Nuclear Magnetic Resonance of Paramagnetic Macromolecules

NATO ASI Series

Advanced Science Institutes Series

A Series presenting the results of activities sponsored by the NATO Science Committee, which aims at the dissemination of advanced scientific and technological knowledge, with a view to strengthening links between scientific communities.

The Series is published by an international board of publishers in conjunction with the NATO Scientific Affairs Division

A Life Sciences **B Physics**	Plenum Publishing Corporation London and New York
C Mathematical and Physical Sciences **D Behavioural and Social Sciences** **E Applied Sciences**	Kluwer Academic Publishers Dordrecht, Boston and London
F Computer and Systems Sciences **G Ecological Sciences** **H Cell Biology** **I Global Environmental Change**	Springer-Verlag Berlin, Heidelberg, New York, London, Paris and Tokyo

PARTNERSHIP SUB-SERIES

1. Disarmament Technologies	Kluwer Academic Publishers
2. Environment	Springer-Verlag
3. High Technology	Kluwer Academic Publishers
4. Science and Technology Policy	Kluwer Academic Publishers
5. Computer Networking	Kluwer Academic Publishers

The Partnership Sub-Series incorporates activities undertaken in collaboration with NATO's Cooperation Partners, the countries of the CIS and Central and Eastern Europe, in Priority Areas of concern to those countries.

NATO-PCO-DATA BASE

The electronic index to the NATO ASI Series provides full bibliographical references (with keywords and/or abstracts) to more than 30000 contributions from international scientists published in all sections of the NATO ASI Series.
Access to the NATO-PCO-DATA BASE is possible in two ways:

– via online FILE 128 (NATO-PCO-DATA BASE) hosted by ESRIN, Via Galileo Galilei, I-00044 Frascati, Italy.

– via CD-ROM "NATO-PCO-DATA BASE" with user-friendly retrieval software in English, French and German (© WTV GmbH and DATAWARE Technologies Inc. 1989).

The CD-ROM can be ordered through any member of the Board of Publishers or through NATO-PCO, Overijse, Belgium.

Series C: Mathematical and Physical Sciences – Vol. 457

Nuclear Magnetic Resonance of Paramagnetic Macromolecules

edited by

Gerd N. La Mar
Department of Chemistry,
University of California,
Davis, California, U.S.A.

Kluwer Academic Publishers

Dordrecht / Boston / London

Published in cooperation with NATO Scientific Affairs Division

Proceedings of the NATO Advanced Research Workshop on
Nuclear Magnetic Resonance of Paramagnetic Macromolecules
Sintra, Portugal
June 4–8, 1994

A C.I.P. Catalogue record for this book is available from the Library of Congress.

ISBN 978-90-481-4522-5

Published by Kluwer Academic Publishers,
P.O. Box 17, 3300 AA Dordrecht, The Netherlands.

Kluwer Academic Publishers incorporates the publishing programmes of
D. Reidel, Martinus Nijhoff, Dr W. Junk and MTP Press.

Sold and distributed in the U.S.A. and Canada
by Kluwer Academic Publishers,
101 Philip Drive, Norwell, MA 02061, U.S.A.

In all other countries, sold and distributed
by Kluwer Academic Publishers Group,
P.O. Box 322, 3300 AH Dordrecht, The Netherlands.

Printed on acid-free paper

This book contains the proceedings of a NATO Advanced Research Workshop held within the programme of activities of the NATO Special Programme on Supramolecular Chemistry as part of the activities of the NATO Science Committee.

Other books previously published as a result of the activities of the Special Programme are:

WIPFF, G. (Ed.), *Computational Approaches in Supramolecular Chemistry.* (ASIC 426) 1994. ISBN 0-7923-2767-5

FLEISCHAKER, G.R., COLONNA, S. and LUISI, P.L. (Eds.), *Self-Production of Supramolecular Structures.* From Synthetic Structures to Models of Minimal Living Systems. (ASIC 446) 1994. ISBN 0-7923-3163-X

FABBRIZZI, L., POGGI, A. (Eds.), *Transition Metals in Supramolecular Chemistry.* (ASIC 448) 1994. ISBN 0-7923-3196-6

BECHER, J., SCHAUMBURG, K. (Eds.), *Molecular Engineering for Advanced Materials.* (ASIC 456) 1995. ISBN 0-7923-3347-0

TABLE OF CONTENTS

PREFACE

The NATO Advanced Research Workshop, "Nuclear Magnetic Resonance of Paramagnetic Macromolecules", took place at the Hotel Tivoli, Sintra, Portugal on June 4-8, 1994. The contents of this monograph represent the lectures at this Workshop. The unique and valuable information content of the NMR spectra of paramagnetic macromolecules has been recognized since the initial report on the spectrum of cytochrome *c* (A. Kowalsky (1965) *Biochemistry*, **4**, 2382). However, the road to developing the methods for first detecting, then assigning the resonances of active site residues, and lastly interpreting the hyperfine shifts and relaxation in terms of molecular and electronic structure, has been a long and tortuous one. Nevertheless, the past decade has witnessed not only a strong resurgence in the interest in NMR of paramagnetic macromolecules, but also remarkable progress in the experimental methodology and understanding of the origins of the paramagnetically influenced spectral parameters. Hence the broad and in-depth potential of NMR studies on paramagnetic macromolecules has been clearly established.

This workshop was organized to assess the current status of NMR studies of paramagnetic macromolecules with respect to the scope and limitations of the needed experimental approaches, the interpretive bases of the hyperfine shifts and relaxation for defining both molecular and electronic structure, and the range of systems which can be studied. The subject matter of the lectures, moreover, was not restricted to NMR studies, but included key contributions on the theoretical framework for interpreting NMR spectral parameters of spin-coupled chromophores, and computational approaches for modeling solution structure of macromolecules with paramagnetic centers.

Two classes of systems have played a key role in the development of the field of NMR in paramagnetic macromolecules, heme proteins and iron-sulfur proteins. The unique role played by these two systems is due, in part, to their diverse function, structure and genetic origin, but largely because they contain well-defined chromophores, each of which can be studied in detail outside the protein matrix. These systems constitute the most successfully studied macromolecules, and a major portion (first eight and last six) of the seventeen contributions deal with heme and/or iron-sulfur proteins. The middle three chapters survey the progress on, and significant promise of, more difficult systems which do not possess a chromophore, but nevertheless have yielded remarkable insight into their structure in spite of the complexities.

The Director is grateful to the members of the Organizing Committee, Ivano Bertini, Bernard Lamotte, Geoffrey R. Moore and Isabel Moura, for their contributions in planning the program, to Anjos L. Macedo for making all local arrangements in Sintra, and to Kendra L. Tanner for valuable and expert assistance in both organizing the Workshop and completing this monograph. Most importantly, the Director and Organizing Committee are indebted to the Scientific and Environmental Affairs Division of NATO, and to Alain Jubier, Director of the Supramolecular Chemistry Program, for the generous financial support of the Advanced Research Workshop.

Gerd N. La Mar, Director
August 20, 1994

PARTICIPANTS

Lucia BANCI • University of Florence, Via G. Caponi 7, I-50121 Florence, Italy

Nursen BAYRAKTAR • Marmara Research Center, Research Institute for Basic Sciences, Department of Chemistry, P.O. Box 21, 41470 Gebze, Turkey

Moshe BELINSKII • School of Chemistry, Tel Aviv University, Ramat-Aviv 69 976, P.O. Box 39040, Tel-Aviv, Israel

Ivano BERTINI • Professor of Inorganic Chemistry, University of Florence, Via G. Caponi 7, I-50121 Florence, Italy

Sergei BORSHCH • IRC, CNRS, 69626 Villeurbanne Cedex, France

Catherine BOUGAULT • Centre d'Etudes Nucléaires de Grenoble DRFMC SESAM SCPM, CEA, 17 rue des Martyrs, 38054 Grenoble Cedex 9, France

Michael CAFFREY • Institut de Biologie Structurale, CNRS CEA, 41, Avenue des Martyrs, 38027 Grenoble Cedex 1 France

Jorge CALDEIRA • Universidade Nova de Lisboa, Departamento de Quimica, Faculdade de Ciencias e Tecnologia, 2825 Monte de Caparica, Portugal

Gerard W. CANTERS • Department of Chemistry, Gorlaeus Laboratory, Leiden University, P.O. Box 9502, 2300 RA Leiden, The Netherlands

Carla M. CARNEIRO • Instituto de Tecnologia Química e Biológica, Universidade Nova de Lisboa, Rua da Quinta Grande 6, Apartado 127, 2780 Oeiras, Portugal

David A. CASE • Scripps Research Institute, Department of Molecular Biology, 10666 N. Torrey Pines Road, La Jolla, CA 92037 USA

Kimber CLARK • University of California, Department of Chemistry, Davis, CA 95616 USA

Christina COSTA • Universidade Nova de Lisboa, Departamento de Quimica, Faculdade de Ciencias e Tecnologia, 2825 Monte de Caparica, Portugal

Helena S. COSTA • Instituto de Tecnologia Química e Biológica, Universidade Nova de LisboaL, Rua da Quinta Grande, 6, Apartado 127, P-2780 Oeiras, Portugal

Isabel B. COUTINHO • Instituto de Tecnologia Química e Biológica, Universidade Nova de Lisboa, Rua da Quinta Grande, 6, Apartado 127, P-2780 Oeiras, Portugal

Sharon DAVY • University of East Anglia, School of Chemical Sciences, Norwich NR4 7TJ, England

Dabney W. DIXON • Georgia State Univiversity, Department of Chem., University Plaza, Atlanta, GA 30303 USA

Antonio DONAIRE • Departamento de Quimica Inorganica, Facultat de Quimica, Universitat de Valencia, C/Dr. Moliner, 50, 46100 Burjassot (Valencia), Spain

Ricardo FRANCO • Universidade Nova de Lisboa, Departamento de Quimica, Faculdade de Ciencias e Tecnologia, 2825 Monte de Caparica, Portugal

Jacques GAILLARD • Département de Recherche Fondamentale sur la Matière Condensée, SESAM-SCPM, Centre d'Etudes Nucleaires de Grenoble, 85X F-38041, Grenoble France

Carlos GERALDES • Departamento de Bioquímica, Universidade de Coimbra, 3049 Coimbra CODEX, Portugal

Jean-Jacques GIRERD • Laboratoire de Chemie Inorganique, CNRS, Institut de Chemie Moleculaire, Bat 420, Univ. de Paris-Sud, 91405 Orsay, France

Brian GOODFELLOW • Universidade Nova de Lisboa, Departamento de Quimica, Faculdade deCiencias e Tecnologia, 2825 Monte de Caparica, Portugal

Halvard HAARKLAU • Department of Chemistry, UNIT AVH, 7055 Trondheim, Norway

Richard C. HOLZ • Utah State University, Department of Chemistry & Biochemistry, Logan, UT 84322-0300 USA

Jürgen HÜTTERMANN • Fachrichtung Biophsik und Physikalische Grundlagen der Medizin, Universität des Saarlandes, Klinikum, Bau 76, 66421 Homburg/Saar, Germany

B. H. (Vincent) HUYNH • Department of Physics, Emory University, Atlanta, GA 30322 USA

Jens-Jakob KARLSSON • Kemisk Laboratorium A, 207, DTH, DK-2800 Lyngby, Denmark

Donald M. KURTZ Jr. • Department of Chemistry, University of Georgia, Athens, GA 30602 USA

Agnete LA COUR • Kemisk Institut, Odense Universitat, 5230 Odense, Denmark

Gerd N. LA MAR • Department of Chemistry, University of California, Davis, CA 95616 USA

Bernard LAMOTTE • Laboratoire de Spectroscopie de Complexes Polymétalliques et de Métalloprotéines, (DRFMC/SESAM/SCPM), Centre d'Etudes Nucléaires de Grenoble, 17 rue des Martyrs, 38054 Grenoble Cedex 9, France

Jean LEGALL • Department of Biochemistry, University of Georgia, Life Sciences Building, Athens, GA 30602 USA

Teresa LEHMANN • Department of Chemistry, University of Minnesota, 207 Pleasant Street S.E., Minneapolis, MN 55455-0431 USA

Gilda H. LOEW • Molecular Research Institute, 845 Page Mill Road, Palo Alto, CA 94304 USA

Ricardo LOURO • Instituto de Tecnologia Química e Biológica, Universidade Nova de Lisboa, Rua da Quinta Grande, 6, Apartado 127, P-2780 Oeiras, Portugal

Claudio LUCHINAT • University of Bologna, Viale Berti Pichat, 10, I-40127 Bologna, Italy

Anjos L. MACEDO • Centro de Tecnologia Química Fina e Biotecnologia, Dep. de Química, Faculdade de Ciências e Tecnologia /UNL, 2825 Monte de Caparica, Portugal

John L. MARKLEY • Department of Biochemistry, University of Wisconsin at Madison, 420 Henry Hall, Madison, Wisconsin 53706 USA

A. Grant MAUK • Department of Biochemistry, University of British Columbia, 2146 Health Sciences Mall, Vancouver, BC, Canada V6T 1Z3

Ann MCDERMOTT • Department of Chemistry, Columbia University, Havemeyer Hall, New York, NY 10027 USA

Li-June MING • Department of Chemistry, University of South Florida, 4202 Fowler Avenue, Tampa, FL 33620 USA

Geoffrey R. MOORE • School of Chemical Sciences, University of East Anglia, Norwich NR4 7TJ, England

Jean-Marc MOULIS • CEA Direction des Sciences du Vivant, Dept de Biol. Molec. et Structurale, CENG, 17 Avenue des Martyrs, 38041 Grenoble Cedex 9, France

Isabel MOURA • Centro de Tecnologia Química Fina e Biotecnologia, Dep. de Química, Faculdade de Ciências e Tecnologia /UNL, 2825 Monte de Caparica, Portugal

José J.G. MOURA • Centro de Tecnologia Química Fina e Biotecnologia, Dep. de Química, Faculdade de Ciências e Tecnologia /UNL, 2825 Monte de Caparica, Portugal

Eckard MÜNCK • Department of Chemistry, Carnegie Mellon University, 4400 Fifth Avenue, Pittsburgh, PA 15213 USA

Akira NAKAMURA • Dept. Macromolecular Science, Faculty of Science, Osaka University, Tuyonaba Osaka 560, JAPAN

Michael J. OSBORNE • University of East Anglia, School of Chemical Sciences, Norwich NR4 7TJ, England

Maurizio PACI • Dipartimento di Scienze e Tecnologie Chimiche, Università di Roma, Tor Vergata, Via Orazio Raimondo, I-00173 Roma, Italy

Nuno PALMA • Universidade Nova de Lisboa, Departamento de Quimica, Faculdade de Ciencias e Tecnologia, 2825 Monte de Caparica, Portugal

Roberta PIERATTELLI • Universita degli Studi di Firenze, Dipartimento di Chimica, Via G. Capponi, 7, 50121 Firenze, Italy

Susana PRAZERES • Centro de Tecnologia Química Fina e Biotecnologia, Dep. de Química, Faculdade de Ciências e Tecnologia /UNL, 2825 Monte de Caparica, Portugal

Lawrence QUE, Jr. • Department of Chemistry, University of Minnesota, 139 Smith Hall, 207 Pleasant St. SE, Minneapolis, Minnesota 55455 USA

Carlos SALGUEIRO • Instituto de Tecnologia Química e Biológica, Universidade Nova de Lisboa, Rua da Quinta Grande, 6, Apartado 127, P-2780 Oeiras, Portugal

Helena SANTOS • Instituto de Tecnologia Química e Biológica, Universidade Nova de Lisboa, Rua da Quinta Grande, 6, Apartado 127, P-2780 Oeiras, Portugal

James D. SATTERLEE • Department of Chemistry, Washington State University, Pullman, Washington 99164 USA

Marco SETTE • Department of Chemistry, University of Rome, Tor Vergata, Via Ricerca Scientifica, 00137 Rome, Italy

Artur M. S. SILVA • Department of Chemistry, University bof Aveiro, 3800 Aveiro, Portugal

Gerard SIMMONEAUX • Université de Rennes I, URA CNRS 415, Laboratoire de Chimie, F-35042 Rennes Cedex, France

Ursula SIMONIS • Department of Chemistry & Biochemistry, San Francisco State Univiversity, 1600 Holloway Ave., San Francisco, CA 94132 USA

Lars SKJELDAL • Agricultural University of Norway, Department of Biochemical Sciences, P.O. Box 5036, N-1432 Ås, Norway

Hilario R. TAVARES • Department of Chemistry, University of Aveiro, 3800 Aveiro, Portugal

Nigel VEITCH • Jodrell Laboratory, Royal Botanic Gardens, Kew, Richmond Surrey TW9 3AB, England

Regitze R. VOLD • Department of Chemistry, University of California, San Diego, La Jolla, CA 92093 USA

F. Ann WALKER • Department of Chemistry, University of Arizona, Tucson, Arizona 85721 USA

Zhigang WANG • Department of Chemistry, University of Minnesota, 207 Pleasant Street S.E., Minneapolis, MN 55455-0431 USA

Victor WRAY • Gesellschaft für Biotechnologische Forschung mbH, Mascheroder Weg 1, D-3300 Braunschweig, Germany

Antonio XAVIER • Instituto de Tecnologia Química e Biológica, Universidade Nova de Lisboa, Rua da Quinta Grande, 6, Apartado 127, P-2780 Oeiras, Portugal

CONTRIBUTORS

Steve L. ALAM • Department of Chemistry, Washington State University, Pullman, Washington 99164 USA

Lucia BANCI • University of Florence, Via G. Caponi 7, I-50121 Florence, Italy

Ivano BERTINI • Professor of Inorganic Chemistry, University of Florence, Via G. Caponi 7, I-50121 Florence, Italy

G. BLONDIN • Laboratoire de Chimie Inorganique, URA CNRS 420, ICMO, Université Paris-Sud, 91405 Orsay, France

E. L. BOMINAAR • Department of Chemistry, Carnegie-Mellon University, 4400 Fifth Avenue, Pittsburgh, PA 15213, USA

Sergei BORSHCH • Institut de Recherche sur la Catalyse, CNRS, 69626 Villeurbanne Cedex, France

David A. CASE • Scripps Research Institute, Department of Molecular Biology, 10666 N. Torrey Pines Road, La Jolla, CA 92037 USA

Hong CHEN • Department of Biochemistry, University of Wisconsin at Madison, 420 Henry Hall, Madison, Wisconsin 53706 USA

J. -L. CHEN • Scripps Research Institute, Department of Molecular Biology, 10666 N. Torrey Pines Road, La Jolla, CA 92037 USA

Zhigang CHEN • Department of Chemistry, University of California, Davis, CA 95616 USA

M. C. COX • School of Chemical Sciences, University of East Anglia, Norwich NR4 7TJ, England

D. CROWE • School of Chemical Sciences, University of East Anglia, Norwich NR4 7TJ, England

G. P. DÄGES • Fachrichtung Biophsik und Physikalische Grundlagen der Medizin, Universität des Saarlandes, Klinikum, Bau 76, 66421 Homburg/Saar, Germany

Jeffrey S. de ROPP • Department of Chemistry, University of California, Davis, CA 95616 USA

Antonio DONAIRE • Departamento de Quimica Inorganica, Facultat de Quimica, Universitat de Valencia, C/Dr. Moliner, 50, 46100 Burjassot (Valencia), Spain

Ping DU • Molecular Research Institute, 845 Page Mill Road, Palo Alto, CA 94304 USA

David P. DUTTON • Department of Chemistry, Washington State University, Pullman, Washington 99164 USA

Raymond GILMOUR • Dept. of Preclinical Veterinary Sciences, Royal (Dick) School of Veterinary Studies, University of Edinburgh, Summerhall, Edinburgh EH9 1QH, UK

C. GIORI • Instituto di Scienze Fiziche, Universitá degli Studi di Parma, Italy

Jean-Jacques GIRERD • Laboratoire de Chemie Inorganique, CNRS, Institut de Chemie Moleculaire, Bat 420, Univ. de Paris-Sud, 91405 Orsay, France

Jürgen HÜTTERMANN • Fachrichtung Biophsik und Physikalische Grundlagen der Medizin, Universität des Saarlandes, Klinikum, Bau 76, 66421 Homburg/Saar, Germany

B. H. (Vincent) HUYNH • Department of Physics, Emory University, Atlanta, GA 30322 USA

H. R. JIMENEZ • Departamento de Quimica Inorganica, Universitat de Valencia, C/Dr. Moliner, 50, 46100 Burjassot (Valencia), Spain

Gerd N. LA MAR • Department of Chemistry, University of California, Davis, CA 95616 USA

Bernard LAMOTTE • Laboratoire de Spectroscopie de Complexes Polymétalliques et de Métalloprotéines, (DRFMC/SESAM/SCPM), Centre d'Etudes Nucléaires de Grenoble, 17 rue des Martyrs, 38054 Grenoble Cedex 9, France

Gilda H. LOEW • Molecular Research Institute, 845 Page Mill Road, Palo Alto, CA 94304 USA

Claudio LUCHINAT • University of Bologna, Viale Berti Pichat, 10, I-40127 Bologna, Italy

Anjos L. MACEDO • Centro de Tecnologia Química Fina e Biotecnologia, Dep. de Química, Faculdade de Ciências e Tecnologia /UNL, 2825 Monte de Caparica, Portugal

John L. MARKLEY • Department of Biochemistry, University of Wisconsin at Madison, 420 Henry Hall, Madison, Wisconsin 53706 USA

A. Grant MAUK • Department of Biochemistry, University of British Columbia, 2146 Health Sciences Mall, Vancouver, BC, Canada V6T 1Z3

Li-June MING • Department of Chemistry, University of South Florida, 4202 Fowler Avenue, Tampa, FL 33620 USA

Geoffrey R. MOORE • School of Chemical Sciences, University of East Anglia, Norwich NR4 7TJ, England

José M. MORATAL • Departamento de Quimica Inorganica, Facultat de Quimica, Universitat de Valencia, C/Dr. Moliner, 50, 46100 Burjassot (Valencia), Spain

J. -M. MOUESCA • Département de Recherche Fondamentale sur la Matière Condensée, Centre d'Etudes Nucléaires de Grenoble, CEA, 85 X 38041 Grenoble, France

Isabel MOURA • Centro de Tecnologia Química Fina e Biotecnologia, Dep. de Química, Faculdade de Ciências e Tecnologia /UNL, 2825 Monte de Caparica, Portugal

José J.G. MOURA • Centro de Tecnologia Química Fina e Biotecnologia, Dep. de Química, Faculdade de Ciências e Tecnologia /UNL, 2825 Monte de Caparica, Portugal

Akira NAKAMURA • Coordination Chemistry Laboratories, Institute for Molecular Science, Okazaki 444, JAPAN

L. NOODLEMAN • Scripps Research Institute, Department of Molecular Biology, 10666 N. Torrey Pines Road, La Jolla, CA 92037 USA

Bjyung-Ha OH • Department of Biochemistry, University of Wisconsin at Madison, 420 Henry Hall, Madison, Wisconsin 53706 USA

Michael J. OSBORNE • University of East Anglia, School of Chemical Sciences, Norwich NR4 7TJ, England

Mario PICCIOLI • Universita degli Studi di Firenze, Dipartimento di Chimica, Via G. Capponi, 7, 50121 Firenze, Italy

Graham PETTIGREW • Dept. of Preclinical Veterinary Sciences, Royal (Dick) School of Veterinary Studies, University of Edinburgh, Summerhall, Edinburgh EH9 1QH, UK

Roberta PIERATTELLI • Universita degli Studi di Firenze, Dipartimento di Chimica, Via G. Capponi, 7, 50121 Firenze, Italy

Larry QUE • Department of Chemistry, University of Minnesota, 139 Smith Hall, 207 Pleasant St. SE, Minneapolis, Minnesota 55455 USA

Natarajan RAVI • Department of Physics, Emory University, Atlanta, GA 30322 USA

H. REINHARD • Fachrichtung Biophsik und Physikalische Grundlagen der Medizin, Universität des Saarlandes, Klinikum, Bau 76, 66421 Homburg/Saar, Germany

G. RIUS • Département de Recherche Fondamentale sur la Matière Condensée, Centre d'Etudes Nucléaires de Grenoble, CEA, 85 X 38041 Grenoble, France

José SALGADO • Departamento de Quimica Inorganica, Facultat de Quimica, Universitat de Valencia, C/Dr. Moliner, 50, 46100 Burjassot (Valencia), Spain

James D. SATTERLEE • Department of Chemistry, Washington State University, Pullman, Washington 99164 USA

G. SCHMIDT • Fachrichtung Biophisik und Physikalische Grundlagen der Medizin, Universität des Saarlandes, Klinikum, Bau 76, 66421 Homburg/Saar, Germany

Lars SKJELDAL • Agricultural University of Norway, Depat. of Biochemical Sciences, P.O. Box 5036, N-1432 Ås, Norway

Andrew T. SMITH • Biochemistry Laboratory, University of Sussex, Brighton, BN1 9QG, UK

Paola TURANO • Universitá degli Studi di Firenze, Dipartimento di Chimica, Via G. Capponi, 7, 50121 Firenze, Italy

Norikazu UEYAMA • Department of Macromolecular Science, Faculty of Science, Osaka University, Osaka 560, JAPAN

Zhigang WANG • Department of Chemistry, University of Minnesota, 207 Pleasant Street S.E., Minneapolis, MN 55455-0431 USA

William M. WESTLER • Department of Biochemistry, University of Wisconsin at Madison, 420 Henry Hall, Madison, Wisconsin 53706 USA

M. T. WILSON • Department of Chemistry and Biological Chemistry, University of Essex, Colchester, Essex CO4 3SQ, U.K.

Bin XIA • Department of Biochemistry, University of Wisconsin at Madison, 420 Henry Hall, Madison, Wisconsin 53706 USA

NEW APPROACHES TO NMR OF PARAMAGNETIC MOLECULES

CLAUDIO LUCHINAT[1] and MARIO PICCIOLI[2]

1:Institute of Agricultural Chemistry
University of Bologna
Viale Berti Pichat, 10
40127 Bologna
2:Department of Chemistry
University of Florence
Via G. Capponi, 7
50121 Florence
Italy

1. Introduction

The frontier of today in the NMR of paramagnetic molecules is that of finding connectivities, both scalar and dipolar, between nuclei which are fast relaxing because of the coupling with the paramagnetic center. A giant step in this direction is represented by the development (and availability) of one dimensional experiments to detect Nuclear Overhauser Effect (NOE) in paramagnetic systems [1-4]. NOE nowadays allows us to detect connectivities among hyperfine shifted signals and between hyperfine shifted signals and diamagnetic signals. Information are obtained by means of difference spectra which must be carefully performed [5,6]. We do not review this aspect here because it has been treated in other articles [7,8]. Here we will give guidelines to record NOESY and COSY spectra in fast relaxing systems and to optimize the experiments to detect cross peaks between signals with different relaxation times. Incidentally, we will define a new spectroscopy called RACTSY. Finally, we discuss the strategy of the sequence-specific assignment and we show selected examples.

2. Through space connectivities

2.1. NOESY

After the first detection of cross peaks in a *2D* experiment on a paramagnetic protein [9], the detection of dipolar connectivities involving signals characterized by large hyperfine shifts and short relaxation times has represented the frontier of the early 90ies [10-15]. The detection of cross peaks among signals having hyperfine shifts of

G.N. La Mar (ed.), Nuclear Magnetic Resonance of Paramagnetic Macromolecules, 1-28.

some tens and hundreds Hertz of line broadening is definitely impressive to those who are not familiar with the peculiarities of paramagnetic systems. Indeed, when the T_1 values of the signals on interest are all similar, the detection of cross peaks is relatively straightforward, provided that T_1 values are not below 1-2 ms and, of course, that the proton-proton distances are not too large (typically geminal or vicinal protons). In the above conditions, the experiments can be easily optimized by using mixing times of the order of the T_1 of the signals of interest, and acquisition times $t_{2max} = t_{1max}$ of the order of the T_2 of signals of interest. The choice of τ_m will ensure the maximum of magnetization transfer due to cross relaxation, the choice of t_2 and t_1 will allow not to acquire noise once the signal is relaxed because of T_2. A shortening of t_{1max} will allow the acquisition of a larger number of transients for each increment and will result in an improvement of the signal-to-noise ratio.

When the aim is that of detecting connectivities between signals with different relaxation behaviors, the choice of the experimental parameters becomes more cumbersome. The cross peak intensity as a function of τ_m is given by

$$M(\tau_m) = \frac{M_0 \sigma_{AB}}{2D}\left[e^{-(R'+D)\tau_m} - e^{-(R'-D)\tau_m}\right] \tag{1}$$

where M_0 is the one dimensional signal intensity, σ_{AB} is the longitudinal cross relaxation between spins A and B coupled through dipolar interaction, τ_m is the mixing time, and R' and D are given by

$$R' = \frac{1}{2}(R_A + R_B) \quad D = \left[\frac{1}{4}(R_A - R_B)^2 + \sigma_{AB}{}^2\right]^{\frac{1}{2}} \tag{2}$$

where R_1 is the relaxation rate $T_1{}^{-1}$. The function [1] reaches its maximum according to

$$\tau_m = -\frac{1}{2D}\ln\frac{(R'-D)}{(R'+D)} \tag{3}$$

Table 1 reports the values of the optimal mixing times as a function of T_{1A} and T_{1B} values [16]. Figure 1 shows that the optimal mixing time is always intermediate between the short (T_{1B}) and the long (T_{1A}) values, but the ratio τ_m/T_{1B} increases with increasing T_{1A}/T_{1B}. If mixing time were adjusted according to the average of T_{1A} and T_{1B}, sizable missettings of mixing time would occur even when the difference between the two T_1 values is small. If the average of the relaxation rates is used to calculate the reciprocal of mixing time,the error is relatively small up to a T_{1A}/T_{1B} ratio larger than 10. The choice of t_{1max} and t_{2max} is related, of course, to the T_2 values of the interested signals. Note that the number of data points in the t_1 and t_2 dimensions may be largely different among systems with similar T_1 and T_2 values, as they obviously depend on the experimental spectral window. Therefore the spreading in chemical shift of the signals of interest is a parameter which needs to be evaluated in setting up the experiment.

For instance, in the case of oxidized High Potential Iron-Sulfur Proteins (HiPIPs), signals have linewidths of 400-500 Hz and are spread over 125 kHz of spectral window at 600 MHz (see later). Under the above conditions, the dwell time is 4 μs and, therefore, 1024 data points give a 4.1 ms acquisition time. When considering reduced HiPIPs, signals with linewidth up to 250-300 Hz are observed in a 18 kHz spectral window, corresponding to a 28 μs of dwell time. Therefore 256 data points are sufficient to achieve a 7.1 ms acquisition time.

TABLE 1. Optimal NOESY mixing times for various T_1 values (ms) of the two signals A and B, calculated using Equation (3). Cross peak intensities for a σ_{AB} value of 1 s^{-1} are shown in parenthesis.

$T_{1A}\backslash T_{1B}$	128	64	32	16	8	4	2	1
128	128 (4.72)	89.0 (3.20)	59.2 (2.02)	38.0 (1.19)	23.7 (0.67)	14.3 (0.36)	8.5 (0.19)	4.9 (0.10)
64	89.0 (3.20)	64.1 (2.36)	44.4 (1.60)	29.6 (1.01)	19.0 (0.59)	11.8 (0.33)	7.2 (0.18)	4.2 (0.09)
32	59.2 (2.02)	44.4 (1.60)	32.0 (1.18)	22.2 (0.80)	14.8 (0.50)	9.5 (0.30)	5.9 (0.17)	3.6 (0.09)
16	38.0 (1.19)	29.6 (1.01)	22.2 (0.80)	16.0 (0.59)	11.1 (0.40)	7.4 (0.25)	4.8 (0.15)	3.0 (0.08)
8	23.7 (0.67)	19.0 (0.59)	14.8 (0.50)	11.1 (0.40)	8.0 (0.29)	5.5 (0.20)	3.7 (0.13)	2.4 (0.07)
4	14.3 (0.36)	11.8 (0.33)	9.5 (0.30)	7.4 (0.25)	5.5 (0.20)	4.0 (0.15)	2.8 (0.10)	1.8 (0.06)
2	8.5 (0.19)	7.2 (0.18)	5.9 (0.17)	4.8 (0.15)	3.7 (0.13)	2.8 (0.10)	2.0 (0.07)	1.4 (0.05)
1	4.9 (0.10)	4.2 (0.09)	3.6 (0.09)	3.0 (0.08)	2.4 (0.07)	1.8 (0.06)	1.4 (0.05)	1.0 (0.04)

2.2. NOE-NOESY

As previously outlined, NOESY experiments (as well as one dimensional NOE) may connect an hyperfine shifted signal to signals which are unaffected by hyperfine interaction. The assignment of the latter is of great interest, as it will be discussed in a following section. In this frame, NOESY is preferred to one dimensional NOE because signals which are connected to hyperfine shifted signals can be easily correlated to other resonances, and this is helpful for their assignment [16]. On the other hand NOE can be performed with much better resolution, and the use of steady state conditions may also allow the detection of signals which arise from second generation NOE. While second generation NOE effects are in general, undesiderable, they can be of great help in the assignment. An experiment, which we termed NOE-NOESY, has been designed to use the advantages of both NOE and NOESY approaches [17]. It can provide information on the connectivities that a given hyperfine shifted signals has with signals in the diamagnetic region, as well as all the dipole-dipole connectivities which involve those signals. NOE-NOESY is based on a NOESY experiment preceded by selective saturation of a hyperfine shifted signal for a time that can vary from a few milliseconds to hundreds of milliseconds.

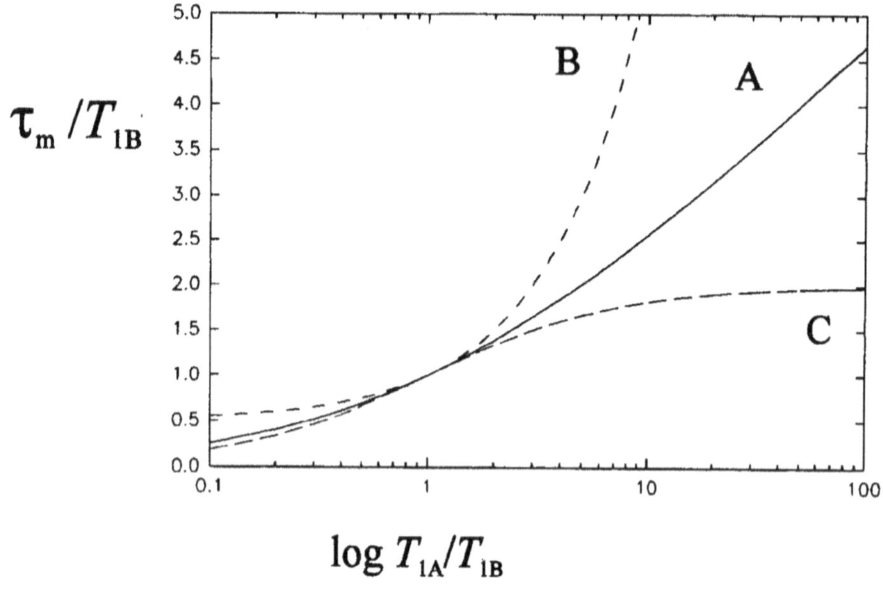

Figure 1. Plot of τ_m/T_{1B} vs T_{1A}/T_{1B} (log scale). Curve A is calculated according to eq (3); curve B is calculated using $\tau_m = ½(T_{1A} + T_{1B})$; curve C is calculated using $\tau_m^{-1} = ½(T_{1A}^{-1} + T_{1B}^{-1})$.

Upon saturation of a hyperfine shifted signal, prior to the beginning of the NOESY sequence, the M_z values of the signals dipole-dipole coupled to it are reduced by the relative NOE and the subsequent NOESY connectivities are reduced accordingly. The experiment is alternated with off-resonance experiments, where the full NOESY intensities are recovered. If the on- and off- resonance experiments are add-subtracted in an interleaved way through phase cycling (according to what is usually done in one dimensional NOE difference spectra) we end up with a NOESY difference spectrum where only cross peaks and diagonal peaks arising from signals experiencing NOE and/or second generation NOE from the saturated signal are observed. The pulse sequence is reported in Figure 2.

The major utility of the NOE-NOESY experiments is thus to "clean" a crowded NOESY map, leaving only those signals which are close to the saturated signal and coupled to it, either directly or through spin-diffusion effects. Depending on the choice of the two acquisition parameters t_0 and τ_m, different information can be obtained. The delay t_0 determines the extent of build up of NOE prior to the NOESY sequence; hence a long t_0 time, will provide steady-state NOEs and development of second generation NOE, while shorter t_0 values will select, in the resulting NOESY difference spectra, only those signals which arise from first generation NOEs. The mixing time τ_m can be chosen according to the guidelines given in section 2.1 in such a way as, for instance, to preferentially select broad, fast relaxing signals which may be unobserved in the standard NOESY experiments because overwhelmed by diamagnetic signals (short τ_m) or to preferentially select the connectivities involving the second coordination sphere (long τ_m) of residues in the proximity of the saturated signals but not much affected by the hyperfine interaction.

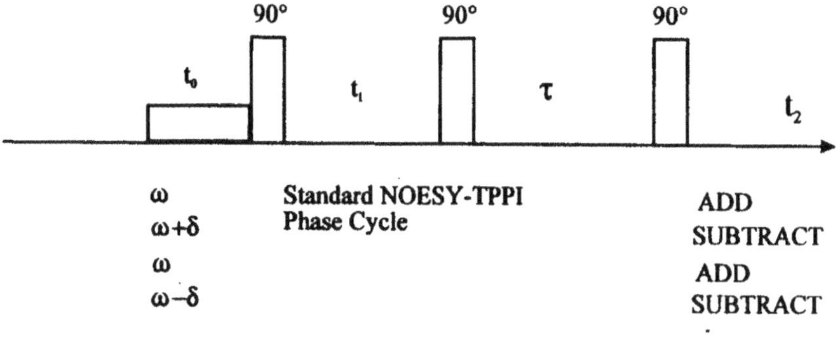

Figure 2. Pulse sequence for the NOE-NOESY experiment.

An interesting property of the NOE-NOESY experiment which is worth being pointed out is that such an experiment is intrinsically asymmetric. In practice, each cross peak can be viewed as a cross peak of a 3*D* spectrum, in which steady state NOE is reached in the first mixing period (t_0) and transient NOE builds during the second mixing period (τ_m). As the two effects are of intrinsically different nature, asymmetric NOESY difference maps can be obtained and their analysis can, in principle, provide additional information. As a practical example of the utility of this pulse sequence we show here the application of the NOE-NOESY experiment to Met-Mb-CN, whose heme pocket is shown in Figure 3.

Figure 4 shows the NOESY experiments at 310 K with 100 ms of mixing time, together with NOE-NOESY experiments performed upon selective saturation of F43 Hε signal (Figure 4B) and of heme 5-CH_3 signal (Figure 4C). The dramatic simplification of the NOE-NOESY map with respect to the standard NOESY experiment permits the detection of even very small cross peaks which arise from signals in the proximity of the saturated signal. Although an extensive assignment through standard 2*D* techniques is available for Met-Mb-CN [1-3,10,18-22], the reported maps have allowed us to assign three additional signals of the βCH_2 and α CH of Phe 43.

2.3 3*D* NOE-NOE

The application of homonuclear 3*D* spectroscopy to paramagnetic systems is still an unexplored field. Indeed, as cross peak intensity is predicted to be always very small, even in the most favorable cases, with respect to diamagnetic systems, the duration of a 3*D* experiments with signal-to-noise ratio suitable to detect the connectivities involving hyperfine shifted signals may be prohibitively long. Moreover, the increase in dimensionality introduces additional delays (at least one additional evolution period) which contribute to the overall loss of information because of T_2 relaxation during the experiment. Under favorable conditions in which such drawbacks can be overcome, homonuclear *3D* spectroscopy can obviously be of great help. Up to now the first and only application of *3D* homonuclear spectroscopy on a paramagnetic metalloprotein is the NOE-NOE experiment, performed on Met-Mb-CN [23].

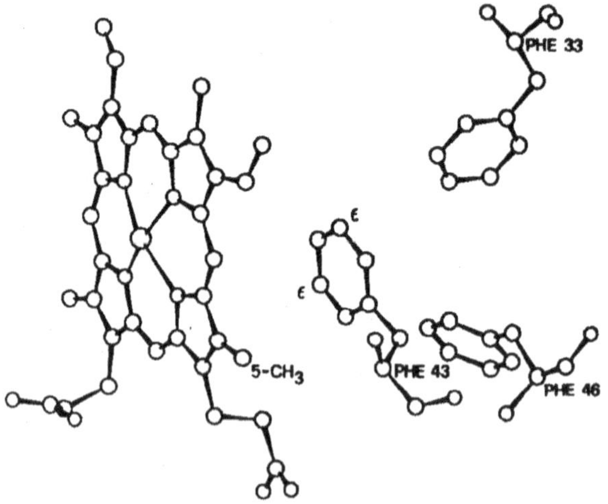

Figure 3. A schematic drawing of the aromatic residues surrounding heme 5-CH_3 in Met-Mb-CN, taken from the X-ray structure of Met-Mb-CO [taken from ref. [24], with permission].

An useful application of *3D* NOE-NOE is achieved by optimizing the first mixing time to the detection of connectivities involving hyperfine shifted signals and using a longer mixing time for the second mixing period. As hyperfine shifted signals are broad and usually well separated, only a few increments in the third dimension are needed. The information that can be obtained are essentially those described in the NOE-NOESY experiment. Of course, on passing from NOE-NOESY to *3D* NOE-NOE all the usual considerations on selectivity and sensitivity, which are related to the increase in dimensionality, hold [25-27]. Additional advantage arises from the fact that, when fast relaxing but not hyperfine shifted signals are present, their selective saturation cannot be performed. In such cases, the *3D* NOE-NOE is the best method to observe the connectivities of interest.

This has been shown, for Mb-CN, in the case of vinyl H-4α signal. Such a signal is affected by hyperfine interaction as the methyl signals, but it is beneath the diamagnetic envelope (5.7 ppm). The connectivities between this signal and other signals affected by the paramagnetic center covered by the diamagnetic envelope may easily be lost in a *2D* NOESY map. Figure 5 shows the ω_1-ω_2 plane at the ω_3 frequency of the vinyl H-4α proton. The two-spin connectivities involving the H-4α proton are several and clearly evident on the lines corresponding to the intersections with the two cross-diagonal planes. Among the several *3D* cross peaks, those involving H-β_{meso}, which is highly affected by the hyperfine interaction and in the

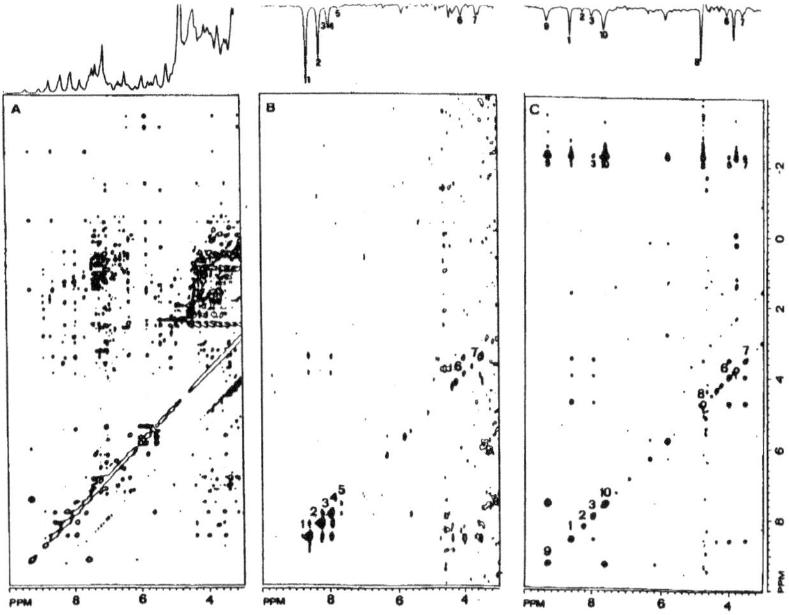

Figure 4. (A) Portion of a NOESY experiment performed on Met-Mb-CN, at 600 MHz and 310 K, with a mixing time of 100 ms. (B and C) Portions of the NOE-NOESY spectra obtained upon saturation of a signal at 12.5 ppm (F43 Hε) (B) and upon saturation of a signal at 26.1 ppm (5-CH_3) (C). A mixing time of 100 ms and saturation time of 300 ms were used for both experiments (B) and (C). Experimental conditions are the same as those given for (A). The reference spectrum of the NOESY and the reference 1*D* NOE difference spectra are also reported. The strong cross peaks observed at -2.2 ppm in the F_1 dimension in (C) directly arise from the irradiated signal 5-CH_3. The spectral window of the experiments in the F_1 dimension has been adjusted in such a way to fold back the row of the saturated signal (originally at 26.1 ppm) in a region were no other cross peak occurs. Folding of the spectrum in F_1 allows a more efficient use of data points in that dimension. From the 1*D* reference spectrum on top of the NOESY experiment (A, inset), it appears that most of the signals considered in (B) and (C) are at least partly overlapped with other signals and some (e.g., signals 6,7,8) are in a very crowded region of the spectrum. Therefore, cross peaks seemingly arising from these signals, even if observed in (A), cannot be unambiguously assigned as they are in (B) and (C). Note also that the relative intensities of signals 1, 2, 3, 9, 10 in the 1*D* NOE reference spectrum, in the diagonal, and in the row of the saturated signal of (C) are sizably different. Such differences can be related to the different relaxation times and cross-relaxation interactions experienced by these signals. The 1*D* NOE is obtained under steady-state conditions. The intensities of the diagonal peaks are the steady state intensities further modulated by the relaxation occurring during τ_m and, to a minor extent, by cross relaxation [taken from ref. [17], with permission].

crowded diamagnetic region (2.7 ppm), are observed. The *3D* cross peak among H-4α H-4β_c and H-4β_t is also clearly detected.

Figure 6 shows the ω_1-ω_2 cross section at the frequency ω_3 corresponding to 5-CH_3 signal. In practice Figure 6 should be compared with Figure 4C in order to discuss the NOE-NOESY versus the *3D* NOE-NOE. The cross peaks observed in the NOE-NOESY between two aromatic signal (Phe 43 Hδ (1) and Phe 46 Hε (3)) and three signals in the aliphatic region (6,7,8) which, on this basis, are assigned to Phe 43 βCH_2 and αCH, are not observed in the cross section plane of the *3D* NOE-NOE. On the other hand, three spin cross peaks involving 4Hα-Hβ_{meso} and 6Hβ-6Hβ' are more apparent in the *3D* NOE-NOE.

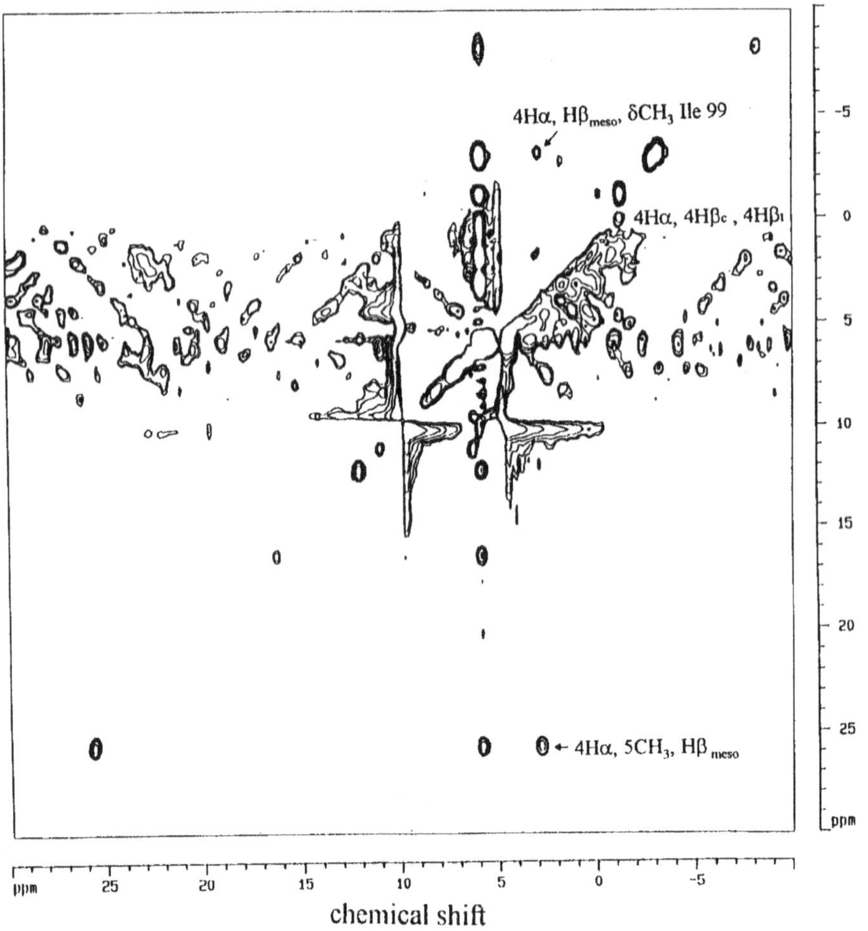

Figure 5. ω_1-ω_2 Cross section at the frequency ω_3 of H-4α signal (5.7 ppm) of a *3D* NOE-NOE spectrum recorded on Met-Mb-CN, at 600 MHz and 310 K [taken from ref. [23], with permission].

Differences rely on the fact that the steady state NOE gives the possibility, by an appropriate choice of saturation time t_0, to transfer the NOE effect from the saturated signal (say I) to the dipole-dipole coupled signals (say S) and, on their turn, to those signals which are coupled to S but not coupled to I.

In the 3D NOE-NOE in which transient NOE is developed during the first mixing time, this effect is less important (second generation NOE contributes to a minor extent).

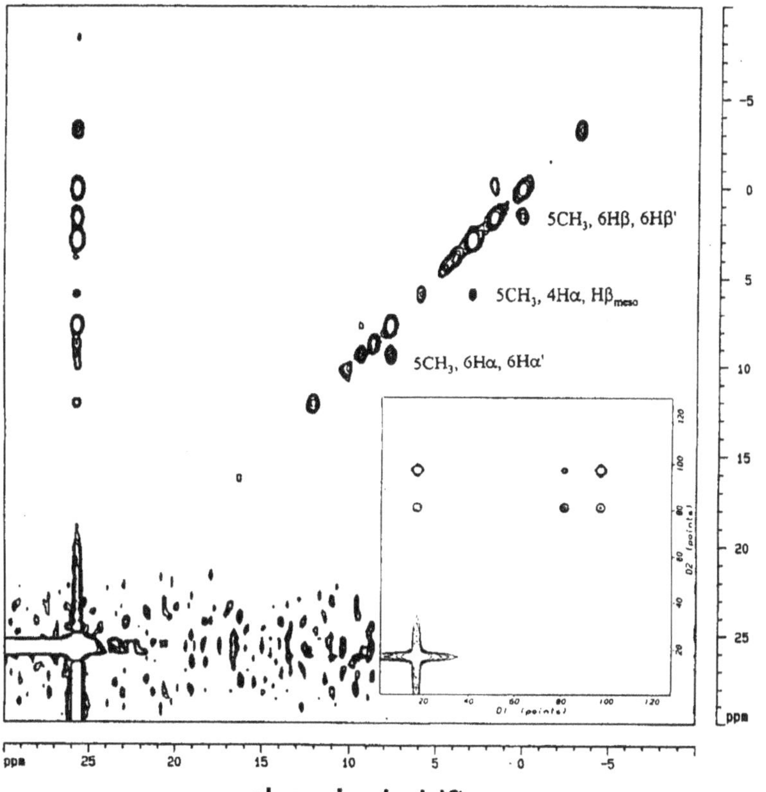

Figure 6. ω_1-ω_2 Cross section at the frequency ω_3 of 5-CH_3 signal of a 3D NOE-NOE spectrum recorded on Met-Mb-CN, at 600 MHz and 310 K. The inset shows the simulated ω_1-ω_2 cross section obtained using the experimental relaxation parameters of the 5-CH_3, H-6α and H-6α' coupled spins, with their geometric properties and with the experimental signal-to-noise ratio [taken from ref. [23], with permission].

Therefore, information which are obtained by the two experiments may be at least partly complementary (as we found in the case of Met-Mb-CN), the NOE-NOESY being powerful in the light of the sequence specific assignment strategy (see later) and 3D NOE-NOE being powerful to assign broad and fast relaxing signals beneath the diamagnetic envelope.

3. Through bond connectivities

3.1. COSY

The applicability of COSY to paramagnetic molecules has been examined in detail from both the experimental and theoretical points of view [22,28-31]. When dealing with COSY experiments on broad and fast relaxing signals, the factors which must be taken into account are essentially the choice of t_{1max} and t_{2max}, the use of filter functions and the type of COSY experiments to be performed.

When relaxation is negligible maximal coherence transfer occurs when $t_1 = t_2 = 1/(2J)$. Therefore, in the F_1 dimension, increasing the number of experiments at the expenses of the number of scans yields an improvement of the S/N ratio only up to $t_{1max} = 1/J$. The appropriate filter under those conditions can be a simple sin or sin^2 function. When relaxation is not negligible, the time at which coherence transfer is maximal (considering, for simplicity, that $R_{2A} = R_{2B} = R_2$, where R_2 is transverse relaxation rate) can be calculated from the coherence transfer as a function of time

$$\begin{aligned} M_A &= \sin \pi J t_1 \sin \pi J t_2 \exp(-t_1 R_{2B}) \exp(-t_2 R_{2A}) \\ M_B &= \sin \pi J t_1 \sin \pi J t_2 \exp(-t_1 R_{2A}) \exp(-t_2 R_{2B}) \end{aligned} \tag{4}$$

by equating the first derivative to zero:

$$t_{peak} = \frac{1}{\pi J} \arctan \frac{\pi J}{R_2} \tag{5}$$

As the ratio $\pi J/R_2$ decreases, t_{peak} moves to the left. An appropriate choice for t_{1max} can be $t_{1max} \cong 2t_{peak}$, again with an appropriate filter function. Either sin or sin^2 or Lorentz-to-Gaussian functions can be used with an appropriate choice of parameters. Exactly the same considerations hold for t_{2max}.

While in diamagnetic systems t_{2max} is usually kept as long as possible, in order to achieve a good resolution in the F_2 dimension (typically 4K data points), when relaxation is not negligible t_2 must be optimized based on the above considerations. Of course, a long acquisition time can always be used in the F_2 dimension with little increase of total experiment time, and then a smaller number of data points can be used when processing the spectrum to observe connectivities from fast relaxing signals.

If $R_{2A} \neq R_{2B}$, t_{1peak} and t_{2peak} are different. If we choose, for example, matched filters with maximal intensities at T_{2A} along t_1 and at T_{2B} along t_2, we optimize the cross peak intensity along the column containing diagonal signal B and along the row containing diagonal signal A, at the expenses of the other cross peak. In other words, the optimal experiment is intrinsically not symmetrical.

As the shape of the cross peak is dependent on the type of COSY experiment, the choice of the COSY experiment and the phasing of the spectrum needs to be considered. In a phase sensitive COSY spectrum, the standard phasing procedure with cross peak components with pure absorption line shape and antiphase structure will provide a considerable cancellation between the cross peak components when the lines are broader than J. This effect is reduced by phasing the cross peak in dispersion mode [29]. Figure 7 show the effects of the two different phase mode at increasing linewidths.

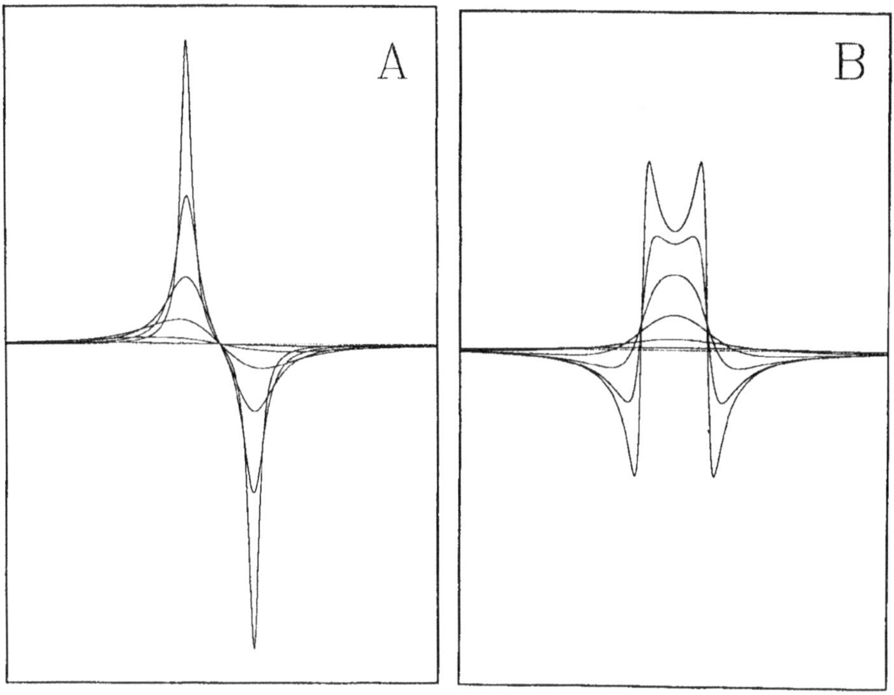

Figure 7. An antiphase doublet with $J = 10/(2\pi)$ Hz and increasing linewidth, phased in absorption (A) and dispersion (B) modes, and $R_2 = 1, 2, 4, 8, 16$, and $32\ s^{-1}$ in order of decreasing signal intensity in (A). From $R_2 = 1$ to $R_2 = 4\ s^{-1}$ the intensity at the center of the doublet in (B) is smaller than that of a single doublet component in (A), and it is bigger from $R_2 =$ 4 to $R_2 = 32\ s^{-1}$ [taken from ref. [31], with permission].

It should be considered that the required accurate phasing over large spectral widths when cross peaks are weak and barely detectable above the noise may not be trivial. For these reasons, the strategy of recording COSY spectra in magnitude mode, which has been largely used in the past years, is still "competitive" with respect to the phase sensitive COSY. A detailed comparison among the various COSY experiments in the case of systems with broad lines has been attempted by Turner [30].

3.2. TOCSY

The use of TOCSY experiments to detect scalar connectivities among broad resonances has not been extensive in the past, due to the satisfactory results obtained by COSY experiments. Indeed, the recently described phenomenon of cross correlation (see next section) suggests that, if they are feasible, TOCSY experiments should be preferred to COSY experiments. The efficiency of the TOCSY experiment is dependent on the effective field strength which acts on each signal. As shown by Bax and coworkers in the articles describing the HOHAHA isotropic mixing [32,33], coherence transfer due to *J*-coupling is maximal if the two *J*-coupled signals experience the same effective field strength. Therefore, if large spreading of chemical shift occurs, *e.g.* of the order of several kHz, as it often happens in paramagnetic proteins, the efficiency of TOCSY experiments is expected to be rather low, unless the spin lock field is relatively close to the middle of the separation in chemical shift among the two signals.

When signals are broad but the chemical shift spreading is not too large, the TOCSY experiments should be considered. Besides the application of TOCSY experiment in small, paramagnetic complexes [34], a nice example of TOCSY experiment on a paramagnetic protein has been performed on the two Fe_4S_4 cluster ferredoxin from *C. pasteurianum* [35]. When the protein is in the oxidized state, the eight βCH_2 protons lie between 20 and 5 ppm (T_1 and T_2 values are of the order of a few milliseconds). It has been shown that, under these conditions, if the spin lock field is applied in the center of the region where the βCH_2 signals are spread (and not on the water signal, as it is usually done when TOCSY are performed in order to achieve information on the diamagnetic part of the protein), for isotropic mixing periods of the order of 10 ms, satisfactory results can be obtained using a spin lock field power about 40% higher than in the standard experiments [35]. The resulting spectrum is shown in Figure 8.

3.3. HETEROCORRELATED EXPERIMENTS

Heterocorrelated NMR spectroscopy has been rarely used to obtain information on residues directly coordinated to the paramagnetic center of a metalloprotein [36]. Applications of heterocorrelated experiments have been reported in the case of Fe_2S_2 ferredoxins [37-39], cytochrome c' and other heme proteins [40-43], cobalt substituted zinc enzymes [44]. We will show here two simple but successful applications of HMQC experiments on two paramagnetic proteins in which the aim is that of investigating residues directly coordinated to the paramagnetic center. These two examples should illustrate the potentialities, largely unexplored, provided by ^{13}C and ^{15}N spectroscopy in paramagnetic systems.

Copper, zinc, superoxide dismutase is a 32000 MW dimeric protein, in which zinc can be easily replaced by cobalt(II) or nickel(II) ions [45]. Such metal substitutions give rise to derivatives showing several hyperfine shifted signals, which can be investigated through NMR [46]. Another chapter of this book will describe theoretical and practical aspects of magnetic coupling inside this dimetallic protein. An extensive review on the NMR work which has been performed on this enzyme has recently appeared [47], here we want to discuss the results obtained on a ^{15}N enriched sample [48]. The schematic drawing of the active site of Cu_2Co_2SOD is reported in Figure 9.

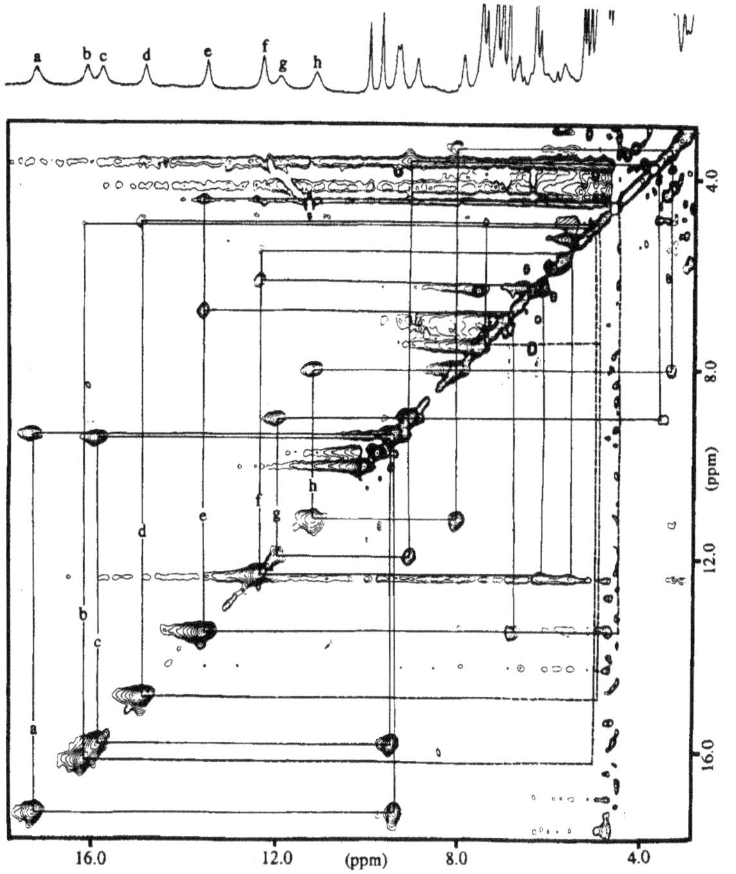

Figure 8. 400 MHz TOCSY spectrum of a 7 mM sample of oxidized *C. pasteurianum* ferredoxin, recorded in D_2O solution. Spin lock and relaxation periods were 10 ms (field strength 20 kHz) and 50 ms, respectively. A trim pulse of 500 μs was applied. 4096x256 points were acquired, each FID consisting of 1024 transients [taken from ref. [35], with permission].

The five distal ring NH belonging to metal coordinated histidines (protons B, C, F, J, and K in Figure 9) have linewidths from about 2 kHz for the NH of His 71 and His 80 (cobalt coordinated residues) to 450 Hz for the NH of His 46, to 250 Hz for the two NH protons of His 48 and His 120. Among the signals of copper coordinated histidines, the larger linewidth of His 46 is due to its proximity to the Co(II) ion. An HMQC experiment on Cu_2Co_2SOD permits the easy detection of ^{15}N-^{1}H connectivities arising from the NH protons of His 48 and His 120, as it can be observed in Figure 10 (signals B and K) [48]. The binding of anions to the copper ion determines changes in ^{15}N hyperfine shifts of such resonances. This can be used to obtain information on the electron delocalization pathways in these systems, as it has been previously done in the case of imidazole axially coordinated to low spin iron(III) porphyrins [49,50].

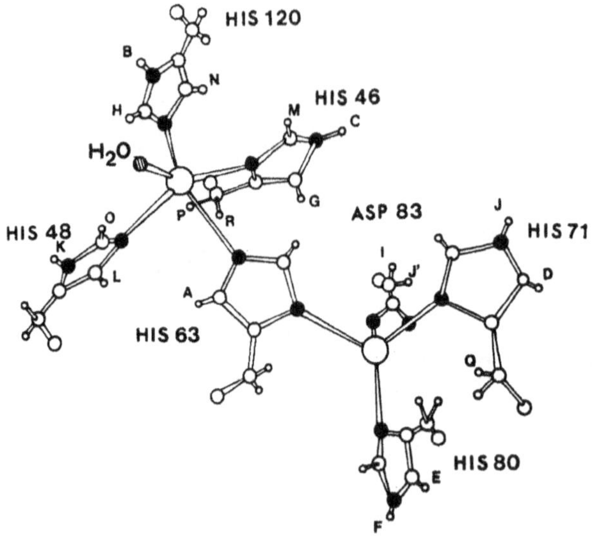

Figure 9. Schematic drawing of the active cavity of copper, zinc superoxide dismutase [51]

It has not been possible to detect any ^{15}N-^{1}H connectivity for the remaining NH metal coordinated residues, but the detection of HMQC peaks involving proton resonances as broad as 250 Hz is a promising result.

It is also interesting to realize how feasible are heterocorrelated experiments on paramagnetic systems at natural abundance. Figure 11 reports the HMQC experiment on the oxidized form of the two Fe_4S_4 cluster ferredoxin from *C. acidi urici*, recorded at natural abundance. All resonances of β carbons and several resonances of the α carbons can be clearly observed [52].

Obviously, access to information concerning heteronuclei can be of great help when analyzing the factorization of the contributions to the hyperfine shift. It can be easily foreseen that the study of relaxation times of ^{13}C and ^{15}N nuclei in proximity of a paramagnetic center will be one of the major issues in future investigations of paramagnetic metalloproteins

4. Relaxation allowed coherence transfer

In paramagnetic molecules Curie spin provides a "chemical shift anisotropy" whose modulation induces nuclear Curie spin relaxation [54-56]. Since the electron magnetic moment is rather large, the latter contribution can be very strong. For protons in paramagnetic molecules Curie spin relaxation is the only sizable form of chemical shift anisotropy.

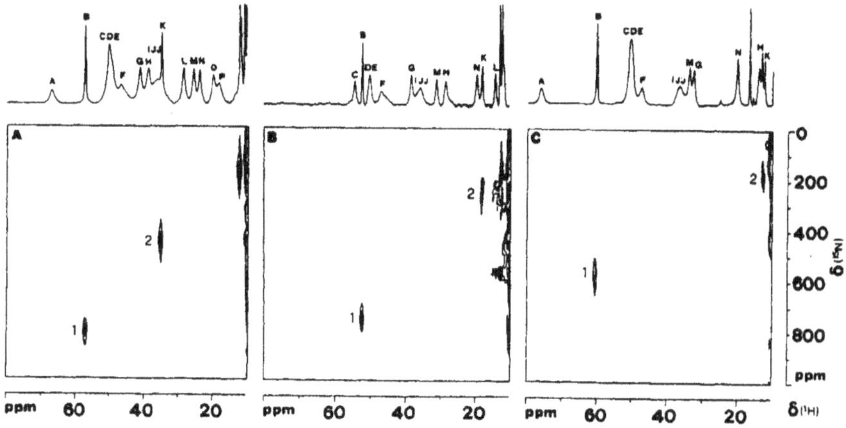

Figure 10. ^{15}N-^{1}H HMQC experiments performed on (A) anion-free Cu_2Co_2SOD, (B) $Cu_2Co_2SOD+N_3^-$, (C) $Cu_2Co_2SOD+CN^-$. ^{15}N enrichment was about 98%. The experiments were performed in magnitude mode, using the four pulse sequence described by Bax et al. [53]. The delay between the first two pulses was set to match the T_2 values of the proton signals (1.5 ms), e.g. shorter than typical $1/(2J_{NH})$ values. 1024 and 100 points were used in the F_2 and F_1 dimensions, respectively. A recycle delay of 135 ms was used. 800 transients each FID were acquired, dwell times were 5 μs in the F_2 dimension and 2.1 μs in the F_1 dimension. Experiments were performed at 298 K on a 600 MHz spectrometer, using a 5mm reverse probe [taken from ref. [48], with permission].

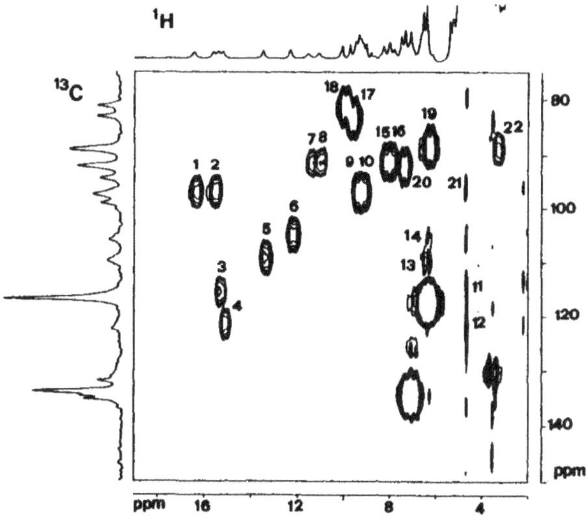

Figure 11. ^{13}C-^{1}H HMQC experiment performed on oxidized ferredoxin from *C. acidi urici* (natural abundance). The labeled cross peaks allow the detection of the ^{13}C resonances of the Cβ of the eight cluster-coordinated cysteines and of six out of eight Cα resonances. Cross peaks 1-16 refer to Cβ-Hβ connectivities, 17-22 to Cα-Hα [taken from ref. [52], with permission].

It is well known that, when chemical shift anisotropy and dipole-dipole coupling are time modulated by the same type of random motion (i.e. by rotation, $\tau_c = \tau_r$), then a cross term appears in the relaxation equations which gives rise to the so called cross-correlation effect. As in paramagnetic molecules Curie spin relaxation is modulated by rotation like the nucleus-to-nucleus dipolar interaction, the appearance of a cross term in the relaxation equations is predicted. In the slow motion limit, the latter has the form:

$$R_{2A}^{CC} = \frac{2}{5}\left(\frac{\mu_0}{4\pi}\right)^2 \frac{\hbar\gamma_A^2\gamma_B B_0 g_e^2 \mu_B^2 S_e(S_e+1)}{(3kT)r_{MA}^3 r_{AB}^3}\left[3\left(\frac{r_{MA}^2 + r_{AB}^2 - r_{MB}^2}{2r_{MA}r_{AB}}\right)^2 - 1\right]\tau_c \qquad (6)$$

where r_{MA} and r_{MB} are the metal-to-nuclei A and B distances, r_{AB} is the nucleus A-to-nucleus B distance.

This recently described phenomenon [57] potentially contains additional information with respect to nucleus-Curie spin and nucleus-nucleus dipolar interactions. Indeed, eq. [6] depends on three distances and it can be extremely useful to obtain structural information on the residues close to the metal ions. Like the "usual" cross correlation terms, recently reviewed and elaborated upon by Bodenhausen and Wimperis [58,59], this effect gives unequal linewidth for the two doublet components. It was shown by Bodenhausen and Wimperis that cross correlation may give rise to the occurrence of cross peaks between two nuclear spins I and S even in the absence of scalar coupling between them.

Such an effect, which has been termed Relaxation Allowed Coherence Transfer (RACT [58]), has practically little influence on the reliability of COSY experiments in diamagnetic systems, as the cross correlation term is small because chemical shift anisotropy for protons is also small. Cross peaks due to RACT have been observed only in special circumstances [58,59].

A rather different situation occurs in paramagnetic molecules when Curie spin relaxation is considered. The latter contribution is, especially at high magnetic fields, the dominant contribution to transverse relaxation, and therefore the occurrence of cross correlation between dipole-dipole coupling and Curie spin relaxation becomes a major source of cross peaks in COSY experiments. It has been shown that, although cross correlation effects are maximal when Curie spin and dipole-dipole interactions are equal, they remains significant up to Curie spin/dipole-dipole relaxation ratios as large as 100 [57]. In other words, detection of relaxation allowed COSY cross peaks in large paramagnetic metalloproteins should be considered as a rule rather than an exception.

This phenomenon gives a sound theoretical ground to the experimental evidences, reported in the last years, of the occurrence of COSY cross peaks involving signals with up to 1000 Hz linewidths [11,60].

Figure 12 (curve A) shows a behavior of the decrease in intensity of the cross peak for an *AX* system scalar-coupled with J_{AX} = 10 Hz as a function of the increased linewidth of the signal. Horizontal dashed lines show the unit intensity and the detectability threshold when the S/N ratio in the experiment is as high as 1000. It appears that for linewidths of the order of 300 Hz the ratio between diagonal peaks and cross peaks is above 1000, which is already an optimistic estimate for the detectability threshold [57]. Curves B-E shows the calculated cross peak intensities plotted as a function of the average of the linewidth of the two signals in such a way to make the comparison easy with the calculated intensity of the "true" scalar cross peak. Only for relatively long internuclear distances (curve E) the dipole-dipole relaxation becomes so small that relaxation-allowed cross peaks intensities become smaller than true scalar cross peaks.

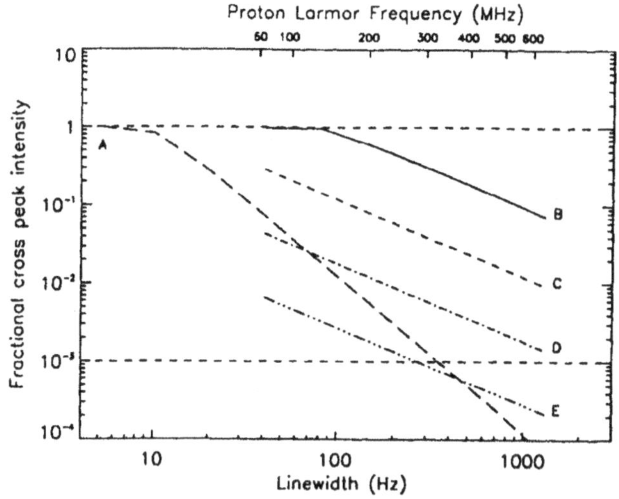

Figure 12. Fractional COSY cross peak intensity for an *AX* system as a function of signal linewidth, in the presence of scalar coupling (J = 10 Hz) (A) or in the absence of scalar coupling and in the presence of Curie relaxation and different *AX* coupling (r_{AX} = 1.6 (B), 2.2 (C), 2.9 (D), 3.7 Å (E)) [taken from ref. [57], with permission].

When COSY spectra are recorded in phase-sensitive mode, the 90° phase shift which occurs between *J*-coupling and RACT-type cross peaks can be used to discriminate between the two effects. This has been recently done in the case of horseradish peroxidase, and the dominance of the RACT contribution has been clearly shown [61]. It can be demonstrated that a refocusing 180° pulse is able to cancel the effect of RACT, while it is well known that the scalar effects are retained. Therefore any pulse sequence of COSY type employing a refocusing -τ -180°- τ - sequence and utilizing the resulting antiphase coherences for subsequent developments will cancel the RACT effect. This is the case of, for instance, ISECR COSY [62] and also of TOCSY (and all its variants) [33,63,64]. While the detection of pure scalar effects with elimination of RACT effects is relatively easy, the selection of RACT with elimination of scalar effects is not trivial. The difficulty is due to the subtle nature of RACT, which is intrinsically different with respect to scalar effect, although it evolves with the same order of coherence transfer. Sequences which can separate the two effects have been recently developed in our laboratory. One such pulse sequence is reported in Figure 13.

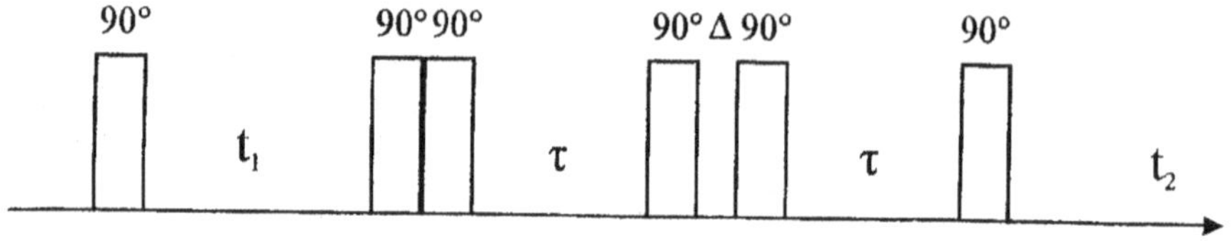

Figure 13. Pulse sequence for detecting RACT cross peaks, while suppressing scalar cross peaks [taken from ref. [31], with permission].

The first 90°-t_1-90° creates a double quantum coherence (both scalar and RACT) which is transformed into antiphase magnetization after the third 90° pulse. The second part of the sequence, τ-90°-Δ-90°-τ- transform the antiphase magnetization into single quantum coherence. The 90°-Δ-90°- scheme acts as a 180° refocusing pulse, and also acts as a *z* filter which retains only single quantum coherences. Within this scheme, prior to the last 90° pulse, both RACT and scalar effects have evolved during t_1 and appear as observable single quantum coherences shifted by 90°. At this point, the last 90° pulse will turn the unwanted effects (scalar effect, in this case) on the *z* axis and will leave the other effect unaffected.

As a practical example of this sequence, we report in Figure 14 the magnitude COSY, RACTSY and ISECR COSY spectra of the thiocyanate adduct of cobalt substituted carbonic anhydrase [31]. The nature of the COSY cross peaks observed in panel A is clarified by the absence of cross peaks in the ISECR COSY experiment (panel C) and by the detection of cross peaks in the RACTSY experiment (panel B). Due to the complexity of the RACTSY sequence, the S/N ratio of the RACTSY spectrum is of much worse quality with respect to the magnitude COSY experiment; nevertheless the origin of the COSY cross peaks is unambiguously proved.

5. Strategies for sequence-specific assignment

The assignment of the hyperfine shifted signals in paramagnetic molecules relied, up to about one decade ago, on the assumption that NMR parameters could be theoretically interpreted and related to distances and/or electronic distributions. Chemical shifts, T_1 values and linewidths have been used to obtain structural information. Chemical modifications like selective deuteriation or substituent replacement have also been used in the assigments. In the absence of chemical modifications, only tentative assignments, on the basis of chemical shifts and relaxation rates could be proposed. For instance, the chemical shifts in Co(II) and

Ni(II) metal substituted proteins have been tabulated as a function of the coordination number and of the aminoacid residue [65]; such type of tables can be used as a reference for tentative identification of metal ligands in proteins of unknown structure.

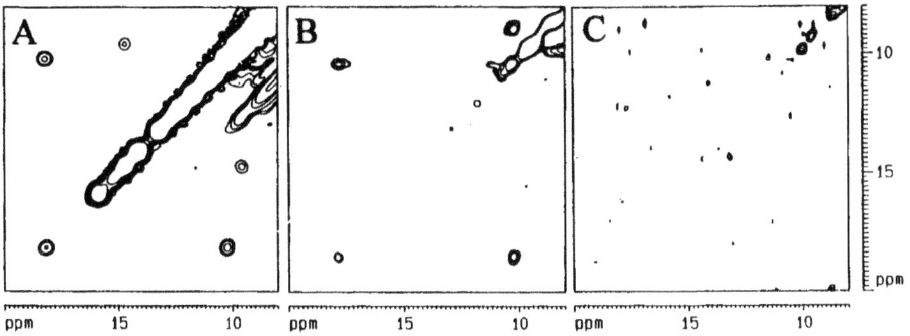

Figure 14. Standard magnitude COSY spectrum of cobalt(II)-substituted carbonic anhydrase (thiocyanate adduct) (A), RACTSY spectrum (B), and ISECR COSY spectrum (C) of the same sample. Conditions: $t_{1max} = t_{2max} = 3.9$ ms. 138 experiments in the F_1 dimensions were performed. magnitude COSY (A) was processed using $\sin^2$ filter in both dimensions, phase-sensitive RACTSY.(B) was processed using $\cos^2$ filter in the F_2 dimension and $\sin^2$ filter in the F_1 dimension, phase sensitive ISECR- COSY (C) was processed using $\cos^2$ filter in both dimensions [taken from ref. [31], with permission].

Since 1983, the application of 1*D* NOE to the study of paramagnetic systems provided the detection of proton-proton connectivities also when the signals of interest were broad and fast relaxing [1]. Heme proteins are the systems in which such studies were first performed [3,66-68]. Among non-heme proteins, cobalt and nickel substituted Cu,Zn, superoxide dismutases provide nice examples [47,69,70].

In the case of heme proteins, the complexity of the prosthetic group permits the detection of an extended network of connectivities which make unambiguous the assignment of the prosthetic group. ^1H NOE, COSY and NOESY experiments provide the necessary intra residue connectivities. Such assignments can thus be independent of the X-ray structure. This has been shown on porphyrin model systems, where the spin density distribution under different redox and coordination states have been also mapped [71-75].

Cu_2Co_2SOD provides a nice example, besides heme proteins. In this case the availability of the X-ray structure of the protein (Figure 9), is crucial for the assignment. Indeed, besides the intra-residue connectivities involving histidine residues, there are inter-residue proton-proton distances smaller than 3.5 Å, which can be monitored through NOESY and NOE experiments [69,76]. The experimental constraints, in the light of the X-ray data, provided the sequence-specific assignment of copper bound histidines [69].

In the case of Met-Mb-CN an extensive assignment not only of the prostethic group but also of almost all (95%) proton signals in a 7.5 Å sphere from the metal ion has allowed Emerson and La Mar to determine the orientation of the magnetic susceptibility tensor in the protein using the experimentally determined pseudocontact shifts of non coordinated aminoacid side chains protons in the heme pocket of the protein [10,19].

In all the above examples, the analysis was essentially restricted to connectivities among hyperfine-shifted signals. However, sequence specific assignment of hyperfine shifted signals can also be performed by detecting connectivities between each of them and one or more amino acids whose spin pattern can be unambiguously identified. This strategy opens new promising pathways.

COSY, NOESY and TOCSY experiments performed on the diamagnetic region, with high resolution and pattern recognition of the various aminoacids, allow us to assign to a specific aminoacid those signals which experience NOE upon saturation of a hyperfine shifted signal. In practice, the "classical" approach for the study of the hyperfine shifted signals is combined with the "classical" approach for the study of diamagnetic systems. The results is the use of the second coordination sphere to assign hyperfine shifted signals.

The connectivities from each hyperfine shifted signal to identified aminoacids may or may not provide unambiguous assignment depending on how "distinctive" they are. For instance, if a signal gives rise to a dipolar connectivity with some diamagnetic signals and, among these signals, only one signal is identified as a CH_3 resonance of a Val residue, this information may not be selective enough to permit the sequence specific assignment because it could well be that two signals of two different residues display this same structural feature.

The absence of the X-ray structure of the investigated protein can be tolerated if the structure of an homologous protein is known. In the latter case, provided that the homology of the residues close to the active site is high, we can obtain a reliable model by replacing, using computer graphics, the amino acids which differ within the two proteins and then submitting the modified structure to a molecular dynamics calculation. Indeed, molecular dynamics calculation have shown to be extremely helpful in this procedure [77].

The reliability of the structural model and the peculiarity of the observed connectivity patterns influence each other. When the X-ray structure is available, even a quantitative analysis of cross peak intensities can be used to discriminate among possible assignments. When this is not the case and only a computer derived model is available, the occurrence of local conformations different from what predicted on the basis of an homologous protein or of a molecular dynamics simulation should always be taken into account. In those cases a more extended network of connectivities is required in order to achieve a convincing assignment.

A successful application of this strategy is provided by high potential iron sulfur proteins (HiPIPs), a class of proteins of about 10,000 MW, whose prosthetic group is a Fe_4S_4 cluster, as depicted in Figure 15 [78,79]. Figure 16 shows the spectrum of reduced and oxidized HiPIPs from *C. vinosum*, whose X-ray structure is known [81], which was the first HiPIP whose Cys protons were sequence-specifically assigned [80].

The pairwise assignment of the βCH_2 of this class of proteins can be easily performed by means of *1D* NOE or NOESY experiments on the entire spectral

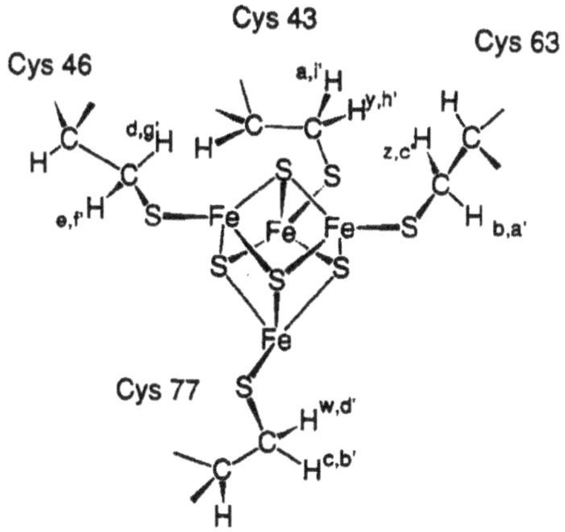

Figure 15. Schematic drawing of the metal cluster of *C. vinosum* HiPIP. Cys β-CH_2 protons are labelled according to their assignment in the ^1H NMR spectra of the reduced and oxidized protein (see Figure 16) [taken from ref. [80], with permission].

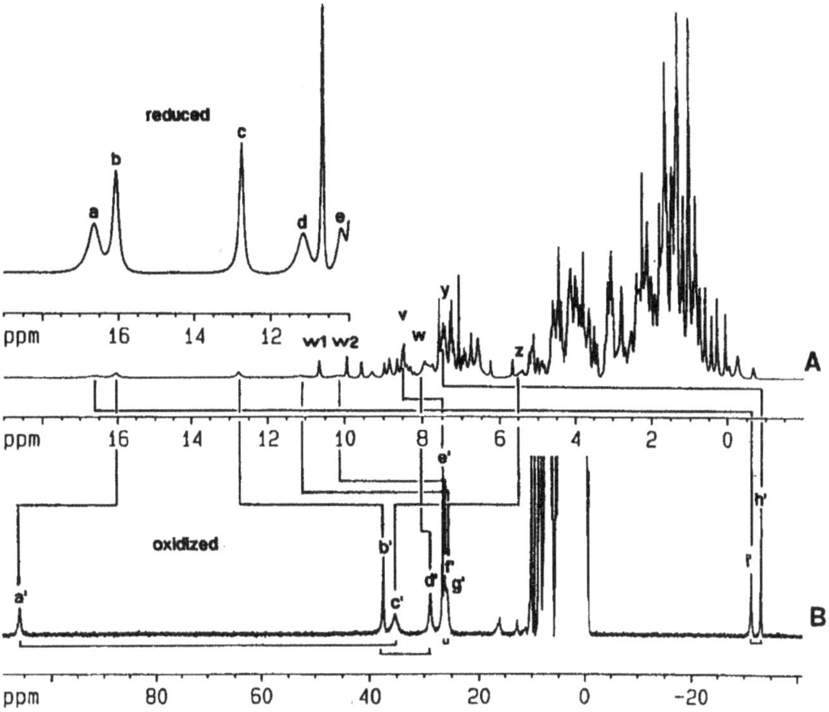

Figure 16. 600 MHz, 300 K ^1H NMR spectra of HiPIP from *C. vinosum* in the reduced (A) state at pH 7.2, and in the oxidized (B) state at pH 5.4. Connectivities established through EXSY experiments are drawn, as well as the geminal connectivities of the oxidized state [taken from ref. [80], with permission].

region, as reported in Table 2. Signals from the protein in the two different oxidation states can be easily related through EXSY experiments, and the information obtained on the two oxidation states are therefore complementary.

TABLE 2. Pairwise identification and sequence specific assignment of cysteine βCH_2 protons in *C. vinosum* HiPIP [80].

Reduced		Oxidized		Sequence Specific Assigment
Signal	δ (ppm)	Signal	δ (ppm)	
a	16.7	i'	-33.01	Cys 43 Hβ2
y	7.57	h'	-31.20	Cys 43 Hβ1
b	16.1	a'	105.82	Cys 63 Hβ2
z	5.52	c'	35.32	Cys 63 Hβ1
c	12.77	b'	37.57	Cys 77 Hβ1
w	8.06	d'	28.90	Cys 77 Hβ2
v (α)	8.52	e' (α)	26.75	Cys 77 Hα
d	11.13	g'	25.91	Cys 46 Hβ2
e	10.16	f'	26.41	Cys 46 Hβ1

The goal is that of assigning each of the four βCH_2 pairs to one of the four Cys residues (namely Cys 43, Cys 46, Cys 63 and Cys 77). Let us consider the cysteine giving rise to signals g' and f' in the oxidized state: signal f' gives rise to a strong NOE with a signal at 10.22 ppm and the corresponding signal in the reduced state is identified through EXSY experiments. In the reduced form, the above mentioned signal (w1) displays a connectivity pattern typical of the NH of a Trp residue. Among the eight βCH_2 Cys protons, Cys 46 Hβ1 is unambiguously the only proton which may experience this feature, because it is close to Trp 80 and therefore is assigned to signal f'. The X-ray structure shows that another Trp has two signals of the six-membered ring (Hε3 or Hζ3) close to the Hα proton of Cys 77. Hα of cysteine 77 in the reduced form has two connectivities (Figure 17D) with two signals of the four membered ring of another Trp (corresponding to the NH signal w2) which is also observed in TOCSY spectrum of Figure 17C. Again, this is an unambiguous feature and leads us to the assignment of this cysteine as Cys 77. Similar considerations hold for the remaining two Cys (see also caption to Figure 17).

Of course, experimental data of this type should be treated with extreme care; for instance, all observed connectivities should be checked at two different temperatures wherever ambiguities due to overlapped signals occur.
The same strategy was then applied to other five HiPIPs, from different bacterial sources, for which no X-ray structure was available [80,82-85].
For each of the investigated HiPIP a structural model was thus generated using as starting structure the one with higher homology among the available X-ray structures of other HiPIPs [86-89]. For *E. halophila* HiPIP II, a refined model was obtained through molecular dynamics simulation, using the structure of *E. halophila* HiPIP I as a starting conformation [82]. In this frame, the utility of MD simulations has been demonstrated by a quantitative analysis of NMR data versus X-ray structure and MD structure on *C. vinosum* HiPIP [77].

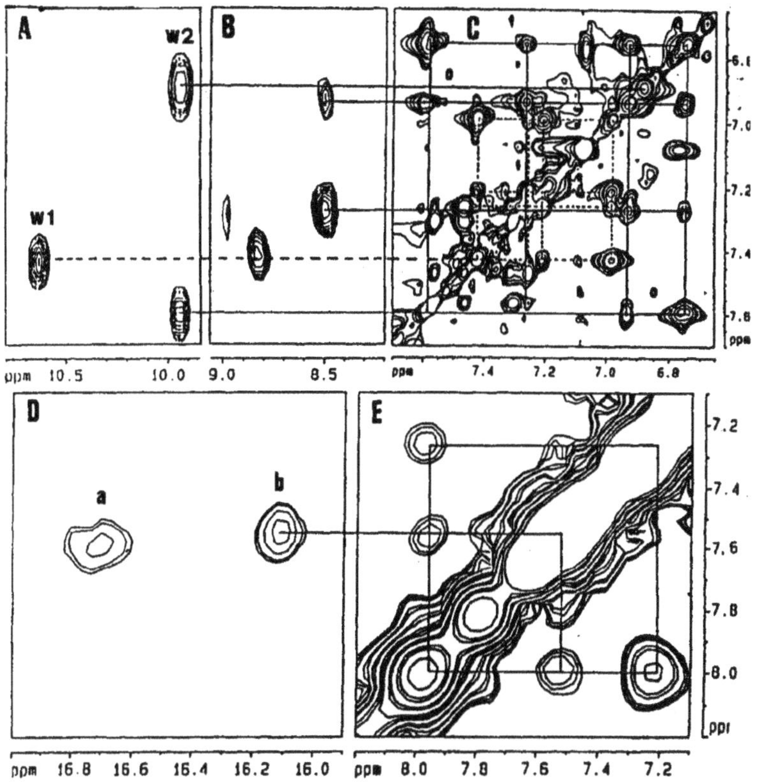

Figure 17. 600 MHz, NOESY (A, B and D), TOCSY (C) and COSY (E) spectra on the aromatic region occur is shown. The patterns of Trp 76 (——) and Trp 80 (- - -) are drawn. The COSY spectrum permits detection of the scalar pattern of Phe 66 (——). The NOESY spectrum allows detection of trough-space connectivities from the Trp 80 and Trp 76 to the respective w1 and w2 NH protons (A), from Phe 66 to signal b (C), and from Hα of Cys 77 and the Trp 76 Hε3 and Hζ 3 protons (B) [taken from ref. [80], with permission].

The above strategy for sequence-specific assignment is being used by several investigators with increasing success [90-95].

Finally, we should underline that it is nowadays possible to achieve the solution structure of a paramagnetic protein, as it is described in another contribution to this book. Therefore, we expect that in the near future there will less and less need to rely on independent structural information to perform the assignment of hyperfine shifted signals.

6. References

1. Johnson, R.D., Ramaprasad, S. and La Mar, G.N. (1983) A method of assigning functionally relevant amino acid residue resonances in paramagnetic hemoproteins using proton NOE measurements, *J. Am. Chem. Soc.* **105**, 7205-7206.
2. Ramaprasad, S., Johnson, R.D. and La Mar, G.N. (1984) Vinyl mobility in myoglobin as studied by time-dependent nuclear Overhauser effect measurements, *J. Am. Chem. Soc.* **106**, 3632-3635.
3. Ramaprasad, S., Johnson, R.D. and La Mar, G.N. (1984) 1H-NMR Nuclear Overhauser Enhancement and Paramagnetic Relaxation Determination of Peak Assignment and the Orientation of Ile-99 FG5 in Metcyanomyoglobin, *J. Am. Chem. Soc.* **106**, 5330-5335.
4. Satterlee, J.D. and Moench, S. (1987) Proton hyperfine resonance assignments using the nuclear Overhauser effect for ferric forms of horse and tuna cytoochrome c, *Biophys. J.* **52**, 101-107.
5. Unger, S.W., Lecomte, J.T.J. and La Mar, G.N. (1985) The Utility of the Nuclear Overhauser Effect for Peak Assignment and Structure Elucidation in Paramagnetic Proteins, *J. Magn. Reson.* **64**, 521-526.
6. Banci, L., Bertini, I., Luchinat, C. and Piccioli, M. (1991) Frontiers in NMR of paramagnetic molecules: ^{1}H NOE and related experiments, in Bertini, I., Molinari, H. and Niccolai, N. (eds) *NMR and biomolecular structure*, VCH, , pp.31-60.
7. Bertini, I., Banci, L. and Luchinat, C. (1989) ^{1}H NMR of paramagnetic metalloproteins, in Oppenheimer, N.J. and James, T.L. (eds) *Nuclear Magnetic Resonance, Part B*, Methods Enzymol., Vol. 177, , pp.246.
8. Lecomte, J.T.J., Unger, S.W. and La Mar, G.N. (1991) Practical Considerations for the Measurements of the Homonuclear Overhauser Effect on Strongly Relaxed Protons in Paramagnetic Proteins, *J. Magn. Reson.* **94**, 112-122.
9. Santos, H., Turner, D.L., Xavier, A.V. and LeGall, J. (1984) Two-Dimensional NMR Studies of Electron Transfer in Cytochrome c3, *J. Magn. Reson.* **59**, 177-180.
10. Emerson, S.D. and La Mar, G.N. (1990) Solution structural characterization of cyanometmyoglobin: resonance assignment of heme cavity residues by two-dimensional NMR, *Biochemistry* **29**, 1545-1556.
11. de Ropp, J.S and La Mar, G.N. (1991) 2D NMR assignment of hyperfine-shifted resonances in strongly paramagnetic metalloproteins: resting state Horseradish Peroxidase, *J. Am. Chem. Soc.* **113**, 4348-4350.
12. Banci, L., Bertini, I., Turano, P., Tien, M. and Kirk, T.K. (1991) Proton NMR investigation into the basis for the relatively high redox potential of lignin peroxidase, *Proc. Natl. Acad. Sci. USA* **88**, 6956-6960.
13. Skjeldal, L., Westler, W.M., Oh, B.-H., Krezel, A.M., Holden, H.M., Jacobson, B.L., Rayment, I. and Markley, J.L. (1991) Two-Dimensional Magnetization Exchange Spectroscopy of Anabaena 7120 Ferredoxin. Nuclear Overhauser Effect and Electron Self-Exchange Cross Peaks from Amino Acid Residues Surrounding the 2Fe-2S Cluster, *Biochemistry* **30**, 7363-7368.
14. Holz, R.C., Que, L.,Jr. and Ming, L.J. (1992) NOESY Studies on the Fe(III) Co(II) Active Site of the Purple Acid Phosphatase Uteroferrin, *J. Am. Chem. Soc.* **114**, 4434-4436.
15. Bertini, I., Luchinat, C., Ming, L.J., Piccioli, M., Sola, M. and Valentine, J.S. (1992) Two-dimensional 1H-NMR studies of the paramagnetic metalloenzyme copper-nickel superoxide dismutase, *Inorg. Chem.* **31**, 4433-4435.
16. Banci, L., Bertini, I. and Luchinat, C. (1994) 2D NMR spectra of paramagnetic systems, in James, T.L. and Oppenheimer, N.J. (eds) *Methods in enzymology*, Academy press, Inc., Florida
17. Bertini, I., Dikiy, A., Luchinat, C., Piccioli, M. and Tarchi, D. (1994) NOE-NOESY: a further tool in NMR of paramagnetic metalloproteins, *J. Magn. Reson. Ser. B* **103**, 278-283.
18. Lecomte, J.T.J. and La Mar, G.N. (1986) The homonuclear overhauser effect in H_2O solution of low-spin hemeproteins. Assignment of protons in the heme cavity of sperm whale myoglobin, *Eur. Biophys. J.* **13**, 373-381.
19. Emerson, S.D. and La Mar, G.N. (1990) NMR determination of the orientation of the magnetic susceptibility tensor in cyanometmyoglobin: a new probe of steric tilt of bound ligand, *Biochemistry* **29**, 1556-1566.

20. Emerson, S.D., Lecomte, J.T.J. and La Mar, G.N. (1988) 1H-NMR Resonance Assignment and Dynamic Analysis of Phenylalanine CD1 in a Low-Spin Ferric Complex of Sperm Whale Myoglobin, *J. Am. Chem. Soc.* **110**, 4176-4182.
21. Lecomte, J.T.J. and La Mar, G.N. (1987) 1H-NMR Probe for Hydrogen Bonding of Distal Residues to Bound Ligands in Heme Proteins: Isotope effect on Heme Electronic Structure of Myoglobin, *J. Am. Chem. Soc.* **109**, 7219-7220.
22. Yu, L.P., La Mar, G.N. and Rajarathnam, K. (1990) ^{1}H NMR Resonance Assignment of the Active Site Residues of Paramagnetic Proteins by 2D Bond Correlation Spectroscopy: Metcyanomyoglobin, *J. Am. Chem. Soc.* **112**, 9527-9534.
23. Banci, L., Bermel, W, Luchinat, C., Pierattelli, R. and Tarchi, D. (1993) ^{1}H 3D NOE-NOE spectrum of met-myoglobin-CN: the first 3D NMR spectrum of a paramagnetic protein, *Magn. Reson. Chem.* **31**, S3-S7.
24. Kuriyan, J., Wilz, S., Karplus, M. and Petsko, G.A. (1986) X-ray structure and refinement of carbon-monoxy (iron II)-myoglobin at 1.5 Å resolution, *J. Mol. Biol.* **192**, 133-154.
25. Griesinger, C., S×rensen, O.W. and Ernst, R.R. (1987) A pratical approach to three-dimensional NMR spectroscopy, *J. Magn. Reson.* **73**, 574-579.
26. Griesinger, C., S×rensen, O.W. and Ernst, R.R. (1989) Three-dimensional Fourier spectroscopy. Application to high-resolution NMR, *J. Magn. Reson.* **84**, 14-63.
27. Oschkinat, H., Griesinger, C., Kraulis, P.J., S×rensen, O.W. and Ernst, R.R. (1988) Three-dimensional NMR spectroscopy of a protein in solution, *Nature* **332**, 374-377.
28. Bertini, I., Capozzi, F., Luchinat, C. and Turano, P. (1991) Applications of COSY to paramagnetic heme-containing systems, *J. Magn. Reson.* **95**, 244-252.
29. Xavier, A.V., Turner, D.L. and Santos, H. (1993) Two-Dimensional NMR of Paramagnetic Metalloproteins, in Vallee, B.L. and Riordan, J.F. (eds) , Methods in Enzymology, Academic Press, San Diego,CA
30. Turner, D.L. (1993) Optimization of COSY and Related Methods. Applications to 1H NMR of Horse Ferricytochrome c, *J. Magn. Reson. Ser. A* **104**, 197-202.
31. Bertini, I., Luchinat, C., Piccioli, M. and Tarchi, D. (1994) COSY spectra of paramagnetic macromolecules, observability, scalar effects, cross correlation effects, relaxation allowed coherence transfer, *Concepts Magn. Reson.* in press
32. Bax, A. and Davis, D.G. (1985) Practical Aspects of Two-Dimensional Transverse NOE Sepctroscopy, *J. Magn. Reson.* **63**, 207-213.
33. Davis, D.G. and Bax, A. (1985) Assignment of Complex ^{1}H NMR Spectra via Two-dimensional Homonuclear Hartman-Hahn Spectroscopy, *J. Am. Chem. Soc.* **107**, 2820-2821.
34. Luchinat, C., Steuernagel, S. and Turano, P. (1990) Application of 2D-NMR techniques to paramagnetic systems, *Inorg. Chem.* **29**, 4351-4353.
35. Sadek, M., Brownlee, R.T.C., Scrofani, S.D.B. and Weed, A.G. (1993) TOCSY Assignment of Broad Resonances in Paramagnetic Proteins, *J. Magn. Reson.* **101**, 309-314.
36. Bertini, I., Turano, P. and Vila, A.J. (1993) NMR of paramagnetic metalloproteins, *Chem. Rev.* **93**, 2833-2932.
37. Oh, B.-H. and Markley, J.L. (1990) Multinuclear magnetic resonance studies of the 2Fe-2S* ferredoxin from Anabaena species strain PCC 7120. 1. Sequence-specific hydrogen-1 resonance assignments and secondary structure in solution of the oxidized form, *Biochemistry* **29**, 3993-4004.
38. Oh, B.-H., Mooberry, E.S. and Markley, J.L. (1990) Multinuclear magnetic resonance studies of the 2Fe-2S* ferredoxin from Anabaena species strain PCC 7120. 2. Sequence-specific carbon-13 and nitrogen-15 resonance assignments of the oxidized form, *Biochemistry* **29**, 4004-4011.
39. Oh, B.-H. and Markley, J.L. (1990) Multinuclear magnetic resonance studies of the 2Fe-2S* ferredoxin from Anabaena species strain PCC 7210. 3. Detection and characterization of hyperfine-shifted nitrogen-15 and hydrogen-1 resonances of the oxidized form, *Biochemistry* **29**, 4012-4017.
40. Santos, H. and Turner, D.L. (1992) 13C and proton NMR studies of horse cytochrome c, *Eur. J. Biochem.* **206**, 721-728.
41. Yamamoto, Y., Nanai, N., Inoue, Y. and Chujo, R. (1988) Natural abundance 13C-NMR study of paramagnetic horse heart Ferricytochrome c Cyanide complex: assignment of hyperfine shifted heme methyl carbon resonances, *Biochem. Biophys. Res. Commun.* **151**, 262-269.
42. Adachi, S. and Morishima, I. (1992) Modification of the Distal Histidyl Imidazole in Myoglobin to N-Tetrazole-Substituted Imidazole and Its Effects on the Heme Environmental Structure and Ligand Binding Properties, *Biochemistry* **31**, 8613-8618.
43. Shiro, Y., Iizuka, T., Makino, R., Ishimura, Y. and Morishima, I. (1989) 15N NMR study on the cyanide (C15N-) complex of cytochrome P-450cam. Effects of d-camphor and putidaredoxin on the iron-ligans structure, *J. Am. Chem. Soc.* **111**, 7707-7711.
44. Bertini, I., Jonsson, B.-H., Luchinat, C., Pierattelli, R. and Vila, A.J. (1994) Strategies of signal assignments in paramagnetic metalloproteins. An NMR investigation of the thiocyanate adduct of the cobalt(II)-substituted human carbonic anhydrase II, *J. Magn. Reson.* in press

45. Valentine, J.S. and Pantoliano, M.W. (1981) , in Spiro, T.G.. (ed) *Copper Proteins*, Protein-metal ionteractions in cuprozinc protein (superoxide dismutase) *Wiley.*, New York, pp.291.
46. Bertini, I., Lanini, G., Luchinat, C., Messori, L., Monnanni, R. and Scozzafava, A. (1985) Investigation of Cu_2Co_2SOD and its anion derivatives. ^{1}H NMR and electronic spectra, *J. Am. Chem. Soc.* **107**, 4391-4396.
47. Bertini, I., Luchinat, C. and Piccioli, M. (1994) Copper zinc superoxide dismutase a paramagnetic protein that provides a unique frame for the NMR investigations, *Progr. Nucl. Magn. Reson. Spectrosc.* **26**, 91-141.
48. Bertini, I., Luchinat, C., Macinai, R., Piccioli, M., Scozzafava, A. and Viezzoli, M.S. (1994) Paramagnetic metal centers in proteins can be investigated through heterocorrelated NMR spectroscopy, *J. Magn. Reson. Ser. B* **B104**, 95-98.
49. Yamamoto, Y., Nanai, N., Inoue, Y. and Chujo, R. (1989) 15N NMR study of iron(III) porphyrin-imidazole complex. Observation of the bound imidazole resonances, *J. Chem. Soc. ,Chem. Commun.* 1419-1421.
50. Goff, H.M. (1981) Iron(III) Porphyrin-Imidazole Complexes. Analysis of Carbon-13 Nuclear Magnetic Resonance Isotropic Shifts and Unpaired Spin Delocalization, *J. Am. Chem. Soc.* **103**, 3714.
51. Tainer, J.A., Getzoff, E.D., Beem, K.M., Richardson, J.S. and Richardson, D.C. (1982) Determination and Analysis of 2 A Structure of Copper Zinc Superoxide Dismutase, *J. Mol. Biol.* **160**, 181-217.
52. Bertini, I., Capozzi, F., Luchinat, C., Piccioli, M. and Vila, A.J. (1994) The Fe_4S_4 centers in ferredoxins studied through proton and carbon hyperfine coupling. Sequence specific assignments of cysteines in ferredoxins from *Clostridium acidi urici* and *Clostridium pasteurianum*, *J. Am. Chem. Soc.* **116**, 651-660.
53. Bax, A., Griffey, R.H. and Hawkins, B.L. (1983) Correlation of Proton and Nitrogen-15 Chemical Shifts by Multiple Quantum NMR, *J. Magn. Reson.* **55**, 301-315.
54. Guéron, M. (1975) Nuclear Relaxation in Macromolecules by Paramagnetic Ions: A Novel Mechanism, *J. Magn. Reson.* **19**, 58-66.
55. Vega, A.J. and Fiat, D. (1976) Nuclear Relaxation Processes of Paramagnetic Complexes. The Slow Motion Case, *Mol. Phys.* **31**, 347-362.
56. Banci, L., Bertini, I. and Luchinat, C. (1991) *Nuclear and electron relaxation. The magnetic nucleus-unpaired electron coupling in solution*, VCH, Weinheim.
57. Bertini, I., Luchinat, C. and Tarchi, D. (1993) Are true scalar proton-proton connectivities ever measured in COSY spectra of paramagnetic macromolecules? *Chem. Phys. Lett.* **203**, 445-449.
58. Wimperis, S. and Bodenhausen, G. (1989) Relaxation-Allowed Cross-Peaks in Two-Dimensional NMR Correlation Spectroscopy, *Mol. Phys.* **66**, 897-919.
59. Wimperis, S. and Bodenhausen, G. (1987) Relaxation-Allowed Transfer of Coherence in NMR between Spins which are not Scalar Coupled, *Chem. Phys. Lett.* **140**, 41-45.
60. Banci, L., Bertini, I., Turano, P. and Vicens Oliver, M. (1992) NOE and two-dimensional correlated ^{1}H NMR spectroscopy of cytochrome c' from *Chromatium vinosum*, *Eur. J. Biochem.* **204**, 107-112.
61. Qin, J., Delaglio, F., La Mar, G.N. and Bax, A. (1993) Distinguishing the Effects of Cross Correlation and J Coupling in COSY Spectra of Paramagnetic Protein, *J. Magn. Reson. Ser. B* **102**, 332-336.
62. Talluri, S. and Scheraga, H.A. (1990) COSY with In-Phase Cross Peaks, *J. Magn. Reson.* **86**, 1-10.
63. Bax, A. and Davis, D.G. (1985) MLEV-17-Based Two-Dimensional Homonuclear Magnetization Transfer Spectroscopy, *J. Magn. Reson.* **65**, 355-360.
64. Griesinger, C., Otting, G., Wüthrich, K.J. and Ernst, R.R. (1988) Clean TOCSY for 1H Spin System Identification in Macromolecules, *J. Am. Chem. Soc.* **110**, 7870-7872.
65. Banci, L. and Piccioli, M. (1994) Cobalt(II) and Nickel(II) Substituted Proteins, *Encycl. of Nuclear Magn. Reson.* in press
66. Pande, U., La Mar, G.N., Lecomte, J.T.J., Ascoli, F., Brunori, M., Smith, K.M., Pandey, R.K., Parish, D.W. and Thanabal, V. (1986) NMR Study of the Molecular and Electronic Structure of the Heme Cavity of Aplysia Metmyoglobin. Resonance Assignments Based on Isotrope Labeling and Proton Nuclear Overauser Effect Measurements, *Biochemistry* **25**, 5638-5646.
67. McLachlan, S.J., La Mar, G.N. and Sletten, E. (1986) Ferricytochrome b5: assignment of heme propionate resonances on the basis of nuclear Overhauser effect measurements and the nature of interprotein contacts with partner redox proteins, *J. Am. Chem. Soc.* **108**, 1285-1291.
68. Thanabal, V., de Ropp, J.S and La Mar, G.N. (1986) Determination of vinyl orientation in resting state and compound I of horseradish peroxidase by 1H nuclear Overhauser effect, *J. Am. Chem. Soc.* **108**, 4244-4245.
69. Banci, L., Bertini, I., Luchinat, C., Piccioli, M., Scozzafava, A. and Turano, P. (1989) ^{1}H NOE studies on dicopper(II) dicobalt(II) superoxide dismutase, *Inorg. Chem.* **28**, 4650-4656.
70. Banci, L., Bertini, I., Luchinat, C. and Piccioli, M. (1990) Transient versus steady state NOE in paramagnetic molecules. Cu_2Co_2SOD as an example, *FEBS Lett.* **272**, 175-180.
71. Lin, Q., Simonis, U., Tipton, A.R., Norvell, C.J. and Walker, F.A. (1992) Models of the cytochromes b. 10. Application of proton COSY to delineate the spin density distribution at the pyrrole .beta. positions of unsymmetrically substituted low-spin iron(III) tetraphenylporphyrins, *Inorg. Chem.* **31**, 4216-4217.

72. Simonis, U., Lin, Q., Tan, H., Barber, R.A. and Walker, F.A. (1993) Two-Dimensional ^{1}H NMR Spectroscopy of Paramagnetic Haem Models of the Cytochromes: Successes, Failures and Unanswered Questions, *Magn. Reson. Chem.* **31**, S133-S144.
73. La Mar, G.N. and Walker, F.A. (1973) Proton Nuclear Magnetic Resonance Line Widths and Spin Relaxation in Paramagnetic Metalloporphyrins of Cr(III), Mn(III), and Fe(III), *J. Am. Chem. Soc.* **95**, 6950-6956.
74. La Mar, G.N. and Walker, F.A. (1979) Nuclear Magnetic Resonance of Paramagnetic Metalloproteins, in Dolphin, D. (ed) *The Porphyrins*, Academic Press, New York, pp.61-157.
75. Shin, K. and Goff, H.M. (1990) Iron(III) porphyrin promoted aerobic oxidation of sulfur dioxide, *J. Chem. Soc., Chem. Commun.* 461-462.
76. Banci, L., Bertini, I., Luchinat, C., Piccioli, M. and Scozzafava, A. (1993) 1D versus 2D ^{1}H NMR experiments in dicopper, dicobalt superoxide dismutase: a further mapping of the active site, *Gazz. Chim. Ital.* **123**, 95-100.
77. Banci, L., Bertini, I., Carloni, P., Luchinat, C. and Orioli, P.L. (1992) Molecular dynamics simulations on HiPIP from *Chomatium vinosum* and comparison with NMR data, *J. Am. Chem. Soc.* **114**, 10683-10689.
78. Bertini, I., Ciurli, S. and Luchinat, C. (1994) The electronic and geometric structures of iron-sulfur proteins studied through electron-nuclear hyperfine coupling, *Angew. Chem.* in press
79. Luchinat, C. and Ciurli, S. (1993) NMR of polymetallic systems in proteins, *Biological Magnetic Resonance* **12**, 357-420.
80. Bertini, I., Capozzi, F., Ciurli, S., Luchinat, C., Messori, L. and Piccioli, M. (1992) Identification of the iron ions of HiPIP from *Chromatium vinosum* within the protein frame through 2D NMR experiments, *J. Am. Chem. Soc.* **114**, 3332-3340.
81. Carter, C.W.J., Kraut, J., Freer, S.T., Xuong, N.-H., Alden, R.A. and Bartsch, R.G. (1974) Two-angstrom crystal structure of Chromatium vinosum high-potential iron protein, *J. Biol. Chem.* **249**, 4212-4215.
82. Banci, L., Bertini, I., Capozzi, F., Carloni, P., Ciurli, S., Luchinat, C. and Piccioli, M. (1993) The iron-sulfur cluster in the oxidized high potential iron sulfur protein from *Ectothiorhodospira halophila, J. Am. Chem. Soc.* **115**, 3431-3440.
83. Bertini, I., Capozzi, F., Luchinat, C. and Piccioli, M. (1993) ^{1}H NMR investigation of oxidized and reduced HiPIP from *R. globiformis*, *Eur. J. Biochem.* **212**, 69-78.
84. Banci, L., Bertini, I., Ciurli, S., Ferretti, S., Luchinat, C. and Piccioli, M. (1993) The electronic structure of $(Fe_4S_4)^{3+}$ clusters in proteins; an investigation of the oxidized HiPIP II from *Ectothiorhodospira vacuolata*, *Biochemistry* **32**, 9387-9397.
85. Bertini, I., Gaudemer, A., Luchinat, C. and Piccioli, M. (1993) Electron self-exchange in HiPIPs. A characterization of HiPIP I from *Ectothiorhodospira vacuolata*, *Biochemistry* **32**, 12887-12893.
86. Backes, G., Mino, Y., Loehr, T.M., Meyer, T.E., Cusanovich, M.A., Sweeney, W.V., Adman, E.T. and Sanders-Loehr, J. (1991) The environment of Fe_4S_4 cluster in ferredoxins and high-potential iron proteins. New information from X-ray crystallography and resonance Raman spectroscopy, *J. Am. Chem. Soc.* **113**, 2055-2064.
87. Carter, C.W.J., Kraut, J., Freer, S.T. and Alden, R.A. (1974) Comparison of oxidation-reduction site geometries in oxidized and reduced Chromatium high potential iron protein and oxidized Peptococcus aerogenes ferredoxin, *J. Biol. Chem.* **49**, 6339-6346.
88. Rayment, I., Wesemberg, G., Meyer, T.E., Cusanovich, M.A. and Holden, H.M. (1992) Three-dimensional structure of the high-potential iron-sulfur protein isolated from the purple phototrophic bacterium *Rhodocyclus tenuis* determined and refined at 1.5 Å resolution, *J. Mol. Biol.* **228**, 672.
89. Benning, M.M., Meyer, T.E., Rayment, I. and Holden, H.M. (1994) Molecular structure of the oxidized High-potential Iron-sulfur protein isolated from *Ectothiorhodospira vacuolata*, *Biochemistry* **33**, 2476-2483.
90. Macedo, A.L., Palma, P.N., Moura, I., LeGall, J., Wray, V. and Moura, J.J.G. (1993) Two-dimensional ^{1}H NMR studies on *Desulfovibrio gigas* ferredoxins. Assignment of the iron-sulfur cluster cysteinyl ligand protons, *Magn. Reson. Chem.* **31**, S59-S67.
91. Busse, S.C., La Mar, G.N., Yu, L.P., Howard, J.B., Smith, E.T., Zhou, Z.H. and Adams, M.W.W. (1992) Proton NMR Investigation of the Oxidized Three-Iron Clusters in the Ferredoxins from the Hyperthermophilic Archae Pyrococcus furiosus and Thermococcus litoralis, *Biochemistry* **31**, 11952-11962.
92. Nettesheim, D.G., Harder, S.R., Feinberg, B.A. and Otvos, J.D. (1992) Sequential resonance assignments of oxidized high-potential iron-sulfur protein from *Chromatium vinosum*, *Biochemistry* **31**, 1234-1244.
93. Cheng, H., Grohmann, K. and Sweeney, W.V. (1992) NMR studies of Azotobacter vinelandii and *Pseudomonas putida* seven-iron ferredoxins. Direct assignment of beta-cysteinil carbon NMR resonances and further proton NMR assignments of cysteinil and aromatic resonances, *J. Biol. Chem.* **267**, 8073-8080.
94. Moratal Mascarell, J.M., Salgado, J., Donaire, A., Jimenez, H.R., Castells, J. and Martinez Ferrer, M.-J. (1993) ^{1}H 2D-NMR characterization of Ni(II)-substituted Azurin from *Pseudomonas aeruginosa*, *Magn. Reson. Chem.* **31**, S41-S46.

95. Moratal Mascarell, J.M., Donaire, A., Salgado, J., Jimenez, H.R., Castells, J. and Piccioli, M. (1993) Two Dimensional ^{1}H NMR spectra of Ferricytochrome c_{551} form *Pseudomonas aeruginosa*, *FEBS Lett*, **343**, 305-308.

THE HYPERFINE COUPLING

IVANO BERTINI and PAOLA TURANO
Department of Chemistry, University of Florence
Via Gino Capponi, 7, 50121 Florence, Italy

1. Introduction: Shift and Relaxation

NMR of paramagnetic compounds monitors the hyperfine coupling of resonating nuclei with unpaired electrons[1-3]. In the spin Hamiltonian formalism, such coupling is expressed as:

$$\aleph = \hat{\mathrm{I}} \cdot \mathbf{A} \cdot \hat{\mathrm{S}} \tag{1}$$

where $\hat{\mathrm{I}}$ and $\hat{\mathrm{S}}$ are spin operators (nuclear and electronic, respectively) and $\mathbf{A}$ is the coupling tensor. The latter contains a constant, A_c, which originates from the unpaired spin density sitting at the resonating nucleus. It is given by:

$$A_c = \frac{\mu_0}{3S} \hbar \gamma_N g_e \mu_B \sum_i \rho_i \tag{2}$$

where μ_0 is the magnetic permeability of vacuum, γ_N is the nuclear magnetogyric ratio, g_e is the free electron g value, μ_B is the electron Bohr magneton, S is the total

G N. La Mar (ed.), Nuclear Magnetic Resonance of Paramagnetic Macromolecules, 29-54.

spin of the molecule, and ρ_i is the spin density at the nucleus due to the *i*th s orbital. This contribution is called contact or Fermi contact coupling. The effect on the resonating nucleus of unpaired spin density on the molecule is dipolar in nature and is represented by the dipolar tensor $\mathbf{A}_{\mathbf{dip}}$. Such tensor is traceless if the system is characterized by an isotropic magnetic susceptibility; otherwise, in the presence of anisotropic magnetic susceptibility, the trace (A_{pc}) can be taken out of the matrix and the remaining tensor is traceless. In the former case in solution the dipolar hyperfine coupling tensor averages zero whereas in the latter it averages the trace, which is termed pseudocontact term.

When we pass to the nuclear shifts, the energy contribution to the $\Delta m = \pm 1$ transitions for the contact term is[4]:

$$\left(\frac{\Delta\nu}{\nu_0}\right)^{con} = -\frac{A_c}{\hbar\gamma_N B_0}\langle S_z\rangle = \frac{A_c}{\hbar}S(S+1)\frac{\bar{g}\mu_B}{3\gamma_N kT} \quad (3)$$

Note that whereas A_c contains g_e, $\langle S_z\rangle$ contains the average of $\mathbf{g}$. B_0 is the external magnetic field.

A better formula outside the spin Hamiltonian formalism is[5]:

$$\left(\frac{\Delta\nu}{\nu_0}\right)^{con} = \frac{A_c}{3\hbar\gamma_N\mu_0\mu_B}\left(\frac{\chi_{xx}}{g_{xx}}+\frac{\chi_{yy}}{g_{yy}}+\frac{\chi_{zz}}{g_{zz}}\right) \quad (4)$$

which, however, is hardly of any use because the values of the components of the magnetic susceptibility tensor χ in solution are generally unknown.

The pseudocontact shift for metal centered unpaired electrons is given by:

$$\left(\frac{\Delta\nu}{\nu_0}\right)^{pc} = \frac{A_{pc}}{\hbar}S(S+1)\frac{\bar{g}\mu_B}{3\gamma_N kT} \quad (5)$$

In this case we have to define the polar coordinates r, θ and φ of the resonating nucleus with respect to the magnetic anisotropic tensor and the pseudocontact hyperfine constant is [6]:

$$A_{pc} = \frac{\mu_0}{4\pi}\frac{\hbar\gamma_N\mu_B}{3\bar{g}r^3}\left\{\left[g_{zz}^2 - \frac{1}{2}\left(g_{xx}^2 + g_{yy}^2\right)\right]\left(3\cos^2\theta - 1\right) + \frac{3}{2}\left(g_{xx}^2 - g_{yy}^2\right)\sin^2\theta\cos 2\varphi\right\} \quad (6)$$

the angle θ being defined as in Fig. 1.

Again, a more precise formula for the pseudocontact shift contains the magnetic susceptibility tensor components[5,7]:

$$\left(\frac{\Delta\nu}{\nu_0}\right)^{pc} = \frac{1}{4\pi}\frac{1}{3N_A r^3}\left\{\left[\chi_{zz} - \frac{1}{2}\left(\chi_{xx} + \chi_{yy}\right)\right]\left(3\cos\theta - 1\right) + \frac{3}{2}\left(\chi_{xx} - \chi_{yy}\right)\sin^2\theta\cos 2\varphi\right\} \quad (7)$$

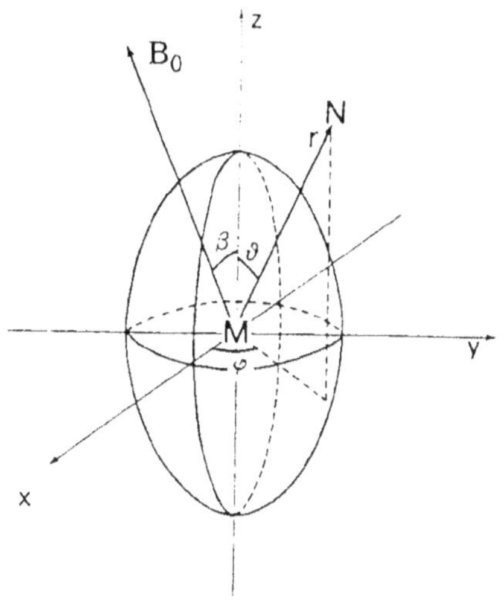

Figure 1. Molecular axis system as defined by the principal directions of an axial tensor. r, θ and ϕ are the nuclear polar coordinates. β defines the direction of the external magnetic field.

Besides the spin densities at the resonating nucleus and at the metal center, there are spin densities all over the molecule that are difficult to handle. Often they are negligibile for protons but not for heteronuclei. When appropriate they are treated parametrically.[2,8]

As far as relaxation is concerned, the general equation holds[3,9]:

$$T_i^{-1} = \overline{E}^2 f_i(\tau_c, \omega_S, \omega_I) \tag{8}$$

where i= 1, 2 and $\overline{E}^2$ is the average of the squared coupling energy. Then there is a function of the correlation time and of the energies of nuclear transitions ω_I, and $\omega_S \pm \omega_I$. The latter transition energies are approximated to ω_S.

In the case of contact coupling we have an average of the squared energy expressed by:[10-13]

$$\overline{E}^2 = \frac{2}{3}\left(\frac{A_c}{\hbar}\right)^2 S(S+1) \tag{9}$$

The average of the squared dipolar energy is given by:[14]

$$\overline{E}^2 = \frac{2}{15}\left(\frac{\mu_0}{4\pi}\right)^2 \frac{\gamma_N^2 g_e^2 \mu_B^2}{r^6} S(S+1) \tag{10}$$

the dipolar energy being expressed by $E = -\left(\frac{\mu_0}{4\pi}\right)\left(\frac{\hat{\mu}_I \cdot \hat{\mu}_S}{r^3}\right)$, where $\hat{\mu}_I$ and $\hat{\mu}_S$ are nuclear and electron magnetic moments, respectively. We have also a dipolar coupling energy between the electron magnetic moment induced by the external field which is due to $\langle S_z \rangle$ and the nuclear magnetic moment.[15,16] This energy is small because small is the induced magnetic moment, but it increases with the external magnetic field and is modulated by rotation which is slow in macromolecules. Therefore, it may be dominating at high magnetic fields for large proteins. Its effect on T_2^{-1} is much larger than on T_1^{-1} because of the non dispersive term in $f(\tau_c, \omega_S, \omega_I)$. This relaxation mechanism is called Curie relaxation. The average of the squared A_{Curie} in solution is given by:

$$\overline{A}^2_{Curie} = \frac{1}{5}\left(\frac{\mu_0}{4\pi}\right)^2 \frac{\hbar^2 \gamma_N^2 g_e^2 \mu_B^2}{r^6} \tag{11}$$

which is the same as $\overline{A}^2_{pc}$.

If the correlation times for, let us say, contact and dipolar relaxation are the same, then cross terms arise when evaluating $\overline{E}^2$. There are cross terms also between other diamagnetic relaxation mechanisms like proton-proton coupling and Curie relaxation. Such cross correlation terms are important in bidimensional NMR spectroscopy of paramagnetic molecules. It has been shown [17,18] that cross peaks in COSY maps between hyperfine broadened signals in paramagnetic macromolecules often arise from relaxation allowed coherence transfer rather than from true scalar coupling. This effect originates from cross correlation between interproton dipolar coupling and Curie relaxation, which acts analogously to chemical shift anisotropy in diamagnetic compounds[19]. Real cross peaks can be recognized from their shape [20].

Finally, it should be kept in mind that nuclear relaxation is caused also by spin densities delocalized onto the molecules[3,21-23]. Although the electron may be only partially delocalized, the contribution to relaxation of even a small fraction of spin density nearby the resonating nucleus may not be negligible because of the short distance. The existence of these ligand centered effects may break the rule that the broader the line the closer the nucleus to the metal ion[24]. However, ligand centered effects are difficult to be quantified because the spin density distribution is generally not known with sufficient accuracy.

Despite the NMR parameters (shift plus relaxation) are the sum of several contributions, often one is dominant, and this simplifies any approach; otherwise more rigorous approaches are needed case by case.

2. The Dependence of A_c on Dihedral Angles

In metal-donor-CH_2 moieties of the type shown in Fig. 2, the spin density ρ may change when dihedral angles are varied.

Figure 2. Schematic view of the metal-donor-C-H dihedral angle θ. A p_z orbital of the donor, orthogonal to the metal-donor-C plane is also shown.

In the case of octahedral nickel(II) complexes with amine ligands, A_c changes with dihedral angle as follows[25-27]:

$$A_c = B_0 + B_2 \cos^2\theta + B_1 \cos\theta \tag{12}$$

Generally, B_1 and B_0 are negligibile. The dihedral angle θ is that between the planes of the metal-nitrogen-carbon and nitrogen-carbon-hydrogen. The meaning of the relationship is that the 1s orbital of hydrogen overlaps with the metal-nitrogen σ bond.

A similar relationship holds in J-J coupling in a H-C-C-H moiety when the C-C bond is a σ bond[28,29].

Typically, the β-CH_2 protons of a histidine bound to a paramagnetic metal ion experience different A_c values depending on their orientation relative to the imidazole plane. Let us define z perpendicular to the imidazole plane and passing for the γ carbon. Then let us define θ as the dihedral angle between the plane passing for z and C_β-C_γ carbons and the plane C_γ-C_β-H. The following relation holds:

$$A_c = B_2 \cos^2\theta \qquad (13)$$

This formula tells us that the largest spin density transmission occurs depending on the overlap between the 1s orbital of hydrogen and the p_z orbital of C_γ. The latter orbital hosts some spin density from the metal. The same reasoning applies to: *i)* the β-CH_2 attached to porphyrin rings when the unpaired electron is delocalized onto the π system of the porphyrin[1,30-32]; *ii)* a CH moiety attached to an sp_2 carbon bearing an unpaired electron in the p_z orbital[33-35]. However, if the angle θ is defined as the angle between the C-C-H plane and the imidazole, porphyrin or sp_2 plane, respectively, the relationship expressed in eq. (13) becomes of the type:

$$A_c = B_2' \sin^2\theta \qquad (14)$$

In this case the definition of the angle θ is analogous to that used for eq. (12). The $\sin^2$ dependence arises from the fact that now the 1s orbital of hydrogen overlaps with a pπ orbital, whose orientation is at 90° with respect to the M-D σ bond.

When the donor atom is a cysteine sulfur, as in iron-sulfur proteins, both σ and π mechanisms are operative. Therefore the formula expressing the relationship between A_c and θ arises from a combination of equations (12) and (14)[36]:

$$A_c = B_0'' + B_1'' \cos\theta + B_2'' \sin^2\theta \qquad (15)$$

where θ is the angle between the Fe-S-C and the S-C-H planes, $B_2'' = B_2' - B_2$, $B_1'' = B_1$, and $B_0'' = B_2 + B_0$. B_2'' is positive or negative depending on whether B_2' is larger than B_2 or viceversa. In the former case the main dependence is of $\sin^2\theta$ type, i.e. the overlap with a pπ orbital of sulfur is dominating. The dependence of the shift for cysteine β-CH_2 protons and Cα carbons in a series of iron-sulfur proteins containing $[Fe_4S_4]^{2+}$ clusters on the dihedral angle θ is reported in Fig. 3. By taking the hyperfine shift as contact, the data indicate that the π overlap is the dominant mechanism[36].

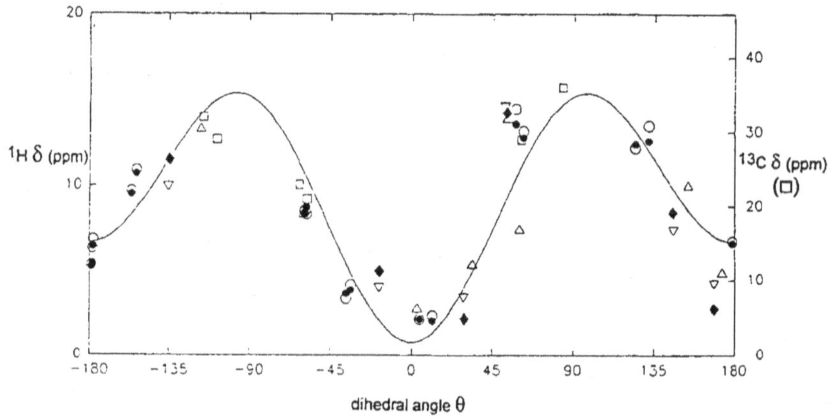

Figure 3. Plot of the hyperfine shifts of cysteine β-CH_2 protons (left hand scale)) for some proteins containing $[Fe_4S_4]^{2+}$ clusters and and of Cα carbons for oxidized *C. acidi urici* ferredoxin (right hand scale), as a function of the dihedral angle θ (i.e. Fe-S-Cβ-H and Fe-S-Cβ-Cα, respectively). The proton hyperfine shifts from different proteins are reported as follows: (●) oxidized *C. acidi urici* ferredoxin; (○) oxidized *C. pasterianum* ferredoxin; (Δ) reduced *C. vinosum* HiPIP; (∇) reduced *E. halophila HiPIP II;* (♦) reduced *E. halophila I.* The curve reports the fitting of the values of *C. acidi urici* ferredoxin obtained by using for the shift an equation. of the type: $\delta = b_0 + b_1\cos\theta + b_2\sin^2\theta$, which is the analogue of eq. (15) for the hyperfine coupling constant. The best fit values for the three parameters are $b_2 = 11.5$, $b_1 = -2.9$, $b_0 = 3.7$.

3. A_c in Low Spin Heme-Iron(III) Systems

Low spin iron(III) is characterized by one unpaired electron. When in tetragonally elongated symmetry, as in heme systems (see Fig. 4), the unpaired electron can reside in two degenerate orbitals, i.e. d_{xz} and d_{yz}. Histidines (and methionines) in axial position split the two orbitals so that the electron is only in the lower energy orbital and gives rise to a spin density pattern on the heme moiety[2,8,37-44].

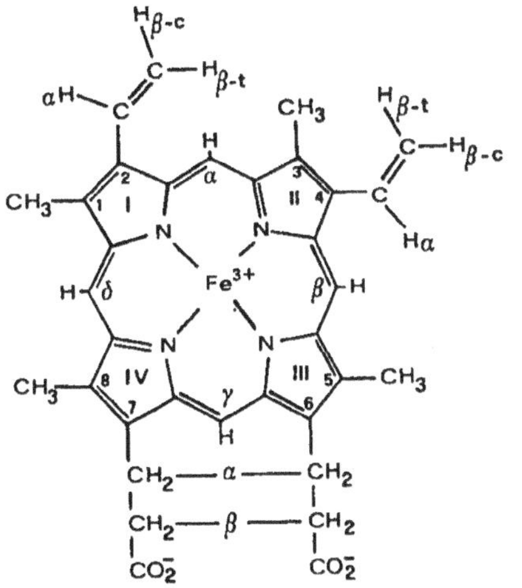

Figure 4. Schematic representation of the heme moiety.

It may happen, however, that the excited level is thermally populated at room temperature and therefore a second spin density pattern sums up to the former.[45] The temperature dependence of the shifts is related to the energy separation. In turn such separation depends on the interaction with the axial ligands. For low spin globins and peroxidases the orientation of the proximal histidine plane defines the electron spin delocalization on opposite pyrroles[46-51]. In the case of cytochrome b_5, which possesses two axial histidine ligands, the heme shift pattern has been drawn to correspond to the His 39 orientation, thus suggesting that the latter residue binds more strongly to the heme iron (Fig. 5)[44]. In the case of *c*-type cytochromes the chirality of the methionine ligand has been proposed to be the determining factor[38,43].

The orbitals d_{xz} and d_{yz} have the correct symmetry to join the π system of the heme. Since the unpaired electron resides in a π orbital no σ spin delocalization is predicted. Under these circumstances the contact shifts for ^{13}C and ^{1}H of heme methyl groups should have constant ratio, being both directly proportional to the unpaired electron density on the pyrrole carbon to which the methyl group is bonded[52,53]. In a number of systems where contact and pseudocontact shifts have been factorized out from the hyperfine shifts, such ratios have been found to be quite variable[54-57]. This means that either σ delocalization is operative or that the contact shifts are capable of detecting the inequivalencies due to the lack of symmetry[57].

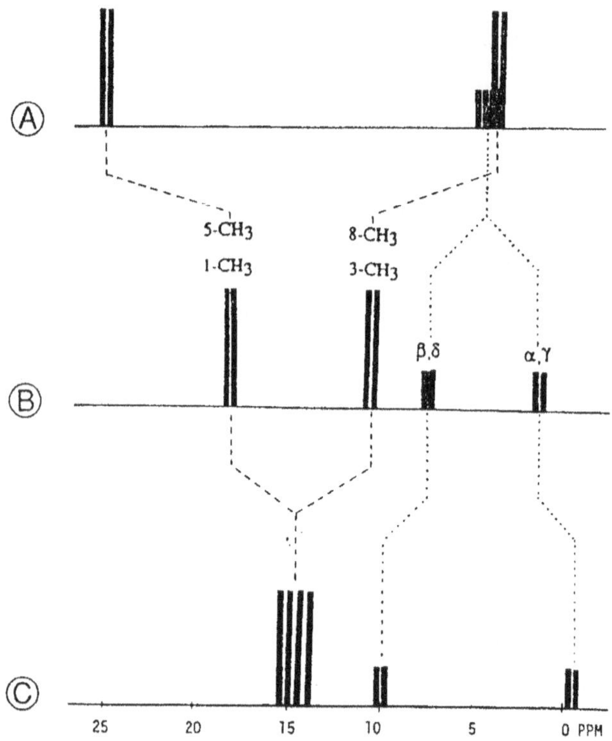

Figure 5. Predicted methyl contact and meso-H rhombic pseudo-contact ^{1}H shift pattern in cyt b_5, as a function of the orientation of the heme relative to the His39 imidazole plane. Panel A corresponds to an orientation of the His plane along the nitrogen atoms of pyrroles I and III; panel C to an orientation along the γ and β meso protons; panel B to an intermediate situation. (Adapted from ref. [44])

The heme π molecular orbitals have nodes at the meso positions, so that the meso protons experience an hyperfine shift which originates predominantly from the pseudo-contact term[44]. For a z axis essentially perpendicular to the heme, as approximately found in proteins, the four meso protons experience all the same upfield shift from the axial term, i.e. the first term of eq. (7). The orientation of the axial ligand is again important in the definition of the rhombic axes x and y: it has been demonstrated that the larger in-plane component of the magnetic susceptibility tensor lies along the axis defined by the orbital hole. The cos 2φ dependence of the rhombic term in eq. (7) cause alternating contributions of the rhombic pseudocontact term to the shift of adjacent meso protons.

4. The Hyperfine Parameters in Magnetically Coupled Dimetallic Systems

When two paramagnetic metal ions, identified by the numbers 1 and 2, are magnetically coupled the wavefunctions are of the type:

$$\left|S',M_{S'}\right\rangle = \sum_{M_{S_{1,2}}} c_{M_{S_{1,2}}} \left|S_1,M_{S_1},S_2,M_{S_2}\right\rangle \tag{16}$$

where $c_{M_{S_{1,2}}}$ are proper coefficients[58].

In general, for a nucleus sensing the metal k, the hyperfine coupling associated to any S'_i level is[59]:

$$A_{i_k} = \frac{A_k \langle S_z \rangle_{i_k}}{\langle S'_z \rangle_i} = A_k C_{i_k} \tag{17}$$

where A_{i_k} is the contact hyperfine constant of the S'_i state, A_k is the contact hyperfine constant of one metal ion (k=1,2) when magnetically uncoupled, $\langle S'_z \rangle_i$ is the expectation value of S'_{zi} and $\langle S_z \rangle_{i_k}$ is the expectation value for the metal k in the coupled system. The ratios C_{i_k} are given, through the Wigner-Eckart theorem, by:

$$C_{i_1} = \frac{\left[S_i'\left(S_i'+1\right)+S_1(S_1+1)-S_2(S_2+1)\right]}{2S_i'\left(S_i'+1\right)} \tag{18a}$$

$$C_{i_2} = \frac{\left[S_i'\left(S_i'+1\right)-S_1(S_1+1)+S_2(S_2+1)\right]}{2S_i'\left(S_i'+1\right)} \tag{18b}$$

Let us suppose that the magnetic coupling constant is larger than kT, so that S_i' is small and we can refer to the ground state only. If the magnetic coupling is antiferromagnetic in nature, $S_2 > S_1$, C_1 of the ground state will be negative and C_2 positive. It follows that a nucleus sensing the larger S_k will have positive C_k and negative $\langle S_z \rangle_k$ whereas a nucleus sensing the smaller S_k will have a positive product of C_k and $\langle S_z \rangle_k$. In other words the contact shifts will be of different sign.

If the nucleus senses also the other metal ion, then the hyperfine constant is the sum for the two ions.

When J is equal or smaller than kT, also the i-excited levels should be considered. The contact shift is given by the sum of the contributions of each level weighted for its population according to the Boltzmann partition function P_i[58-60].

$$\left(\frac{\Delta v}{v_0}\right)^{con} = -\frac{1}{\hbar\gamma_N B_0}\sum_i A_{i_k} \langle S_z' \rangle_i P_i \tag{19}$$

The A_i values are obtained from eq. (18) through A_k, which is generally assumed or taken from mononuclear systems[3,59-61].

4.1 THE CASE OF $[Fe_2S_2]^+$

The dimetallic center $[Fe_2S_2]^+$ is present in reduced two iron - two sulfur ferredoxin.[8,62] It contains one Fe^{3+} and one Fe^{2+}, both high spin[63]. The S_i' levels are characterized by the following energies:

$$E_i = \frac{1}{2} J S_i' (S_i' + 1) \tag{20}$$

Magnetic Mössbauer on ^{57}Fe at 4 K shows that Fe^{3+} is associated with a negative value of A_{g_k} relative to the hyperfine coupling with ^{57}Fe (mean value of -47.4 MHz) while Fe^{2+} is associated with a positive value (mean value of +21.1 MHz)[63]. Fig. 6 shows the hyperfine shifted signals of the ^1H NMR spectrum of reduced spinach ferredoxin[59,64,65], together with their experimental temperature dependence. Equation (19) allows us to calculate the contact shifts for the protons around room temperature. In the case of spinach ferredoxin tha data are consistent with antiferromagnetic coupling of about 200 cm $^{-1}$[66] and a hyperfine constant of both uncoupled metal ions of 1.8 MHz [59,62].

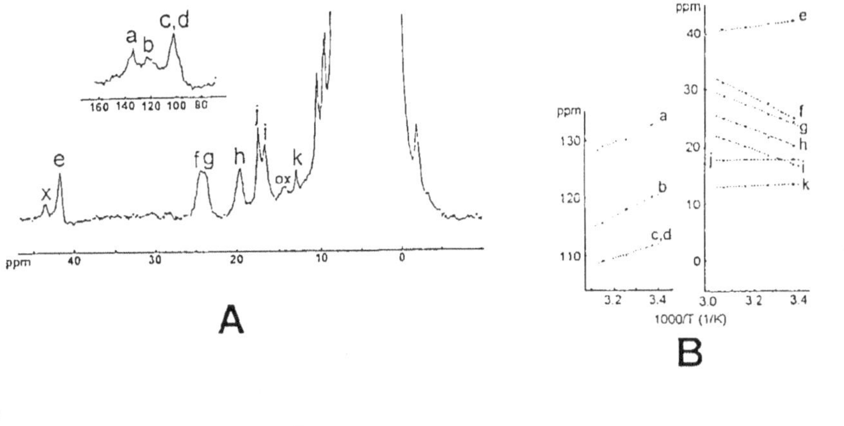

Figure 6. (A) 300 MHz and 297 K ^1H NMR spectrum of reduced $[Fe_2S_2]^+$ spinach ferredoxin.Signals from residual oxidized protein (ox) and from minor components are indicated. (B) Temperature dependence of the ^1H NMR shifts. (Taken from ref. [62], with permission)

4.2 THE CASE OF Cu_2Co_2SOD

The active site of cobalt-substituted superoxide dismutase is shown in Fig. 7[67-69]. In this case we have a copper(II) (S=1/2) and a high spin cobalt(II) (S=3/2) antiferromagnetically coupled. Magnetic susceptibility measurements have shown that J= 33 cm^{-1} [70]. The coupling energy is therefore small compared to kT, so that at room temperature all the magnetically coupled levels are almost equally populated. The summation over all levels in eq. (19) approximates the A_c value of the uncoupled system. The result is that the contact shifts of one nucleus does not depend on magnetic coupling[71].

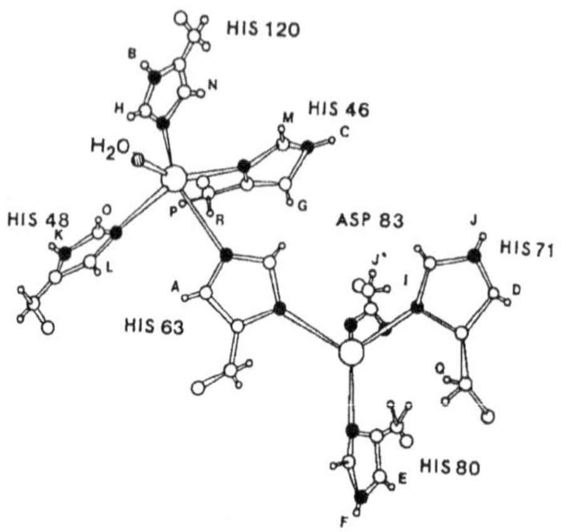

Figure 7. Schematic drawing of the metal sites in superoxide dismutase. The numbering of the residues is that of human isoenzyme.

4.3 GENERAL REMARKS ON DINUCLEAR CENTERS

In the case of dimetallic systems it is possible to predict the contact shift and its temperature dependence from the energies of the magnetically coupled levels and from the wavefunctions of the levels [3,62,72]. Possibly, the same rules apply to dipolar contributions to the hyperfine coupling, provided that magnetic exchange does not affect the g values of each center. There are rules to calculate from the g_{ll} values of each metal the g_{ll} (l=x, y, z) of each S_i', when uncoupled[73]. Once these values are known, the A_{pc} for each S_i' level would be obtained by applying eq. (6). At each temperature the pseudocontact shift should be given by an equation analogous to eq.

(19), where now A_{i_k} is the pseudocontact hyperfine coupling constant. However, no attempts to estimate this contribution have been reported in the literature.

Finally we should note that in the case of different metal ions, one of the two will have shorter electron relaxation than the other and magnetic coupling will allow the latter to relax faster. For T_1 and T_2 equations can be written which are analogous to eq. (19) for the shift [3,58,59]. However, the largest effect here is the change in τ_S upon establishment of magnetic coupling when the ions are different. An equation similar to that derived by Bloembergen for contac relaxation of nuclei can be used to describe how the slow relaxing metal ion increases its electronic relaxation rates through coupling with a fast relaxing metal ion[3]:

$$\tau_{S_{M1}}^{-1}(J) = \tau_{S_{M1}}^{-1}(0) + \frac{2}{3}\left(\frac{J}{\hbar}\right)^2 S_2(S_2+1)\frac{\tau_{S_{M2}}}{1+\left(\omega_{S_{M1}}-\omega_{S_{M2}}\right)^2\tau_{S_{M2}}^2} \qquad (21)$$

where $\tau_{S_{M1}}^{-1}(J)$ is the actual value for the relaxation rate of the slow relaxing metal ion in the coupled system, $\tau_{S_{M1}}^{-1}(0)$ is the relaxation rate in the absence of coupling, and $\tau_{S_{M2}}^{-1}$ is the relaxation time of the fast relaxing metal ion , which is assumed not to be altered when interacting with the other metal ion. Eq. (21) holds within the Redfield limit.

As far as NMR is concerned, shortening of the electron relaxation rates means sharpening of the NMR signals. That is why the copper(II) center of superoxide dismutase can be investigated in the Cu_2Co_2SOD. In Cu_2Zn_2SOD the proton signal of histidines bound to copper(II) are broad beyond detection because copper(II) has long electron relaxation times. In Cu_2Co_2SOD the ^1H NMR spectra with signals of histidine protons of both copper and cobalt domains can be easily recorded (see Fig. 8)[24,69,74]. The T_1 values of histidine protons range from 1 to 5 ms at 200 MHz, those of copper domain being longer. The τ_S values are 10^{-9} and 10^{-11} s for copper and cobalt, respectively, when uncoupled.

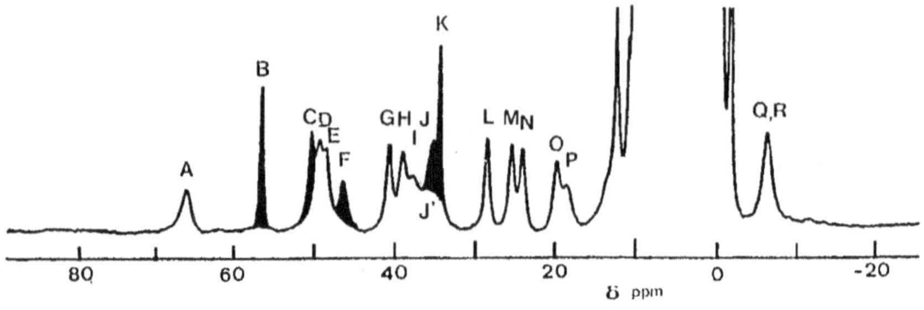

Figure 8. 300 MHz ^{1}H NMR spectrum of Cu_2Co_2SOD. Signal assignment follows the proton labeling of Fig. 7.

5. The Hyperfine Coupling in Polymetallic Systems

In the case of polymetallic systems, the wavefunctions are obtained by solving a Hamiltonian of the type[3]:

$$\aleph = \sum_{i \neq j} J_{ij} \hat{S}_i \cdot \hat{S}_j \tag{22}$$

where i and j are the metal ions taken pairwise.

Then eq. (18) holds and allows us to interpret any kind of contact hyperfine coupling obtained through ENDOR, Mössbauer and NMR[62].

We will here report the case of $[Fe_4S_4]^{3+}$ present in the protein called HiPIP II from *E. halophila*. The Mössbauer spectra show that there are two Fe^{3+} ions and two $Fe^{2.5+}$ ions[75]. The mean values for the hyperfine coupling constant with ^{57}Fe are +21.4 MHz and -31.5 MHz, respectively. The two average hyperfine values have opposite signs, analogously to that observed in the reduced $[Fe_2S_2]^+$ ferredoxin. However, in the present system the negative hyperfine coupling is associated with the

$Fe^{2.5+}$ ions and with Fe^{3+} in $[Fe_2S_2]^+$ ferredoxin. According to eq. (22) these values are consistent with a subspin for the $Fe^{2.5+}$ ions larger than that of the Fe^{3+} ions giving rise to a S=1/2 ground state. The latter could be either $\left|\frac{9}{2},4,\frac{1}{2}\right\rangle$ or $\left|\frac{7}{2},3,\frac{1}{2}\right\rangle$, where the first figure is the subspin of the $Fe^{2.5+}$ ions, the second is the subspin of the Fe^{3+} pair, and the last is the total spin [62,76,77]. The ^1H NMR spectrum of oxidized *E. halophila* HiPIP II is shown in Fig. 9A. According to the above considerations, the spectrum displays the four signals of the β-CH_2 of coordinated cysteines sensing the larger subspin downfield, and the four signals of the cysteine β-CH_2 sensing the smaller subspin shifted upfield [78,79].

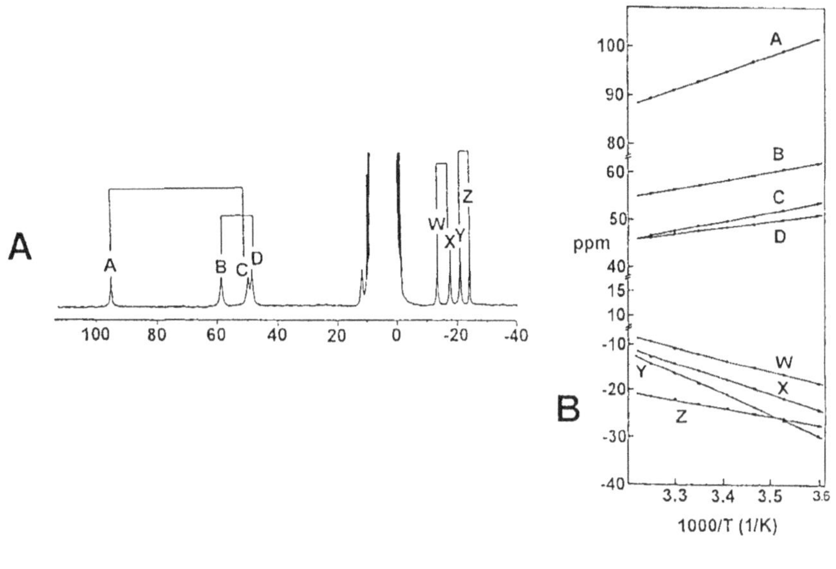

Figure 9. (A) 600 MHz and 298 K ^1H NMR spectrum of oxidized $[Fe_4S_4]^{3+}$ HiPIP from *E. halophila*. The cysteine β-CH_2 geminal connectivities are indicated. (B) Temperature dependence of the hyperfine shifted resonances. The signals are labeled as in (A). (Taken from ref. [62], with permission)

The presence of $Fe^{2.5+}$ requires electron delocalization. However, even in a Heisenberg exchange scheme a choice of a larger $J(Fe^{3+}\text{-}Fe^{3+})$ (J_{12}) and a smaller $J(Fe^{2.5+}\text{-}Fe^{2.5+})$ (J_{34}), the other J values being intermediate, gives rise to the

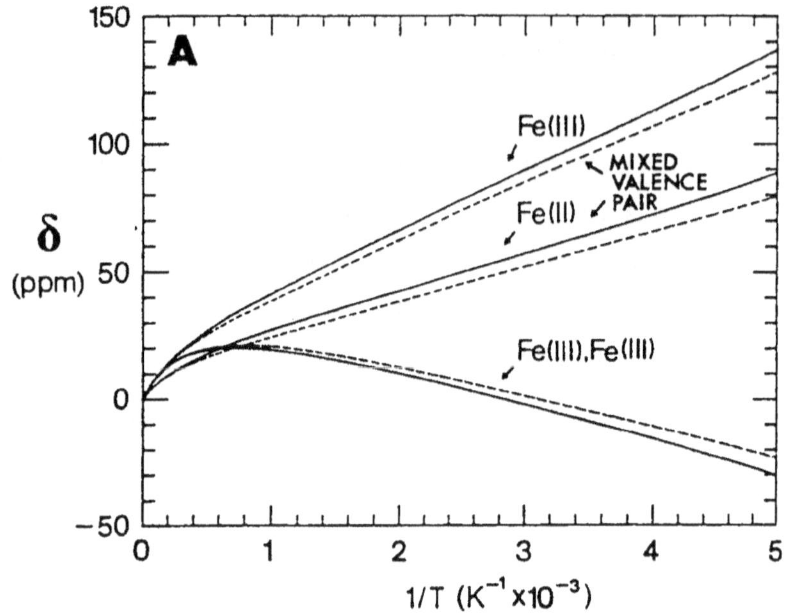

Figure 10. Calculated temperature dependence oh the ^{1}H NMR shifts of oxidized HiPIP using J_{ij}=300 cm^{-1}, J_{12}=400 cm^{-1}, (—) B=0, J_{34}=200 cm^{-1} and (- -) B= 300 cm^{-1}, J_{34}=300 cm^{-1}. (Adapted from ref. [81])

expected spin states. This is an example of the so-called spin frustration mechanism. Similar results can be obtained using a larger $J(Fe^{2.5+}$-$Fe^{2.5+})$ and introducing in the Hamiltonian a term which takes into account the electron delocalization between the two metal centers[80]. This new term is of the type BT_{12}, where T_{12} is the electron transfer operator and B the double exchange constant between the metal centers 1 and 2, for which the electron delocalization is required. A debate is presently occurring for the extimation of the actual values of the double exchange constant and on the possibility of determining it. Figure 10 shows the expected calculated temperature dependence of the shifts in the two cases [81] and Figure 9B the experimental one.

The protein matrix presumably determines which iron pair among the possible spin pairs is the mixed valence one. Other HiPIPs show different spectra (Fig.

11) with two pairs of cysteine β-CH_2 being less shifted from the diamagnetic position [82-86]. We have interpreted this observation as due to two different localization of the two $Fe^{2.5+}$ ions in fast equilibrium.(Fig. 12).

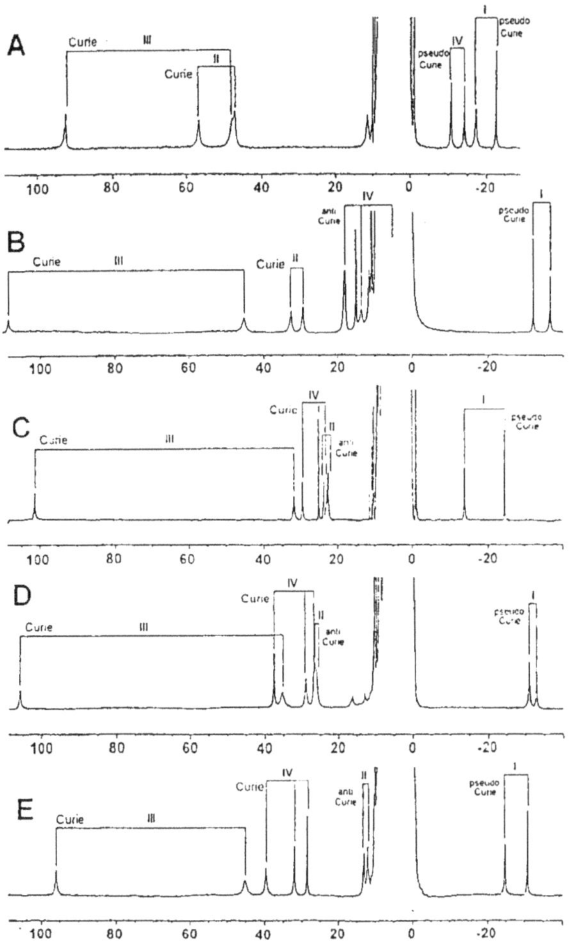

Figure 11. 600 MHz ^1H NMR spectra of oxidized $[Fe_4S_4]^{3+}$ HiPIP from: (A) *E. halophila* (iso-II); (B) *R. globiformis*; (C) *E. vacuolata* (iso II); (D) *C. vinosum*; (E) *R. gelatinosus*. The cysteine β-CH_2 connectivities are also indicated. For cysteine IV the α-CH is also assigned. (Taken from ref. [62], with permission)

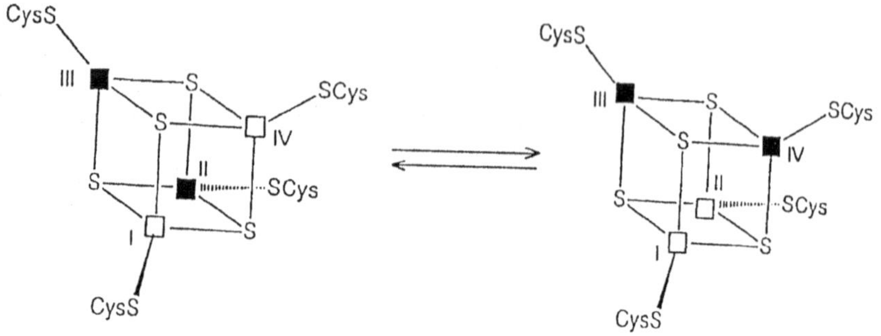

Figure 12. Possible low symmetry situation in the oxidized $[Fe_4S_4]^{3+}$ cluster in HiPIP as obtained in the presence of chemical equilibrium between two situations of higher symmetry.

It should be noted that the EPR spectrum of *E. halophila* is easily interpretable as due to a single species [75] whereas those of the other HiPIPs are not (Fig. 13)[62,86-88].

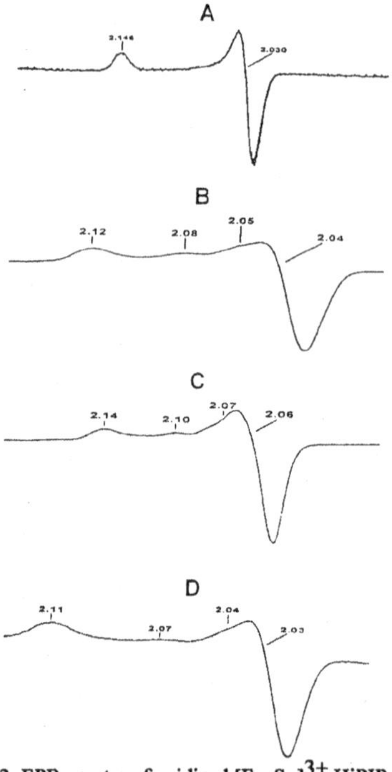

Figure 13. EPR spectra of oxidized $[Fe_4S_4]^{3+}$ HiPIP from : (A) *E. halophila* (iso II); (B) *C. vinosum*; (C) *E. vacuolata* (iso II); (D) *R. gelatinosus*. (Taken from ref. [62], with permission)

References

1. La Mar, G.N. (1973) , in La Mar, G.N., Horrocks, Jr.,W.D. and Holm, R.H. (eds) *NMR of Paramagnetic Molecules*, Academic Press, New York
2. Bertini, I. and Luchinat, C. (1986) *NMR of paramagnetic molecules in biological systems*, Benjamin/Cummings, Menlo Park, CA.
3. Banci, L., Bertini, I. and Luchinat, C. (1991) *Nuclear and electron relaxation. The magnetic nucleus-unpaired electron coupling in solution*, VCH, Weinheim.
4. McConnell, H.M. and Chesnut, D.B. (1958) Theory of isotropic hyperfine interactions in p-electron radicals, *J. Chem. Phys.* **28**, 107-117.
5. Kurland, R.J. and McGarvey, B.R. (1970) Isotropic NMR shifts in transition metal complexes: calculation of the Fermi contact and pseudocontact terms, *J. Magn. Reson.* **2**, 286-301.
6. McConnell, H.M. and Robertson, R.E. (1958) Isotropic nuclear resonance shifts, *J. Chem. Phys.* **29**, 1361-1365.
7. Horrocks, W.D.J. and Hall, D.D. (1971) Direct evaluation of dipolar nuclear magnetic resonance shifts from single-crystal magnetic measurements. Paramagnetic anisotropy of bis(2,4-pentanedionato)bis(pyridine)-cobalt(II), *Inorg. Chem.* **10**, 2368-2370.
8. Bertini, I., Turano, P. and Vila, A.J. (1993) NMR of paramagnetic metalloproteins, *Chem. Rev.* **93**, 2833-2932.
9. Kowalewski, J., Nordenskiöld, L., Benetis, N. and Westlund, P.-O. (1985) Theory of nuclear spin relaxation in paramagnetic systems in solution, *Progr. Nucl. Magn. Reson. Spectrosc.* **17**, 141-185.
10. Bloembergen, N. (1957) Comments on "Proton relaxation times in paramagnetic solutions", *J. Chem. Phys.* **27**, 575-596.
11. Bloembergen, N. (1957) Proton relaxation times in paramagnetic solutions, *J. Chem. Phys.* **27**, 572-573.
12. Solomon, I. and Bloembergen, N. (1956) Nuclear magnetic interactions in the HF molecule, *J. Chem. Phys.* **25**, 261-266.
13. Koenig, S.H. (1982) A classical description of relaxation of interacting pairs of unlike spins: extension to T_{1p}, T_2, and T_{1poff}, including contact interactions, *J. Magn. Reson.* **47**, 441-453.
14. Solomon, I. (1955) Relaxation processes in a system of two spins, *Phys. Rev.* **99**, 559-565.
15. Guéron, M. (1975) Nuclear relaxation in macromolecules by paramagnetic ions: A novel mechanism, *J. Magn. Reson.* **19**, 58-66.
16. Vega, A.J. and Fiat, D. (1976) Nuclear relaxation processes of paramagnetic complexes The slow motion case, *Mol. Phys.* **31**, 347-355.
17. Bertini, I., Luchinat, C., Piccioli, M. and Tarchi, D. (1994) COSY spectra of paramagnetic macromolecules, observability, scalar effects, cross correlation effects, relaxation allowed coherence transfer, *Concepts Magn. Reson.* in press
18. Bertini, I., Luchinat, C. and Tarchi, D. (1993) Are true scalar proton-proton connectivities ever measured in COSY spectra of paramagnetic macromolecules? *Chem. Phys. Lett.* **203**, 445-449.

19. Wimperis, S. and Bodenhausen, G. (1989) Relaxation-allowed cross-peaks in two-dimensional NMR correlation spectroscopy, *Mol. Phys.* **66**, 897-919.
20. Qin, J., Delaglio, F., La Mar, G.N. and Bax, A. (1993) Distinguishing the effects of cross correlation and J coupling in COSY spectra of paramagnetic protein, *J. Magn. Reson. Ser. B* **102**, 332-336.
21. Gottlieb, H.P.W., Barfield, M. and Doddrell, D.M. (1977) An electron spin density matrix descriptin of nuclear spin-lattice relaxation in paramgnetic molecules, *J. Chem. Phys.* **67**, 3785-3794.
22. Waysbort, D. and Navon, G. (1978) The effect of spin delocalization on the proton magnetic relaxation in transition metal hexaaquo ions, *J. Chem. Phys.* **68**, 3704-3707.
23. Nordenskiöld, L., Laaksonen, L. and Kowalewski, J. (1982) Applicability of the Solomon-Bloembergen equation to the study of paramagnetic transition-metal-water complexes. An ab initio SCF-MO study, *J. Am. Chem. Soc.* **104**, 379-382.
24. Banci, L., Bertini, I., Luchinat, C., Piccioli, M., Scozzafava, A. and Turano, P. (1989) ^{1}H NOE studies on dicopper(II) dicobalt(II) superoxide dismutase, *Inorg. Chem.* **28**, 4650-4656.
25. Fitzgerald, R.J. and Drago, R.S. (1968) Contact-shift studies, delocalization mechanisms, and extended Hückel calculations of Nickel(II)-Alkylamine complexes, *J. Am. Chem. Soc.* **90**, 2523-2527.
26. Ho, F.F.-L. and Reilley, C.N. (1969) Conformational studies of chelated ethylenediamines by nuclear magnetic resonance. Paramagnetic nickel(II) complexes of N-Alkylethylenediamines, *Anal. Chem.* **41**, 1835-1841.
27. Carney, M.J., Papaefthymiou, G.C., Spartalian, K., Frankel, R.B. and Holm, R.H. (1988) Ground spin state variability in $[Fe_4S_4(SR)_4]^{3-}$. Synthetic analogues of the reduced clusters in ferredoxins and other iron-sulfur proteins: cases of extreme sensitivity of electronic state and structure to extrinsic factors, *J. Am. Chem. Soc.* **110**, 6084-6095.
28. Karplus, M. (1959) Contact Electron-Spin coupling of nuclear magnetic moments, *J. Chem. Phys.* **30**, 11-15.
29. Karplus, M. (1963) Vicinal proton coupling in nuclear magnetic resonance, *J. Am. Chem. Soc.* **85**, 2870-2871.
30. Yamamoto, Y., Nanai, N., Inoue, Y. and Chujo, R. (1988) Natural abundance 13C-NMR study of paramagnetic horse heart Ferricytochrome c Cyanide complex: assignment of hyperfine shifted heme methyl carbon resonances, *Biochem. Biophys. Res. Commun.* **151**, 262-269.
31. Santos, H. and Turner, D.L. (1992) ^{13}C and proton NMR studies of horse cytochrome c, *Eur. J. Biochem.* **206**, 721-728.
32. Santos, H. and Turner, D.L. (1986) ^{13}C and proton NMR studies of horse ferricytochrome c, *FEBS Lett.* **194**, 73-77.
33. Karplus, M. and Fraenkel, G.K. (1961) Theoretical interpretation of carbon-13 hyperfine interactions in electron spin resonance spectra, *J. Chem. Phys.* **35**, 1312-1323.
34. Heller, C. and McConnell, H.M. (1960) Radiation damage in organic crystals. II. Electron spin resonance of $(CO_2H)CH_2CH(CO_2H)$ in b-succinic acid, *J. Chem. Phys.* **32**, 1535-1539.

35. Stone, E.W. and Maki, A.H. (1962) Hindered internal rotation and ESR spectrsocopy, *J. Chem. Phys.* **37**, 1326-1333.
36. Bertini, I., Capozzi, F., Luchinat, C., Piccioli, M. and Vila, A.J. (1994) The Fe_4S_4 centers in ferredoxins studied through proton and carbon hyperfine coupling. Sequence specific assignments of cysteines in ferredoxins from *Clostridium acidi urici* and *Clostridium pasteurianum*, *J. Am. Chem. Soc.* **116**, 651-660.
37. La Mar, G.N. and Walker, F.A. (1979) Nuclear magnetic resonance of paramagnetic metalloproteins, in Dolphin, D. (ed) *The Porphyrins*, Academic Press, New York, pp.61-157.
38. Satterlee, J.D. (1986) NMR spectroscopy of paramagnetic haem proteins, *Annu. Rep. NMR Spectrosc.* **17**, 79-178.
39. Shulman, R.G., Glamur, S.H. and Karplus, M. (1971) Electronic structure of cyanide complexes of hemes and heme proteins, *J. Mol. Biol.* **57**, 93-115.
40. Traylor, T.G. and Berzinis, A.P. (1980) Hemoproteins models: NMR of imidazole chelated protohemin cyanide complexes, *J. Am. Chem. Soc.* **102**, 2844-2846.
41. Keller, R.M. and Wüthrich, K. (1978) Evolutionary change of the heme electronic structure: ferricytochrome c-551 from *pseudomonas aeruginosa* and horse heart ferricytochrome c, *Biochem. Biophys. Res. Commun.* **83**, 1132-1139.
42. La Mar, G.N., Viscio, D.B., Smith, K.M., Caughey, W.S. and Smith, M.L. (1978) NMR studies of low-spin ferric complexes of natural porphyrin derivatives. 1. Effect of peripheral substituents on the p electronic asymmetry in biscyano complexes, *J. Am. Chem. Soc.* **100**, 8085-8092.
43. Moore, G.R. (1985) 1H NMR studies of thr haem and coordinated methionine of Class I and Class II cytochromes c, *Biochim. Biophys. Acta* **829**, 425-429.
44. Lee, K.-B., La Mar, G.N., Mansfield, K.E., Smith, K.M., Pochapsky, T.C. and Sligar, S.G. (1993) Interpretation of hyperfine shift patterns in ferricytochromes b_5 in terms of angular position of the heme: a sensitive probe for peripheral heme protein interactions, *Biochim. Biophys. Acta* **1202**, 189-199.
45. Turner, D.L. (1993) Evaluation of ^{13}C and ^{1}H Fermi contact shifts in horse cytochrome c. The origin of the anti-Curie effect, *Eur. J. Biochem.* **211**, 563-568.
46. La Mar, G.N., Davis, N.L., Parish, D.W. and Smith, K.M. (1983) Heme orientational disorder in reconstituted and native sperm whale Myoglobin. Proton nuclear magnetic resonance characterizations by heme methyl deuterium labelling in the met-cyano protein, *J. Mol. Biol.* **168**, 887-896.
47. Satterlee, J.D., Erman, J.E. and de Ropp, J.S (1987) Proton hyperfine resonance assignments in cyanide-ligated cytochrome c peroxidase using the nuclear Overhauser effect, *J. Biol. Chem.* **262**, 11578-11583.
48. Thanabal, V., de Ropp, J.S and La Mar, G.N. (1987) ^{1}H NMR study of the electronic and molecular structure of the heme cavity in Horseradish Peroxidase. Complete heme resonance assignments based on saturation transfer and nuclear overhauser effects, *J. Am. Chem. Soc.* **109**, 265-272.
49. Banci, L., Bertini, I., Turano, P., Tien, M. and Kirk, T.K. (1991) Proton NMR investigation into the basis for the relatively high redox potential of lignin peroxidase, *Proc. Natl. Acad. Sci. USA* **88**, 6956-6960.
50. de Ropp, J.S, La Mar, G.N., Wariishi, H. and Gold, M.H. (1991) NMR study of the active site of resting state and cynide-inhibited Lignin Peroxidase from

Phanerochaete chrysosporium. Comparison with Horseradish Peroxidase, *J. Biol. Chem.* **266**, 15001-15008.

51. Banci, L., Bertini, I., Pease, E., Tien, M. and Turano, P. (1992) ^{1}H NMR investigation of manganese peroxidases from *Phanerochaete chrysosporium* A comparison with other peroxidases, *Biochemistry* **31**, 10009-10017.
52. McConnell, H.M. (1956) Indirect hyperfine interactions in paramagnetic resonance spectra of aromatic free radicals, *J. Chem. Phys.* **24**, 764-766.
53. McConnell, H.M. (1957) Vector model for indirect proton hyperfine interactions in p-electron radicals, *Proc. Natl. Acad. Sci. USA* **43**, 721-723.
54. Yamamoto, Y., Iwafune, K., Nanai, N., Osawa, A., Chujo, R. and Suzuki, T. (1991) NMR study of Galeorhinus japonicus myoglobin ^{1}H NMR study of molecular structure of the heme cavity, *Eur. J. Biochem.* **198**, 299-306.
55. Yamamoto, Y., Nanai, N. and Chujo, R. (1990) Mapping paramagnetic metal-centered dipolar field in haemoprotein using haem methyl carbon and the attached proton resonance, *J. Chem. Soc. ,Chem. Commun.* **22**, 1556-1557.
56. Yamamoto, Y., Komori, K., Nanai, N., Chujo, R. and Inoue, Y. (1992) Determination of the principal axes of the magnetic susceptibility tensor for horse heart oxidized cytochrome c in solution, *J. Chem. Soc. ,Dalton Trans.* 1813-1819.
57. Banci, L., Bertini, I., Pierattelli, R. and Vila, A.J. (1994) ^{1}H ^{13}C HETCOR investigations on heme-containing systems, *Inorg. Chem.* in press
58. Dunham, W.R., Palmer, G., Sands, R.H. and Bearden, A.J. (1971) On the structure of the iron-sulfur complex in the two-iron ferredoxins, *Biochim. Biophys. Acta* **253**, 373-384.
59. Banci, L., Bertini, I. and Luchinat, C. (1990) The ^{1}H NMR parameters of magnetically coupled dimers - The Fe_2S_2 proteins as an example, *Struct. Bonding* **72**, 113-135.
60. Banci, L., Bertini, I., Briganti, F. and Luchinat, C. (1991) The electronic structure of paramagnetic polynuclear metal clusters in proteins studied through ^{1}H NMR spectroscopy, *New J. Chem.* **15**, 467-477.
61. Bertini, I., Luchinat, C., Owens, C. and Drago, R.S. (1987) NMR Proton relaxation in bimetallic complexes containing Co(II), *J. Am. Chem. Soc.* **109**, 5208-5212.
62. Bertini, I., Ciurli, S. and Luchinat, C. (1994) The electronic and geometric structures of iron-sulfur proteins studied through electron-nuclear hyperfine coupling, *Angew. Chem.* in press
63. Dunham, W.R., Bearden, A.J., Salmeen, I., Palmer, G., Sands, R.H., Orme-Johnson, W.H. and Beinert, H. (1971) The two-iron ferredoxins in spinach, parsley, pig adrenal cortex, *Azotobacter Vinelandii*, and *Clostridium Pasterianum*: studies by magnetic field Mössbauer spectroscopy, *Biochim. Biophys. Acta* **253**, 134-152.
64. Salmeen, I. and Palmer, G. (1972) Contact-shifted NMR of spinach ferredoxin: additional resonances and partial assignments, *Arch. Biochem. Biophys.* **150**, 767-773.
65. Bertini, I., Lanini, G. and Luchinat, C. (1984) ^{1}H NMR spectra of reduced spinach ferredoxin, *Inorg. Chem.* **23**, 2729-2730.
66. Palmer, G., Dunham, W.R., Fee, J.A., Sands, R.H., Izuka, T. and Yonetani, T. (1971) The Magnetic Susceptibility of spinach ferredoxin from 77-250K: A

measurement of the antiferromagentic coupling between the two iron atoms, *Biochim. Biophys. Acta* **245**, 201-207.

67. Rotilio, G., Finazzi Agro', A., Calabrese, L., Bossa, F., Guerrieri, P. and Mondovì, B. (1971) Metal sites of copper proteins. Ligands of copper in hemocuprein, *Biochemistry* **10**, 616-621.
68. Valentine, J.S. and Pantoliano, M.W. (1981) , in Sigel, H. (ed) *Metal Ions in Biological Systems (Vol. 3)*, Dekker, New York, pp.291-358.
69. Bertini, I., Banci, L., Luchinat, C. and Piccioli, M. (1990) Spectroscopic studies on Cu_2Zn_2SOD: a continuous advancement of investigation tools, *Coord. Chem. Rev.* **100**, 67-103.
70. Morgenstern-Badarau, I., Cocco, D., Desideri, A., Rotilio, G., Jordanov, J. and Dupre', N. (1986) Magnetic susceptibility studies of native cupro-zinc superoxide dismutase and its cobalt-substituted derivatives. Antiferromagnetic coupling in the imidazolate-bridged copper(II)-cobalt(II) pair, *J. Am. Chem. Soc.* **108**, 300-302.
71. Bertini, I., Banci, L. and Luchinat, C. (1988) NMR of paramagnetic systems: magnetically coupled dimetallic systems. Cu_2Co_2-superoxide dismutase as an example, in Que, L.,Jr. (ed) *Metal clusters in proteins*, Am. Chem. Soc., Washington, DC, pp.70-84.
72. Scaringe, R.P., Hodgson, D.J. and Hatfield, W.E. (1978) The coupled representation matrix of the pair Hamiltonian, *Mol. Phys.* **35**, 701-713.
73. Kent, T.A., Huynh, B.H. and Munk, E. (1980) Iron-sulfur proteins: spin coupling model for three-iron cluster, *Proc. Natl. Acad. Sci. USA* **77**, 6574-6576.
74. Banci, L., Bertini, I., Luchinat, C., Piccioli, M. and Scozzafava, A. (1993) 1D versus 2D ^1H NMR experiments in dicopper, dicobalt superoxide dismutase: a further mapping of the active site, *Gazz. Chim. Ital.* **123**, 95-100.
75. Bertini, I., Campos, A.P., Luchinat, C. and Teixeira, M. (1993) A Mössbauer investigation of oxidized Fe_4S_4 HiPIP II from *E. halophila*, *J. Inorg. Biochem.* **52**, 227-234.
76. Mouesca, J.M., Rius, G. and Lamotte, B. (1993) Single-crystal proton ENDOR studies of the $[Fe_4S_4]^{3+}$ cluster: determination of the spin population distribution and proposal of a model to interpret the ^1H NMR paramagnetic shifts in high potential ferredoxins, *J. Am. Chem. Soc.* **115**, 4714-4731.
77. Rius, G.J. and Lamotte, B. (1989) Single-crystal ENDOR study of a ^{57}Fe-enriched iron-sulfur $[Fe_4S_4]^{3+}$ cluster, *J. Am. Chem. Soc.* **111**, 2464-2469.
78. Banci, L., Bertini, I., Briganti, F., Luchinat, C., Scozzafava, A. and Vicens Oliver, M. (1991) ^1H NOE studies of oxidized high potential iron sulfur protein II from *Ectothiorhodospira halophila*, *Inorg. Chim. Acta* **180**, 171-175.
79. Krishnamoorthi, R., Markley, J.L., Cusanovich, M.A., Przysiecki, C.T. and Meyer, T.E. (1986) Hydrogen-1 nuclear magnetic resonance investigation of high-potential iron-sulfur protein from *Ectothiorhodospira halophila* and *Ectothiorhodospira vacuolata*: a comparative study of hyperfine-shifted resonances, *Biochemistry* **25**, 60-67.
80. Noodleman, L. (1988) A model for the spin states of high-potential $[Fe_4S_4]^{3+}$ proteins, *Inorg. Chem.* **27**, 3677-3679.

81. Bertini, I., Briganti, F., Luchinat, C., Scozzafava, A. and Sola, M. (1991) ^{1}H NMR spectroscopy and the electronic structure of the high potential iron-sulfur protein from *Chromatium vinosum*, *J. Am. Chem. Soc.* **113**, 1237-1245.
82. Bertini, I., Capozzi, F., Luchinat, C., Piccioli, M. and Vicens Oliver, M. (1992) NMR is a unique and necessary step in the investigation of iron-sulfur proteins: the HiPIP from *R. gelatinosus* as an example, *Inorg. Chim. Acta* **198-200**, 483-491.
83. Bertini, I., Capozzi, F., Ciurli, S., Luchinat, C., Messori, L. and Piccioli, M. (1992) Identification of the iron ions of HiPIP from *Chromatium vinosum* within the protein frame through 2D NMR experiments, *J. Am. Chem. Soc.* **114**, 3332-3340.
84. Bertini, I., Capozzi, F., Luchinat, C. and Piccioli, M. (1993) ^{1}H NMR investigation of oxidized and reduced HiPIP from *R. globiformis*, *Eur. J. Biochem.* **212**, 69-78.
85. Nettesheim, D.G., Harder, S.R., Feinberg, B.A. and Otvos, J.D. (1992) Sequential resonance assignments of oxidized high-potential iron-sulfur protein from *Chromatium vinosum*, *Biochemistry* **31**, 1234-1244.
86. Banci, L., Bertini, I., Ciurli, S., Ferretti, S., Luchinat, C. and Piccioli, M. (1993) The electronic structure of $(Fe_4S_4)^{3+}$ clusters in proteins; an investigation of the oxidized HiPIP II from *Ectothiorhodospira vacuolata*, *Biochemistry* **32**, 9387-9397.
87. Dunham, W.R., Hagen, W.R., Fee, J.A., Sands, R.H., Dunbar, J.B. and Humblet, C. (1991) An investigation of *Chromatium vinosum* high-potential iron-sulfur protein by EPR and Mössbaur spectroscopy; evidence for a freezing-induced dimerization in sodium cloride solution, *Biochim. Biophys. Acta* **1079**, 253-262.
88. Beinert, H. and Thomson, A.J. (1983) Three-iron clusters in iron-sulfur proteins, *Arch. Biochem. Biophys.* **222**, 333-361.

ASSIGNMENT STRATEGIES AND STRUCTURE DETERMINATION IN CYANIDE-INHIBITED HEME PEROXIDASES

GERD N. LA MAR, ZHIGANG CHEN and JEFFREY S. DE ROPP
Department of Chemistry and NMR Facility
University of California, Davis
Davis, California 95616 USA

Abstract

A strategy is described for locating and assigning all of the hyperfine-shifted and/or relaxed resonances in the active site of the low-spin, cyanide-inhibited complex of horseradish peroxidase. The serious problems in spectral resolution and dynamic range due to the large size (44 kDa) of the protein can be overcome by taking advantage of the strong temperature dependence of the hyperfine shift, and by use of DEFT or WEFT pulse sequences in conjunction with standard 2D experiments. Detailed analysis of COSY experiments reveals that cross-correlation contributes significantly to cross peak intensity and dominates for geminal protons. It is shown that many COSY cross peaks, in fact, are inter-residue cross correlation peaks. Hence COSY has limited use in mapping scalar connectivities, even in its phase-sensitive form. TOCSY, on the other hand, particularly when ROESY is suppressed, provides an effective method for mapping essentially all spin systems. The sequence-specific assignment of active site residues via standard backbone NOESY connectivities in ^2H_2O is greatly facilitated by the very slow exchange rates of the peptide protons for both the proximal and distal helices. The combination of 2D NMR methods lead to the complete assignment of the active site protons <6.5 Å from the iron. The resulting dipolar shifts for non-coordinated residues are shown to be quantitatively described by the orientation of the magnetic axes which reflect a tilt of the Fe-CN unit away from the heme normal. It is likely that similar methods will allow detailed structural studies in a variety of comparably sized cyanide-inhibited heme peroxidases.

1. Introduction

One of the first paramagnetic enzymes studied by ^1H NMR was horseradish peroxidase, HRP, a member of the plant super class of heme peroxidases which carry out one electron oxidation of a variety of substrates at the expense of hydrogen peroxide [1,2]. In general, these proteins consist of one protohemin prosthetic group embedded in a single polypeptide chain of ~300 amino acids which may or may not be extensively glycosylated. The resting heme peroxidase enzyme contains a high-spin ferric hemin which reacts with peroxides to produce two oxidizing equivalents in compound I, one each on the iron and either a porphyrin or a nearby amino acid cation radical [1]. The first one-electron oxidation of a substrate by compound I yields compound II, in which the radical is abolished. Thus the complete catalytic cycle of heme peroxidases includes only paramagnetic states of the chromophore. The paramagnetism and the relatively large size, (34-48 kDa for the classical plant and fungal peroxidases, and up to 155 kDa for mammalian peroxidases such as myeloperoxidase, MPO), would not make these enzymes ideal candidates for ^1H NMR studies. However, valuable ^1H NMR studies on several of the paramagnetic states of the prototype plant peroxidase, HRP, and the fungal peroxidase, cytochrome *c* peroxidase, C*c*P, have been reported [3]. C*c*P has yielded high resolution X-ray crystal

G.N. La Mar (ed.), Nuclear Magnetic Resonance of Paramagnetic Macromolecules, 55-74.

structures for several derivatives, providing the first quantitative picture of the catalytic site of a heme peroxidase, and leading to a reaction mechanism that has been generalized to other classical peroxidases [4,5]. The proximal side possesses a coordinated His whose ring NH serves as a strong hydrogen bond donor to the side chain of a proximal Asp, and thereby imparts imidazolate character to the axial ligand which stabilizes compounds I and II. The key distal catalytic residues are proposed to be a His which serves as a general base, and an Arg that facilitates heterolytic bond cleavage for the activating peroxide. These four residues, the proximal His, Asp and distal Arg, His appear highly conserved in the plant and fungal peroxidases [6]. A schematic representation of the heme pocket of HRP, based on the *CcP* crystal structure, sequence homology between *CcP* and HRP, and ^{1}H NMR data [7], is shown in Figure 1.

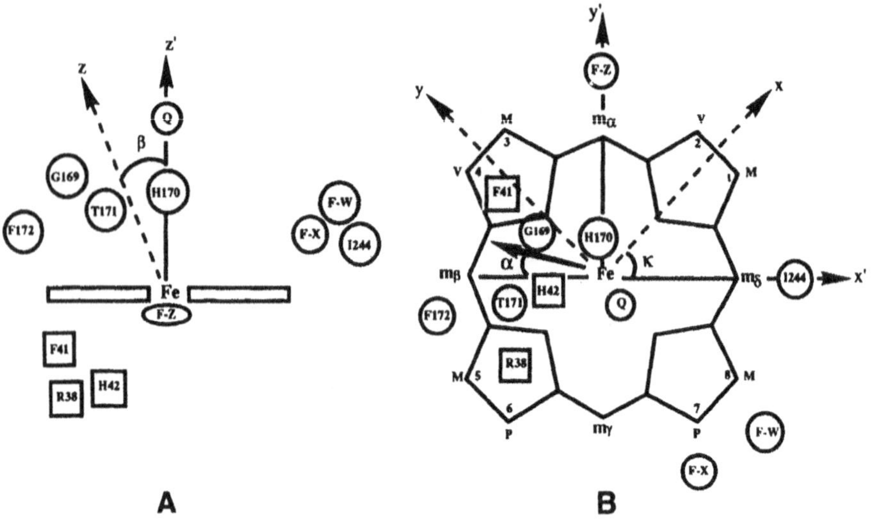

Figure 1 Schematic representation of the heme cavity of HRP based on the crystal structure of *CcP*, sequence homology to *CcP*, and HRP-CN ^{1}H NMR data. (A) edge on view with major magnetic z-axis, with angle β to heme normal z', and (B) face-on view showing rhombic magnetic axes, x, y, oriented by $\kappa \sim \alpha + \gamma$ with respect to pseudosymmetry axes, x', y'. Amino acids are identified by their one-letter code and sequence position, with proximal and distal residues depicted as circles and squares, respectively. Three Phe not sequence specifically assigned are designated F-W, F-X, F-Z, and an as yet unidentified proximal residue in contact with His 170 is labeled residue Q.

While HRP is the most extensively studied heme peroxidase, the failure to yield suitable single crystals [8] has placed heavy emphasis on spectroscopic characterization of the active site. Of the various methods, ^{1}H NMR has provided the most comprehensive insight of both the electronic and molecular structure of the heme pocket [3,7]. The enzyme, beside being readily available, is remarkably stable over a range of temperatures and pH. The most informative derivative of heme peroxidases for ^{1}H NMR study is the low-spin ferric (S = 1/2) cyanide-inhibited form, *i.e.*, HRP-CN, which exhibits relatively narrow and weakly relaxed resonances for the active site with excellent chemical shift dispersion due to the hyperfine fields for both the coordinated and non-coordinated residues. Thus HRP-CN has served as a benchmark molecule for both testing NMR methodology and evaluating the interpretive basis of the resulting hyperfine shifts. The initial assignments were based on isotope labeling the reversibly extractable heme [9], followed by the use of steady-state nuclear Overhauser effect, NOE, to complete the heme assignments and tentatively identify several active site residues based on sequence and structural homology to C*c*P [10-12]. More recently, we have shown that modern 2D NMR methods are directly applicable to HRP-CN and significantly extend the prospects of solution molecular structure determination [7,13]. The NOESY experiment was found to be particularly effective [7,13,14], and shown to be superior to steady-state NOEs in most cases. More surprisingly, COSY experiments, particularly in the magnitude form, MCOSY, were found to exhibit cross peaks for protons expected to be spin coupled [13]. Hence, it appeared as if the two crucial 2D NMR experiments needed to define both molecular and electronic structure in the active site, *i.e.*, scalar or bond correlation to identify residues, and dipolar correlation to characterize spatial relationships among the residues, were applicable to low-spin heme peroxidases.

However, a close examination of the status of NMR studies of peroxidases reveals three main problems which must be resolved to make such studies useful for characterizing the active site. First, until recently, both 1D and 2D NMR methods have been successfully applied primarily to the relatively few resolved residues [10,13,14]. The 500 MHz ^1H NMR spectrum of HRP-CN in ^2H_2O is shown in Figure 2A; not shown is a single proton peak from His 170 $C_\varepsilon H$ at -28 ppm [11]. The unresolved envelope which contains the bulk of the protons is well off-scale at the vertical scale needed to clearly detect the resolved hyperfine shifted peaks from single protons and methyl groups. The problems in extending the heme pocket assignments to the main envelope are resolution (the protein contains ~$3x10^3$ protons in the diamagnetic window and only about 20 resolved signals), and dynamic range (the cross peaks in 2D spectra between relaxed resonances in the diamagnetic window are much weaker and less numerous than those between protons remote from the active site) [3]. Second, the MCOSY experiments previously used [13] to map spin connectivities in heme peroxidases are now known to exhibit cross peaks that arise from cross correlation (a cross term between ^1H-^1H dipolar and Curie spin relaxation) that does not relate to spin connectivity [7,15,16]. This mechanism has been proposed to dominate in high-spin resting state HRP [15]. More recently, Qin *et al* have shown [16] that coherence (spin connectivity) and cross-correlation contribute to MCOSY cross peaks for HRP, but are readily distinguished experimentally on the basis of their differential phase properties relative to the diagonal in the phase-sensitive COSY or P-COSY experiment. The initial examination of the axial His 170 COSY cross peaks has shown [7] that they are dominated by cross correlation even in low-spin HRP-CN, where Curie spin relaxation is much less important than in high-spin HRP. Hence a general investigation of the scope and limits of MCOSY experiments for low-spin cyanide inhibited peroxidases is appropriate. The dominant influence of cross correlation for spatially proximate protons in COSY, and its interference in assessing scalar connectivity, have led us to evaluate rotating-frame scalar correlation experiments (TOCSY [17,18]) for which cross-correlation is abolished. Third, while NOESY has been shown to be very effective in detecting dipolar contacts in the heme cavity of HRP-CN [13,14], the large size and relatively ineffective paramagnetic relaxation lead to spin diffusion, where many cross peaks detected at the conventionally used mixing time of 20 ms [3] arise from secondary rather than primary NOEs [14]. The focus of this report is a survey of approaches to surmount these problems [7].

Lastly, partial assignments of HRP-CN [7,9-14], C*c*P-CN [19] and LiP-CN [20,21] (LiP = lignin peroxidase), have shown that remarkably different patterns of hyperfine shifts, particularly for the catalytically relevant proximal His and distal His and Arg, can be observed. This is despite the fact that the X-ray structures of C*c*P [4] and LiP [22], as well as the solution structure of HRP [7], reflect a remarkably conserved structure for their heme cavity residues. Hence it is desirable to develop a semiquantitative model for the hyperfine shift pattern which will aid in interpreting NMR spectral differences in terms of structural differences among both native genetic variants and point mutants of heme peroxidases. In the following sections, we survey approaches to answer a number of the experimental problems discussed above and show that these methodologies can lead to the needed assignment of active site residues in HRP-CN. The identity of some of the residues located in these studies can be guided by the results of molecular modeling of HRP, while those residues assigned directly by NMR provide important constraints in selecting among alternate models [23]. These NMR methods, moreover, should be applicable to a broad range of isoelectronic heme peroxidases. With the molecular structure of the heme cavity of HRP-CN in hand, we show that the hyperfine shift pattern for amino acid residues can be interpreted in terms of the orientation of the paramagnetic susceptibility tensor, and that differences in hyperfine shift patterns among genetic variants can result solely from modulation of the orientation of the tensor by distal steric interactions with the bound cyanide.

2. Experimental

2.1. SAMPLE PREPARATION

Protein solutions were prepared from lyophilized HRP (>98% isozyme C) purchased from Boehringer-Mannheim and dissolved into 99.9% ^2H_2O. Excess solid KCN was added to the solution to generate HRP-CN; solution pH was adjusted to 7.0 (value not corrected for isotope effect) by addition of small amounts of ca. 0.2 M ^2HCl or NaO^2H. Protein concentration was 3 mM in a 5 mm NMR tube.

2.2. NMR DATA COLLECTION

All spectra were collected on a GE-NMR Ω-500 operating at 11.75T. Data were collected over the temperature range 25° to 55°C with chemical shifts referenced to residual water, which in turn had been calibrated against internal 2,2-dimethyl-2-silapentane-5-sulfonate, DSS. WEFT [7,24] and DEFT [7,25] 1D spectra were collected with 160 msec and 55 msec relaxation delays respectively and repetition rates of 5 s^{-1}. 2D spectra were collected with 1024 or 2048 complex points in t_2 over a 20-31 kHz bandwidth (in some scalar correlation experiments the downfield methyls were folded in). The residual solvent was saturated during the predelay, and all pulse sequences utilized single 90° or 180° pulses (as opposed to composite). Data sets were collected at a repetition rate of 5 s^{-1}. Magnitude COSY (MCOSY) [26] spectra were collected with 512 blocks in t_1 and 300-800 scans/block. Phase-sensitive COSY (P-COSY) [27], TOCSY [17,18] and NOESY [28] spectra were collected with 512 hypercomplex blocks in t_1. All NOESY data, except rise curves, used a 20 ms mixing time; TOCSY mixing times varied from 6 to 22 ms [7]. WEFT-NOESY experiments utilized a 160 ms relaxation delay; DEFT-NOESYs a 55 ms relaxation delay [7,29]. In all phase-sensitive 2D experiments the delay time between pulse and first data point acquired was adjusted to produce minimal baseline curvature.

2.3. NMR DATA PROCESSING

All 2D data sets were either processed on a SPARC-2 workstation with GE-NMR Ω 6.0 or on a SGI 4D/35 with Biosym Felix 1.1 or 2.10. MCOSY data sets were processed with 0°-shifted sine-bell-squared apodization in both dimensions. P-COSY,

TOCSY, and NOESY data sets were processed with 30° to 60°-shifted sine-bell-squared apodization in both dimensions, and phase corrected and baseline leveled in both dimensions [3]. All data sets were zero-filled as necessary to 1024 x 1024 or 1024 x 2048 points. No data sets were symmetrized.

3. Results

3.1. DYNAMIC RANGE/RESOLUTION

Improvements in dynamic range and resolution in a paramagnetic protein are based on the two distinguishing characteristics of active site residue protons [3], effective spin-lattice relaxation given by:

$$T_1^{-1} \propto R_{Fe}^{-6} \quad (1)$$

where R_{Fe} is the distance to the iron, and the temperature dependence of the hyperfine shift contribution, δ_{hfs}, to the observed shift, δ_{obs}, as approximated by the Curie law:

$$\delta_{hfs} \propto T^{-1} \quad (2)$$

The normal 500 MHz ^1H NMR hyperfine spectrum of HRP-CN in ^2H_2O is shown in Figure 2A. The intense diamagnetic envelope 0-10 ppm obscures many active site resonances. The use of a DEFT pulse sequence to suppress all peaks with $T_1 \geq 150$ ms [7], leads to the trace in Figure 2C, which reveals several effectively relaxed peaks (T_1s ≤ 50 ms) that are completely obscured in the normal spectrum, each of which is readily identified as a previously assigned heme peak or as an active site amino acid residue proton (see below). Particularly noteworthy is the detection of the previously assigned heme 1-CH_3 and 5-CH_3 resonances [9,13], as labeled in Figure 2C. Thus DEFT (or WEFT) pulse sequences are remarkably effective for locating signals under the diamagnetic envelope based on their effective relaxation by the iron.

An alternate method for identifying active site residue signals under the diamagnetic envelope is to use the temperature sensitivity of the chemical shift (*i.e.*, Eq. (2)) [3,30]. The MCOSY map for the upfield window of HRP-CN in ^2H_2O, where the proposed Arg 38 peaks resonate, is shown in Figure 3. The dashed lines map out an eight-spin system at 55°C for which several frequencies are degenerate in the previously reported 50°C MCOSY map [13] and hence failed to allow detection of one of the frequencies. The previous characterization of a seven spin system was consistent with that of an Arg in ^2H_2O. The present identification [7] of an eighth spin implies that the peptide NH for Arg 38 must exchange slowly in ^2H_2O (see below). Such variable temperature MCOSY (or NOESY) maps are invaluable for not only locating temperature sensitive (and hence active site) proton signals, but providing a ready means for resolving accidental degeneracies. The systematic search for all temperature-sensitive chemical shifts by 2D NMR in a related, but smaller, cyano-metmyoglobin has been shown to lead to the identification of all protons in the active site [30]. A similar approach applied to HRP-CN appears possible.

The most effective approach for resolving relaxed cross peaks within the diamagnetic envelope is the combination of DEFT or WEFT and NOESY pulse sequences in a single experiment, in which cross peaks from slowly relaxed protons remote from the active site are strongly suppressed when compared to the cross peaks from relaxed residues near the active site. The DEFT (or WEFT) NOESY, in combination with variable temperature experiments, effectively improves both dynamic range and resolution for unresolved active site residues [7]. Portions of the normal NOESY and DEFT-NOESY maps of HRP-CN which illustrate the enhanced resolution/dynamic range are shown in Figure 4. For example, two cross peaks in the

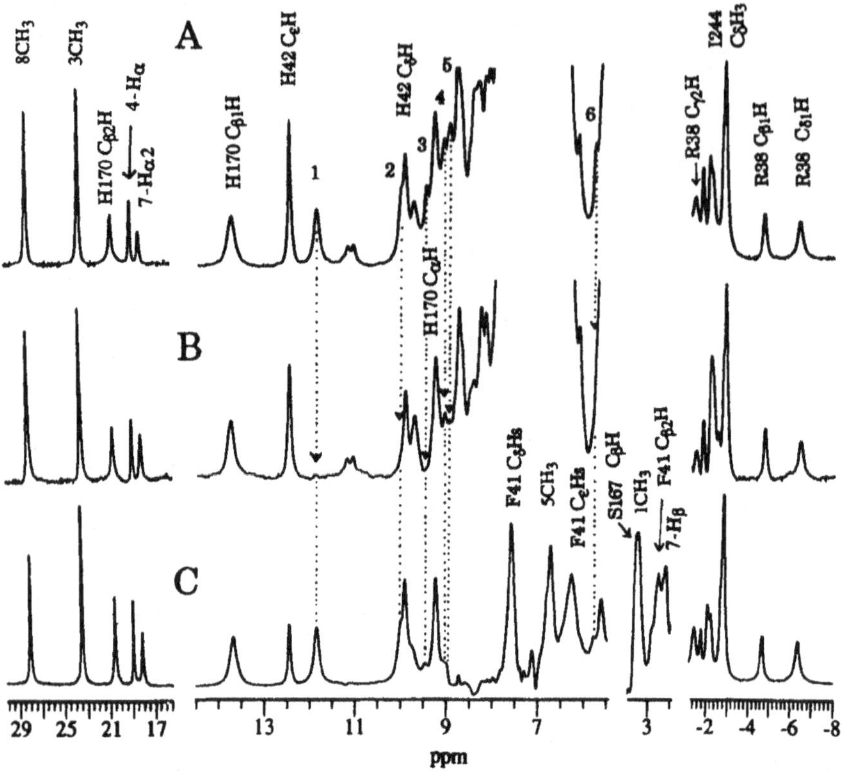

Figure 2 500 MHz ^{1}H NMR trace for isozyme C of HRP-CN in ^2H_2O at 55°C at pH 7.0, with peaks labeled as assigned previously [7]; note upfield His 170 $C_\varepsilon H$ peak at -28 ppm is not shown. (A) Normal trace of HRP-CN one month after dissolution in ^2H_2O; partially resolved, slowly exchanging labile peptide NH peaks are marked by numbers 1-6, and are due to His 170, Gly 169, Phe 172, His 42, Thr 171 and Arg 38, respectively. (B) normal trace of HRP-CN prepared from apo-HRP soaked in ^2H_2O prior to reconstitution of the heme [12]; peptide NHs whose intensity is lost or reduced are marked by vertical arrows. (C) DEFT NMR trace illustrates the suppression of slowly relaxing protons in the diamagnetic envelope and thereby allows the direct detection of strongly relaxed resonances in the 2-4 and 5-10 ppm windows.

aromatic window in the normal NOESY map (labeled Y-1 and Y-2), which arise from an immobile Phe side chain remote from the heme cavity, are completely suppressed in the DEFT-NOESY map, while weakly detected or poorly resolved cross peaks (*i.e.*, Phe 41) in the normal NOESY map are better resolved and strongly enhanced in the DEFT-NOESY map. In fact, all of the cross peaks to the low-field of the water ridge in the DEFT-NOESY map have been assigned to heme pocket residues [7]. Similarly useful selective enhancement of cross peaks for active site residues is achieved in other portions of the DEFT or WEFT-NOESY maps [7].

3.2. CROSS PEAK ORIGINS IN COSY

The MCOSY map in Figure 3 exhibits cross peaks expected for an Arg, a vinyl and a portion of an Ile, but are these cross peaks due to spin coupling? Figure 5 illustrates the predicted cross peak phases expected for J spin coupling (coherence) versus cross-

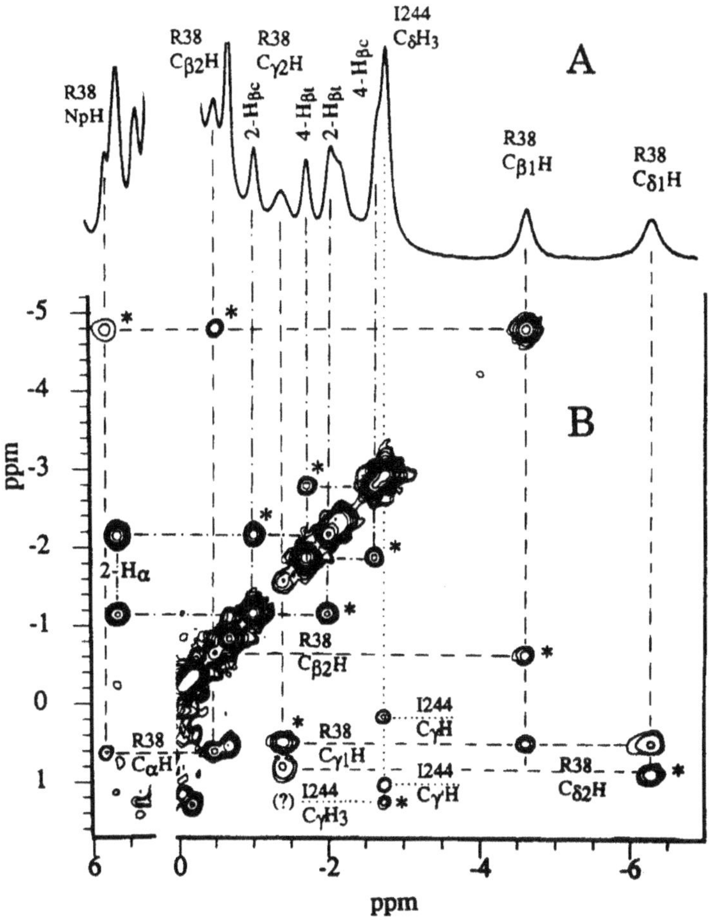

Figure 3 (A) Upfield portion of the 500 MHz ^1H NMR trace of HRP-CN in ^2H_2O at 55°C at pH 7.0 where the Arg 38 and Ile 244 signals resonate. (B) Magnitude COSY, MCOSY, spectrum of HRP-CN illustrating the Arg 38 (dashed lines), Ile 244 (dotted lines) and heme vinyl (dash-dot lines) cross peaks. The cross peaks marked with asterisks arise solely or predominantly from cross-correlation.

correlation in phase-sensitive COSY, P-COSY, as discussed in detail elsewhere [16]. The simulations in Figure 5 show the expected diagonal and cross peaks for the limit of narrow lines compared to the splitting, Figures 5A-5C, and in the limit of broad lines compared to the splitting, Figures 5A'-5C'. Only the latter case is applicable to HRP-CN. The cross peaks due to either scalar (J coupling) or cross-correlation are each composed of antiphase components, but those for cross correlation are in-phase (Figures 5C, 5C') with the in-phase components of the diagonal (Figure 5A, 5A'), while those due to J coupling are 90° out-of-phase with respect to the diagonal (Figures 5B, 5B'). The splitting of the diagonal and cross peak components is J_{AX} for the scalar cross peaks, but Δ_A or Δ_X for cross correlation, where Δ_A or Δ_X can be much larger than J ($\Delta_A = (T_{2A}^{X=\alpha} - T_{2A}^{X=\beta})\{2\pi(T_{2A}^{X=\alpha} T_{2A}^{X=\beta})\}^{-1}$, $\Delta_X = (T_{2X}^{A=\alpha} - T_{2X}^{A=\beta}) \bullet \{2\pi(T_{2X}^{A=\alpha} T_{2X}^{A=\beta})\}^{-1}$) [16]. It should be noted that the convention for phasing the

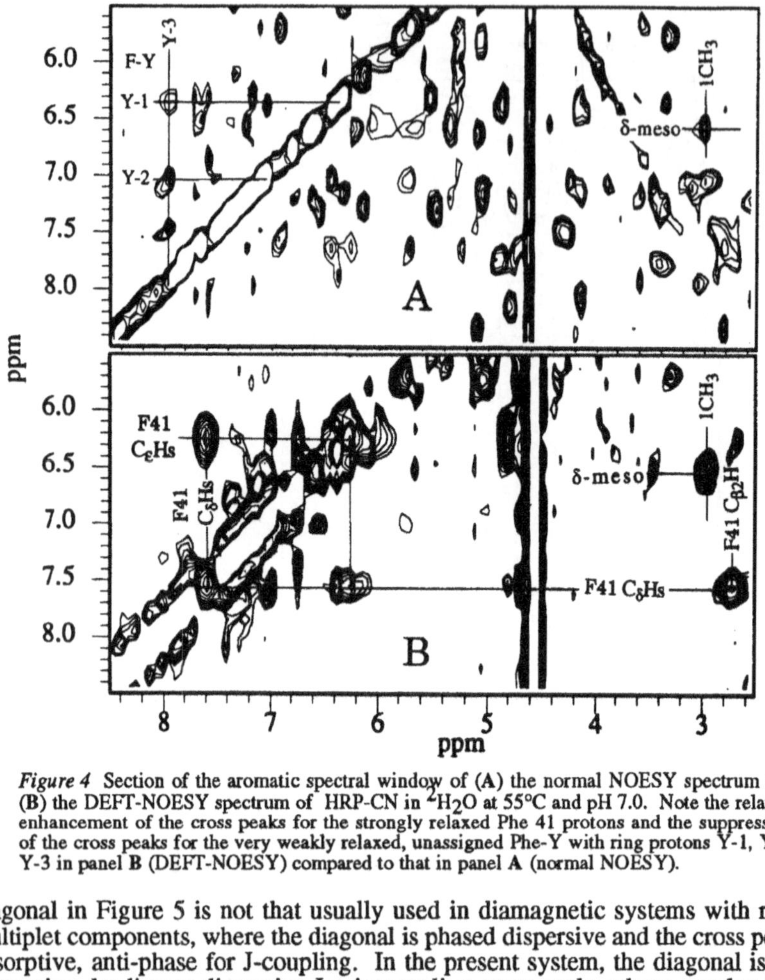

Figure 4 Section of the aromatic spectral window of (A) the normal NOESY spectrum and (B) the DEFT-NOESY spectrum of HRP-CN in ^2H_2O at 55°C and pH 7.0. Note the relative enhancement of the cross peaks for the strongly relaxed Phe 41 protons and the suppression of the cross peaks for the very weakly relaxed, unassigned Phe-Y with ring protons Y-1, Y-2, Y-3 in panel **B** (DEFT-NOESY) compared to that in panel **A** (normal NOESY).

diagonal in Figure 5 is not that usually used in diamagnetic systems with resolved multiplet components, where the diagonal is phased dispersive and the cross peaks are absorptive, anti-phase for J-coupling. In the present system, the diagonal is phased absorptive, leading to dispersive J spin coupling cross peaks whose overlap leads to the "apparent" absorptive peak shown in Figure 5B'.

The upfield portion of the ^1H NMR spectrum where the Arg 38 signals resonate is shown in Figure 6A. The slice of the MCOSY map through Arg 38 $C_{\beta 1}H$ with its surprising intense cross peak to $C_{\beta 2}H$ is reproduced in Figure 6B [13]. In Figure 6C, we show the same slice for the P-COSY experiment. However, the phase of the cross peak to $C_{\beta 2}H$ is that shown in Figure 5C', and hence the cross peak is strongly dominated by cross correlation [7,16]. In fact, all geminal protons in HRP-CN are found to exhibit COSY cross peaks dominated by cross correlation [7]. Moreover, even many vicinal proton COSY cross peaks are largely due to cross correlation. Our conclusions are that conventional COSY experiments have very limited use for mapping spin correlations in low-spin heme peroxidases, although they may provide some additional structural information if their assignments can be determined by less ambiguous methods. It is likely that many COSY cross peaks reported for a variety of heme peroxidases are similarly due primarily to cross correlation rather than

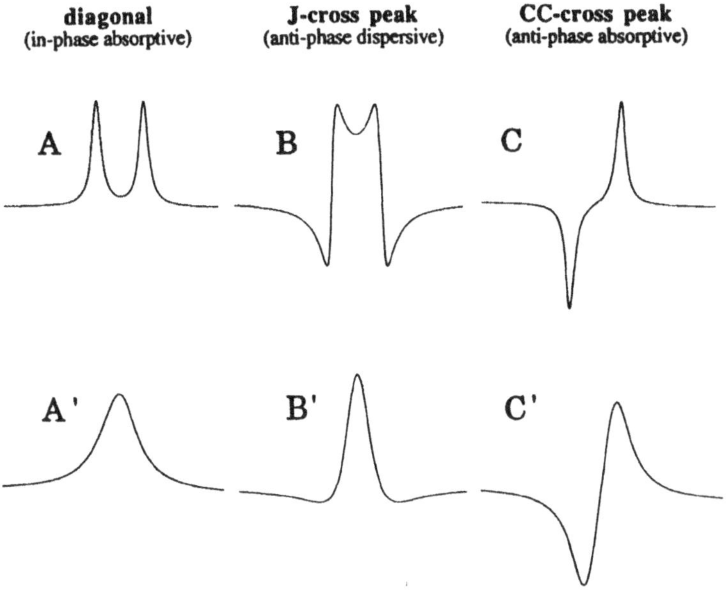

Figure 5 Simulation of the diagonal (**A**, **A'**) and cross peak phases for J-coupling or coherence (**B**, **B'**) and cross-correlation (**C**, **C'**) in a phase-sensitive COSY (P-COSY) experiment. **A**-**C** show the case of linewidth smaller than the splitting, while **A'**-**C'** give the case of linewidth larger than the splitting of its components [16]. Note that the cross peaks from cross-correlation are phase-shifted 90° with respect to those expected from coherence or J-coupling.

coherence. The serious problem of misinterpretations that can result from using COSY maps as the basis for establishing scalar connectivities will be considered in section 3.5, after we have considered alternate methods for mapping coherence in HRP-CN.

3.3. SCALAR CORRELATION VIA TOCSY

Moving to the rotating frame abolishes cross correlation. However, broad lines (short T_2s) and small coupling lead to only weak TOCSY cross peaks; such cross peaks are optimally detected with very short mixing times [3]. Moreover, the large size of HRP-CN, and resulting effective ^{1}H-^{1}H cross relaxation, lead to strong ROESY responses [7]. The opposite absorptive (and potentially self-cancelling) phases for these two responses [18], together with the required short mixing times and large chemical shift dispersion in paramagnetic systems, can make detection of TOCSY cross peaks very difficult, and such studies for peroxidases have not been reported previously. The portion of the 500 MHz ^{1}H NMR spectrum where the His 170 NH, C_αH and C_βH signals resonate is shown in Figure 7A. The normal TOCSY slice through $C_{\beta 1}$H of His 170 (Figure 7B) leads to the expected TOCSY (positive phase) cross peak to C_αH, but a ROESY (negative phase) cross peak for the geminal $C_{\beta 2}$H for a 9 ms duration, 12 KHz spin lock field with the spectrometer frequency set at 15.5 ppm. Note that this places the carrier much closer to the geminal $C_\beta H_2$ pair than the typical condition of having the carrier at the residual solvent, yet still a TOCSY response cannot be obtained for the geminal signals. Increasing the spin lock field to 24 KHz in Figure 7C gives similar results. Note that the ROESY cross

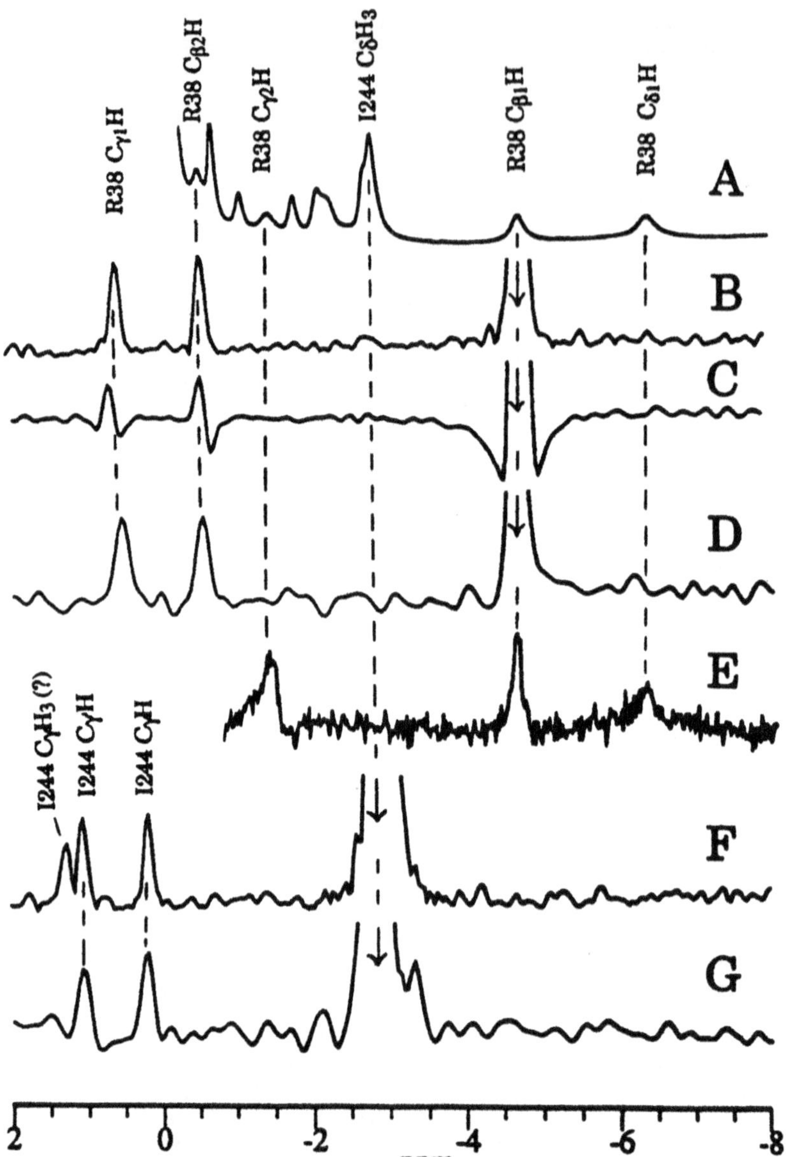

Figure 6 Comparison of scalar correlation methods for the upfield region of HRP-CN in ^2H_2O, pH 7.0 at 55°C. (A) reference spectrum with assignments. (B) MCOSY slice through Arg 38 $C_{\beta 1}H$. (C) PCOSY slice through Arg 38 $C_{\beta 1}H$. (D) CLEAN-TOCSY slice with 12 ms mixing time through Arg 38 $C_{\beta 1}H$. (E) CLEAN-TOCSY slice with 12 ms mixing time through Arg 38 $C_{\gamma 1}H$. (F) MCOSY slice through Ile 244 $C_\delta H_3$. (G) CLEAN-TOCSY slice with 12 ms mixing time through Ile 244 $C_\delta H_3$.

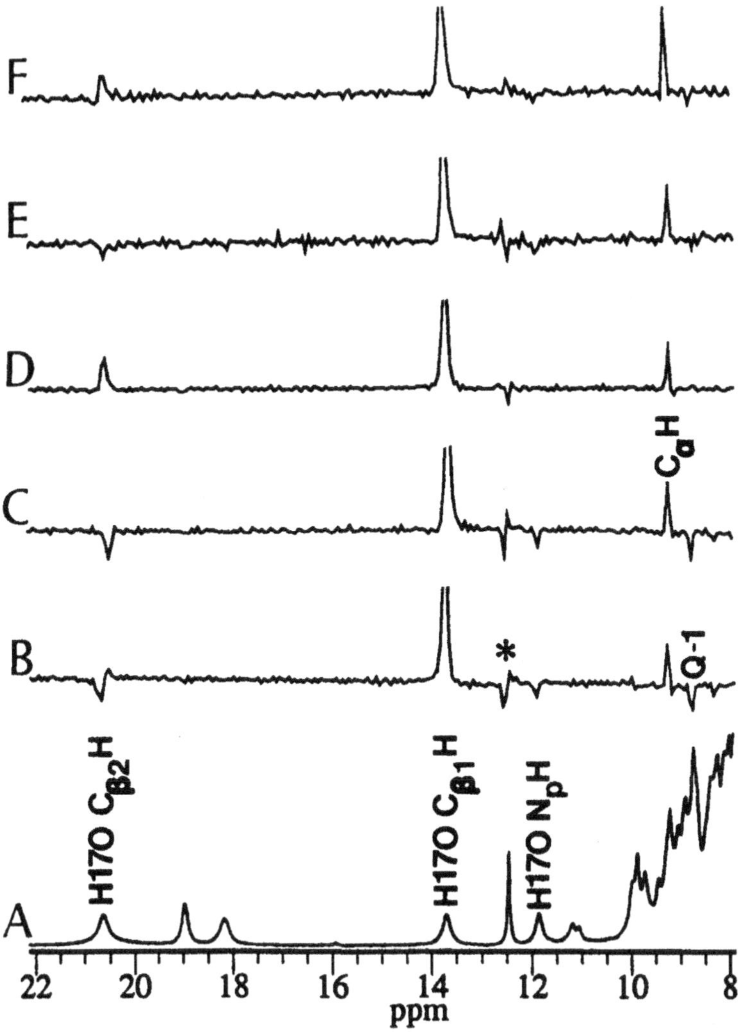

Figure 7 TOCSY slices through the diagonal of His 170 $C_{\beta1}H$ of HRP-CN in ^2H_2O, pH 7.0 at 55°C under the given conditions. (A) Portion of the reference spectrum. (B) TOCSY with 9 ms, 12 kHz spin-lock, carrier at 15.5 ppm. (C) TOCSY with 9 ms, 24 kHz spin-lock, carrier at 15.5 ppm. (D) CLEAN-TOCSY with 12 ms, 12 kHz spin-lock, carrier at 15.5 ppm. (E) CLEAN-TOCSY with 12 ms, 12 kHz spin-lock, carrier at residual solvent (4.5 ppm). (F) CLEAN-TOCSY with 12 ms, 24 kHz spin-lock, carrier at solvent. The symbol * denotes an artifact.

peaks to non-spin-coupled His 170 N_PH and to Q-1 of (unassigned) residue Q are present, in agreement with the NOESY data in Figure 8C.

However, when the CLEAN-TOCSY [18] sequence, designed to suppress ROESY peaks, is used, even the weaker 12 KHz spin lock field produces a TOCSY response in the geminal partner, as shown in Figure 7D. In addition, the ROESY cross peaks to His 170 N_PH and Q-1 are eliminated. Thus it is essential to use CLEAN-TOCSY for peroxidases to suppress cross relaxation; note also that the CLEAN-TOCSY sequence allows the use of a 25% longer mixing time, an important advantage in detecting weak cross peaks in paramagnetic systems. It is more convenient experimentally to collect data with the carrier on the residual solvent as is shown in Figure 7E for a CLEAN-TOCSY with a 12 KHz spin lock field. This field is not sufficient to spin lock the entire paramagnetic bandwidth necessary and no TOCSY response is seen from $C_{\beta 2}H$, but increasing the spin lock field to 24 KHz as in Figure 7F restores the TOCSY cross peak response to $C_{\beta 2}H$ and provides unambiguous evidence of scalar correlation to both geminal and vicinal protons. Hence CLEAN-TOCSY is capable of mapping out the scalar connectivities for residues in the active site of HRP-CN [7].

The advantage of TOCSY over MCOSY is clearly illustrated for the proposed Ile 244 (or Ile 180 [23]) residue, whose δ-methyl is partially resolved in the upfield spectral window of Figure 6A. The MCOSY slice for the relaxed $C_\delta H_3$, which is in dipolar contact with the heme 1-CH_3 and 8-CH_3 [11,13], reveals *three* cross peaks (Figure 6F), as also shown in the MCOSY map in Figure 3, a situation which is incompatible with the possible scalar connectivities of a methyl group. The P-COSY slice exhibits insufficient resolution to clearly establish the phase properties of these three cross peaks (not shown). However, the CLEAN-TOCSY slice through the Ile $C_\delta H_3$ in Figure 6G shows scalar cross peaks to only two of the three frequencies, clearly identifying the γ-methylene protons of the Ile [7]. The origin of the third MCOSY cross peak is cross correlation to the likely Ile $C_\gamma H_3$ (see below). CLEAN-TOCSY also resolves the ambiguities in the Arg 38 spin system, which contains some of the broadest (shortest T_2) peaks in HRP-CN. Figure 6D shows the CLEAN-TOCSY slice through Arg 38 $C_{\beta 1}H$, producing clear scalar correlation cross peaks to both the geminal $C_{\beta 2}H$ and vicinal $C_{\gamma 1}H$, eliminating the ambiguities present in the traces in Figures 6B and 6C. The CLEAN-TOCSY slice through $C_{\gamma 1}H$ of Arg 38 (Figure 6E) produces definite scalar cross peaks to its geminal ($C_{\gamma 2}H$) and vicinal ($C_{\beta 1}H$, $C_{\delta 1}H$) protons [7].

A section of the CLEAN-TOCSY map of HRP-CN with a 12 ms mixing time is shown in Figure 8B. Clearly observed are the $C_{\beta 1}H$-$C_\alpha H$ and N_PH-$C_\alpha H$ cross peaks for His 170, both direct and remote cross peaks for the four protons Q-1 to Q-4 of an as yet unassigned residue Q, as well as several characteristic N_PH-$C_\alpha H$ cross peaks for residues with slowly exchanging N_PHs, of which we label those from Gly 169 (see below), His 42 and Phe 172. The cross peaks for the relatively strongly relaxed His 170 and Gly 169 resonances are undetected at longer (22 ms) mixing times. In all, complete or partial spin systems for 14 residues in the active site of HRP-CN have been identified by CLEAN-TOCSY [7].

3.4. SEQUENCE-SPECIFIC ASSIGNMENTS IN ^2H_2O

The standard sequence-specific assignment of residues in a folded protein proceeds by tracing the backbone (N_PH, $C_\alpha H$) NOESY cross peaks in ^1H_2O solution [31]. This approach has serious limitations in proteins as large as HRP because of the severe spectral congestion from the ~300 NHs in the amide region. In fact, any hope of achieving such assignments demands the introduction of ^{15}N or $^{15}N/^{13}C$ labeling into the protein and the use of heteronuclear 2D/3D methods [32]. Because of the inefficiency in refolding recombinant HRP [33], the isotope labeling approach is not yet practical. However, both the normal 1D spectrum in Figure 2A, as well as the CLEAN-TOCSY spectrum in Figure 8B, indicate that several peptide NHs can be observed for months to years after dissolution of HRP-CN in ^2H_2O, even at neutral pH. The resolved His 170 NpH peak had been assigned earlier and shown to be exchangeable in ^2H_2O only in the apo-protein [12]. Figure 2B shows the ^1H NMR trace of HRP-CN in ^2H_2O prepared from apo-HRP previously soaked in ^2H_2O. Significant intensity is lost in the 7-10 ppm region, but most prominent are six NpH peaks marked by numbers, one of which is the His 170 NpH. The NpH at 10.12 ppm exhibits a NOESY cross peak to His 170 NpH (Figure 8C), and hence must arise from the residue adjacent to His 170 in the sequence (*i.e.* Gly 169 or Thr 171).

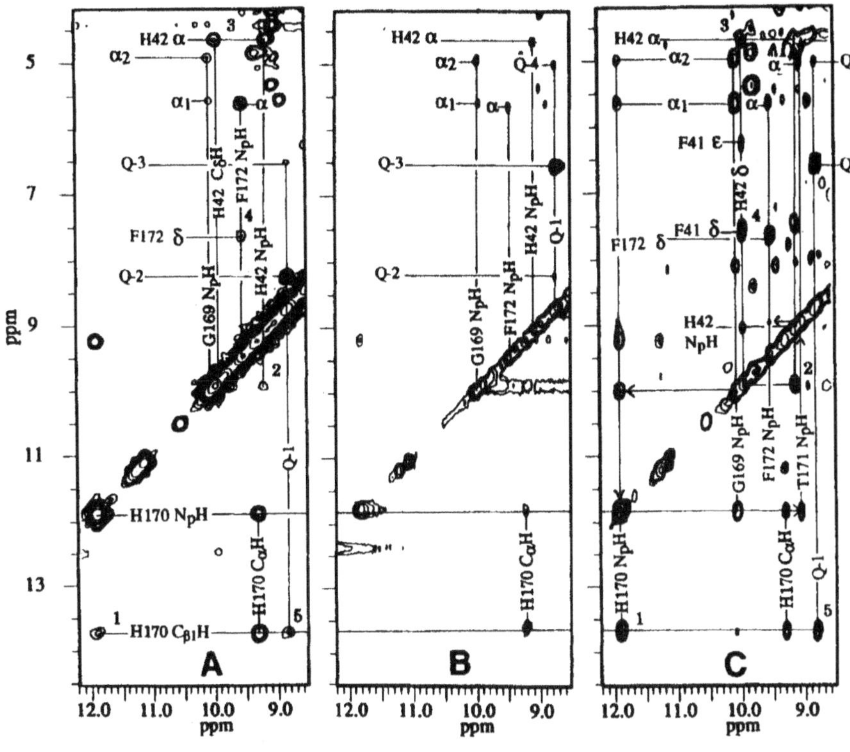

Figure 8 Portion of the aromatic spectral window of the (A) MCOSY; (B), CLEAN-TOCSY with 12 ms mixing time, and (C) NOESY with 20 ms mixing time, for HRP-CN in ^2H_2O at 55°C and pH 7.0. Cross peaks are labeled by one letter amino acid code. The cross peaks marked with numbers in (A) are due to cross-correlation as identified: 1, H170NpH-H170C$_{\beta1}$H; 2, H42NpH-H42C$_\delta$H; 3, H42C$_\delta$H-H42C$_\alpha$H; 4, F172 NpH-F172 C$_\delta$Hs and 5, Q-1-H170 C$_{\beta1}$H. Note that the numbered cross peaks in the MCOSY spectrum (A) are not present in the TOCSY map (B), but correspond to very intense NOESY cross peaks (C).

However, the two TOCSY peaks in Figure 8B from the NpH at 10.12 ppm to two non-labile protons, together with intense TOCSY and NOESY cross peaks between these two non-labile protons, identify this residue as Gly 169 and establishes the direction of the helix. The backbone NH connectivity can be followed to Phe 172 (not shown). In fact, numerous N_i-N_{i+1}, α_i-N_{i+1}, α_i-β_{i+3} cross peaks indicative of an α-helix are readily observed [7].

The peak at 5.67 ppm was observed in Figure 3 as part of the Arg 38 spin system in ^2H_2O. The loss of that signal in Figure 2B confirms the presence of the labile NpH in the eight-spin system [7]. The slowly exchanging His 42 NpH is assigned by its expected strong NOE to the His 42 ring CδH (Figure 8C). Hence, both the proximal and distal helices of HRP-CN exhibit remarkable dynamic stability in retaining the backbone peptide protons after many months to years in ^2H_2O [7]. The distal helix signature NH-NH NOESY cross peaks have not yet been identified, but are likely detectable in ^2H_2O at higher field. The fact that the slowly exchanging peptide NHs are localized in the active site opens up the prospects for extensive standard sequence-specific assignment of residues by homonuclear 2D NMR methods [31] by taking advantage of the resolution in the fingerprint region due to the selective retention of NHs in the active site.

3.5. MIS(?)-INFORMATION IN MCOSY SPECTRA

Inspection of the MCOSY map in Figure 8A reveals intra-residue cross peaks between numerous unambiguously assigned proton resonances which are known not to be spin-coupled [7]. Several such prominent cross peaks are: Cβ_1H-NpH for His 170, CδH-NpH, CδH-CαH for His 42, and NpH-CδHs for Phe 172. A similar non-spin-coupled MCOSY cross peak for NpH-Cβ_1H of Arg 38 is observed in Figure 3. Most dramatic, however, is the observed *inter*-residue MCOSY cross peak in Figure 8A between His 170 Cβ_1H and proton Q-1 of residue Q, which is homologous to Phe 202 in C*c*P, but has not been unambiguously assigned in HRP-CN [7]. Note that none of these cross peaks are found in the TOCSY map in Figure 8B. Clearly, taking these cross peaks for evidence of spin connectivity can lead to disaster.

Cross correlation between protons is strongly dependent on the interproton distance [15]. Inspection of the NOESY map in Figure 8C reveals that indeed, all cross-correlation MCOSY cross peaks in Figure 8A correspond to proton pairs which exhibit intense NOESY cross peaks. We therefore conclude that MCOSY cross peaks for which the protons also exhibit intense NOESY cross peaks likely arise from cross correlation, and cannot be taken as evidence for spin connectivity [7]. Hence, MCOSY cannot serve as a basis for mapping scalar correlation in any heme peroxidase. While MCOSY cross peaks provide mis-information if interpreted in terms of coherence or spin connectivity, the correct interpretation, together with NOESY cross peak intensity, should provide additional valuable information on the relative spatial disposition of various protons in the heme cavity.

3.6. DIPOLAR CORRELATION

NOESY experiments appear to work well for low-spin heme peroxidases, since the paramagnetic relaxation is relatively weak and the cross relaxation, because of the molecular size, is strong [7,13,14,19-21]. In fact, the paramagnetic relaxation, to some extent, quenches spin diffusion in the active site, allowing the detection of primary NOEs for longer mixing times than would be possible in an isostructural diamagnetic system [3]. Initial studies of cross peak intensity for heme and His 170 resonances in HRP-CN have already shown that interproton distances can be determined from the linear portion of rise curves, and that these distances correlate well with the known fixed intramolecular distances within the heme and axial His [14].

The linear portion of rise curves from geminal methylene protons and strongly paramagnetically relaxed protons, however, can be very short, dictating the use of mixing times ≤ 3 ms, for which it may be difficult to obtain the sensitivity to quantitate cross peak volume, unless the protein is available in plentiful supply.

The majority of NOESY spectra of low-spin heme peroxidases have utilized a mixing time of 20 ms [13,19-21], which appears optimal for detecting cross peaks of interest. However, such a long mixing time can lead to significant spin diffusion (secondary NOEs) [3,14]. Such secondary NOEs, when not explicitly considered, lead to misinformation on the proximity of pairs of protons. The slices through the heme 3-CH_3 peak as a function of mixing time for HRP-CN in ^2H_2O are shown in Figure 9. The proton assignment for several cross peaks are labeled, and the system, 1°, 2° as used to designate primary and secondary NOEs (*i.e.*, those whose intercept of zero

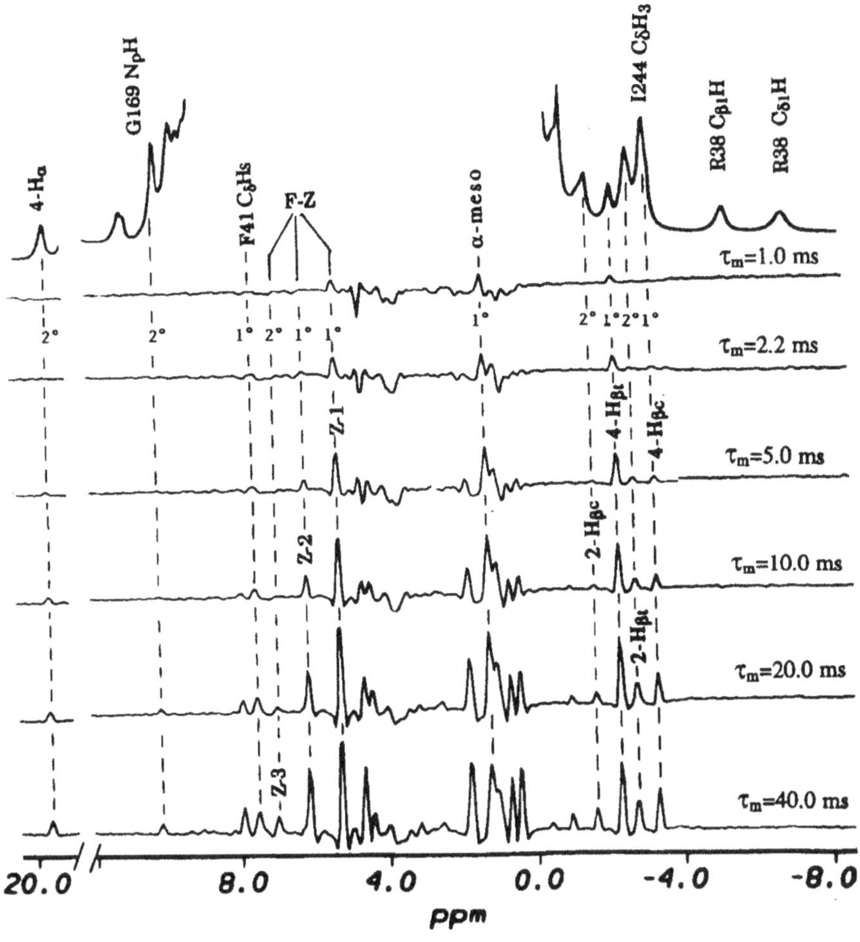

Figure 9 Slices through the heme 3-CH_3 peak in the NOESY spectrum of HRP-CN in ^2H_2O at 35°C as a function of mixing time, τ_m. Cross peaks to selected assigned resonances are labeled 1°, 2° for primary and secondary NOE, respectively.

intensity is at $\tau = 0$ and $\tau > 0$, respectively). The rise curves for several of the protons of interest are shown in Figure 10. The intra-heme 3-CH_3-4$H_{\beta c}$, 3-CH_3-4$H_{\beta t}$ cross peaks are primary and confirm the *cis* orientation of the 4-vinyl group [14]; the 3-CH_3-4H_α NOE in Figure 9 is clearly a secondary NOE via the 4H_βs. The 3-CH_3-2$H_{\beta t}$, 3-CH_3-2$H_{\beta c}$ cross peaks have been shown to be secondary NOEs via the α-meso-H [14], but still confirm the *trans* orientation of the 2-vinyl group.

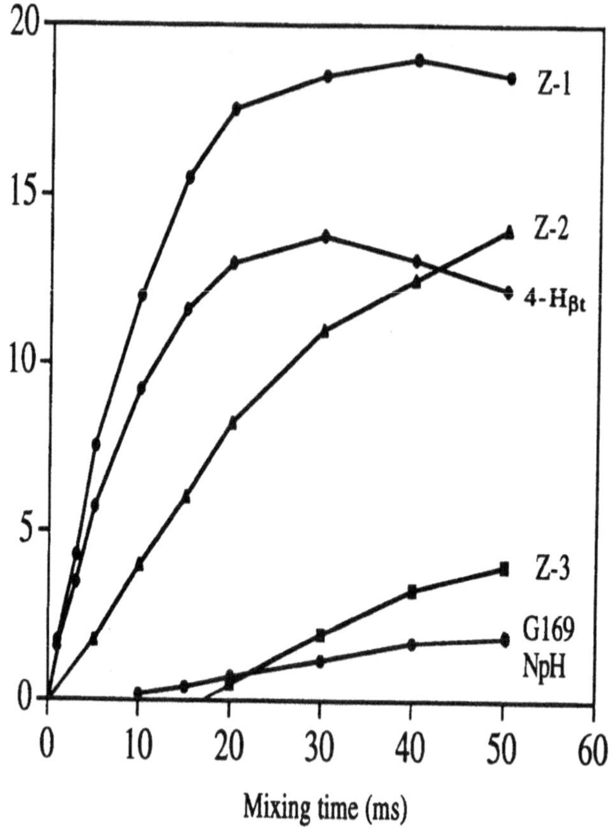

Figure 10 Plot of the intensities (arbitrary scale) of cross peaks between 3-CH_3 and several resonances of interest as a function of mixing time, τ_m. Note that the NOEs from 3-CH_3 to Z-1, Z-2 and 4-$H_{\beta t}$ are primary, while those to Z-3 and G169 NpH are secondary.

The cross peaks to two protons, labeled Z-1 and Z-2 in Figure 9, of a Phe (likely Phe 152) in contact with both 3-CH_3 and 2-vinyl are primary, with 3-CH_3 much closer to Z-1 than Z-2; the third Phe ring proton, Z-3, exhibits only a secondary NOE. The Gly 169 NpH-3-CH_3 NOE is secondary, and occurs via a relatively strong NOESY cross peak Gly 169 NpH-4$H_{\beta t}$ [7], as expected from the CcP [4] crystal structure. The cross peak to the resonance at 7.62 ppm in Figure 9 is primary, and with only weak temperature dependence to its shift, identifies the strongly relaxed signal at 7.62 ppm in Figure 2C as the Phe 41 ring C_δHs. This signal, as well as the Phe 41 C_εHs, also exhibit NOESY cross peaks to His 42 C_δH (Figure 8C) [7]. Numerous primary NOEs between heme and assigned residues, and between pairs of assigned residues have been identified [7] for which further detailed work should yield estimates of the internuclear separations. A particularly important set of cross peaks occur between the residue Q and the C$_\beta$Hs of His 170. The contact is similar to that expected for Phe 202 in C*c*P [4] or Phe 204 in LiP [22]. This Phe, however, is not conserved in HRP [6], and the relevant region of the sequence is particularly difficult to place in the preliminary X-ray crystal structure of HRP isozyme E5 [8]. Molecular modeling of HRP has led to the proposal of Leu 250 as the origin of residue Q [23].

3.7. AN INTERPRETIVE BASIS OF HYPERFINE SHIFTS

The heme contact shift pattern in all low-spin ferric hemoproteins is dominated by the orbital ground state [34], which is controlled by the orientation of the proximal His imidazole plane in cyanide ligated ferri-hemoproteins [35,36]. The heme contact shift pattern in HRP-CN of large shifts for the substituents on pyrroles B and D (3-CH_3, 4-H_α, 7H_αs, 8-CH_3) and small shifts for substituents on pyrroles A, C (1-CH_3, 2H_α, 5-CH_3, 6H_αs), is consistent with the orientation of the axial His in C*c*P-CN [4], and has been concluded to be very similar in HRP-CN [7,11]. The non-coordinated residues, on the other hand, exhibit only dipolar shifts given by [35,36]:

$$\delta_{dip} = \frac{1}{3N}\left[\Delta\chi_{ax}\,(3\cos^2\theta - 1)\,r^{-3} + \frac{3}{2}\Delta\chi_{rh}\,(\sin^2\theta\cos 2\Omega)\,r^{-3}\right]R(\alpha, \beta, \gamma) \qquad (3)$$

where $\Delta\chi_{ax}$, $\Delta\chi_{rh}$ are the axial and rhombic anisotropies of the magnetic susceptibility tensor, θ, Ω, r are the proton coordinates in some arbitrary iron-centered protein coordinate system (usually taken as a set of X-ray crystal coordinates), and $R(\alpha, \beta, \gamma)$ are the Euler rotation angles that rotate the iron-centered crystal coordinates into the magnetic axes. The tilt of the major magnetic (z) axis with respect to the heme normal is β (Figure 1A), and the location of the rhombic magnetic axes in the heme plane is given by $\kappa \sim \alpha + \gamma$ (Figure 1B). 2D NMR has identified some fifteen dipolar shifted signals which could be sequence-specifically assigned (and hence provide experimental δ_{dip}) and shown [7] to have spatial disposition relative to the heme and to each other that are essentially identical to those conserved residues in C*c*P [4]. Hence the C*c*P crystal data serve as coordinates (*i.e.*, geometric factors $(3\cos^2\theta-1)r^{-3}$, $(\sin^2\theta\cos 2\Omega)r^{-3}$ in Eq. (3)) for analyzing the HRP-CN dipolar shifts in terms of $R(\alpha, \beta, \gamma)$ and $\Delta\chi_{ax}$, $\Delta\chi_{rh}$. A least-square computer search for a minimum between observed and calculated dipolar shifts afforded both the anisotropies, which are only slightly smaller than in metMbCN, and the orientation of the axes [36]. The excellent correlation between observed and predicted dipolar shifts for 15 resonances is illustrated in Figure 11. The orientation of the magnetic axes is described by a tilt from the heme normal by ~20° in the direction of the β-meso-H. For metMbCN, the tilts of the z axis and Fe-CN vectors were found coincident [35], so that the tilt of the z-axis (β) in HRP-CN [36] is attributed to tilt of the Fe-CN from the heme normal. The direction and magnitude of the tilt in HRP-CN is similar to what is observed in the crystal structure of C*c*P-CN [37].

It is therefore concluded that the hyperfine shifts for non-coordinated residues in HRP-CN reflect the orientation of the magnetic axes as determined by Fe-CN tilt. A decrease in the tilt of the major magnetic axis from the heme normal predicts shift changes [36] that correlate directly with the differences observed for conserved catalytic residues upon comparing HRP-CN with LiP-CN [20,21] and MnP-CN (MnP = manganese peroxidase) [38]. Hence, it is concluded [36] that the significant differences in hyperfine shift pattern for a variety of cyanide-inhibited heme peroxidases with largely conserved active site structure are due to varied degrees of tilt of the Fe-CN unit. It is likely that the significant perturbations to the hyperfine shift pattern of distal point mutants of C*c*P-CN or HRP-CN can be interpreted in terms of altered distal constraints on the bound ligand. Such investigations of HRP-CN mutants are in progress.

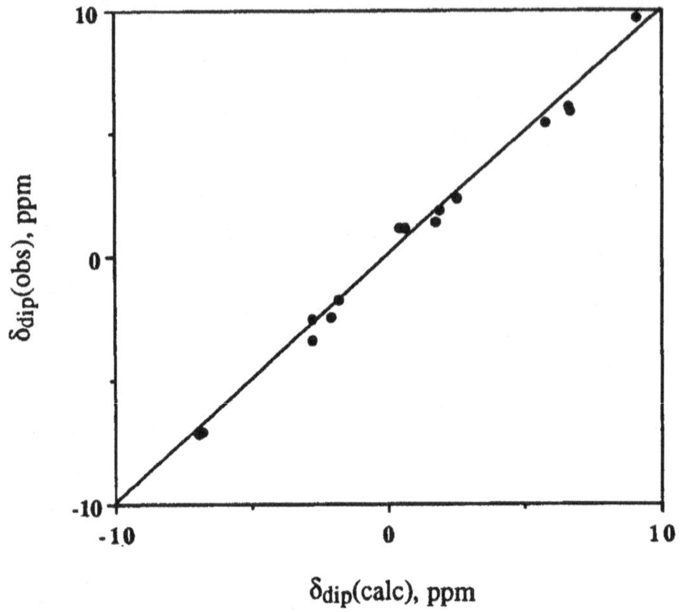

Figure 11 A plot of observed, δ_{dip}(obs), versus calculated, (δ_{dip}(calc) as in Eq. (3)), dipolar shifts for active site residues of HRP-CN based on magnetic axes R(20°, 23.2°, 70°), $\Delta\chi_{ax} = 0.90 \times 10^{-33}$ m^{-3}, and $\Delta\chi_{rh} = 0.60 \times 10^{-33}$ m^{-3}, as obtained from a five parameter least-square search [36]. The solid line represents the ideal correlation.

4. Conclusions

The combination of 1D and 2D NMR methods enables systematic structure elucidation of the active site of plant and fungal peroxidases. These limited structural studies can provide important experimental support for computer modeled structures and such modeling, in turn, can guide the assignment strategy for residues in the active site [23]. Moreover, changes in hyperfine shift patterns are readily interpreted in terms of the position of the heme with respect to the axial His, and the degree of Fe-CN tilt from the heme normal. The latter effect is likely to provide important information on distal steric and hydrogen bonding effects on the bound ligand.

5. Acknowledgments

This research was supported by a grant from the National Institutes of Health, GM 26226. The authors are indebted to Drs. Marco Sette, Griselda Hernández and K. Vyas for experimental assistance, and to Dr. Jun Qin for valuable discussions.

6. References

1. Everse, J., Everse, K.E. and Grisham, M.B. (eds) (1991) *Peroxidases in Chemistry and Biology*, CRC Press, Boca Raton, FL, Vol. II.

2. Ortiz de Montellano, P.R. (1992) Catalytic sites of hemoprotein peroxidases, *Annu. Rev. Pharmacol. Toxicol.* **32**, 89-107.

3. La Mar, G.N. and de Ropp, J.S. (1993) NMR methodology for paramagnetic proteins, in L.J. Berliner and J. Reuben (eds), *Biological Magnetic Resonance*, Plenum Press, New York, 12, pp. 1-78.

4. Poulos, T.L. and Kraut, J. (1980) The stereochemistry of peroxidase catalysis, *J. Biol. Chem.* **255**, 8199-8205.

5. Finzel, B.C., Poulos, T.L., and Kraut, J. (1984) Crystal structure of yeast cytochrome *c* peroxidase refined at 1.7-A resolution, *J. Biol. Chem.* **259**, 13027-13036.

6. Welinder, K.G. and Gajhede, M. (1993) Structure and evolution of peroxidases, in K.G. Welinder, S.K. Rasmussen, C. Penel, and H. Greppin (eds), *Plant Peroxidases: Biochemistry & Physiology*, University of Geneva Press, Geneva, Switzerland, pp. 35-42.

7. Chen, Z., de Ropp, J.S., Hernandez, G., and La Mar, G.N. (in press) 2D NMR approaches to characterizing the molecular structure and dynamic stability of the active site for cyanide-inhibited horseradish peroxidase, *J. Am. Chem. Soc.*

8. Morita, Y., Funatsu, J., and Mikami, B. (1993) X-ray crystallographic analysis of horseradish peroxidase E5, in K.G. Welinder, S.K. Rasmussen, C. Penel, and H. Greppin (eds), *Plant Peroxidases: Biochemistry and Physiology*, University of Geneva Press, Geneva, Switzerland, pp. 1-4.

9. de Ropp, J.S., La Mar, G.N., Smith, K.M., and Langry, K.C. (1984) Proton NMR studies of the electronic and molecular structure of ferric low-spin horseradish peroxidase complexes. *J. Am. Chem. Soc.* **106**, 4438-4444.

10. Thanabal, V., de Ropp, J.S., and La Mar G.N. (1987) ^{1}H NMR study of the electronic and molecular structure of the heme cavity in horseradish peroxidase. Complete heme resonance assignments based on saturation transfer and nuclear Overhauser effects, *J. Am. Chem. Soc.* **109**, 265-272.

11. Thanabal, V., de Ropp, J.S., and La Mar, G.N. (1987) Identification of the catalytically important amino acid residue resonances in ferric low-spin horseradish peroxidase with nuclear Overhauser effect measurements, *J. Am. Chem. Soc.* **109**, 7516-7525.

12. Thanabal, V., de Ropp, J.S., and La Mar, G.N. (1988) Proton NMR characterization of the catalytically relevant proximal and distal hydrogen-bonding networks in ligated resting state horseradish peroxidase, *J. Am. Chem. Soc.* **110**, 3027-3035.

13. de Ropp, J.S., Yu, L.P., and La Mar, G.N. (1991) 2D NMR of paramagnetic metalloenzymes: cyanide-inhibited horseradish peroxidase, *J. Biomolec. NMR* **1**, 175-190.

14. Sette, M., de Ropp, J.S., Hernandez, G., and La Mar, G.N. (1993) Determination of interproton distances from NOESY spectra in the active site of paramagnetic metalloenzymes: cyanide-inhibited horseradish peroxidase, *J. Am. Chem. Soc.* **115**, 5237-5245.

15. Bertini, I., Luchinat, C., and Tarchi, D. (1993) Are true scalar proton-proton connectivities ever measured in COSY spectra of paramagnetic macromolecules, *Chem. Phys. Lett.* **203**, 445-449.

16. Qin, J., Delaglio, F., La Mar, G.N., and Bax, A. (1993) Distinguishing the effects of cross correlation and J coupling in COSY spectra of paramagnetic proteins, *J. Magn. Reson, Series B* **102**, 332-336.

17. Braunschweiler, L. and Ernst, R.R. (1983) Coherence transfer by isotropic mixing: application to proton correlation spectroscopy, *J. Magn. Reson.* **53**, 521-528.

18. Griesinger, C., Otting, G., Wüthrich, K., and Ernst, R.R. (1988) Clean TOCSY for ^{1}H spin system identification in macromolecules, *J. Am. Chem. Soc.* **110**, 7870-7872.

19. Satterlee, J.D. and Erman, J.E. (1991) Proton NMR assignments of heme contacts and catalytically implicated amino acids in cyanide-ligated cytochrome *c* peroxidase determined from one- and two-dimensional nuclear Overhauser effects, *Biochemistry* **30**, 4398-4405.

20. de Ropp, J.S., La Mar, G.N., Wariishi, H., and Gold, M. (1991) NMR study of the active site of resting state and cyanide-inhibited lignin peroxidase from *phanerochaete chrysosporium*, *J. Biol. Chem.* **266**, 5001-5008.

21. Banci, L., Bertini, I., Turano, P., Tien, M., and Kirk, T. K. (1991) Proton NMR investigation into the basis for the relatively high redox potential of lignin peroxidase, *Proc. Natl. Acad. Sci. U.S.A.* **88**, 6956-6960.

22. Poulos, T.L., Edwards, S.L., Wariishi, H., and Gold, M.H. (1993) Crystallographic refinement of lignin peroxidase at 2Å, *J. Biol. Chem.* **268**, 4429-4440.

23. Smith, A. T., Du, P., and Loew, G. H. (1994) Homology modeling of horseradish peroxidase, in G. N. La Mar (ed), Nuclear Magnetic Resonance of Paramagnetic Macromolecules, Kluwer Academic Publisher, Dordrecht, this volume.

24. Patt, S.L. and Sykes, B.D. (1972) Water-eliminated Fourier transform NMR spectroscopy, *J. Chem. Phys.* **56**, 3182-3184.

25. Becker, E.D., Ferretti, J.A., and Farrar, T.C. (1975) Dynamic range in Fourier transform proton magnetic resonance, *J. Magn. Reson.* **19**, 114-117.

26. Bax, A. (1982) *Two-dimensional nuclear magnetic resonance in liquids*, Delft University Press, Dordrecht, Holland.

27. Marion, D. and Wüthrich, K. (1983) Application of phase-sensitive two-dimensional correlated spectroscopy (COSY) for measurements of ^{1}H-^{1}H spin-spin coupling constants in proteins, *Biochem. Biophys Res. Comm.* **113**, 967-974.

28. Macura, S. and Ernst, R.R. (1980) Elucidation of cross relaxation in liquids by two-dimensional NMR spectroscopy, *Mol. Phys.* **41**, 95-117.

29. Kao, L.-F. and Hruby, V.J. (1986) Suppression or differentiation of solvent resonance by a combination of DEFT with a two-dimensional sequence, *J. Magn. Reson.* **70**, 394-407.

30. Qin, J. and La Mar, G.N. (1992) Complete sequence-specific ^{1}H NMR resonance assignment of hyperfine-shifted residues in the active site of a paramagnetic protein: application to *Aplysia* cyano-metmyoglobin, *J. Biomolec. NMR* **2**, 597-618.

31. Wüthrich, K. (1986) *NMR of Proteins and Nucleic Acids*, Wiley & Sons, New York.

32. Clore, G.M. and Gronenborn, A.M. (1991) Two, three, and four-dimensional NMR methods for obtaining larger and more precise three-dimensional structures of proteins in solution, *Annu. Rev. Biophys. Biophys. Chem.* **20**, 29-63.

33. Smith, A.T., Sanders, S.A., Sampson, C., Bray, R.C., Burke, J.F., and Thorneley, R.N.F. (1993) Folding and activation of recombinant horseradish peroxidase from *E. coli* and analysis of protein variants produced by site-directed mutagenesis, in K.G. Welinder, S.K. Rasmussen, C. Penel, and H. Greppin (eds), *Plant Peroxidases: Biochemistry & Physiology*, University of Geneva Press, Geneva, Switzerland, pp. 159-168.

34. Shulman, R. G., Glarum, S. H. and Karplus, M. (1971) Electronic structure of cyanide complexes of hemes and heme proteins, *J. Mol. Biol.* **57**, 193-115.

35. Rajarathnam, K., La Mar, G.N., Chiu, M.L., and Sligar, S.G. (1992) Determination of the orientation of the magnetic axes of the cyano-met complexes of point mutants of myoglobin by solution ^{1}H NMR: Influence of His E7 → Gly and Arg CD3 →Gly substitutions, *J. Am. Chem. Soc.* **114**, 9048-9058.

36. La Mar, G. N., Chen, Z., Vyas, K. and McPherson, A. D. (1994) An interpretive basis of the hyperfine shifts in cyanide-inhibited horseradish peroxidase based on the magnetic axes and ligand tilt: Influence on substrate binding and extension to other peroxidases, *J. Am. Chem. Soc.*, submitted.

37. Edwards, S. and Poulos, T.L. (1990) Ligand binding and structural perturbations in cytochrome c peroxidase, *J. Biol. Chem.* **265**, 2588-2595.

38. Banci, L., Bertini, I., Pease, E.A., Tien, M., and Turano, P. (1992) ^{1}H NMR investigation of manganese peroxidase from *phanerochaete chrysosporium*. A comparison with other peroxidases, *Biochemistry* **31**, 10009-10017.

HOMOLOGY MODELING OF HORSERADISH PEROXIDASE

ANDREW T. SMITH
Biochemistry Laboratory, University of Sussex
Brighton, BN1 9QG, UK

PING DU and GILDA H. LOEW
Molecular Research Institute
845 Page Mill Road, Palo Alto, CA 94304 USA

Abstract

Three dimensional (3D) models of horseradish peroxidase (HRP) and their Cyano complexes were constructed using the known structures of three other members of the plant peroxidase superfamily: cytochrome c peroxidase (CCP), lignin peroxidase (LiP), and *Arthromyces ramosus* peroxidase (ARP). Six models of the cyano complexes of the full protein, differing mainly in the conformation of a large insertion between helices F and G were obtained. Comparisons of the calculated distances between specific heme moieties and nearby residues in these models with those derived from 2D NMR data allowed the selection of one of these models as the most plausible and the identification of the Ile 180 as Ile-X residue responsible for a cross peak with the heme 8-CH3 group and Leu 250 as the aliphatic residue responsible for cross peaks with the proximal histidine 170 Cβ-protons deduced from NMR studies.

1. Introduction

Horseradish peroxidase (HRP) is a member of the plant peroxidase superfamily [1]. It is able to utilize hydrogen peroxide to catalyze the one electron oxidation of a wide range of aromatic phenols and anilines. It has been proposed [2] that such aromatic hydrogen donor substrates, as well as phenylhydrazine [3], bind to peripheral sites near the δ-meso and 8-CH3 heme edge about 8-11Å from the iron, while other substrates, typified by benzhydroxamic acid, bind to a distal heme site. Although crystallographic work is underway [4], no high resolution structure is currently available. NMR studies of the paramagnetic cyano HRP complex including wildtype and HRP protein variants have proven particularly powerful [5,6,7] and have provided information about the nature and probable peripheral site at which aromatic donors interact with the enzyme, as well as key residues near the distal and proximal sides of the heme. On the distal side, Arg 38, His 42 and Phe 41 have been identified in earlier studies [7] and confirmed by additional evidence in subsequent studies [5,6] with the main evidence for Phe 41 based on the disappearance of the single aromatic cross peak in the F41V mutant [5]. These distal residues are conserved and appear to be in the same positions as the corresponding residues in the crystal structure of CCP, forming a common catalytic pocket [8]. More recently, an additional residue on the distal site, Phe 152, was assigned on the basis of its interaction with the 3-CH3 and 2-vinyl group of the heme [5]. On the proximal side, His 170 was identified early as the heme-iron ligand [7]; but two other assignments of residues Leu 237 and Tyr 185 have subsequently been questioned. Specifically, the

G.N. La Mar (ed.), Nuclear Magnetic Resonance of Paramagnetic Macromolecules, 75-93.

assignment of Tyr 185 was questioned by Veitch & Williams [9] based on its location in an unconserved region with CCP and because they proved that the signal arose from a phenylalanine side chain via 2D NMR spectra. In addition, the study by de Ropp et al. [6] concluded that an unidentified isoleucine-X instead of Leu 237 should be close to the 8-methyl group of the heme as part of the peripheral binding site for aromatic substrates. In very recent work reported at this NATO workshop [10], observed cross peaks with the 4Hβ vinyl group of the heme were assigned to the NH group of Gly 169 and observed cross peaks with the Hβ1 and Hβ2 protons of the proximal His 170 were assigned to two protons of an unidentified aliphatic residue.

While NMR studies of the paramagnetic HRP-CN complexes have provided useful information of the position of local residues near the paramagnetic heme unit, these techniques cannot, by themselves provide global 3D structure of the protein. However, when NMR results are combined with the technique of homology modeling used to construct 3D models of the protein, they can contribute to the selection of the most plausible 3D model. Thus, homology modeling provides the missing link between local and global structural information that can be derived from NMR studies and thus enhances its usefulness.

In the work reported here, we have used the known crystal structures of three members of the superfamily, cytochrome c peroxidase (CCP) [11], *Arthromyces ramosus* peroxidase (ARP) [12] and ligninase (LiP) [13] as templates to construct 3D models of HRP. By comparing several candidate models obtained with the distances between specific substituents of the heme and nearby residues deduced from NMR studies, the most plausible model could be selected and the unknown residues in the peripheral binding site identified.

2. Methods

In this section we describe the steps used to: 1) construct the initial 3D models of HRP; 2) refine the initial models; 3) construct and refine their corresponding cyano complexes, and 4) further refine the most plausible model.

2.1 CONSTRUCTION OF INITIAL FULL PROTEIN 3D MODELS OF HRP-C

2.1.1 *The Core Framework*

The structures of three members of the plant peroxidase superfamily [1] were obtained from the Brookhaven Protein Data Bank [14]. These included cytochrome c peroxidase (CCP) [11], lignin peroxidase (LiP) [13] and *Arthromyces ramosus* peroxidase (ARP) [12]. Since the level of sequence identity between CCP and the other two peroxidases was weak (~20% identity), the first step chosen was to identify structurally conserved regions (SCRs) by pairwise structural comparisons of the three known structures. This process was carried out using the HOMOLOGY module of INSIGHT II [15]. Structurally similar regions were identified by calculating the RMS deviation of the Cα distance matrices. The parameter used in this search were: a contingency threshold of 4.2Å and an orientation threshold of 4.5Å. Sequence alignments for the three known structures were then adjusted manually so that SCRs common to all three structures were brought into alignment. Gaps were introduced outside of the SCRs and of the conserved helical regions that occur in all three of the known structures. Ten SCRs were found. Seven of these have RMS deviations for the Cα positions less than 1.5Å in all three structures.

The sequence of horseradish peroxidase, isoenzyme Cla [16] was then brought into alignment manually with those of the three known structures taking into account the consensus alignments of the plant peroxidase superfamily [17] and placing a high weighting on residues with an established structural or functional role in the known structures [1]. These residues included the distal residues Arg 38, His 42, assigned catalytic significance in the acid/base mechanism of Compound I formation; Asn 70 involved in the formation of a conserved hydrogen bond link to the distal His 42; the proximal His 170 ligand and nearby Asp 247, and other residues involved in the formation of a buried salt bridge (Asp 99 and Arg 123). Additional residues implicated by sequence alignment [17] to be required as Ca(II) ligands, for both the distal (Asp 43, Asn 47, Ser 52, Asp 50 and Gly 48) and proximal (Thr 171, Asp 222, Thr 225, I 228 and Asp 230) calcium binding sites were aligned with their counterparts in the known structures. In the resulting alignment, only the positioning of helix A of HRP was ambiguous with respect to the known structures. Thus, several alignments of helix A were evaluated in the initial model building studies until the distance between a disulfide forming Cys pair (Cys 11 and 91) was reasonable. Superposition of the three known structures on the basis of the 29 identical matches in the derived alignments gave overall RMS values as follows: CCP/ARP 5.73Å; CCP/LiP 5.93Å; and ARP/LiP 1.37Å. These overall RMS scores could not be improved significantly by further adjustments to the alignments of the secondary structure elements, indicating a best alignment was achieved (Figure 1).

To build the core framework of the HRP model, the atomic coordinates of the SCRs of ARP were used for the substitution of the HRP residues in the alignment. Side chain conformations were preserved whenever possible. In case of steric contact, the side chain was adjusted to avoid direct overlap of atoms.

2.1.2 *Small Insertions, Regions For Which Template Was Not Available In The Known Structures*

Small insertions (3-4 residues) were made using the procedure embedded in the QUANTA/CHARMm (Molecular Simulations, Inc.) protein design module. A Cα distance matrix was set up using a portion of the Brookhaven protein data bank.This subset included 16 representative high resolution structures plus 12 additional globular heme protein structures. These structures were examined for loop regions containing the correct number of residues and spanning the correct distance. Candidate loops were chosen from the 10 most significant matches. The criteria used for their selection were: (i), no steric repulsion with conserved secondary structural elements; (ii) ability to allow formation of the appropriate disulphide bonds, and (iii) conservation of Gly and Pro residues in insertion sequences containing these residues at a particular loop position. Insertion of these residues was followed by 50 cycles of steepest descent constrained minimization of their Cα carbon chain.

2.1.3 *Prediction Of Secondary Structure For Regions With No Template*

Application of a neutral network program for homology derived prediction of secondary structure [18] gave some insight as to the secondary structure of untemplated regions. This method led to the prediction of the secondary structure elements in HRP known to be conserved among the three known structures. Using this procedure, two additional helices unique to HRP were also predicted. These were confirmed by use of another secondary structure predicting algorithm [19]. The first of these include 5 residues, Ala 129 - Leu 133, corresponding to helix D' in schematic diagrams of the low resolution tertiary structure of HRP E5 [4] that has 70% identity with HRP-C [5]. Helix D' was constructed *de novo* within QUANTA and energy minimized (50 cycles of steepest descent). This helix was docked in the core molecule with its most hydrophobic face inwards so as to span the gap in the core molecule. Connectivities were made by regularizing the last residue at each end of the helix.

The most challenging aspect in the construction of full protein models for HRP-C was the identification of a 34 residue insertion between helices F-G characteristic of class III peroxidases. The following strategy was adopted for this long insertion (
Phe 179 - Asp 212). A secondary structure analysis of this sequence using the same algorithms that predicted helix D' led to prediction of a helix labeled F' within this region. A search of homologous sequence fragments was then carried out for the 34 residue insertions by probing the sequences of the proteins in PDB. The most significant finding was 36.8% identity in a 19 residues overlap (no gaps) to rhinovirus 14 (pdb code 2rm2). Other matches of reduced significance were also found, but only 2rm2 contained a helix at precisely the point predicted by the secondary structure prediction of HRP. Two of the three Pro residues in the F-G insertion were also conserved. Accordingly, the Cα backbone for the homologous region of 2rm2 was copied, HRP side chains were added in their default positions and the 2rm2 Cα backbone regularized (50 cycles of steepest descent minimization).

To construct a full protein 3D model,the F-G insertion element, an 8 residue helix with an extended loop on one side and a reverse turn on the other, was manually docked to the core molecule. Three candidate sites for this element were chosen such that an internal disulfide bond could be made between one partner, Cys 177, in a conserved region and the other, Cys 209, in the long insertion. An additional criterion used was that the correct connection of the two ends of this insertion with the corresponding core residues could be made using the loop searching strategy (2-3 residues at each end). In order to constrain the F-G region further before energy optimization, an additional variant of the loop was constructed for each model giving rise to three more initial 3D models for the full protein. These variants contained an additional false disulfide bond created by mutating Met 181 to Cys and Leu 211 to Cys and using the Cys-Cys distance as a constraint during local loop minimization using AMBER [20]. The rationale for including these variants was based on the observation that Tomato peroxidase contains an additional disulfide bond in this region and it was thought that the corresponding side chains of Met 181 and Leu 111 in HRP may also be spatially close to each other. This false disulfide bond was later removed and the Cys residues were mutated back to Met 180 and Leu 211.

2.2 STRUCTURE REFINEMENT OF THE SIX INITIAL 3D MODELS OF HRP-C

A multistep procedure was carried out in parallel for the refinement of the six initial models (Figure 2) using the AMBER program package. These models include polar hydrogens explicitly and treat nonpolar hydrogens as united atoms with the bound heavy atoms. A constant dielectric of 1.0 and nonbonded cutoff of 9Å were used for energy calculations. These steps included:

Full unconstrained protein energy minimization with 1000 steps of steepest descent.

Addition of the 246 crystallographic water molecules of ARP to the six minimized models of HRP-C. The coordinates of the oxygen atoms of these waters were taken from the structure of ARP by first allowing maximum superimposition between the heme units of ARP and the models. Two waters were removed from the six models since they are in close contact with the protein (≤0.5Å), resulting in 244 structural waters. Hydrogen atoms were added to the waters with the OH bond length of 0.96Å and HOH bond angle of 105°. The orientations of the water molecules were assigned arbitrarily. Fixing the oxygen atom positions, the hydrogen atom positions were refined by energy minimization with 200 steps steepest descent and 400 step of a conjugate gradient method, followed by an MD simulation of 0.5ps at room temperature and a second energy minimization using the same procedure as for the initial one.

Refinement of all side chains with the backbone atoms fixed by energy minimization with 200 steps of steepest descent and 600 steps conjugate gradient method, followed by 1.0ps MD at room temperature and then a second optimization using the same procedure as for the first.

Refinement of regions outside the SCRs, while fixing the rest of the protein, using two cycles of energy minimization and short MD simulations followed by a final energy optimization. All energy minimization were performed using 400 steps steepest descent and 1600 steps of the conjugate gradient method and 0.5ps MD simulations performed at room temperature.

Further refinement of the six full protein 3D models with no constraint after removal of the disulfide bond in models 4-6 between residues 180 and 211 and mutating these cystines to Met 180 and Leu 211 corresponding to the wild type HRP-C. This refinement was done using two cycles of energy minimization and short MD simulations, followed by a final energy optimization. All energy minimizations were performed using 400 steps steepest descent and 1600 steps of the conjugate gradient method and 0.5ps MD simulations performed at room temperature.

2.3 CONSTRUCTION AND REFINEMENT OF CYANO COMPLEXES OF 6 MODEL HRPS

A cyano anion was added as the sixth ligand of the heme iron to each of the six 3D models of HRP in order to have a direct comparison with the NMR results, which were performed for the paramagnetic low spin cyano complex of HRP. Using a united atom model, the six CN-HRP complexes were first minimized with 200 steps of steepest descent and 1800 steps of conjugate gradient methods. A 20ps room temperature MD simulation of each minimized model was then performed to dynamically characterize the residues near the heme units and assess their possible roles in interacting with the heme or bound substrates as suggested by 2-D NMR studies [5,6]. The first 10ps were required for equilibration. The remaining 10ps was used for analysis of distances between specific groups of the heme unit and nearby residues to identify and confirm assignments of residues as the origin of cross peaks in the 2D NMR studies of HRP-CN complexes [5,6,10].

The comparisons of these distances with the 2D NMR studies [5,6,10] of HRP-CN, and particularly of the distances of isoleucine residues to the 8- CH3 group and aliphatic residues to the Cβ protons of the distal His 170 imidazole allowed the identification of the unassigned Ile-X and aliphatic residues and the selection of the best model from among the six candidates.

2.4 FURTHER REFINEMENT AND ANALYSIS OF THE SELECTED MODEL

The most plausible model selected was further refined. All hydrogen atoms were explicitly included and an energy minimization was performed with 200 steps of steepest descent and 1800 steps of conjugate gradient methods. The side chains were then refined by 1.0ps MD followed by an energy minimization of 1000 steps while fixing the backbone atoms.

The full protein model was further refined without constraints by two cycles of energy optimization followed by short MD simulations and a final energy optimization. All energy minimizations were performed with 200 steps of steepest descent and 800 steps of conjugate gradient methods. The short MD simulations were performed for 1.0ps at room temperature

The refined model was used for a 50.0ps MD at room temperature. The first 10.0ps were required for equilibration and the last 40ps were used for analysis of the atomic distances of interest.

3. Results and Discussion

The multiple sequence alignment between CCP, ARP, LiP, and HRP is shown in Figure 1. The α helices of the known structures, labeled A-J, are preserved in the HRP model. Two new helices, D' and F', were predicted for HRP from secondary structure prediction algorithms. Helix D' is located at a short insertion between helices D and E. Helix F' is in the middle of the 33 residue insertion between helices F and G. The SCRs determined from the structural alignment of CCP, ARP, and LiP are also highlighted. These regions include all known secondary structures and some loop regions.

Figure 2 shows the 3D structures of the six refined models of HRP-CN with backbone Cα chains and the two Ca(II) binding sites explicitly displayed. We see in this figure that the main difference in these models is in the orientation and position of the 33 amino acid insertion between the F and G helices.

Table 1 gives the residues identified as ligands of the Ca(II) binding sites and distances between the ligand atom and the Cα in the HRP model 1 and in the X-Ray structures of ARP and LiP. CCP has no such sites. The binding sites are the same in all six HRP models and are highly conserved in the three peroxidases. As shown in Figures 3a, 3b and Table 1 in each site, the Ca(II) has at least six oxygen ligands with Asp 43 providing two carbonyl oxygen as ligands in Site I and Asp 222 in Site II.

TABLE 1: Two Ca (II) binding sites identified
Residues Identified as Ligands of the Ca(II)

SITE I	1	2	3	4	5	6
HRP Model 1	Asp43	Asn47	Gly48	Asp250	Ser52	H_20
Distances	2.3,2.4,2.8	2.5,2.6	2.4	2.3	2.4	2.7
LiP	Asp48	-	Gly66	Asp68	Ser70	-
Distances	2.6,2.4	-	2.7	2.6	2.2	-
ARP	Asp57	-	Gly75	Asp77	Ser79	-
Distances	2.6,2.5	-	2.6	2.6	2.4	-

SITE II	1	2	3	4	5	6
HRP Model 1	Thr171	Asp222	Thr225	Pro226	Asp230	H_20
Distances	2.4	2.3,2.4	2.4	2.7	2.3	2.5
LiP	Ser177	Asp194	Thr196	Ile199	Asp201	-
Distances	2.3,2.8	3.0,3.0	2.8,2.4	2.4	2.4,3.9	-
ARP	Ser185	Asp202	Thr204	Val207	Asp209	-
Distances	2.5,2.7	2.5,2.6	2.5,2.7	2.5	2.6	-

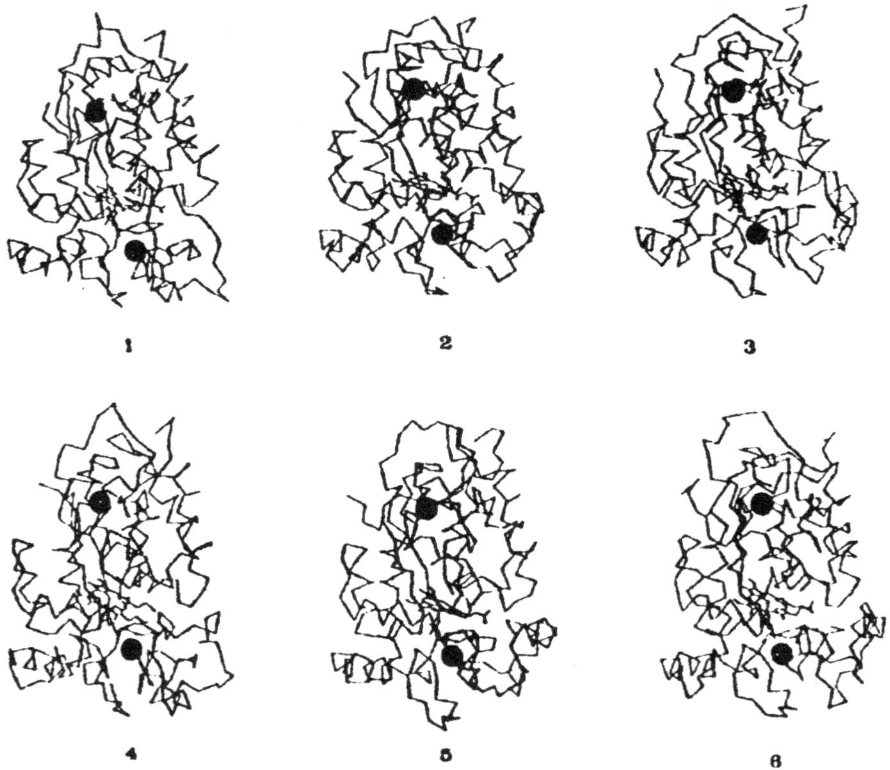

Figure 2. Cα trace of the six initial models of HRP. The heavy atoms of the heme are displayed. The van der Waals surface of the Ca(II) ions are shown in black spheres. The F-G insertion that differs in the six models is in the lower right corner.

```
                                        A--------------------------A              B----------------B
CCP       *2  *  TPLVHVASV*11 *EKGRSYEDFQ*21 *KVYNAIALKL*31 *REDDEYDNYI*41 *GYGPVLVRLA*51 *WHISGTWD
ARP       *11 *     TCPGGQ*17 *STSNSQCCVW*27 *FDVLDDLQTN*37 *FYQGSK  CE*45 *SPVRKILRIV*55 *FHDAIGFSPA
LiP       *2  *     TCANGK*8  *TVGDASCCAW*18 *FDVLDDIQAN*28 *MFHGGQ  CG*36 *AEAHESIRLV*46 *FHDSIAISPA
HRP MODEL *1  *      qltpt*6  *fydnSCPNVS*16 *NIVRDTIVNE*26 *LRSDP     *31 *RIAASILRLH*41 *FHDCFVN
                                     SCR1                                     SCR2

                                     B'------B'                 C---------------C                D--
CCP       *59 *     KHDNTG*65 *GSYGGTYRFK*75 *KEFND      *80 *PSNAGLQNGF*90 *KFLEPIHKEF*100*P     WISS
ARP       *65 *LTAAGQFGGG*75 *GADGSIIAHS*85 *NIELA     F*91 *PANGGLTDTI*101*EALRAVGINH*111*G     VSF
LiP       *56 *MEAKGKFGGG*66 *GADGSIMIFD*76 *TIETA     F*82 *HPNIGLDEVV*92 *AMQKPFVQKH*102*G     VTP
HRP MOdEL *   *        *48 *GCDASILLDN*58 *TTSFRTEKDA*68 *FGNANSARGF*78 *PVIDRMKAAV*88 *ESACPRTVSC
                                SCR3                     SCR4

               ------------------D                          D'-D'             E------------E
CCP       *105*GDLFSLGGVT*115*AVQEM QGPK*124*IPWRCGRVDT*134*  PEDTTPDNG*143*RLPDADKDAG*153*YVRTFFQRLN
ARP       *115*GDLIQFATAV*125*GMSNCPGSPR*135*LEFLTGRSNS*145*   SQPSPPS*152*LIPGPGNTVT*162*AILDRMGDAG
LiP       *106*GDFIAFAGAV*116*ALSNCPGAPQ*126*MNFFTGRKPA*136*   TQPAPDG*143*LVPEPFHTVD*153*QIIARVNDAG
HRP MODEL *98 *ADLLTIAAQQ*108*SVTLA GGPS*117*WRVPLGRRDS*127*LQAFLDLANA*137*NLPAPFFTLP*147*QLKDSFRNVG
                          SCR5                                          SCR6

             F---------------F                                            F'----F'
CCP       *163* MNDREVVAL*172*MGAHALGKTH*182*LK            *   *            *   *            *184*      NSG
ARP       *172* FSPDEVVDL*181*LAAHSLASQE*191*GL            *   *            *   *            *193*      NSAI
LiP       *163*EFDELELVWM*173*LSAHSVAAVN*183*DV            *   *            *   *            *185*      DPTV
HRP MODEL *157*LNRSSDLVAL*167*SGGHTFGKNQ*177*CRflmdrlyn*187*fsntglpdpt*197*lnttylqtlr*207*glcplnGNLS
                    SCR7
```

```
                          G------------G                                                    H------------H  I
CCP        *187*YEGPWG AAN*196*NVFTNEFYLN*206*LLNEDWKLEK*216*NDANNEQWDS*226* KSGYMMLPT*235*DYSLIQ   D
ARP        *197*FRSPLD STP*206*QVFDTQFYIE*216*TLLKGTTQPG*226*PSLGFAEELS*236*PFPGEFRMRS*246*DALLAR  D
LiP        *189*QGLPFD STP*198*GIFDSQFFVE*208*TQFRGTLFPG*218*SGGNQGEVES*228*GMAGEIRIQT*238*DHTLAR  D
HRP MODEL  *217*ALVDFD1RTP*227*TIFDNKYYVN*237*LEEQK      *    *         *242*     GLIQS*247*DQELFSSPNA
                SCR8                                                                      SCR9

           ----------------I J---------------------------J
CCP        *242*PKYLSIVKEY*252*ANDQDKFFKD*262*FSKAFEKLLE*272*DGITFPKDAP*282*SPFIFKTLEE*292*QGL
ARP        *253*SRTACRWQSM*263*TSSNEVMGQR*273*YRAAMAKMSV*283*LGFDRNALTD*293*CSDVIPSAVS*303*NN    AAPVI
Lip        *245*SRTACEWQSF*255*VGNQSKLVDD*265*FQFIFLALTQ*275*LGQDPNAMTD*285*CSDVIPLSKP*295*IPGNGPFSFF
HRP MODEL  *257*TDTIPLVRSF*267*ANSTQTFFNA*277*FVEAMDRMGN*287*ITPLTGTQGQ*297*IRLNCRVVNS*307*NS
                 SCR10

CCP        *    *           *    *           *    *          *    *
ARP        *310*PGGLTVDDIE*320*VSCPSEPFPE*330*IATASGPLPS*340*LAPAP
LiP        *305*PPGKSHSDIE*315*QACAETPFPS*325*LVTLPGPATS*335*VARIPPHKA
HRP MODEL  *    *           *    *           *    *          *    *
```

Figure 1. Multiple sequence alignment of CCP, ARP, LiP, and HRP. The α helices are labeled with letters A to J above the alignment. Helices D' and F' are predicted for the HRP model and the rest are from the known structures of CCP, ARP and LiP. The SCRs derived from these known structures are labeled with --- below the alignment.

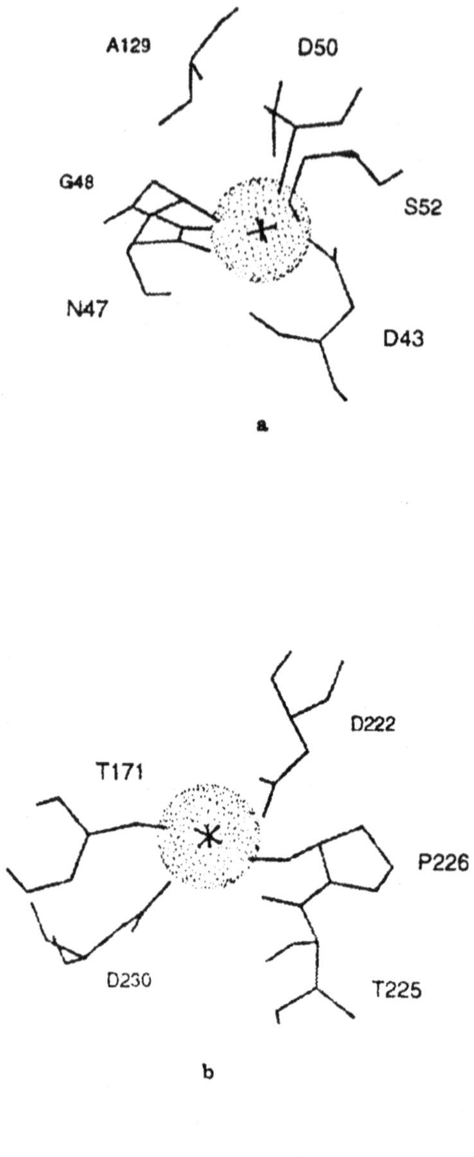

Figure 3. Ca(II) binding site I (a) and II (b) in the six models of HRP.

Table 2 specifies the four pairs of disulfide bond partners and the extent to which they are in conserved regions in HRP. These partners are the same in each of the 6 models and each of the models was initially built to satisfy these disulfide bonds.

TABLE 2: Four (4) disulfide bonds identified

Disulfide Bond Partners	Degree of Conservation
Cys 44----Cys 49	Highly conserved in LiP/ARP
Cys 11----Cys 91	Required small adjustment for HRP
Cys 97----Cys 301	Required small adjustment for HRP
Cys 177----Cys 209	One partner (Cys 177) in highly conserved region Other partner (Cys 209) in long insertion

Nine candidate glycosylation sites, identified as surface accessible Asn-X-Ser/Thr Motifs for each of the 6 models, were identified. These are Asn 13, Asn 57, Asn 158, Asn 186, Asn 198, Asn 214, Asn 255, Asn 268, and Asn 286. These are the same residues for all models, but some are in the variable F-G loop, resulting in different conformations. All but Asn 286 are known to be glycosylated [21].

Table 3 shows the heme unit and the three residues; Arg 38, Phe 41 and Phe 152; in the distal site that have been assigned as the origin of cross peaks to the heme in the 2D-NMR studies [5,6]. Shown in this table are the mean distances calculated from the 10 ps MD simulation in the six model structures of HRP-CN. We see from this table that all models are consistent with the assignment [6] of heme-Arg 38 cross peaks connecting the heme 5-CH3 with the CβH of Arg 38 and the CβH of the propionate group with the CβH and CδH of Arg 38, although the three distances vary between the models. Similarly, all models are consistent with the NMR assignment [5] of Phe 41 connected to 3-CH3 of the heme and Phe 152 connected to the 3-CH3 as well as the C2-B vinyl protons. These assignments involve highly conserved distal residues validated by the six proposed models. However, they cannot be used to select the most plausible candidate among them.

Turning to the proximal side of the heme unit, as shown in Table 4, there are two isoleucines, 180 and 244, that are candidates for the Ile-X identified as the origin of the cross peaks with the 8-CH3 group of the heme. Table 4 lists the distances between the carbon atom of the 8-CH3 group and the C atoms of the isoleucine side chains in the six model structures. Isoleucine 180 is in the long insertion between the F and G helices that is the most variable in the six structures, while Ile 244 is in a conserved region in helix H. Thus, it is not surprising that the distances between the Ile 180 side chain atoms and the 8-CH3 group vary among the models. What is striking, however, is that only model 1 gives a short distance (3.65Å) from the Cγ1 atom of Ile 180 to the 8-CH3 group, consistent with the NMR assignment. All other models have the distance >12Å. Ile 244, on the other hand, is more than 9Å away from 8-CH3 in all six models. Thus, Ile 180 is predicted to be the isoleucine group interacting with the 8-CH3 of heme and this assignment allows model 1 to be selected as the most plausible and the only one that is consistent to the 2-D NMR studies of HRP.

Table 5 shows five candidate aromatic residues in the vicinity of the peripheral binding site and gives the distances from the center of the aromatic side chains of these five residues to the 8-CH3 carbon atom of the heme unit in the six models. These distances show considerable variability among the 6 models since these aromatic residues are, in general, not in highly conserved regions. Three phenylalanines, Phe 221, Phe 142, and Phe 143, are within 10Å from the 8-CH3 group in model 1, serving as possible candidates for the Phe A and Phe B residues observed in NMR studies [5]. Other aromatic residues, such as Tyr 185 and Phe 68, also have short distances to 8-CH3 in models 2, 3, 5, and 6. However, as noted only for model 1 is an isoleucine residue close to heme found. Thus, model 1 is the only model that provides a consistent explanation of the 2-D NMR data.

TABLE 3: Mean Distance from Distal Residues to the Specific Heme Groups

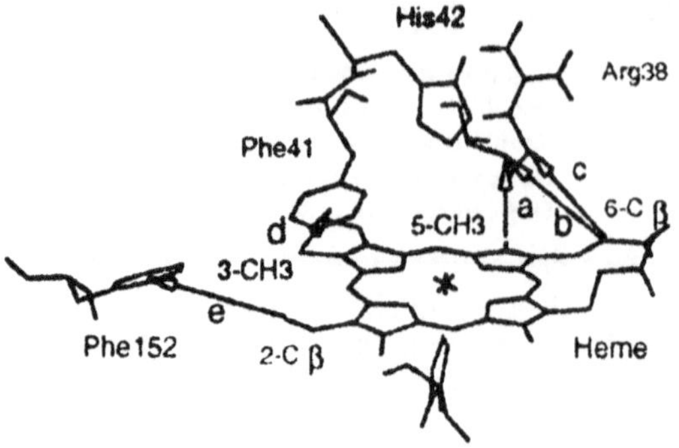

		MODELS					
Heme	Residue	1	2	3	4	5	6
5-CH3	Arg Cβ	4.0	4.6	5.2	5.4	4.0	5.1
6-Cβ	Arg Cβ	6.1	4.9	7.6	5.7	5.9	3.7
6-Cβ	Arg Cδ	8.0	5.4	5.2	4.0	5.1	5.2
3-CH3	Phe 41	3.8	4.6	3.6	4.9	4.8	4.0
C-2-Cβ	Phe 152	4.4	4.6	4.3	6.7	4.9	4.4
3-CH3	Phe 152	6.1	8.5	7.2	7.7	7.0	5.1

TABLE 4: Mean Distance from Ile180 and Ile244 to the 8-CH3 Heme Group

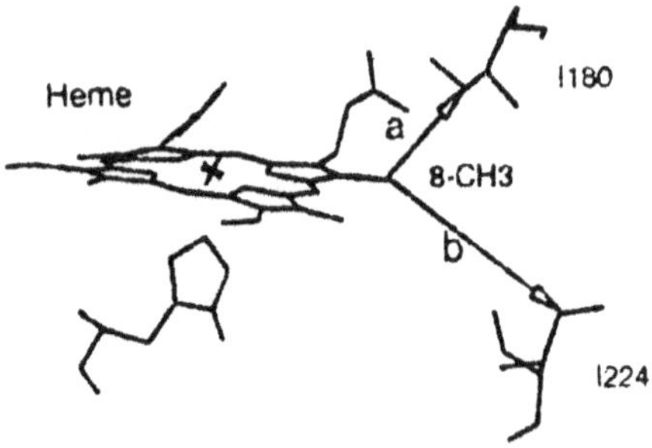

		MODELS					
Heme	Residue	1	2	3	4	5	6
a) 8-CH3	Ile 180 Cγ	3.6	13.7	13.7	17.2	12.6	15.0
b) 8-CH3	Ile 244 Cδ	9.3	9.4	9.4	12.2	12..4	11.0

TABLE 5: Distance (Å) between 8-methyl heme and Five Candidate Peripheral Site Aromatic Residues (Phe A and Phe B)

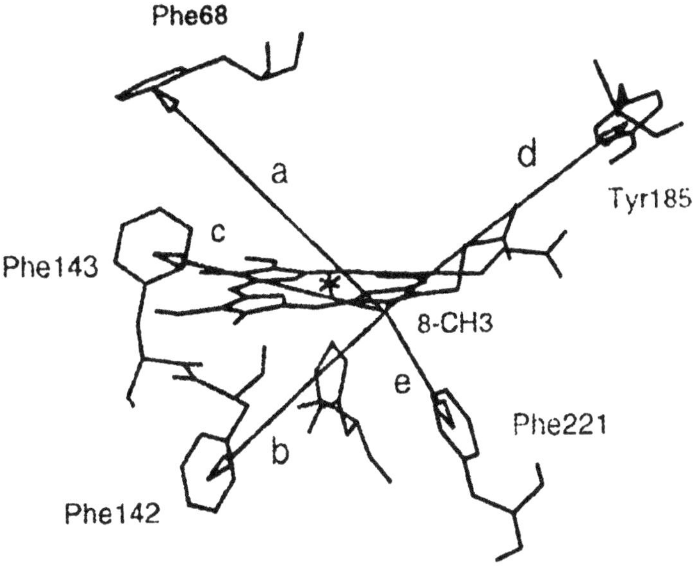

	MODELS					
RESIDUE	1	2	3	4	5	6
Phe 221	7.7	6.8	6.5	5.0	9.7	6.9
Phe 143	8.8	11.3	13.5	9.6	9.5	10.6
Phe 142	9.1	11.5	12.1	14.4	11.6	11.5
Phe 68	14.5	8.3	7.7	10.7	7.3	10.7
Tyr 185	12.1	5.1	4.8	15.9	4.5	7.0

Additional support for model 1 comes from the identification of an unidentified aliphatic residue found in a very recent NMR study [10] to couple with the 2 protons on the β-C of His 170. In these same studies, Gly 169 was assigned as the origin of a cross peak to the Cβ of the 4-vinyl group of the heme. As shown in Table 6, the N atom of Gly 169 is very close to the Cβ of the 4 vinyl group and Leu 250 is close to His 170. Although the distance from Gly 169 to the heme does not distinguish among the six models, the four distances from the Cβ atom of His 170 to the side chain C atoms of Leu 250 are the shortest in model 1, ranging between 5 and 7Å. Thus, model 1 provides an explanation for the observed coupling to an unknown aliphatic residue of His 170 and identifies it as Leu 250. Taken together then, model 1 provides the most consistent explanation of all the 2D-NMR data and allows the assignment of Ile 180 as Ile-X and of Leu 250 as the aliphatic residue interacting with His 170.

TABLE 6. Mean distance between Gly 169 and the heme and between His 170 and Leu 250 in the six initial models of HRP

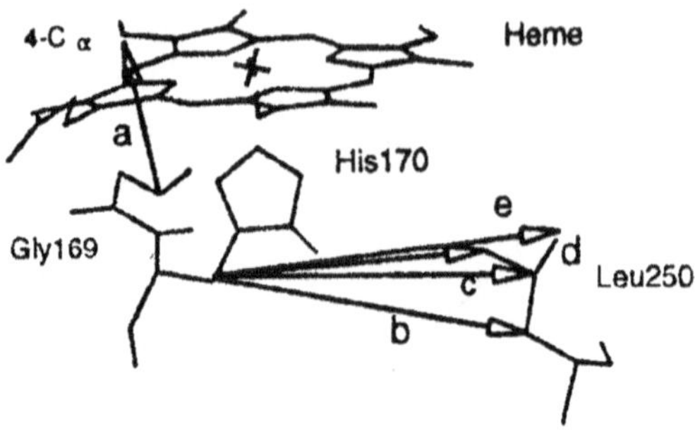

Heme	Residues		1	2	3	4	5	6
4-Cβ (vinyl)	N Gly169	(a)	4.9	5.4	4.8	5.4	4.9	4.8
2-Cβ (His 170)	Cβ Leu 250	(b)	6.4	8.9	10.6	9.2	8.5	9.0
	Cγ Leu 250	(c)	6.4	8.7	11.3	8.9	8.3	8.6
	Cδ1 Leu 250	(d)	5.2	7.4	11.5	7.4	6.9	7.6
	Cδ2 Leu 250	(e)	6.9	9.5	11.1	9.5	9.2	8.9

Having selected model 1 as the most plausible model, a more detailed analysis of this model was carried out by performing a 50ps MD simulation with all H atoms explicitly included. Figure 4 shows the calculated B factors for the backbone Cα atoms from this simulation. In general, the N-terminal domain (residues 1-179) are relatively stable during the simulation. The B factors are relatively large for the C-terminal domain between residues 190 and 308. The largest fluctuations occur in the F-G insertion, indicating more flexibility of this region coupled to the rest of the protein. This large flexibility is consistent to its poorly defined electron density in a low resolution X-ray structure [4].

Figure 5 shows a comparison of the preferred Model of HRP with the three known structures of CCP, ARP and LiP. This figure illustrates the overall similarity of the model HRP to the known structure in the distal domain. Large deviations, however, were found in the proximal domain. These deviations include the F-G insertion which plays a key role in determining the residues that define the peripheral binding site. Figure 6 shows residues within 10Å of the 8-CH3 of the heme that may form the peripheral binding site in HRP-C. This site is located in a similar position as the substrate binding site we have identified near the heme unit in LiP [22]. The residues involved in both are shown in Table 7.

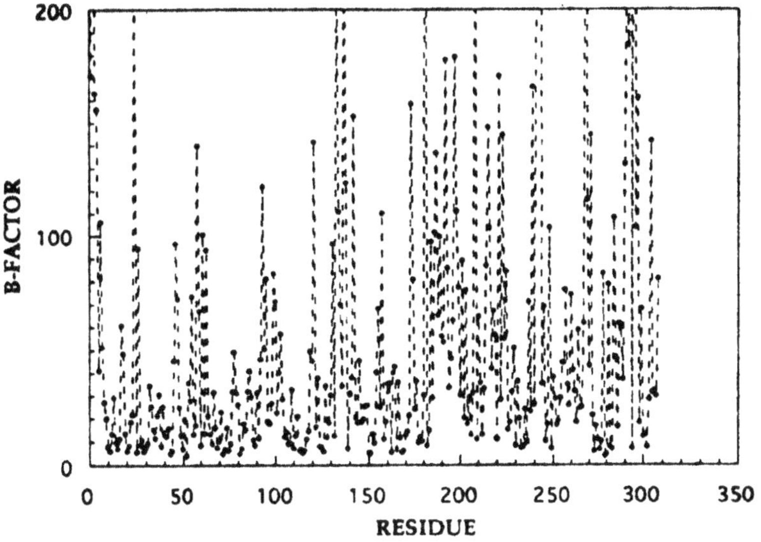

Figure 4. B-factors calculated for model 1 of HRP in a 40ps MD simulation.

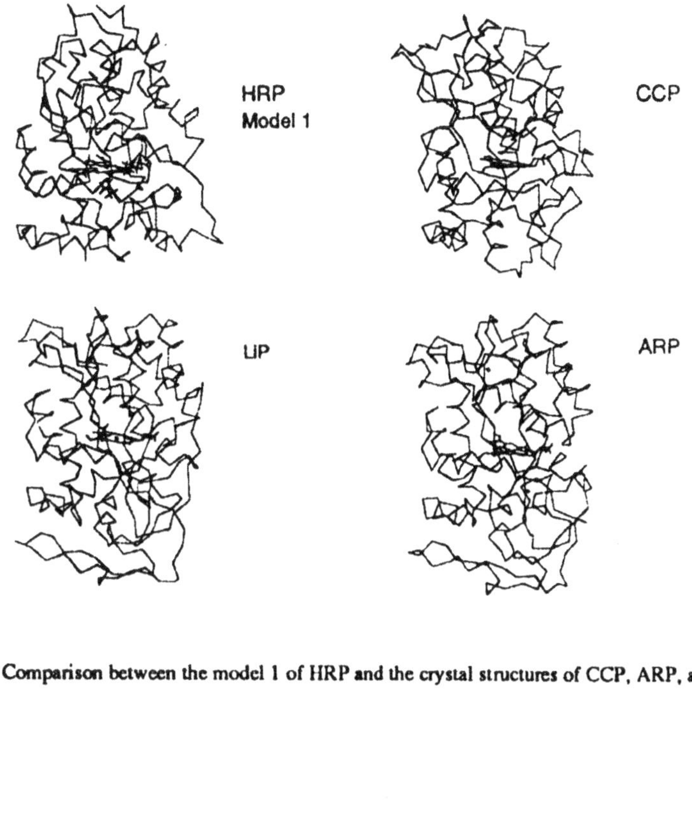

Figure 5. Comparison between the model 1 of HRP and the crystal structures of CCP, ARP, and LiP.

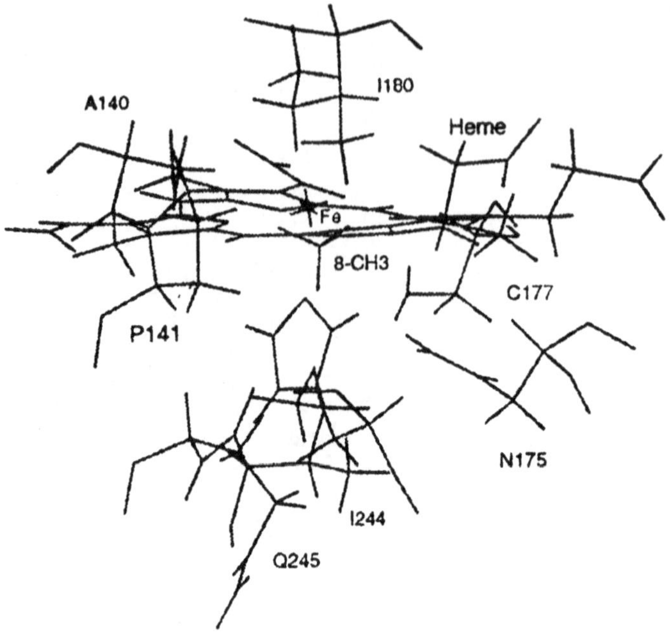

Figure 6. Peripheral substrate binding site of model I of HRP.

TABLE 7. Peripheral binding site Model 1

HRP-C	LiP
Ala 140	His 82
Pro 141	Ile 85
Heme	Pro 147
Asn 175	Phe 148
Cys 177	Heme
Ile 180	Asp 183
Ile 244	Val 184
Glu 245	Asp 185
	Arn 221
	Glu 222
	Gly 223

In order to determine the extent to which the dynamic behavior of the refined model including H atoms explicitly, is consistent with known NMR data, we have calculated H-H distances for all the residues identified as responsible for the observed cross peaks.

The calculated mean distances and standard deviation between atoms of the distal residues, Arg 38, Phe 41 and Phe 152, and proximal residues, Gly 169, Leu 250, and peripheral residues, Ile 180, Ile 244, Phe 221, Phe 143, Phe 142 to the H atoms of specific heme group from the 50 ps MD simulation are listed in Table 8. As can be seen, during the MD simulation, the hydrogen atoms of the Arg 38 side chain remain close to the heme hydrogens of the 5-CH3 and 6-Cβ groups with small standard deviations. The Phe 41 and Phe 152 H atoms are close to the 3CH3 and 2 vinyl protons. The H atom of isoleucine 244 remain too far from those of the 8-CH3 group to be viable candidates for Ile-X. By contrast, several hydrogen atoms bound to the $C\gamma_2$, $C\gamma_1$, and $C_{\delta 1}$ atoms of Ile180 are within 4Å from the hydrogen atoms of the 8-CH3 group, confirming direct interaction between these atoms, and strengthening the assignment of Ile 180 as Ile-X. Also given in Table 8 are the distances between the centers of the three Phe residues that are candidates for PheA and PheB and the H atoms of the 8-CH3 group. Although the average distances to the H atoms have increased in the longer MD simulations of the all atom version of model I, Phe 221 especially, can be considered to be a viable candidate for these unassigned aromatic residues.

TABLE 8. MD averaged proton-proton distances (Å) in refined Model 1 between heme and candidate residues in NMR spectra

Heme Proton	Residue Proton	<r>			sd
DISTAL					
H1 (5-CH3)	H1 (Cβ) @R38	4.1			0.8
H2 (6-Cβ)	H1 (Cβ) @R38	3.6			0.4
H1 (6-Cβ)	H2 (Cδ) @R38	4.7			0.3
H1 (3-CH3)	$H_{\varepsilon 2}$ @ Phe 152	6.6			0.5
H_t, H_c (2-Cβ)	$H_{\varepsilon 2}$ @ Phe 152	3.5	4.8		0.7
H1 (3-CH3)	$H_{\delta 2}$ @ Phe 41	3.1			0.7
H1 (2-Cβ)	$H_{\varepsilon 1}$ @ Phe 41	4.8			0.9
PROXIMAL					
H1 (4-Cβ)	HN @ Gly 169	3.1			0.3
H1 (Cβ) @ His 170	H1 (Cδ1) @ Leu 250	6.9			0.8
PERIPHERAL					
H1 (8-CH3)	H ($C\gamma_1$) @I180	2.7	4.3		0.5
H1 (8-CH3)	H ($C\delta_1$) @I180	4.0	3.4	3.8	0.8
H1 (8-CH3)	H ($C\gamma_2$) @I180	4.0	5.5	5.0	0.6
H1 (8-CH3)	H ($C\delta_1$) @I244	8.9			0.7
H1 (7-Cα)	H ($C\delta_1$) @I244	9.8			0.6
H1 (8-CH3)	Phe 221	8.2			0.3
H1 (8-CH3)	Phe 143	10.8			0.4
H1 (8-CH3)	Phe 142	10.9			0.5

4. Conclusions

3D (3D) models of HRP-C have been constructed using the known peroxidase structures CCP, LiP and ARP as templates. Six models were obtained, differing mainly in the long insertion between the F and G helices. Comparisons of calculated distances in the six model structures of the HRP-CN complexes with 2D-NMR cross peak assignments between the heme and nearby residues in the distal, proximal and peripheral regions of the heme allowed the clear selection of one model as the most likely, and the assignment of Ile 180 as rather than Ile 244 as Ile-X, and Leu 250 as the unidentified aliphatic residue interacting with His 170.

When all the H atoms were explicitly considered in the preferred model and it was further refined, the MD averaged proton-proton distances between the heme and candidate residues in the distal binding site confirmed the assignments of Arg 38 and Phe 41 and Phe 152 cross peaks in the 2D-NMR and of Gly 169, Ile 180 and Leu 250 in the proximal and peripheral regions. Less clear is the tentative assignment identification of Phe 221, 142, and 143 as possible candidates for Phe A or Phe B in the more variable peripheral regions of the heme. Site specific mutations of Phe 142 and Phe 143 reported at this NATO workshop [23] indicate that Phe 143 is neither Phe A nor Phe B while Phe 142 is not Phe A. Thus, it is still possible that Phe 221 could be Phe A and Phe 142, Phe B. Site specific mutations of Phe 221 as well as the two candidates for Ile-X, Ile 180, or 244 and for the unknown aliphatic residue assigned as Leu 250 should help to further elucidate these assignments.

While the predicted 3D model of HRP cannot be expected to have the accuracy of a high resolution X-Ray structure, the model helps explain and assign unknown residues and is consistent with all the known NMR data and appears to be reasonably stable under MD simulations. Thus, it enhances the usefulness of these NMR studies and allows not only local but global 3D structural information to be deduced from them. In addition, the model provides guide to mutations that can be made to further validate it. Finally, it provides binding sites that can be used in studies of explicit substrate-enzyme complexes, with a variety of substrates in each candidate binding site and with mutant forms of the enzyme. This new capability that the model provides should be very useful in the continued elucidation of the relationship between structure and function in this ubiquitous family of metabolizing heme proteins.

5. Acknowledgments

Support from this work from a NATO Collaborative Research Grant and from National Science Foundation Grant # DMB9096181 is gratefully acknowledged.

6. References

1. Welinder, K. G. (1992) Superfamily of plant, fungal and bacterial peroxidases. *Curr. Opin. Struc. Biol.* **2**, 388-393.
2. Sakurada, J., Takahashi, S. and Hosoya, T. J. (1986) Nuclear magnetic resonance studies on the spatial relationship of aromatic donor molecules to the heme iron of horseradish peroxidase. *J. Biol. Chem.* **261**, 9657-9662.
3. Ortiz de Montellano, P. R. (1987) Control of the Catalytic activity of prosthetic heme by the structure of hemoproteins. *Acc. Chem. Res.* **20**, 289-294.
4 Morita, Y., Funatsu, J. and Mikami, B. (1993). In Welinder, K. G., Rasmussen, S. K., Penel, C., Greppin, H. A. (eds.) *Plant peroxidases: Biochemistry and Physiology*, University of Geneva. p 1-4.
5. Veitch, N. C., Williams, R. J. P., Bray, R. C., Burke, J. F., Sanders, S. A., Thorneley, R. N. F. and Smith, A. T. (1992) Structural studies by proton NMR spectroscopy of plant horseradish peroxidase C, the wild type recombinant from *Escherichia coli* and two protein variants, Phe 41->Val and Arg 38-> Lys. *Europ. J. Biochem.* **207**, 521-531.

6. de Ropp, J. S., Yu, L. P. and La Mar, G. N. (1991) 2D NMR of paramagnetic metalloenzymes cyanide-inhibited horseradish peroxidase. *J. Biomol.* NMR **1**, 175-190.
7. Thanabal, V., de Ropp, J. S. and La Mar, G. N. (1987) Identification of the catalytically important amino acid residue resonances in ferric low-spin horseradish peroxidase with NOE measurements. *J. Am. Chem. Soc.*, **109**, 7516-7525.
8. Finzel, B. C., Poulos, T. L. and Kraut, J. J. (1984) Crystal structures of yeast cytochrome c peroxidase refined at 1.7Å resolution. *J. Biol. Chem.* **259**, 13027-13036.
9. Veitch, N. C. and Williams, R. J. P. (1990) Two dimensional 1H-NMR studies of horseradish peroxidase C and its interaction with indole-3-propionic acid. *Eur. J. Biochem.*, **159**, 351-362.
10. La Mar, G. N., Chen, L., de Ropp, J. S. (1994). Presented at NATO Workshop - Nuclear Magnetic Resonance of Paramagnetic Molecules (Sintra, Portugal).
11. Poulos, T. L., Freer, S. T., Alden, R. A., Edwards, S. L., Skogland, U., Takio,V., Erikson, B., Xuong, Ng H., Yonetani, T., Kraut, J. (1980). The crystal structure of yeast cytochrome c peroxidase. *J. Biol. Chem.* **255**, 575-580.
12. Kurishima, N., Fukiyama, J., Matsubara, H., Hatanaka, H., Shibano, Y., Amachi, T. (1994). Crystal structure of the fungal peroxidase from *Arthromyces Ramosus* of 1.9Å resolution. Structural comparison with the lignin and cytochrome c peroxidase. *J. Mol. Biol.* **235**, 331-344.
13. Poulos, T. L., Edwards, S. L., Wariishi, H., Gold, M. H. (1993). Crystallographic refinement of lignin peroxidase at 2Å. *J. Biol. Chem.* **268**, 4429-4440.
14. Bernstein, F. C., Koetze, T. F., Williams, G. J. B., Meyer, E. F., Brice, M. D., Rodgers, J. R., Kennard, O., Schimanovichi, T. and Tasumi, M. (1977). The Protein Data Bank.: A Computer based archival file for macromolecular structures. *J. Mol. Biol.* **112**, 535-542.
15. Insight II Version 2.3.0 Biosym Technologies, Inc. (1993).
16. Welinder, K. G. (1976) FEBS Lett. **72**, 19-23.
17. Welinder, K. G., Gajhede, M. (1993). In Welinder, K. G., Rasmussen, S. K., Penel, C., Greppin, H. A. (eds.) *Plant peroxidases: Biochemistry and Physiology*, University of Geneva. pp. 35-42.
18. Rost, B. and Sander, C. (1993) Prediction of protein structure at better than 70% accuracy. *J. Mol. Biol.* **232**, 584-599.
19. Garnier, J., Osguthorpe, D. J., Robson, B. (1978). Analysis of the accuracy and implications of simple methods for predicting the secondary structures of globular proteins. *J. Mol. Biol.* **120**, 97-120.
20, Pearlman, D. A., Case, D. A., Caldwell, J. C., Seibel, G. L., Singh, U. C., Weiner, P., Kollman, P. A. (1991). AMBER 4.0, University of California, San Francisco.
21, Welinder, K. G. (1979) Amino acid sequence studies of horseradish peroxidase. *Eur. J. Biochem.* **96**, 483-502.
22. Du, P., Loew, G. H., (1993) Molecular Dynamic Simulations of Lignin Peroxidase Complexes for the Prediction of Substrate Binding Sites. Plant Peroxidases: Biochem. and Physiol. (K. G. Welinder, et al., eds) Univ. of Geneva, p 27-30.
23, Veitch, N. C. (1994). Presented at NATO Workshop on Nuclear Magnetic Resonance of Parmagnetic Molecules (Sintra, Portugal).

NMR STUDIES OF PARAMAGNETIC SYSTEMS TO CHARACTERISE SMALL MOLECULE:PROTEIN AND PROTEIN:PROTEIN INTERACTIONS

G. R. MOORE [1], M. C. COX [1], D. CROWE [1], M. J. OSBORNE [1],
A. G. MAUK [2] AND M. T. WILSON [3]
[1]*Centre for Metalloprotein Spectroscopy and Biology,*
School of Chemical Sciences, University of East Anglia,
Norwich NR4 7TJ, U.K.
[2]*Department of Biochemistry and Molecular Biology,*
University of British Columbia, Vancouver,
British Columbia, V6T 1Z3, Canada.
[3]*Department of Chemistry and Biological Chemistry,*
University of Essex, Colchester, Essex CO4 3SQ, U.K.

1. Introduction

Intermolecular interactions are a feature of all biochemical processes, and whilst some involve the interaction of small molecules with biopolymers or membranes, others involve the interaction of two or more biopolymers. In all cases, a thorough description of the molecular details of the interaction requires kinetic, thermodynamic and structural investigations. In general, however, there are a limited number of possible kinetic and thermodynamic schemes. Consider the association of molecule A with molecule B:

$$A + B \underset{k_{-1}}{\overset{k_1}{\rightleftharpoons}} AB \qquad (1)$$

There could be a high binding affinity characterised by a low dissociation constant, K_d, and a low rate of dissociation, k_{-1}; or a weaker binding affinity with k_{-1} approaching relatively high rates. In some cases the rate of association, k_1, is greater than that expected from simple collision theory because of an electrostatic attraction between the molecules. Some intermolecular reactions take place in a transient complex which is characterised by an extremely high K_d; for example, electron self-exchange reactions of redox proteins [1 and references therein]. In all these situations investigations employing NMR spectroscopy have contributed to defining specific systems. In the present article we consider how to structurally characterise intermolecular associations which are relatively weak, and therefore in fast exchange on the NMR time-scale. Such interactions are not usually easy to characterise by other structural methods, such as X-ray crystallography, making the NMR approach particularly important.

G.N. La Mar (ed.), Nuclear Magnetic Resonance of Paramagnetic Macromolecules, 95-122.

There are numerous studies of small molecule:protein interactions characterised by NMR methods such as saturation-transfer and transfer-NOE [2,3 and references therein]. Typically these involve diamagnetic molecules interacting with proteins. The study of paramagnetic small molecules interacting with proteins has a long history [4,5 and references therein], partly because such an approach offers a way of defining protein structures, for example lysozyme [6], and partly because reactions between inorganic redox reagents and electron-transfer proteins can be investigated by such procedures, for example cytochromes [7]. However, the paramagnetic small molecule approach has other applications, as we illustrate in the present article.

Here we consider two systems: bovine superoxide dismutase [SOD] and eukaryotic cytochrome *c*. SOD catalyses the disproportionation of the superoxide ion, thus;

$$2O_2^- + 2H^+ \rightarrow H_2O_2 + O_2 \qquad [2]$$

It consists of two identical subunits, each containing one copper and one zinc ion and 151 amino acids [8]. Various mechanistic proposals have been made and two of their common features are that there is an electrostatic facilitation of the dismutase reaction that enhances the number of productive collisions between SOD and O_2^-, and that O_2^- binds to the Cu(II) ion [8-10]. The purpose of the present study is to use the binding of paramagnetic anions to SOD to investigate anion binding sites close to the metal centres.

Eukaryotic cytochrome *c* transfers electrons between a variety of proteins located in the intermembrane space of mitochondria, either attached to membranes or free in the aqueous phase [11 and references therein]. These partner proteins include cytochrome b_5, cytochrome *c* peroxidase and cytochrome *c* oxidase. The structures of the reactive complexes they form with cytochrome *c* are of considerable interest for defining mechanisms of electron-transfer but there are many difficulties in determining these structures. This stems from the relatively weak intermolecular binding interactions, which result, in part, from the electrostatic nature of the interactions [11,12]. Many suggestions have been made regarding the structures of protein complexes of cytochrome *c* from computer modelling calculations [e.g. see 13-16] and some of these models have been tested by a variety of chemical and mutagenesis approaches [e.g. see 14,16]. Only one complex, that with yeast cytochrome *c* peroxidase, has been crystallographically determined [17], but whether the structures obtained accurately describe the functional solution state[s] is not known. The aims of our study are to characterise the solution structures of interprotein complexes of cytochrome *c*, and to develop a procedure for studying such electrostatically-stabilised complexes that is generally applicable.

2. Small Molecule:Protein Interactions

2.1 GENERAL ASPECTS OF EXTRINSIC PARAMAGNETIC PROBES

The use of extrinsic paramagnetic reagents to map out interaction sites on protein surfaces is well established [2-4,6,7]. Fig. 1 illustrates the general approach.

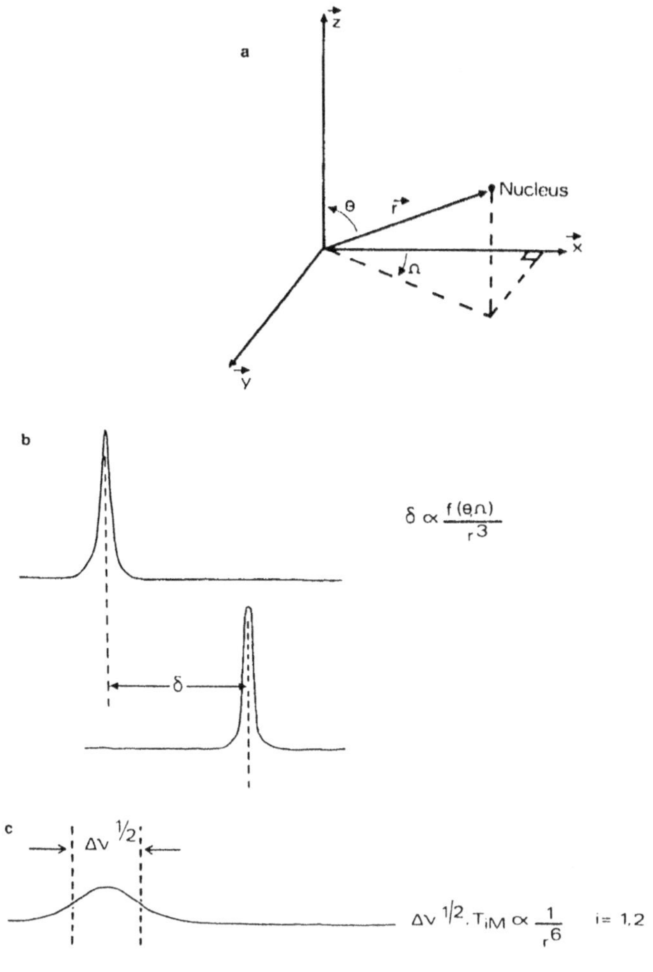

Figure 1. The effects of NMR shift and relaxation probes. (a) The coordinate system for the calculation of paramagnetic effects is centred on the unpaired electron and has the z axis aligned along the principal symmetry axis of the metal ion-protein complex. (b) If the electron relaxation time is short and the electron distribution anisotropic, the effect of probe binding is to shift the resonances of nearby nuclei from their unperturbed positions. (c) If the electron relaxation time is long, the dominant effect is on the relaxation times of nearby nuclei whose resonances are broadened. The perturbations to the chemical shift (d) and relaxation times [T_{iM}, where i = 1 or 2, the spin-lattice and spin-spin relaxation times, respectively] are related to the geometry of the complex by the geometric relationships given above.

The bound extrinsic paramagnetic metal is at the centre of the coordinate system with the nucleus to be monitored at a distance *r* away. If the extrinsic metal is a shift probe, the resonance of the nucleus will be shifted, and if the metal is a relaxation probe, the resonance will be broadened. In general, a paramagnet with long electron relaxation times is a relaxation probe, while a paramagnet with both short electron relaxation times and an anisotropic g-tensor is usually a shift probe [2,3]. When the paramagnetic perturbations arise from dipolar effects only, the spectral perturbation can be related to the conformation of the protein around the bound metal. Such analyses generally require the binding constant of the paramagnetic reagent: protein complex to be known because the occupancy of the binding site is a key parameter. Thus, it is possible to observe a small effect on a resonance resulting from weak binding at a site near to the group whose resonance is observed, and this could be mistaken for strong binding at a site far away from the monitored group. Numerous procedures exist for determining the relevant association constants; for reagents such as $[Cr(CN)_6]^{3-}$ a suitable method is to use ^{59}Co NMR of the corresponding Co(III) compound [18]. $[Cr(CN)_6]^{3-}$ is a shift probe to go alongside $[Cr(CN)_6]^{3-}$ but in general transition metal complexes are not suitable shift probes for the kind of studies reported here [7]. Some lanthanide complexes are suitable however [2,6].

2.2 $[Cr(CN)_6]^{3-}$ BINDING TO SUPEROXIDE DISMUTASE

^{59}Co NMR of $[Cr(CN)_6]^{3-}$ binding to oxidised SOD, unbuffered at pD 7.2 and 298K, has shown that there is one $[Co(CN)_6]^{3-}$ binding site per subunit with a K_d of 4mM [19]. The ^{13}C NMR spectrum of oxidised SOD at pD 7.2 and 25°C is given in Fig. 2.

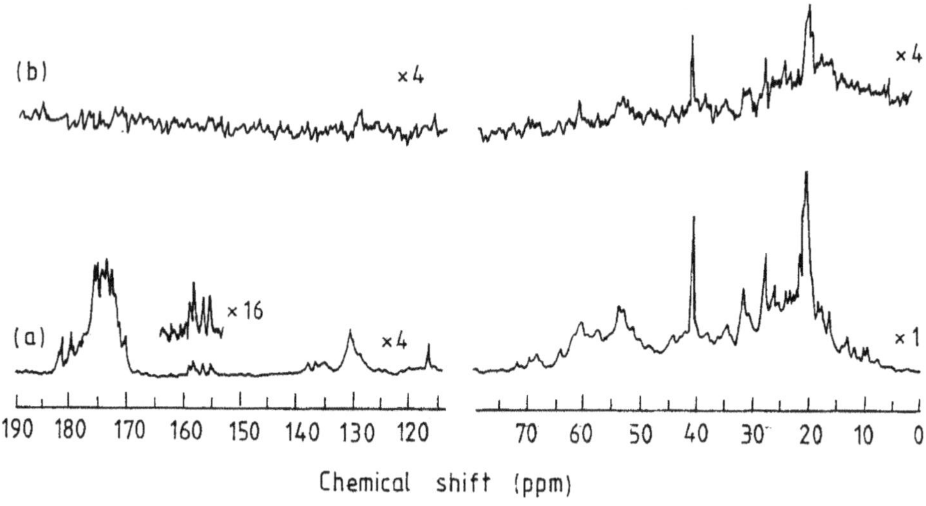

Figure 2. 75.4 MHz ^{13}C NMR spectra of Cu(II)-Zn(II) SOD at pD 7 and 25°C. The spectra were obtained with a solution of 6 mM SOD in D_2O. The spectra are proton decoupled and are the sum of 40,000 scans. They were obtained with a pulse width of 12 μs [75° pulse], an acquisition time of 0.44s and a recycle delay of 1.0s. (a) The spectrum of SOD before the addition of $[Cr(CN)_6]^{3-}$. (b) The paramagnetic difference spectrum (PDS) produced by the addition of 15 mM $[Cr(CN)_6]^{3-}$.

Generally resonances at about 150 to 160 ppm in ^{13}C spectra can be assigned to quarternary carbons of Arg and the C-4 carbon of Tyr [3]. In the spectrum of SOD four resonances are observed in this region with chemical shifts of 159.0, 158.2, 156.6 and 155.3 ppm. Bovine SOD contains four Arg residues and one Tyr residue [8,10] and thus five carbon resonances should be present in this region. However, the X-ray structure shows Arg 141 is close to the copper [10] and its resonances are therefore likely to be considerably broadened by the paramagnetic effect of Cu(II).

The addition of $[Cr(CN)_6]^{3-}$ caused some ^{13}C resonances to broaden. These appear as positive peaks in the PDS (Fig. 2) with chemical shifts of 40.4, 27.6, 23.8 and 20.1 ppm. The first three peaks have chemical shifts corresponding to the ε, δ and γ carbons respectively of lysine. Their relative intensities in the PDS decrease in the order $\varepsilon > \delta > \gamma$ and this is consistent with $[Cr(CN)_6]^{3-}$ binding to one or more lysine $-NH_3^+$ groups. The intense peak at 20.1 ppm probably comes from the methyl groups of Val or Thr residues. These occur in small peptides at 20.0 ppm and 19.6 or 18.6 ppm respectively.

The effect of $[Cr(CN)_6]^{3-}$ upon the ^1H NMR spectra of Cu(II)-Zn(II) SOD and Cu(I)-Zn(II) SOD is shown in Figs. 3 and 4 respectively. Resonances in the aromatic and aliphatic regions of the spectra were broadened and appear in the PDS.

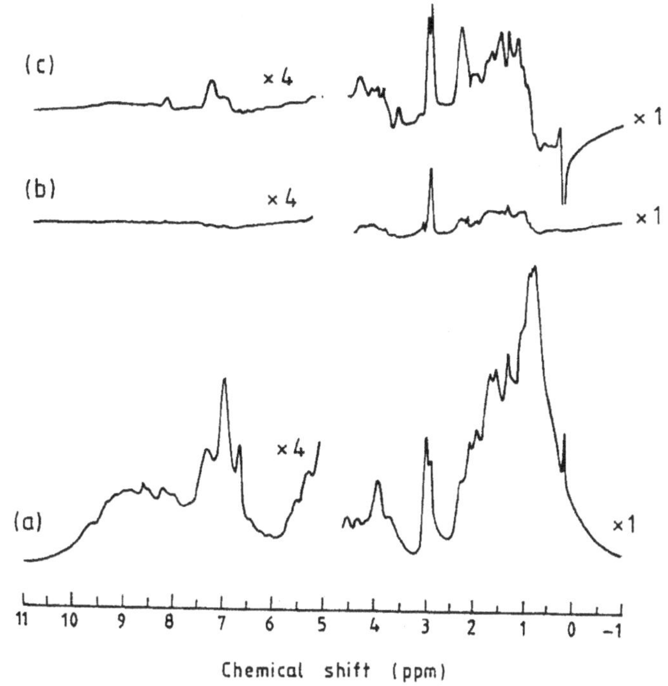

Figure 3. 300 MHz ^1H NMR spectra of 3 mM oxidised SOD at pD 6.7 and 25°C.
(a) is the normal spectrum in the absence of $[Cr(CN)_6]^{3-}$; (b) and (c) are the PDS produced on the addition of 0.1mM and 0.43mM $[Cr(CN)_6]^{3-}$, respectively.

The intense peaks at ~2.9 ppm can be assigned to the εCH_2 proteins of lysine residues. The intensity of these peaks indicates more than one lysine is affected. The appearance of a peak at 1.25 ppm in both sets of PDS at low $[Cr(CN)_6]^{3-}$ concentration is consistent with the suggestion from ^{13}C NMR that a Thr residue is close to the $[Cr(CN)_6]^{3-}$ binding site; the ^1H chemical shift of an unshifted Thr methyl group is 1.23 ppm [3,4].

The peak in the ^1H PDS of oxidised SOD at 7.32 ppm (Fig. 3) and reduced SOD at 7.36 (Fig. 4) are probably from the same group. These peaks do not appear until relatively high concentrations of $[Cr(CN)_6]^{3-}$ but they become more intense than a single proton resonance. Their chemical shift and intensity leads to their assignment to one or more Phe residues. Bovine SOD does not contain Trp and the aromatic resonances of the sole Tyr have been identified to be at 7.01 and 6.67 ppm [20].

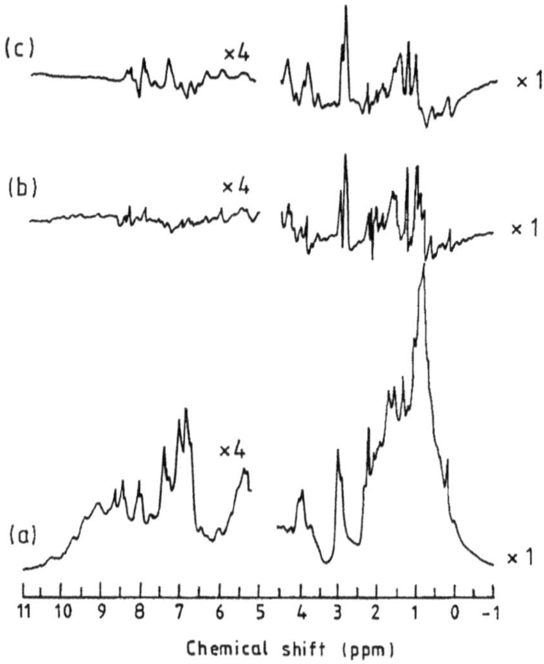

Figure 4. 300 MHz ^1H NMR spectra of 3 mM reduced SOD at pD 7 and 25°C. (a) is the normal spectrum in the absence of $[Cr(CN)_6]^{3-}$; (b) and (c) are the PDS produced on the addition of 0.1 mM and 0.43 mM $[Cr(CN)_6]^{3-}$, respectively.

Most of the resonances in the region 6 to 9 ppm of the PDS of Fig. 4 that are absent from the PDS of Fig. 3 come from groups that are ligands to the metal ions. These are considerably broadened in spectra of Cu(II)-Zn(II) SOD by the unpaired electron of the Cu(II). The peak at 7.94 ppm has been assigned to a non-ligand His that titrates with pH with a pK of 6.7. Its appearance in the PDS is probably due to the direct, weak binding of $[Cr(CN)_6]^{3-}$ to the protonated form of the His. This binding is likely to be very weak and not detected by the ^{59}Co experiment carried out at pH 7.2 [19]. Similar weak binding of $[Cr(CN)_6]^{3-}$ to the cationic form of His has been observed by NMR for lysozyme [21] and cytochrome *c* [7]. The absence of this peak from the PDS of Fig. 3 is probably because the pH of that sample was at the His pK_a leading to a broadening of the His resonances by an intermediate rate of proton exchange. The peaks at 8.45, 8.36 and 6.02 ppm are affected by relatively low concentrations of $[Cr(CN)_6]^{3-}$. The peaks at 8.00, 7.71, 6.82, 6.76 and 6.41 ppm appear in the PDS at higher $[Cr(CN)_6]^{3-}$ concentrations (Fig. 4). All except the peak at 8.00 ppm have previously been assigned to histidine ligands of the metal ions with those at 8.45, 8.36 and 6.41 ppm from Zn(II) ligands and most of the remainder from Cu(II) ligands [19,20]. The pattern of relaxation effects show that resonances of histidine ligands to Zn(II) are more affected than are those to Cu(I), indicating the bound anion is closer to Zn(II) than to Cu(I).

Implications for the dismutase reaction. The identification of a region of the SOD surface of high positive charge close to the metal site supports the proposal that the O_2^- is attracted to the reaction site by electrostatic forces. Although it is likely that the bound anion is mobile, it is nevertheless of interest that it spends most of its time closer to the zinc than to the copper. If this reflects the electrostatic field of the protein then O_2^- is unlikely to be directed to the copper by it. However, it is possible that steric factors prevent the $[Cr(CN)_6]^{3-}$ from approaching closer to the copper.

2.3 $[Cr(CN)_3]^{3-}$ AND $[Cr(ox)_3]_3]^{3-}$ BINDING TO CYTOCHROME *c*

Similar studies with cytochrome *c* to those described for SOD in section 2.4, have led to the identification of a number of regions of the cytochrome *c* surface where anions bind [7,18,22]. Fig. 5 summarises the structural interpretation of the NMR data [7].

Site 1 is positioned at the top left of cytochrome *c* and includes residues 65 and 89. Lysines 5, 86, 87, and 88 may provide the binding groups. Site 2 lies to the right of the haem on the front right edge and includes the amino acids Val 11, Ala 15, and Thr 19. Lysines 7, 25, and 27 may provide the binding groups. Site 3 lies on the front face to the left of the haem. It is binding at this site that is predominantly responsible for the perturbation of the resonance of haem methyl 3. Resonances of the amino acids Ile 81, Phe 82, and Ala 83 are affected by binding at this site. Lysines 13, 72, and 86 may provide the binding groups. Resonances of Val 28 are affected by binding at both sites 2 and 3 and resonances of Ile 85 by binding at both sites 1 and 3. Sites 1 and 3 have a higher affinity for simple substitution-inert anions than does site 2. $[Fe(edta)(H_2O)]^-$ binds preferentially at the lower end of site 2, close to the acidic residue Glu 21, which may actually coordinate to the extrinsic non-haem iron [23].

These data for the interaction of paramagnetic molecules, that are either redox reagents or analogues of redox reagents, with cytochrome *c* have allowed the electron transfer function of cytochrome *c* to be investigated [7,23]. They also underpin the competitive paramagnetic difference spectroscopy approach described in section 3.5 for the investigation of interprotein complexes of cytochrome *c*.

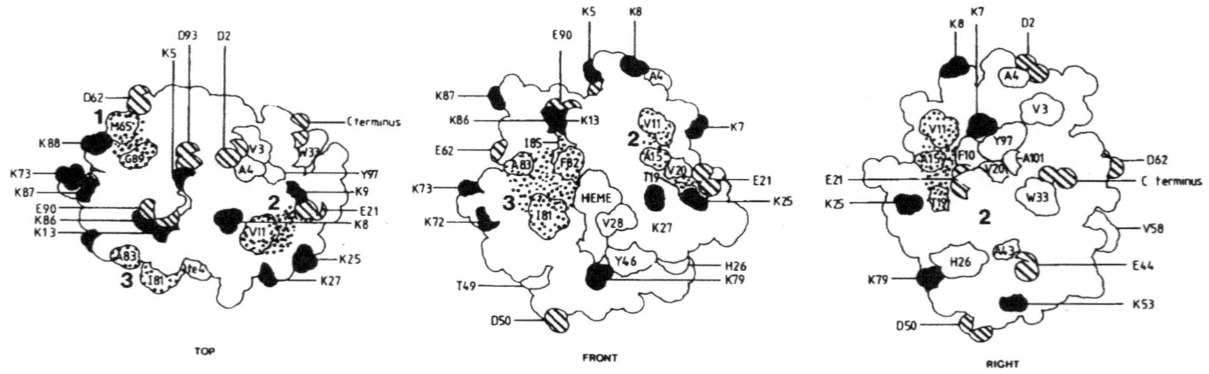

Figure 5. Representation of the three major anion binding sites on cytochrome *c*. The space-filling diagrams for tuna cytochrome *c* illustrate the positions of negatively charged groups [striped] and positively charged groups [solid]. Residue numbers in the figure and text refer to tuna cytochrome *c*. Important residues with assigned NMR resonances are shown in outline. The stippled residues are those affected by low concentrations of probes of the type $[Cr(CN)_6]^{3-}$ and $[Cr(ox)_3]^{3-}$, and the numbering scheme refers to binding sites defined in the text.

3. Protein:Protein Interactions

3.1 INTRODUCTION

The development of 2D and 3D NMR methods and their application to proteins uniformly labelled with ^{13}C and/or ^{15}N [e.g. 24,25 and references therein] has opened the field of interprotein interactions to a variety of novel procedures. For example, one protein in a complex may be labelled and the other unlabelled, thus allowing heteronuclear experiments to focus on one component only. Such an approach is being pursued with complexes of high binding affinity so that the complex is in slow exchange between free and bound components on the NMR time-scale; an example is the complex formed between the DNase domain of colicin E9 and its inhibitor protein [26]. Complexes of cytochrome *c* have not been studied by such a procedure.

The major structural questions with complexes of cytochrome *c* are:

(a) What is the stoichiometry of the complex?

(b) Is there a conformation change of either protein on complex formation?

(c) Is there a unique structure for the complex?

(d) What regions of the protein surfaces make intermolecular contacts?

Generally there are a variety of procedures for answering question (a), including titration by electronic spectroscopy [27] and NMR [28]. Under conditions resembling physiological conditions cytochrome *c* forms bimolecular complexes with cytochrome b_5 and cytochrome *c* peroxidase but has two binding sites with different affinities on intact cytochrome *c* oxidase [11 and references therein]. The answer to question (b) is that there does not appear to be a substantial conformation change, if any at all, to cytochrome *c* when it binds to cytochrome b_5 [28,29]. It is questions (c) and (d) that are difficult to answer and it is these which are addressed in the remainder of the paper.

3.2 ^1H NMR SPECTROSCOPY

3.2.1 Amino acid resonances

The formation of an interprotein complex generally leads to broadening of ^1H NMR peaks as a consequence of the increased tumbling time of the complex compared to the individual proteins. 1D NMR experiments rarely allow specific amino acids to be monitored because of a lack of resolution, and many 2D NMR experiments, such as TOCSY, are handicapped by the relatively rapid relaxation of protons within the complex. In the case of cytochrome *c*, increasing ionic strength leads to a rapid rate of exchange between free and bound forms, thus overcoming the linewidth increase, but in so doing much of the specific information about the complex is lost. If ^1H NMR alone is to be useful it may need to be accompanied by the uniform deuteration of one of the proteins within the complex in order to reduce the linewidth contribution to the remaining proton resonances from ^1H-^1H dipole interactions.

Specific examples of where ^1H NMR alone has been useful are given below. For trimethyllysine residues of cytochromes *c*, interactions with cytochrome b_5 [28] and cytochrome b_2 [30] could be monitored and valuable information obtained. Also, in ^1H NMR spectra of cytochrome *c* complexed with cytochrome b_5, NOE experiments to monitor Phe 82 of cytochrome *c* were productive [29]. These were important in showing that the predicted conformation change of cytochrome *c* within the complex, that involved Phe 82 [15], did not actually occur.

3.2.2 *Hyperfine shifted resonances*

Gupta and Yonetani [31] first observed that protein complexation with cytochrome *c* caused shifts to the haem methyl 3 and 8 resonances at ~32 ppm and ~34 ppm. These workers studied the system yeast cytochrome *c* peroxidase interacting with cytochrome *c* and used the chemical shift changes to propose an interaction region for the peroxidase on cytochrome *c*. Subsequent experiments revealed that other redox proteins binding to cytochrome *c* produced similar chemical shift perturbations. Proteins studied included bovine cytochrome b_5 and yeast cytochrome b_2 [28,30]. However, it became clear from studies of the effect of polyglutamate binding to cytochrome *c*, and from the effects of chemical modification of lysine residues of cytochrome *c* (Fig. 6) that the perturbations could not be interpreted in terms of a specific alignment for the proteins within the complex [32].

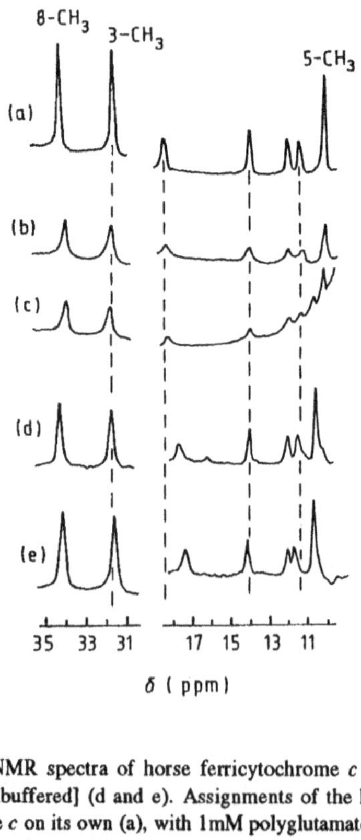

Figure 6. 270MHz ^{1}H NMR spectra of horse ferricytochrome *c* at pD 6.2 [5mM phosphate buffer] (a-c) or at pD7 [unbuffered] (d and e). Assignments of the haem methyl resonances are indicated. 3mM cytochrome *c* on its own (a), with 1mM polyglutamate (b) and 1.5mM cytochrome *c* peroxidase (c); and 3mM lysine-modified cytochromes *c*, (d) acetimidylated and (e) maleylated.

The large hyperfine shifts of the ferricytochrome *c* haem methyl resonances are a consequence of the presence of an unpaired electron in the haem group, and the central conclusion from Fig. 6, that the haem methyl shifts are sensitive to the charged nature of groups at the surface of the ferricytochrome, reflects variation in the distribution of the unpaired electron. Minor changes in the charge distribution of the protein surface leads to changes in the distribution of the unpaired electron and hence to chemical shift perturbations.

Although the chemical shift perturbations cannot be used to produce detailed structural models, they can be used to show complexation takes place and, in favourable cases, they allow binding constants and stoichiometries to be determined: e.g. for cytochrome *c* and cytochrome b_5 [28 and Section 3.6]. Fig. 7 shows that the complexation between ferricytochrome *c* and a monomeric form of bovine cytochrome oxidase can be detected by this procedure. The molecular mass of the monomeric cytochrome oxidase is ~160 kD [33] and its intrinsic haem resonances have linewidths of 400-1900 Hz [34]. In the experiment of Fig. 7 there is an excess of cytochrome *c* over the oxidase and since the system is in rapid exchange between free and bound forms, the linewidths of the cytochrome *c* resonances are not appreciably broadened.

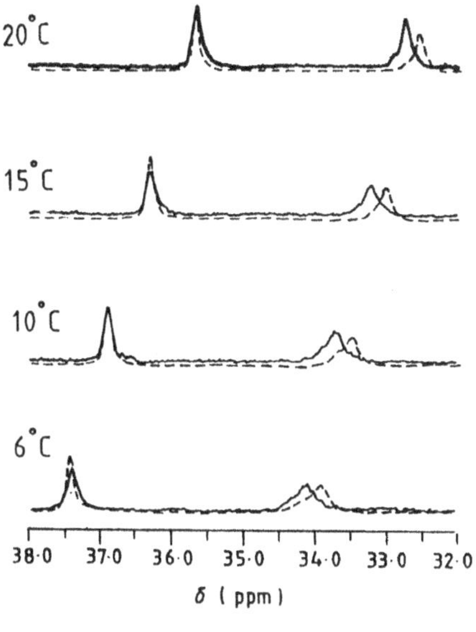

Figure 7. The haem methyl 3 and 8 resonances of 400MHz ^{1}H NMR spectra of horse ferricytochrome *c* at pD 7.4 and various temperatures. The broken line represents cytochrome *c* on its own and the solid line a mixture of 1.0 mM cytochrome *c* and 0.2 mM monomeric cytochrome oxidase [20 mM phosphate; 0.3 M NaCl; 0.1% triton X-100]

3.3 ^{13}C NMR SPECTROSCOPY

^{13}C NMR spectroscopy has the advantage over ^{1}H NMR of a greater chemical shift dispersion and thus a reduced overlap problem. Furthermore, ^{13}C NMR shifts are more sensitive to minor structural perturbations. Therefore whilst ^{1}H NMR may not provide much information regarding the involvement of amino acids in interprotein interactions, ^{13}C NMR may. This is illustrated by the cytochrome b_5 - cytochrome *c* complex. Part of the methyl region of the spectra of free and complexed cytochromes *c* and b_5 is shown in Fig. 8. The improved resolution of the ^{13}C NMR spectra allows groups to be monitored directly. The environment of Ile 81 of horse cytochrome *c* is slightly affected by complexation, and the environments of Ile 75 and Ile 85 are also slightly perturbed. The shift of the resonance at 16.9 ppm, which is the chemical shift of the γ-CH_3 of Ile 85, indicates that it is probably the δ-CH_3 resonance of Ile 85 at 11.9 ppm that is perturbed by complexation. Further analysis of the ^{13}C NMR spectra requires a full assignment of the cytochrome b_5 spectrum.

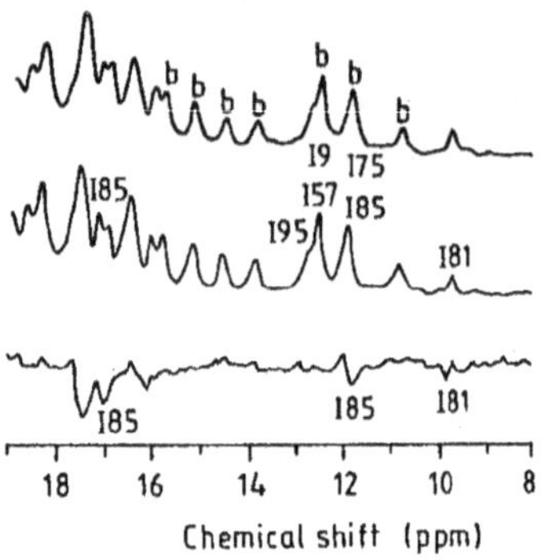

Figure 8. Aliphatic region of the 100 MHz ^{13}C NMR spectra of ferricytochromes *c* and b_5 in 50 mM phosphate [pH 7.6] at 25°C. *Top*, mixture spectrum; *middle*, summed spectrum; and *bottom*, difference spectrum [summed - mixture]. Resonances labelled "b" are resonances of cyt b_5 that have not yet been assigned.

3.4 ^{15}N NMR SPECTROSCOPY AND NH EXCHANGE RATES

As with ^{13}C NMR spectroscopy, ^{15}N NMR signals are generally more sensitive to the environments of their corresponding nuclei than are ^{1}H NMR signals to the environments of their corresponding nuclei. To date, most of the ^{15}N NMR studies of interprotein interactions have monitored the peptide nitrogen resonances. For this reason we consider this procedure together with the rates of peptide NH $\rightarrow$ ND exchange. However, we note investigating lysine side chain nitrogen resonances may be informative, particularly in view of the approach we describe in section 3.6.

The most complete study of a protein complex of cytochrome *c* involving its peptide resonances is the work of Jeng *et al* [35] on the cytochrome *c*-cytochrome *c* peroxidase complex. These authors found that the NH $\rightarrow$ ND exchange rates for some peptide resonances were slower in bound cytochrome *c* than in free cytochrome *c*. Only 33 NH resonances could be monitored, of which 16 were substantially affected. Both the identity of some of the affected resonances, and the extent of the perturbations seen were informative. Many of the perturbed resonances came from groups on the front side of the protein, (see Figs. 5 and 10), where its redox partners are known to bind [11-17], but some came from further away indicating that either there was an additional interaction site or that there had been an alteration in the dynamic properties of the cytochrome *c* on complexation. The process of NH → ND exchange requires some opening of the protein structure and a breaking of hydrogen-bonds. If complexation reduces the extent to which the protein can open and close its structure then rates of NH exchange will generally be reduced, and not just for groups at the protein interface. A similar comment has been made concerning protection experiments involving chemical modification of cytochrome *c* peroxidase residues in its complex with cytochrome *c* [11]. In one case tryptophan residues far from the binding site were protected from photo-oxidation [36]. However, in the study of Jeng *et al* [35] what was surprising was that the decrease in NH → ND exchange rates on complexation was much smaller than anticipated. This could only come about if the complex was highly mobile; a point we return to in section 4.

Studies of ^{15}N NMR spectra of the peptide resonances of cytochrome *c* in protein complexes have not been reported but from work with DNase-inhibitor protein complexes [26] we anticipate that many resonances will be affected if the experiment is carried out under conditions where the cytochrome *c* is bound for much of the time. This is because the NH ^{15}N chemical shifts reflect the strength of the corresponding peptide hydrogen-bond interactions and these will be altered by complexation as the protein dynamics are altered (see above).

3.5 COMPETITIVE PARAMAGNETIC DIFFERENCE NMR

The problems introduced by non-specific broadening resulting from protein:protein complex formation can be partly overcome by the use of paramagnetic reagents to amplify specific effects resulting from complex-formation by two macromolecules [28]. Fig. 9 illustrates the idea behind the experiment. In Fig. 9(i) three resonances of protein A are non-specifically broadened as a result of the formation of a complex between proteins A and B. The specific effects of a small relaxation reagent bound to protein A in the absence of protein B are illustrated in Fig. 9(ii). One resonance (α) is unaffected, one (β) is weakly affected, and one (γ) is strongly affected. The effect of the relaxation reagent on the resonances of protein A in the complex formed between proteins A and B is illustrated in Fig. 9(iii). Only resonance β is affected. A comparison of the difference spectra (f) and (i) reveals which resonances are protected from the paramagnetic reagent by complex-formation. In this case resonance γ is protected. The protected resonances come from groups of protein A that are shielded by the bound protein B and thus specific information is obtained about the interaction between the two proteins.

There are a number of complications to this experiment. Perhaps the most important is that in a solution of a mixture of two proteins and a small reagent, a number of different complexed species may exist, with the amounts of each depending on the initial concentrations of the uncomplexed species and their relative association constants. If the presence of protein B results in the loss of a resonance of protein A from a PDS it could be because protein B sequesters the reagent, or because protein B and the reagent compete for the same binding site on protein A. In general, it is not possible to distinguish between these two mechanisms unless the appropriate association constants are known. However, when there are several binding sites for the small reagent on protein A, a test of which mechanism operates is possible. Eley and Moore [28] describe a number of other potential problems, and indicate ways in which they can be addressed. In general, for proteins such as cytochrome *c* and b_5 and cytochrome *c* peroxidase, competition between protein partners and reagents such as $[Cr(ox)_3]^{3-}$, $[Cr(CN)_6]^{3-}$, $[Cr(NH_3)_6]^{3+}$ and $[Cr(en)_3]^{3+}$ is readily interpreted in terms of the structures of the various intermolecular complexes [28,37].

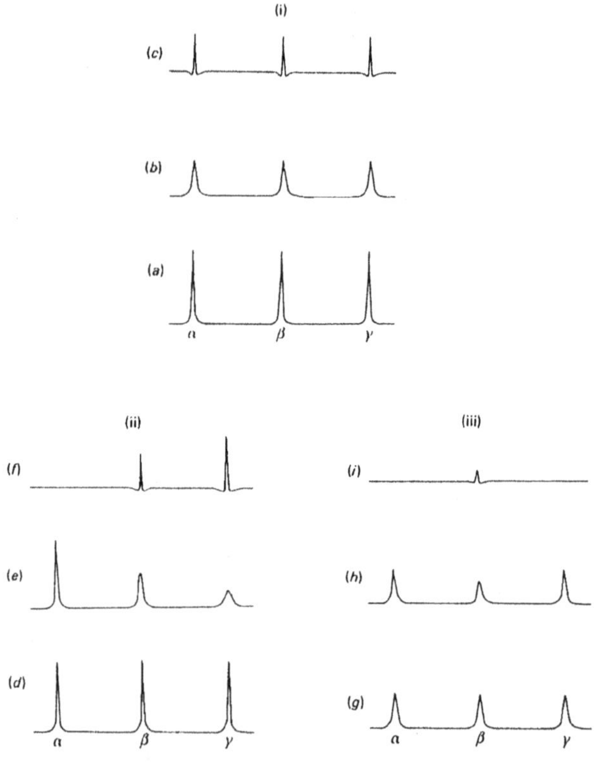

Figure 9. Simulated spectra illustrating the effects of (i) correlation time broadening, (ii) paramagnetic broadening and (iii) a combination of the two effects. (i)(a) Three resonances in the spectrum of a macromolecule (A) with linewidths at half peak height of 25 Hz. (b) The effect of binding A to a second macromolecule (B) is to broaden the resonances non-specifically: the linewidths are 50 Hz. (c) The difference spectrum (a)-(b). (ii)(d) The spectrum of A. (e) The effect of adding a specific small paramagnetic relaxation reagent. Resonance α is unaffected, but the linewidths of resonances β and γ increase to 50 Hz and 100 Hz respectively. (f) The difference spectrum (d)-(e). (iii)(g) The spectrum of A bound to B. (h) The effect of adding the relaxation reagent. Only resonance β is affected, its linewidth increases to 75 Hz. (i) the difference spectrum (g)-(h).

Fig. 10 compares the site on cytochrome *c* for cytochrome b_5 determined by the competitive NMR PDS approach, employing $[Cr(ox)_3]^{3-}$ and $[Cr(CN)_6]^{3-}$ to compete with cytochrome b_5, with the site obtained from kinetic studies of chemically modified cytochromes [38,39]. There is reasonable agreement between the two sets of data, particularly in the dominance given to the top left side of the protein. However, the kinetic experiments also suggest that Lys 25 and Lys 27 have important roles, which conflict with the NMR data. Both sets of data agree reasonably well with the predicted model proposed by Salemme [13] and are consistent with the ^{13}C data described in section 3.2.

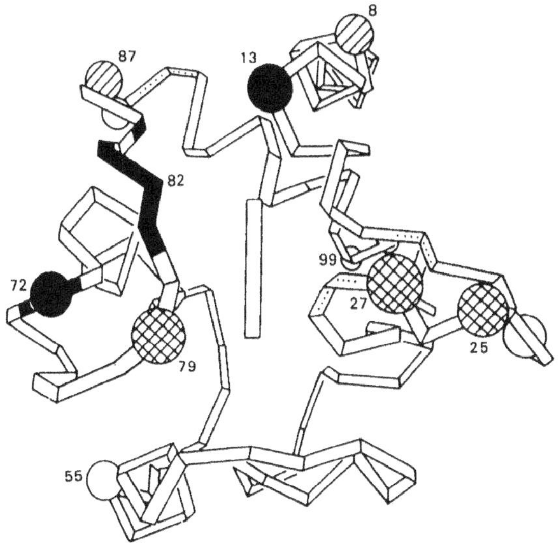

Figure 10. Ribbon diagram with a comparison of NMR and lysine-modification data [38,39] for residues of cytochrome *c* that illustrates the binding site for cytochrome b_5. The circles represent the lysine εNH_3^+ groups that have been shown by chemical modification to be involved in the interaction of cytochromes b_5 and *c*; shading indicates their degree of implication in binding, with an open circle indicating no involvement. The ribbon represents the peptide backbone; groups shown by NMR to be in the interaction region are indicated by ■; ..... indicates groups not shielded by cytochrome b_5.

The site on ferricytochrome b_5 for cytochrome *c* binding has also been probed by the competitive PDS approach using $[Cr(en)_3]^{3+}$ [37]. $[Cr(en)_3]^{3+}$ was found to be able to bind close to the haem of cytochrome b_5 in its complex with cytochrome *c* with the binding of cytochrome *c* to cytochrome b_5 not significantly inhibiting the binding of $[Cr(en)_3]^{3+}$. The data indicate that $[Cr(en)_3]^{3+}$ probably binds to cytochrome b_5 close to at least one of its haem propionates, in both the absence and presence of cytochrome *c*. There are a number of situations that could give rise to this result.

(i) $[Cr(en)_3]^{3+}$ could dissociate the cytochrome complex and bind to free cytochrome b_5. The constant broadening and shift of cytochrome *c* resonances argue against this in the present case.

(ii) Cytochrome *c* could bind to cytochrome b_5 at the latter's haem edge but the addition of $[Cr(en)_3]^{3+}$ causes a shift in the position of the bound cytochrome *c* so that although an interprotein complex is maintained the cytochrome b_5 haem edge is free to bind $[Cr(en)_3]^{3+}$. That is, binding of $[Cr(en)_3]^{3+}$ induces formation of an alternative docking alignment.of cytochrome *c* with cytochrome b_5.

(iii) Cytochrome *c* and $[Cr(en)_3]^{3+}$ may bind to different regions of the cytochrome b_5 surface. Cytochrome b_5 is a highly anionic protein [net charge -9 at pH 7], and its cation-binding surface is large enough to bind both cytochrome *c* and $[Cr(en)_3]^{3+}$ simultaneously. Whitford [40] has shown that there are at least three $[Cr(en)_3]^{3+}$ binding sites on cytochrome b_5 with one of them being the haem edge site previously described [39].

At present it is not clear whether scheme (ii) or (iii) best describes the ternary complex. However, these studies do indicate some of the problems with extracting structural information from such paramagnetic NMR data.

3.6 NMR PROBES OF LYSINES

One important aspect of protein:protein complexation for which the experiments described so far provide no direct information, is the number of cytochrome *c* lysine residues involved in binding cytochrome b_5. This deficiency arises because the proton resonances of the 19 lysines of horse cytochrome *c* are complex multiplets, and both the ^1H and ^{13}C resonances occur in crowded regions of the spectrum. Thus, these Lys residues are, in effect, NMR invisible. To make them visible in NMR, we chemically modified them to produce the N-acetimidylated derivative. The derivative retained the structure and key functional properties of native cytochrome *c* [29]. Subsequently, we have turned to reductive methylation to yield the N^{ε}-dimethyl lysine derivative of cytochrome *c* [41]. This modification also retains a positive charge but it has the advantage over N^{ε}-acetimidylation of introducing two methyl groups per lysine without a significant increase in the size of the modified residue, and of being more readily obtained in an isotopically enriched form in which the dimethyl groups contain ^{13}C.

Fig. 11 compares the region of the ^1H NMR spectra of modified and native ferricytochrome *c* where the added methyl resonances occur. The wide chemical shift dispersion of these resonances is, in part, a result of the paramagnetism of the haem because the dispersion is not so great for the ferrocytochromes. 2D-NMR spectra show that the modifications do not substantially alter the protein structure, and kinetic and thermodynamic measurements show that the functional properties of the cytochrome are not substantially altered either.

Fig. 12 shows the 2D ^{13}C-^1H COSY spectrum of the N^{ε}–dimethyl lysine derivative of horse ferricytochrome *c* uniformly labelled with ^{13}C in the lysine methyl groups. 15 of the 19 lysines are resolved in the 2D spectrum. Of the remaining 4 lysines, at least two, those with ^1H chemical shifts of 2.67 and 2.72 ppm, do not appear in the 2D spectrum. It is probable that the other two lysines do not have appreciable cross-peak intensity either. The reasons for this are not clear but it may be associated with relatively rapid relaxation in either or both of the ^{13}C and ^1H dimensions. Nevertheless, the resolution of 15 lysines in the 2D spectrum is clearly superior to the resolution in either 1D spectrum.

Assignment of the methyl resonances to specific lysines is not straightforward, even though many of the lysine β–ε ^{1}H assignments are known [42]. Perhaps the best approach is to study variant cytochromes produced by site-directed mutagenesis, and this work will be undertaken shortly with yeast iso-1 cytochrome *c*. A related study with calmodulin shows what is possible [43]. In the absence of site-directed variants we have employed $[Cr(CN)_6]^{3-}$ binding to provide tentative assignments.

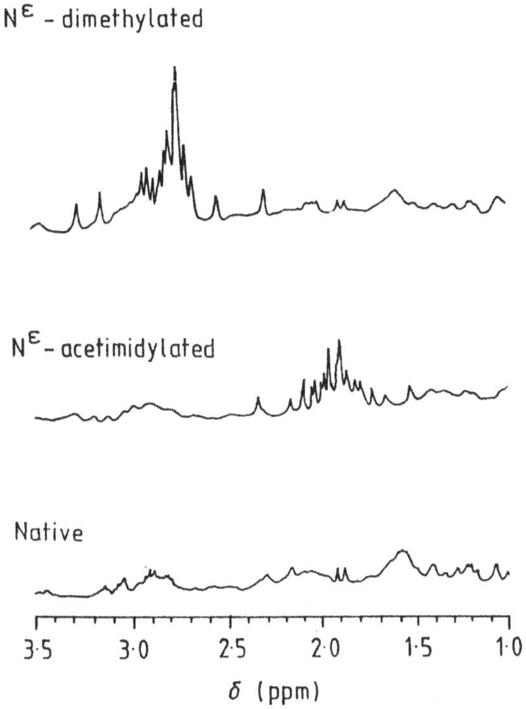

Figure 11. 400 MHz ^{1}H NMR spectra of native and derivatised horse ferricytochrome *c* at pD 7 and 25°C.

Fig. 13 shows the effect of adding $[Cr(CN)_6]^{3-}$ to the ^{1}H-^{13}C COSY spectrum of ^{13}C labelled Nε-dimethyllysine cytochrome *c*. Some cross-peaks are largely unaffected whilst others are broadened. From parallel studies of ^{1}H NMR spectra analogous to those described in secton 2.3, used to map out the interaction sites shown in Fig. 5, some of the lysine cross-peaks have been associated with different binding regions [Table 1]. This, together with a comparison of the spectra of dimethylated yeast iso-1 and horse cytochromes *c* provide some possible assignments [Table 1].

3.6.1 Nε-dimethyllysine cytochrome c and cytochrome b_5

In order to examine the effect that complexation has on the methyl lysine resonances, a titration was carried out, where small aliquots of a solution of lipase cleaved bovine

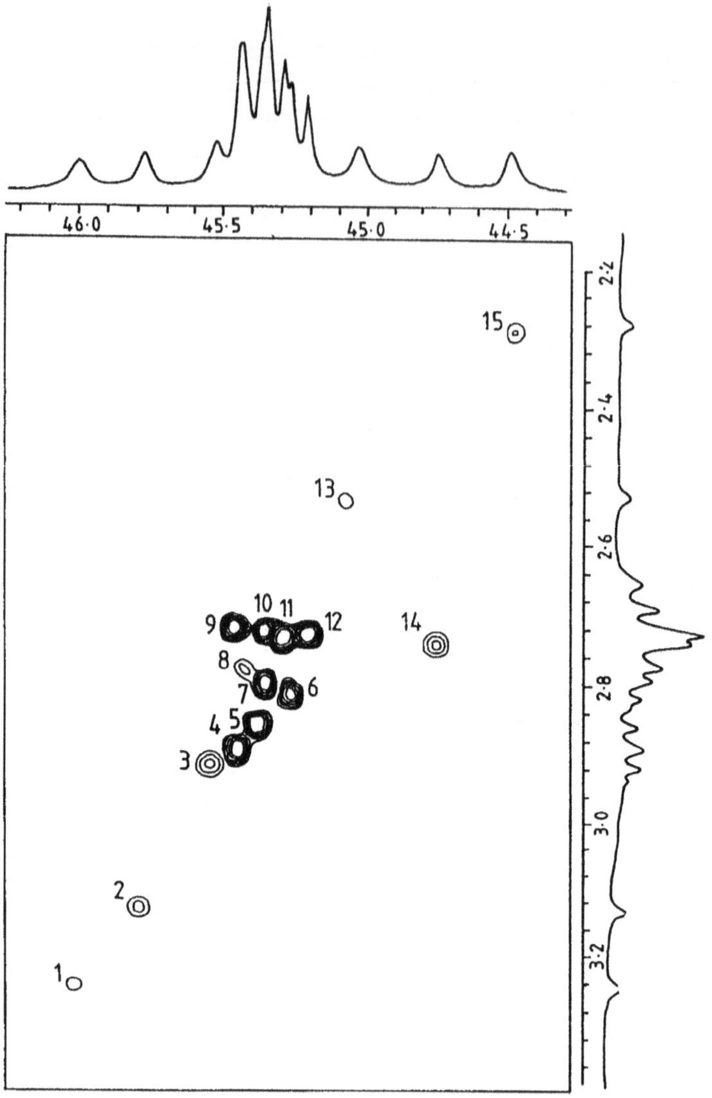

Figure 12. The 2D ^{1}H-^{13}C COSY spectrum of ^{13}C labelled N^{ε}-dimethyllysine ferricytochrome *c* acquired with a 400MHz ^{1}H NMR spectrometer. The 1D spectra are the corresponding ^{1}H and ^{13}C spectra. The ^{1}H NMR spectrum was measured with unlabelled protein. All samples were in 5mM phosphate buffer at pD 7 and the spectra were measured at 25°C. The acquisiton parameters for the COSY spectrum were: spectral widths of 4000Hz and 1000Hz respectively for the ^{13}C and ^{1}H dimensions; 512 t_1 increments of 2048 data points each; and 16 scans per increment.

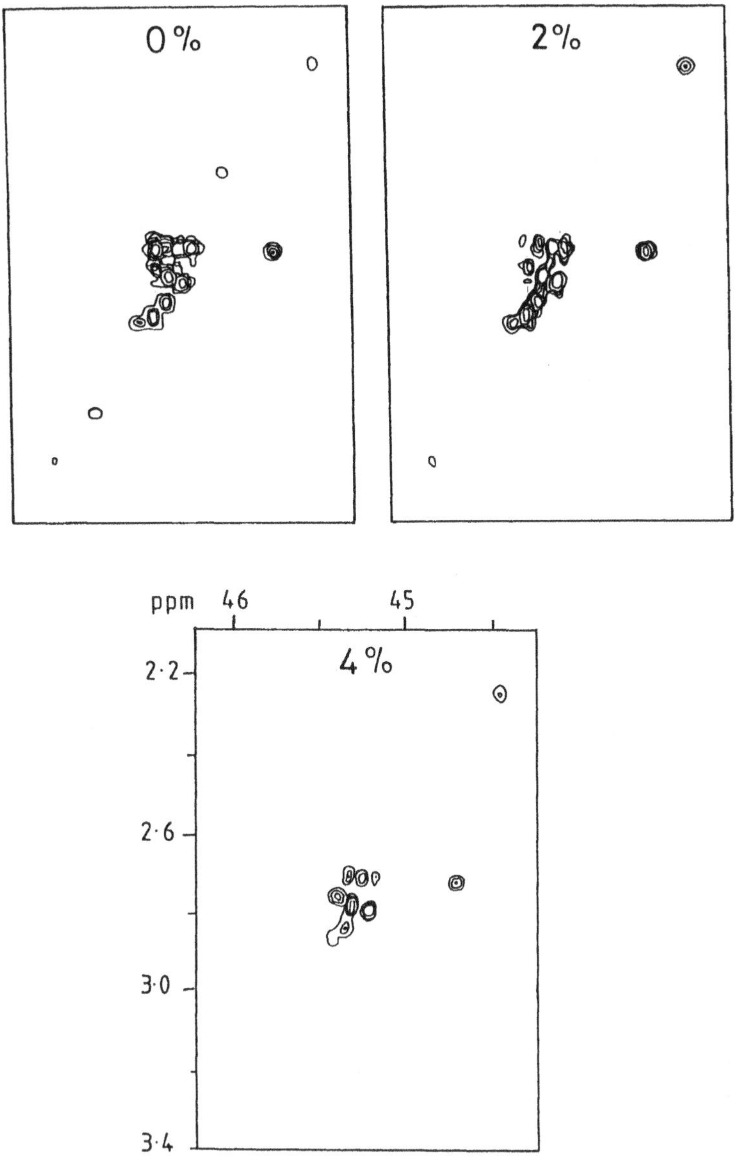

Figure 13. The effect of $[Cr(CN)_6]^{3-}$ on the ^1H-^{13}C COSY spectrum of ^{13}C labelled N$^{\varepsilon}$-dimethyl lysine ferricytochrome *c*. For experimental details see Fig. 12. Mole percent of added $[Cr(CN)_6]^{3-}$ was 0%, 2% and 4%, as indicated.

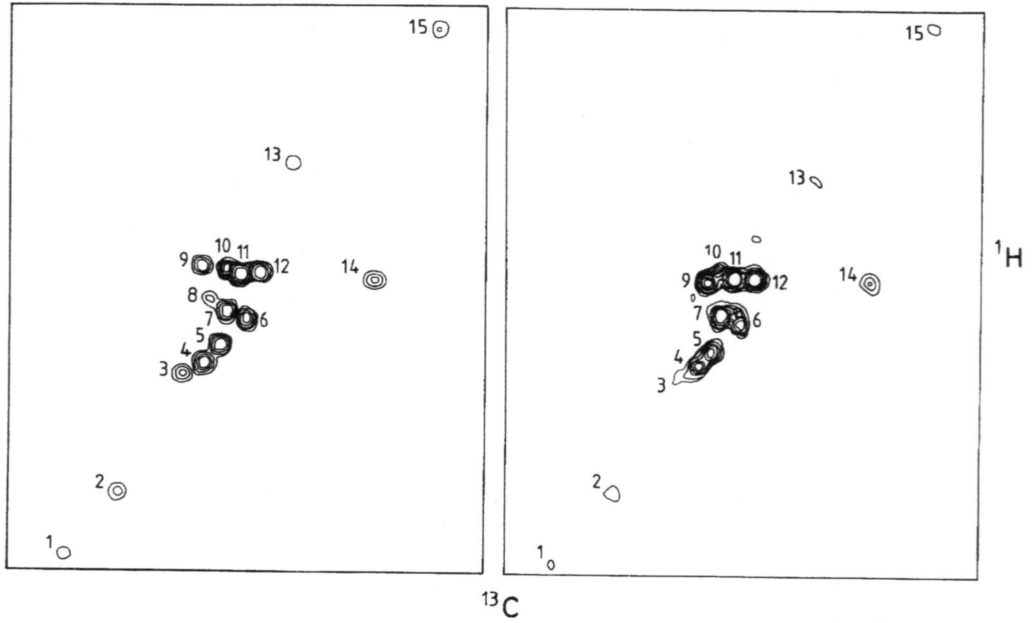

Figure 14. ^{1}H-^{13}C COSY spectra of 3mM ^{13}C labelled N^{ϵ}-dimethylferricytochrome *c* in the absence (*left*) and presence (*right*) of 3mM ferricytochrome b_5. Other experimental details are given in Fig. 12.

TABLE 1. Effect of $[Cr(CN)_6]^{3-}$ and cytochrome b_5 on dimethyl lysine resonances of N^{ε}-dimethyl lysine ferricytochrome *c*

Lysine resonance	$[Cr(CN)_6]^{3-}$ binding site[1] [see Fig. 5]	Tentative[2] lysine assignment	Maximum on binding ^{13}C	Δδ [ppm] cytochrome b_5 ^{1}H
1	site 2	7/25/27	0.03	0.02
2	site 3	13/72/86	u	u
3	u		0.02	0.01
4	site 2	7/25/27	u	u
5	u		0.03	0.02
6	u		u	u
7	u		u	u
8	u		?	?
9	site 3	13/72/86	0.05	0.03
10	u		?	?
11	u		0.02	0.00
12	site 2	7/25/27	u	u
13	site 3	13/72/86	0.09	0.03
14	u	39/53/60	u	u
15	u	39/53/60	u	u

1. u = unaffected by cytochrome b_5 or by 6% $[Cr(CN)_6]^{3-}$
   ? = affected by cytochrome b_5 but resolution insufficient to quantitate effect.
2. From $[Cr(CN)_6]^{3-}$ binding studies and comparison of spectra of derivatised yeast and horse ferricytochromes *c*.

microsomal cytochrome b_5 were added to a sample of ^{13}C labelled methylated cytochrome *c*. The ratio of cytochrome b_5 to cytochrome was gradually increased over the course of the titration. As with native cytochrome *c*, the haem methyl resonances shifted during the experiment. Some of the dimethyl lysine ^{13}C-^{1}H COSY peaks also shifted throughout the titration. Fig. 14 shows the ^{1}H-^{13}C COSY spectrum of a mixture of the two proteins where the ratio of cytochrome b_5-cytochrome *c* concentrations were 1:1, and Fig. 15 shows a plot of the change in ^{13}C chemical shifts of peaks 9 and 13 against the molar ratio of cytochrome b_5 to cytochrome *c*. In all at least 6 of the ^{13}C-^{1}H methyl resonances are affected by the presence of cytochrome b_5.

There are a number of points to be made concerning these data. First, the observation that at least six lysine residues are affected by complex formation is different to predictions based on computer-graphics molecular-modelling experiments, all of which predict four or five lysines to be involved [13-16]. Second, and perhaps more importantly, the chemical shift perturbations on complex formation are extremely small. This was unexpected but it may be explicable for the reason given below. The chemical shift perturbations will be a weighted average of their values in the complexed and uncomplexed states, and it may well be that in a dynamic complex important lysine residues are not always involved in intermolecular interactions, even though the cytochrome *c* remains bound to cytochrome b_5 for most of the time. In effect, the expected chemical shift perturbations are spread over more lysine residues than are involved in intermolecular intractions at any moment of time. This would be consistent with the observation that more lysines are seen to be perturbed than expected.

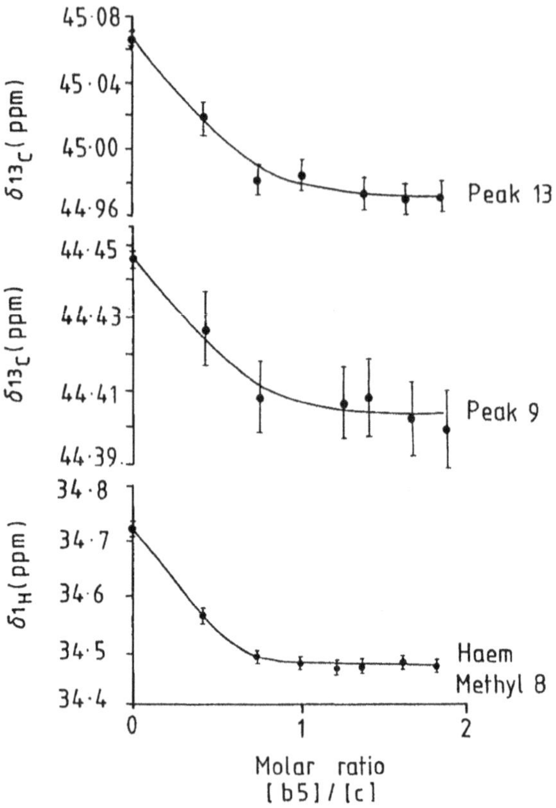

Figure 15. The change in ^{13}C chemical shifts of two dimethyllysine peaks of N^{ϵ}-dimethyllysineferricytochrome *c* as a function of the mole ratio of ferricytochrome b_5 to ferricytochrome *c* and the comparable ^{1}H chemical shift of a haem methyl resonance of N^{ϵ}-dimethyllysineferricytochrome *c*.

3.6.2 N^{ϵ}-dimethyllysine ferricytochrome c and cytochrome c peroxidase.

Similar experiments to those described above for ferricytochrome b_5 binding to N^{ϵ}-dimethyllysine ferricytochrome *c* were carried out for the binding of cytochrome *c* peroxidase. Of the 15 resolved lysines in the 2D ^{1}H-^{13}C COSY spectrum, at least 7 were shifted by the binding of cytochrome *c* peroxidase. These were dimethyl lysine peaks: 1,2,3,9,10,13 and 15. Peaks 2, 13 and 15 were also substantially broadened, probably because their motion is constrained by intermolecular contacts. There is some similarity with the dimethyl lysine peaks shifted by cytochrome b_5 (Table 1), and in both cases dimethyl lysine peaks 9 and 13 were strongly affected. However, the most suprising feature of these data is the difference in response to cytochrome b_5 and cytochrome *c* peroxidase. Interpretation of this requires firm assignment of the dimethyl lysine peaks.

4. Binding surfaces on proteins

Ion binding and protein:protein interactions occur at the surfaces of proteins, and therefore it is important to know what these look like. However, because of the dynamic properties of surfaces it is far harder to define precisely their structure than it is to define the internal structure of a protein. Indeed, since many groups on a surface possess a number of conformations that are almost energetically equivalent, it is an oversimplification to describe the surface as though it has a unique structure. An additional problem is that X-ray structures may be distorted at the surface by crystal packing effects, and NMR structures may be poorly defined at the surface by the lack of sufficient NOE constraints.

4.1 THE BINDING SURFACE ON CYTOCHROME *c*

With cytochrome *c*, most attention has been focused on the lysines, all of which are located on the surface, because it is these groups that are responsible for binding the physiological oxidoreductases [11 and references therein]. These lysines are distributed in a highly asymmetric manner with the majority clustered on the front face near the haem. There are few carboxylates on the front face. Only 7 out of 16 lysines of tuna cytochrome *c* are crystallographically well defined [7,44]. Of the remaining 9 lysines, some have blurred images in the electron density maps and some have different positions in what should be identical X-ray structures. Four of the 7 well-defined lysines are involved in interactions with other resides that presumably reduce their conformational flexibility. The surface carboxylates are generally well defined, probably because their side chains are shorter than that of lysine.

Bearing in mind the conformational flexibility of lysine, and the fact that its side chain from α-C to ε-N is about 10 Å long when fully extended, we can ask whether there are discrete anion binding sites on the surface of cytochrome *c*, as is suggested by the representation of NMR results given in Figs. 5 and 10. Fig. 16 is a diagrammatic representation of the problem. There are three binding sites on the protein, two of about equal affinity and one of lesser affinity. These are assumed to be the anion binding sites described in Fig. 5. Fig. 16 one binding site is separate but the other two sites are either indistinct [case III] or are only partially distinct [case II]. In case III the two sites have almost completely merged and it is incorrect to describe them as separate sites, though for the sake of simplicity we shall do so.

In case I, for an ion to move between sites A and B it would have to dissociate from the protein, but in the other cases it need not fully dissociate from the protein in order to move between sites, i.e., the ion could migrate on the protein surface. This might be facilitated by, for example, the two sites sharing a binding group. We do not have sufficient data to determine which case best describes anion binding to cytochrome *c*, but the most probable situation is as follows. Site 2, on the right side of the haem [Fig. 5] resembles site C in Fig.16, and sites 1 and 3 [Fig. 5] resemble sites A and B in Fig.16, case I or II. Thus, there may be some movement of bound anions between sites 1 and 3 but not to the extent suggested by Fig. 16, case III. The overall impression is that anion binding to cytochrome *c* has a considerable degree of dynamism.

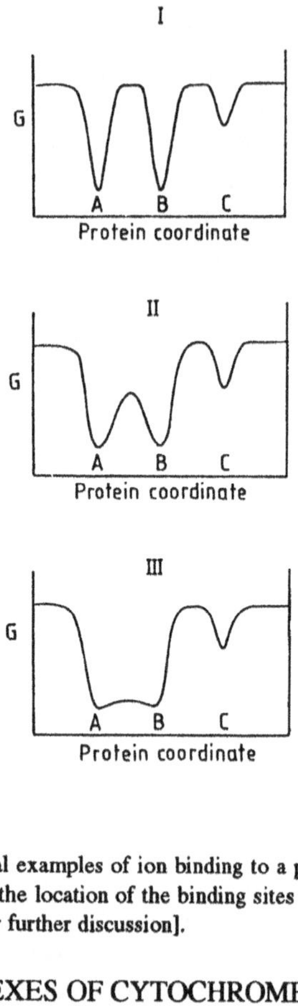

Figure 16. Three hypothetical examples of ion binding to a protein. G is the binding energy; the protein coordinate represents the location of the binding sites on the protein. The stoichiometry is three ions/protein [see text for further discussion].

4.2 PROTEIN COMPLEXES OF CYTOCHROME *c*

Similar dynamic characteristics to those described above for anion binding to cytochrome *c* have been ascribed to protein complexes of cytochrome *c*, where the phrase "rolling-ball model" was introduced to describe the relationship of two globular proteins forming a complex in which the proteins could move relative to each other [28,29]. The observations of mobility described by Eley & Moore [28] from their competitive PDS experiments, by Jeng *et al* [35] from their NH exchange experiments, and in section 3.6 from the ^{13}C NMR studies of lysine-modified cytochrome *c* are consistent with each other. Thus cytochrome *c*, in its complexes with cytochrome *c* peroxidase and cytochrome b_5 at least, exhibit considerable mobility. If a constrained system is required then the approach adopted by Moench *et al* [45], namely to study cross-linked complexes of cytochrome *c*, could be adopted. However, although chemically interesting, such complexes are not physiologically relevant.

Leaving aside the question of the mobility of interprotein complexes, the overall impression gained from the studies reported here is that the general location of the cytochrome b_5 binding site on cytochrome *c* proposed by Salemme [13] is correct. There is also considerable overlap between this region and that for cytochrome *c* peroxidase, as expected from modelling and crystallographic studies [17 and references therein].

The data we present are not consistent with models of cytochrome *c* complexes in which there is a substantial change in the cytochrome structure. In particular, the proposal of Wendoloski *et al* [15] that Phe 82 undergoes a conformational change that leads to its relocation close to the cytochrome b_5 haem in the cytochrome *c*:cytochrome b_5 complex is not supported our work [29]. There are a number of possible reasons for this discrepancy but the major one may be that the NMR studies are probing the collision complex, whereas the molecular dynamics calculations are investigating the subsequent activated complex.

5. Conclusions

A variety of NMR approaches have been described for studying the interactions of small molecules with proteins and for studying interprotein interactions. The use of small paramagnetic reagents as analogues of other small molecules (e.g. $[Cr(CN)_6]^{3-}$ to mimic O_2^- binding to superoxide dismutase) or as a competitor for proteins (e.g. $[Cr(CN)_6]^{3-}$ competing with cytochrome b_5) have been shown to be valuable procedures. These need to be incorporated into multinuclear studies of proteins where many resonance assignments have been obtained. Good definition of interaction sites requires the prior assignment of protein resonances. If resonances of binding groups such as lysines cannot be resolved in spectra of native proteins then chemical modification to produce functional, isotopically enriched groups, should be employed.

A combination of these methods ha bee shown to be effective in characterising the interaction between cytochrome *c* and cytochrome b_5, and it appears likely that the cytochrome *c*:cytochrome peroxidas conplex can be similarly defined.

Throughout the work described in the present article the emphasis has been on systems interacting relatively weakly. With proteins where there is a strong binding interaction, so that the components are in slow exchange between free and bound forms on the NMR time-scale, more conventional NMR experiments based on NOEs may be sufficient to define the structure. A major problem with the relatively weak binding systems is how are the structures of their surfaces to be described? Their inherent dynamism may mean that the search for unique complexes, which underpins many of the reported NMR studies of such systems, is misguided.

6. Acknowledgements

We thank Drs. T. Alleyne, E. Borghi, A.M. Burch, W.D. Funk, M.R. Mauk, S.E.J. Rigby and Professor J.V. Bannister for their assistance with some aspects of the project described in this manuscript. We also thank the Commission of the European Comunity, NATO [travel grant 870145] and the NIH [grant GM33804 to AGM] for their support of this work, and the Biotechnology and Biology Science Research Council and the Engineeering and Physical Sciences Research Council for their support of the UEA Centre for Metalloprotein Spectroscopy and Biology through their Biomolecular Sciences Panel.

7. References

1. Moore, G. R. and Pettigrew, G. W. (1990) *Cytochromes c: Evolutionary, Structural and Physiochemical Aspects*, Springer-Verlag, Berlin
2. Dwek, R. A. (1973) *NMR in Biochemistry*, Clarendon Press, Oxford
3. Wüthrich, K. (1976) *NMR in Biological Research:Peptides and Proteins*, North-Holland Publishing, Amsterdam
4. Jardetzky, O. and Roberts, G. C. K. (1981) *NMR in Molecular Biology*, Academic Press, New York
5. Feeney, J. and Birdsall, B. (1993) NMR studies of protein-ligand interactions in G. C. K. Roberts [ed], *NMR of Macromolecules*, IRL Press, Oxford pp. 183-215
6. Dobson, C. M. and Williams, R. J. P. (1977) NMR Studies of the Interaction of Lanthanide Cations with Lysozyme, in B. Pullman and N. Goldblum (eds), *Metal-Ligand Interactions in Organic Chemistry and Biochemistry*, D.Reidel Publishing, Dordrecht, pp 255-282
7. Moore, G. R., Eley, C. G. S. and Williams, G. (1983) Electron Transfer Reactions of Class I cytochromes *c*, *Advances in Inorg. and Bioinorg. Mech.* **3**, 1-96
8. Bannister, J. V. and Rotilio, G. (1984) A decade of superoxide dismutase activity, in J. V. Bannister and W. H. Bannister (eds) *The Biology and Chemistry of Active Oxygen*, Elsevier, New York, pp 146-189
9. Cudd, A. and Fridovich, I. (1982) Electrostatic interactions in the reaction mechanism of bovine erythrocyte superoxide dismutase, *J. Biol. Chem.*, 257, 11443-11447
10. Tainer, J. A., Getzoff, E. D., Beem, K. M., Richardson, J. S. and Richardson, D. C. (1982) Determination and Analysis of the Structure of Copper, Zinc Superoxide Dismutase, *J. Mol. Biol.*, **160**, 181-217
11. Pettigrew, G. W. and Moore, G. R. (1987) *Cytochromes c: Biological Aspects*, Springer-Verlag, Berlin
12. Koppenol, W. H. and Margoliash, E. (1982) The asymmetric distribution of charges on the surface of horse cytochrome *c*, *J Biol Chem*, **257**, 4426 - 4437
13. Salemme, F. R. (1976) An Hypothetical Structure for an Intermolecular Electron Transfer Complex of Cytochromes *c* and b_5, *J. Mol. Biol.*, **102**, 563-568
14. Mauk, M. R., Mauk, A. G., Weber, P. C. and Matthew, J. B. (1986) Electrostatic Analysis of the Interaction of Cytochrome *c* with Native and Dimethyl Ester Heme Substituted Cytochrome b_5, *Biochemistry*, 25, 7085-7091
15. Wendoloski, J. J., Matthew, J. B., Weber, P. C. and Salemme, F. R. (1987) Molecular Dynamics of a Cytochrome *c* -Cytochrome b_5 Electron Transfer Complex, *Science*, 238, 794-797
16. Rodgers, K.K., Pochapsky, T. C. and Sligar, S. G. (1988) Probing the Mechanisms of Macromolecular Recognition: The Cytochrome *c* -Cytochrome b_5 Complex, *Science*, 240, 1657- 1659
17. Pelletier, H. and Kraut, J. (1992) Crystal structure of a complex between electron transfer partners, cytochrome *c* peroxidase and cytochrome *c*, *Science* **258**, 1748-1755
18. Ragg, E. and Moore, G. R. (1984) Association constants for metal hexacyanide binding to cytochrome *c*, *J. Inorg. Biochem.* **21**, 253-261

19. Bannister, J. V., Borghi, E. and Moore, G. R. unpublished data
20. Bertini, I., Capozzi, F., Luchinat, C., Piccioli, M. and Viezzoli, M. S. (1991) Assignment of active-site protons in the ^{1}H NMR spectrum of reduced human Cu/Zn superoxide dismutase, *Eur. J. Biochem.* **197**, 691-697
21. Campbell, I. D., Dobson, C. M., Williams, R. J. P. and Xavier, A. V. (1973) Resolution Enhancement of Protein NMR Spectra Using the Difference Between a Broadened and a Normal Spectrum, *J. Mag. Reson.* **11**, 172-181
22. Eley, C. G. S., Moore, G. R.,Williams, G. and Williams, R. J. P. (1982) NMR studies of the electron-exchange between cytochrome *c* and iron hexacyanides, *Eur. J. Biochem.* **124** 295-303
23. Williams, G.,. Eley, C. G. S., Moore, G. R., Robinson, M. N. and Williams, R. J. P.(1982) The reaction of cytochrome *c* with [Fe(edta)]$^-$, *FEBS Letters* **150**, 293-299
24. Bax, A. and Grzesiek, S. (1993) Methodological Advances in Protein NMR, *Accounts Chem. Res.* **26**, 131-138
25. Markley, J. L. and Kainosho, M. (1993) Stable isotope labelling and resonance assignments in larger proteins in G. C. K. Roberts [ed], *NMR of Macromolecules*, IRL Press, Oxford pp. 101-152
26. Osborne, M. J., Lian, L-Y., James, R., Kleanthous, C. and Moore, G. R. unpublished data
27. Mauk, M. R., Reid, L. S. and Mauk, A. G. (1982) Spectrophotometric Analysis of the Interaction between cytochrome *c* and cytochrome b_5, *Biochemistry* **21**, 1843-1846
28. Eley, C. G. S. and Moore, G. R. (1983) ^{1}H NMR investigation of the interaction between cytochrome *c* and cytochrome b_5, *Biochem. J.* **215**, 11-21
29. Burch, A. M., Rigby, S. E. J., Funk, W. D., MacGillivray, R. T. A., Mauk, M. R., Mauk, A. G. and Moore, G. R. (1990) NMR characterization of surface interactions in the cytochrome b_5,-cytochrome *c* complex, *Science*, **247**, 831-833
30. Thomas, M-A., Delsuc, M. A., Beloil, J-C. and Lallemand, J. Y. (1987) ^{1}H NMR investigation of yeast cytochrome *c*. Interaction with the corresponding specific reductase, *Biochem. Biophys. Res. Commun.* **145**, 1098-1104
31 Gupta, R. K. and Yonetani, T. (1973) NMR studies of the interaction of cytochrome *c* with cytochrome *c* peroxidase, *Biochim. Biophys. Acta.* **292**, 502-508
32. Boswell, A. P., McClune, G. J., Moore, G. R., Williams, R. J. P., Pettigrew, G. W., Inubishi, T.,Yonetani, T. and Harris, D. E. (1981) NMR study of the interaction of cytochrome *c* with cytochrome *c* peroxidase, *Biochem. Soc. Trans.*, **8**, 637-638
33. Puettner, I., Carafoli, E. and Malatesta, F. (1985) Structural and Functional Properties of Cytochrome *c* Oxidase Isolated from Sharks, *J. Biol. Chem.*, **255**, 2722-2729
34 Rigby, S. E. J., Alleyne, T., Wilson, M. T. and Moore, G. R. (1989) A ^{1}H NMR study of bovine cytochrome oxidase: paramagnetically shifted resonances of haem *a*, *FEBS Lett.* **257**, 155-158
35. Jeng, M-F., Englander, S. W., Pardue, K., Rogalskyj, J. S. and McLendon, G. (1994) Structural dynamics in an electron-transfer complex, *Nature Structural Biology* **1**, 234 - 238
36. Bosshard, H. R., Banziger, J., Hasler, T. and Poulos, T. L. (1984) The cytochrome *c* -cytochrome *c* peroxidase complex-the role of histidine residues, *J. Biol. Chem.* **259**, 5683-5690
37. Hartshorn, R. T., Mauk, A. G., Mauk, M. R. and Moore, G. R. (1987) NMR study of the interaction between cytochrome b_5 and cytochrome *c*: observation of a ternary complex formed by the two proteins and $[Cr(en)_3]^{3+}$, *FEBS Letters* **213**, 391-395
38. Ng, S., Smith, M. B., Smith, H. T. and Millet, F. (1977) Effect of modification of individual cytochrome *c* lysines on reaction with cytochrome b_5, *Biochemistry* **16**, 4975-4978
39. Smith, M. B., Stonehuerner, J., Ahmed, A. J., Staudenmayer, N. and Millet, F. (1980) Use of specific trifluoroacetylation of lysine residues in cytochrome c_1 to study the reaction with cytochrome b_5, cytochrome c_1 and cytochrome oxidase, *Biochim. Biophys. Acta* **592**, 303-313
40. Whitford, D. (1992) The identification of cation-binding domains on the surface of microsomal cytochrome b_5 using ^{1}H NMR paramagnetic difference spectroscopy, *Eur. J. Biochem.* **203**, 211-223
41. Cox, M. C., Osborne, M. J., Thurgood, A. G. P. and Moore, G. R. (1991) Structural characterisation Of interprotein complexes involving cytochrome *c*, Abtsr. ICBIC-5, *J. Inorg. Biochem.* **43**, 106
42. Feng, Y., Roder, H., Englander, S. W., Wand, A. J. and DiStefano, D. L. (1989) Ptoton NMR Asignments of Horse ferricytochrome *c*, *Biochemistry* **28**, 195 - 203

43. Zhang, M. and Vogel, H. J. (1993) Determination of the side chain pK_a values of the lysine residues in calmodulin, *J. Biol. Chem.* **268**, 22420 - 22428
44. Takano,T. and Dickerson, R. E. (1981) Conformation change of cytochrome *c*, *J. Mol. Biol.* **153**, 79 - 115
45. Moench, S. J., Chroni, S., Lou, B-S., Erman, J. E. and Satterlee, J. D. (1992) Proton NMR comparison of noncovalent and covalently cross-linked complexes of cytochrome *c* peroxidase with horse, tuna and yeast ferricytochromes *c*, *Biochemistry* **31**, 3661-3670

RECOMBINANT PERDEUTERATED PROTEIN AS AN EFFICIENT METHOD FOR MAKING UNAMBIGUOUS HEME PROTON RESONANCE ASSIGNMENTS: CYANIDE-LIGATED *GLYCERA DIBRANCHIATA* MONOMER METHEMOGLOBIN COMPONENT IV AS AN EXAMPLE

STEVE L. ALAM
DAVID P. DUTTON
JAMES D. SATTERLEE
Department of Chemistry
Washington State University, Pullman, WA 99164-4630, USA

Abstract

In order to overcome the difficulties of selectively assigning the heme proton resonances in paramagnetic low-spin heme proteins, a method involving perdeuteration of the globin has been developed. This method allows rapid proton assignments of the heme prosthetic group to be made, however its use is restricted to proteins for which a suitable expression system exists. As an example of this method, the process of making complete heme proton assignments of the cyanide-ligated *Glycera dibranchiata* monomer hemoglobin Component IV in both the naturally protonated, native protein and the recombinant, perdeuterated protein is presented. There are many potential uses of this method aside from heme-containing proteins.

1. Introduction

Making unambiguous proton hyperfine resonance assignments in magnetically anisotropic low-spin ferriheme proteins is often a difficult task because many of these resonances occur in the very dense spectral region between 10 ppm and -2 ppm. In the specific case of cyanide-ligated *Glycera dibranchiata* monomer methemoglobin Component IV (GMH4CN), these normal difficulties are exacerbated by an inherently low proton spectral dispersion [1,2]. The total proton hyperfine shift region of GMH4CN spans only about 60% that of sperm whale metmyoglobin-cyanide (SWMbCN), a protein of comparable size to GMH4CN. This means that many more of the hyperfine-shifted proton resonances are difficult to locate in this protein because they are buried in the dense spectral region dominated by the overlapping resonances of ~1400 diamagnetic protons.

Additionally, there are two further complicating features of GMH4CN proton NMR spectra. The first of these is the fact that apparently homogeneous protein solutions of GMH4CN display a second set of spectra from a minor

G.N. La Mar (ed.), *Nuclear Magnetic Resonance of Paramagnetic Macromolecules*, 123-140.

protein form [2,3]. The minor form accounts for about 15% of the total amount of GMH4CN in solution and is a potentially confusing and complicating factor when proton assignments are being sought. The second complication is the presence of internal dynamic processes that cause temperature dependent line broadening of several of the hyperfine-shifted proton resonances [1].

The combination of these factors has made even the relatively selective task of identifying just the heme and axial ligand proton resonances in the GMH4CN spectrum very complicated. Using a brute force approach over the course of two years of shared instrument time we were able to identify a self-consistent group of proton hyperfine-shifted resonances that we felt could be confidently assigned as the complete heme and proximal histidine resonances of GMH4CN [1]. Of the 29 protons that occur on the heme and proximal histidine our initial assignments were, to varying degrees, ambiguous for only two. We have subsequently attempted to resolve those ambiguities using a perdeuterated, expressed version of native GMH4CN [4].

Achieving unambiguous assignment of the heme and proximal histidine resonances is of significance to us for two primary reasons. First, is so that we may have a rapid assay of heme pocket integrity for the mutant recombinant proteins that are currently being produced. We have long felt that the hyperfine-shifted proton spectrum would be an especially good indicator of the overall similarity of heme pockets between parent and mutant GMH4 proteins. We have felt this because of the demonstrated sensitivity of proton hyperfine shifts to even relatively small changes in structural and magnetic environments [1,3,5]. Second, is that a complete set of heme and proximal histidine proton assignments is the first step in subsequently making assignments of other hyperfine-shifted protons that belong to amino acids situated within the heme pocket.

After considering several strategies we decided to explore expression of fully deuterated monomer Component IV globin (ie GMG4, the apoprotein). Following purification, subsequent constitution of the perdeuterated globin with naturally protonated heme to form the perdeuterated holoprotein, and ligation with KCN, 1D and 2D proton NMR experiments have been used to unambiguously detect and assign all heme and proximal histidine resonances. Prior reports of aspects of this work have already appeared [1,4], and in this manuscript we wish to elaborate details of the methodology.

2. Experimental

2.1. SAMPLE PREPARATION

2.1.1. *Native GMH4 and Sperm Whale Myoglobin*

Native oxidized Component IV monomer hemoglobin (metGMH4) was prepared from *Glycera dibranchiata* erythrocytes as previously described (1,7,8), but with the modification that for these experiments Components II and III were rapidly eluted from the CM-Sepharose (Sigma) cation exchange column, without care for their separation, by a rapidly applied linear gradient which ranged from 0 to 10 mM KCl, at pH 6.8. Component IV was subsequently eluted from the column using a very shallow gradient that ran up to 20 mM KCl. Myoglobin (Sperm Whale, Sigma) was used without further purification.

2.1.2. *Samples For Proton NMR Spectroscopy*
Sample preparation, regardless of the protein, followed the protocol previously described [1,2,3,7]. In general, a protein sample was re-oxidized with a small amount of potassium ferricyanide (Mallinckrodt) following collection from the CM-Sepharose column and concentration. Next, the sample was treated with excess KCN (Mallinckrodt) and extensively washed in an Amicon pressure ultrafiltration cell. Sample buffers included: (a) H_2O-buffer, consisting of 100 mM potassium phosphate/100 mM KCl/20 mM KCN at pH = 6.82 in 90% H_2O/10%D_2O (Isotec, 99.9%); or (b) D_2O-buffer, consisting of the same salt concentrations dissolved in D_2O (99.9%, Isotec), pH' 6.41. The pH of these solutions was monitored with a calibrated Fisher combination electrode and a Fisher Accumet 925 meter. All samples of the native proteins had concentrations between 3 and 4 mM. The perdeuterated GMH4CN sample was 280 μM, as determined spectrophotometrically.

2.2. RECOMBINANT DNA METHODS

Recombinant native GMG4 was expressed in *E. coli* strain BL21(DE3) pLysS using a cDNA created from the previously cloned cDNA for a related monomer hemoglobin, rec-gmg [5,9,10]. This bacterial strain carries the DE3 immunity region of phage 21 which has been inserted into the int gene, and contains the LacI gene, the LacUV5 promoter, the beginning of the LacZ gene and the gene for the T7 RNA polymerase. Transcription of the T7 RNA polymerase is directed by the LacUV5 promoter, which in turn is induced by isopropyl-β-D-thiogalactopyranoside (IPTG).

Detailed experimental protocols have been given elsewhere [5]. Briefly, the native GMG4 sequence-containing plasmid pET3d:GMG4 was created by ligating the excised fragment of cDNA coding for rec-gmg from the plasmid pIBI76:GMG4 and ligating it into the multiple cloning region of Pst1-cut M13mp19 phage so that the 3' end is next to the BamH1 site. The correct orientation of the insert was proved by Sanger dideoxy-sequencing methods using the protocols from the kit supplied by US Biochemicals. The three mutations needed to transform the rec-gmg cDNA into GMG4 cDNA were made using single-stranded site-directed mutagenesis [11]. The three mutations were made by simultaneously annealing the mutagenic oligonucleotides (all about 25 bases in length) to a uracil-containing template, then generating the complementary strand with T4 DNA polymerase. The reaction mixture was next transformed into JM101(f+) and the secreted phage were purified. Mutations at all three positions were confirmed by dideoxy-DNA sequencing. Next, the replicating form of phage M13mp19:recGMG4 was isolated and directionally subcloned into the T7 expression plasmid pET3d [12-14] using the restriction enzymes Nco1 and BamH1, resulting in the plasmid pET3d:GMG4. Expression (Figure 4) was carried out using either of the following culture media: (A) 2XYT (1.6% Bacto-tryptone, 1% yeast extract and 0.5% NaCl); or (B) tryptonephosphate (2% Bacto-tryptone, 0.2% sodium hydrogen phosphate, 0.1% potassium dihydrogen phosphate, 0.8% NaCl, 1.5% yeast extract and 0.2% glucose). The general method for expression consisted of diluting an overnight culture 1/100 into fresh medium, monitoring the growth progress by optical spectroscopy until the OD_{600} reached 0.8, then reducing the temperature to 21 °C and initiating expression by adding IPTG to the desired concentration. At the end of the induction period the cells were collected, lysed by sonication, and the supernatant collected and dialyzed against 10 mM potassium phosphate buffer, pH 6.0. A solution of hemin chloride (Porphyrin Products) was added at this point to constitute the holoprotein, GMH4. The GMH4 was separated from other proteins in the lysate by ion exchange chromatography on a CM-Sepharose column which was equilibrated with the dialysis buffer. GMH4 was eluted with a shallow concentration gradient of this buffer ranging from 10-20 mM. The GMH4 fraction was brought to 1.2 M ammonium sulfate and stirred overnight to precipitate additional contaminants. The solution was clarified by centrifugation and desalted by passage down a Sephadex G-50 column that had been equilibrated with 100 mM potassium phosphate/100 mM potassium chloride buffer, pH 6.8. The general expression, isolation and purification scheme is given in Figure 5.

Perdeuterated recombinant GMG4 (DrecGMG4) was isolated by growing the bacteria and inducing globin expression in a deuterated algal hydrolysate in D_2O (Martek Corp). In this case it was necessary to adapt the bacteria for growth in D_2O medium prior to actual induction. Additional details have been previously described [4]. Constitution and purification of the expressed perdeuterated GMG4 to form DrecGMH4 followed the steps described above and in Figure 5. It should be noted that the protein was handled in H_2O solutions following lysis and prior to constitution with heme (steps 5-9, Figure 5). Otherwise the protein was maintained in D_2O solutions.

2.3. NMR METHODS

All experiments were homonuclear proton NMR spectroscopy. A variety of 1D and 2D methods were carried out as described more extensively, elsewhere [1,4]. Briefly, the experiments consisted of : standard 1D-NOE difference spectroscopy; SuperWEFT [15]; NOESY [16]; phase-sensitive and magnitude COSY[17]; and phase-sensitive TOCSY [18]. NOESY paramaters are typical of the instrumental conditions: 1024 data points in t_2; 512 t_1 increments; 256 transients per t_1 block; a spectral width of 20 kHz; 30 msec mixing time; hypercomplex acquisition [19]. Zero-filling to final data matrix sizes of 2048 x 2048 points was frequently carried out. The spectra were acquired on a Varian VXR500s spectrometer operating in quadrature detection mode at a nominal frequency of 500 MHz. Typically, the residual water resonance was presaturated during the relaxation delay period using the decoupler channel. Observed shifts were referenced to external DSS by assigning the residual water resonance a shift of 4.70 ppm.

3. Results and Discussion

3.1. BACKGROUND

The marine annelid *Glycera dibranchiata* contains erythrocytes from which three major monomer hemoglobins have been consistently isolated [6-8]. These monomer hemoglobins are labeled Components II, III and IV. The holoproteins are abbreviated GMH2, GMH3, and GMH4, respectively. They each resemble myoglobin in size and structure [6-8], although the sequence homology between any of the Component Monomer Hemoglobins and SWMb is low (~18%) [9,10], there is apparently a strong global structural similarity [20]. The feature of primary interest in these proteins is that it has now been established that all three Monomer Hemoglobin Components lack a distal histidine [9,10,20,21]. Instead, each has a significantly altered heme ligand binding site in which the distal amino acid situated in the normal histidine position is leucine. This substitution has obvious implications for ligand binding and the kinetic data currently available on our protein preparations suggests that ligand binding is, indeed, anomalous [22,23]. Although work in this laboratory is progressing on all three Components, the focus of this paper will be on Component IV.

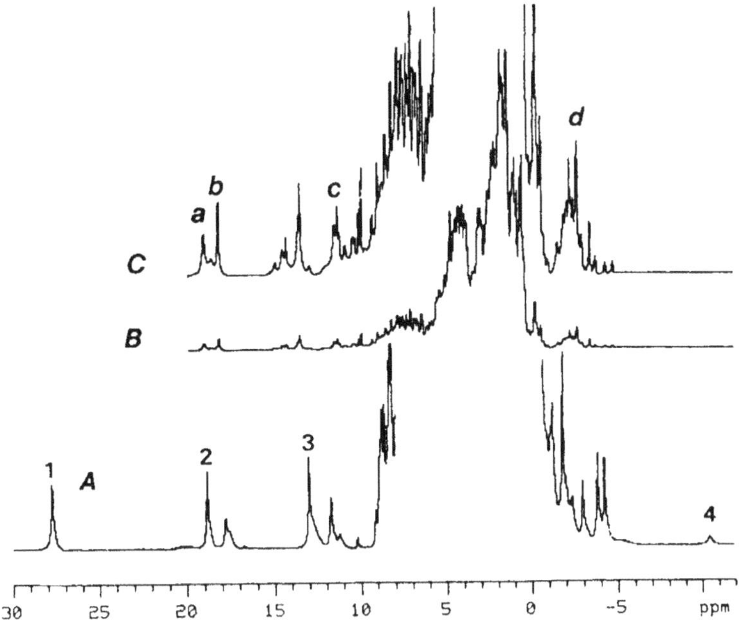

Figure 1. Proton NMR spectra at 500 MHz. (A) Sperm whale metmyoglobin-CN with the following resonances labeled: 1) heme $5CH_3$; 2) heme $1CH_3$; 4) I99 CγH. (B) Native metGMH4CN. (C) 7x vertical expansion of GMH4CN spectrum shown in (B) with the following resonance assignments: (a) heme $8CH_3$; (b) heme $3CH_3$; (c) an envelope of overlapping resonances containing (among others): heme 4αH; H90βH and heme 7α1H; (d) an envelope of overlapping heme resonances (among others) containing: heme 6βH, 7β1H, 7β2H, 6α2H, 4βHcis and 4βHtrans. Conditions were identical for each sample: pH' = 6.8, 20 °C, D_2O-buffer.

3.2. PROTON NMR SPECTRA OF NATIVE GMH4CN

Monomer Hemoglobin Component IV has been the subject of several NMR studies in the past [1-5,24,25]. The proton NMR spectrum of cyanide-ligated GMH4CN is generally similar to that of other low-spin met-heme proteins, with several obvious hyperfine-shifted resonances outside the 10 to -2 ppm envelope. This is illustrated in Figure 1 which compares the proton NMR spectra of GMH4CN and SWMbCN taken under identical conditions. However, closer inspection reveals that, compared to SWMbCN, there are significant differences. Lower spectral dispersion (Fig. 1), and variable temperature linewidths (*c.f* Fig. 2-peaks at 14-15 ppm) are characteristic of GMH4CN. In particular, preliminary analysis of variable temperature behavior, such as that shown in Figure 2, indicates that there is substantially more local motional mobility of active-site amino acid side chains than occurs in other met-heme proteins. The sources of this dynamic behavior have been tentatively identified as two phenylalanines.

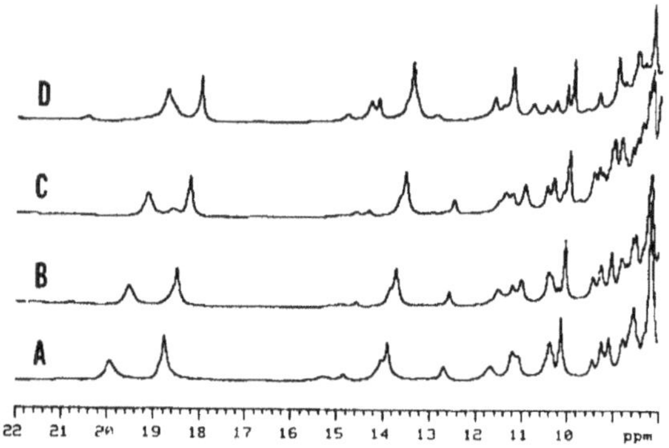

Figure 2. 500 MHz proton NMR spectra of the high frequency hyperfine shift region of native GMH4CN at (A) 10, (B) 15, (C) 20, and (D) 25 °C. The sample was in D_2O buffer at pH' = 6.93.

Initial attempts to overcome the lower hyperfine-shift dispersion in order to advance the status of resonance assignments focused on taking advantage of the differential relaxation properties of the hyperfine-shifted protons. Experiments like the WEFT spectrum shown in Figure 3 identified several rapidly relaxing resonances both in the hyperfine shift region and under the diamagnetic envelope (labeled u-z in Fig. 3). For the hyperfine shift region several broad non-heme resonances (~21 ppm and 15-11 ppm) relax as fast, or faster than the assigned heme resonances (see Fig. 1). In the region covered by the diamagnetic envelope, as more precise experiments evolved, it became clear that the resonances labeled u-z in Figure 3 were primarily due to amino acid side-chains.

Subsequently a full suite of homonuclear proton 1D NOE, NOESY, COSY and TOCSY experiments were required for obtaining complete heme and proximal histidine assignments [1]. This was a lengthy task because each of these experiments was carried out at multiple temperatures and various pHs. Also, sets of experiments were carried out both in H_2O and D_2O solutions. In addition, NOE build-up experiments were carried out at multiple temperatures. Together these experiments have led to a self-consistent set of resonance assignments for all of the heme and proximal histidine protons of Monomer Hemoglobin Component IV, with two of the set remaining tentative [1]. A Figure illustrating the results of a NOESY experiment is presented later.

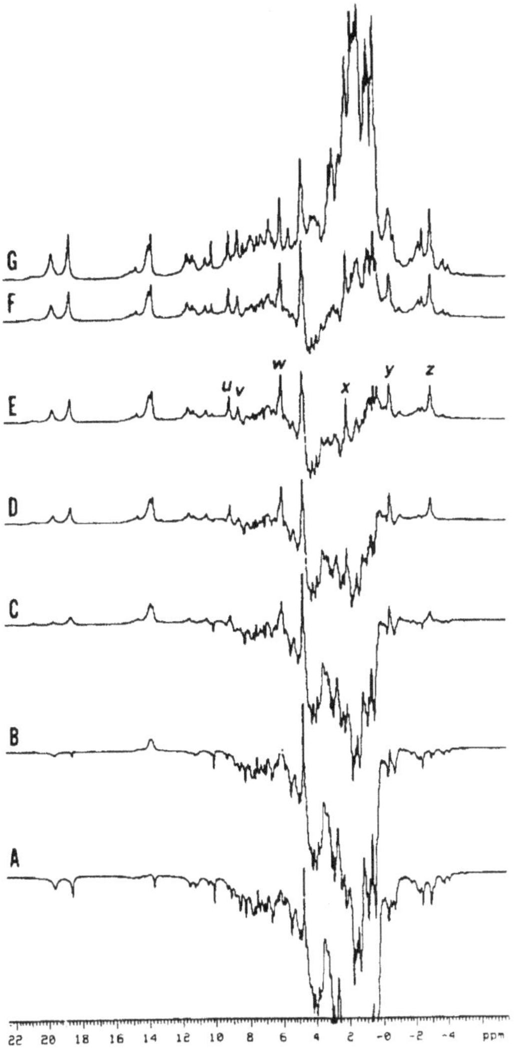

Figure 3. 500 MHz Proton superWEFT calibration spectra for native GMH4CN at 20 °C, pH' = 6.81 in D_2O buffer. Interpulse delay times are (A) 20 ms, (B) 40 ms, (C) 60 ms, (D) 80 ms, (E) 100 ms, (F) 120 ms, (G) 200 ms. Labeled peaks correspond to the most obvious, fast-relaxing non-heme resonances occurring in the diamagnetic envelope.

3.3. RECOMBINANT GMH4

3.3.1. *Preparation of Recombinant Native GMH4CN*

In parallel with the NMR work described above we had also been successful in cloning the *Glycera dibranchiata* Monomer Globin Component IV (GMG4) gene

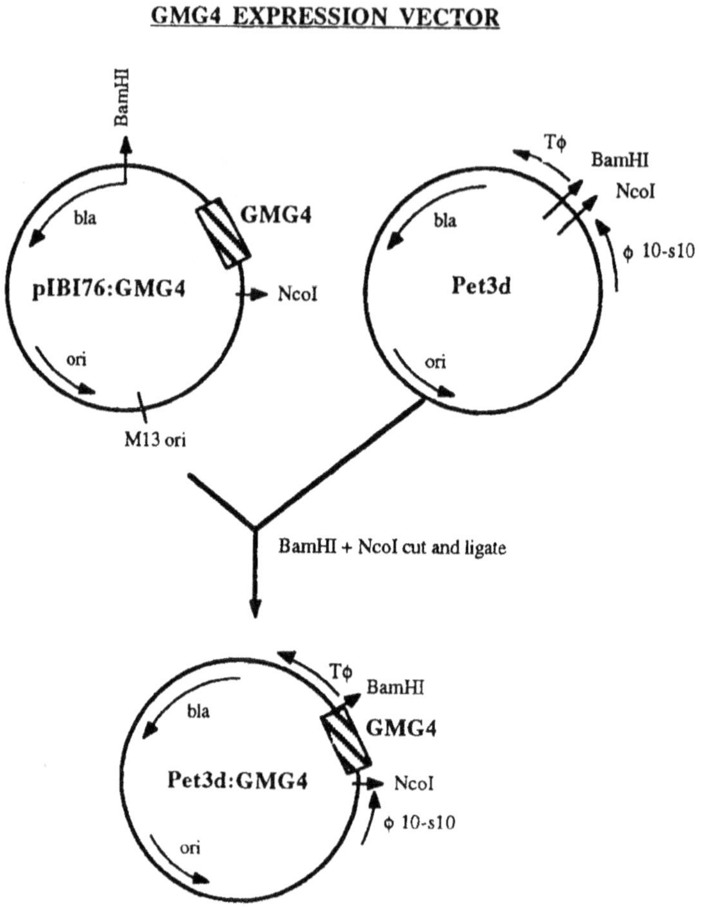

Figure 4. A GMG4 expression vector was created by excising the GMG4-containing sequence from the plasmid pIBI76:GMG4 and ligating it into pET3d. The resultant plasmid pET3d:GMG4 was then used to transform *E. coli* strain BL21(DE3) pLysS. Expression was induced by IPTG. Inductions were run at 21 °C so that expressed GMG4 would remain soluble in the bacterial cytosol.

into a system for mutagenesis, and high level expression based upon the pET3d plasmid [12-14]. Figure 4 shows a plasmid diagram and outlines the expression scheme now commonly in use with GMG4 which involves the *E. coli* strain BL21(DE3)pLysS. Recombinant native GMG4 has been isolated from this system to the extent of ~40 mg of protein per liter of induced growth medium. The recombinant globin can be constituted with hemin chloride to form the holoprotein, GMH4, whose characterization is described below.

When considering whether an expression system could play a central role in resolving the ambiguities in the Monomer Hemoglobin Component IV proton hyperfine-resonance assignments, two ideas emerged. First attempts focused on exploiting a heme-protein expression system that, unlike our BL21/pET3d system, possessed the property of having heme biosynthesis occur at a rate comparable to polypeptide synthesis so that a complete holoprotein was produced by the bacteria. The intention was to express this protein in such a manner that uniformly ^{13}C enriched heme could be harvested and used to constitute holoGMH4 from recombinantly expressed GMG4. Heteronuclear correlation experiments could then lead to the heme proton (and carbon) assignments. This effort was unsuccessful.

3.3.2. *Recombinant Perdeuterated Native GMG4*

The second idea was to create perdeuterated Component IV globin from our BL21/pET3d expression system using a >90% deuterated medium. This effort was successful and resulted in a purified sample of recombinant perdeuterated GMH4CN (DrecGMH4CN) following constitution of the perdeuterated globin with naturally protonated hemin chloride and ligation with KCN.

The idea to perdeuterate a protein is not new. The essence of the idea was contained in two 1968 papers which reported studies carried out on proteins harvested from bacteria and algae grown on deuterated medium [26,27]. More recently, Forsen and co-workers [28] have prepared selectively protonated calbindin D_{9k} using a bacterial expression system grown on 99% D_2O medium, with added protonated amino acids. Also, random fractional deuteration to the extent of 50-75% deuterium has been used to simplify NMR spectra [29,30]. The difference between our work (presented here) and the previous work is that we have expressed a perdeuterated protein. Instead of using protonated amino acids added to the bacterial growth medium for the purpose of generating a selectively protonated protein, we have used the naturally protonated heme group as our "selective" proton spectroscopic probe.

3.4. CHARACTERIZATION OF NATIVE AND RECOMBINANT GMH4

Prior to expressing recombinant perdeuterated GMG4 it was necessary to express, purify and constitute the naturally protonated holoprotein: recombinant GMH4. The flow chart in Figure 5 illustrates the general procedure. Although Figure 5 specifically depicts how the naturally protonated recombinant native GMH4 was prepared and purified, the recombinant perdeuterated protein was similarly prepared, as outlined in section 2.2 and the Figure 5 caption.

For purposes of characterizing the expressed protein, our experience to-date has shown that one-dimensional proton NMR spectra of the high-spin paramagnetic ferric forms of expressed ferriheme proteins are excellent initial indicators of heme pocket integrity and sensitive to even small perturbations of the heme active site. Thus, the virtually identical pattern of proton hyperfine shifts of the recombinant native metGMH4 (recGMH4) and native metGMH4 (GMH4)

PURIFICATION AND CONSTITUTION OF RECOMBINANT GMG4

(1) Grow BL21(DE3)pLysS pET3d:GMG4 to OD600=0.8
at 37 degrees{AMP=50ug/ml CAM=25ug/ml}

↓

(2) Switch to 21 degrees for 10 minutes to cool down to induction temperature

↓

(3) Add IPTG to a final concentration of 0.4mM and induce at 21 degrees
for 3-5 hours

↓

(4) Harvest cells by centrifugation at 4500 RPM for 20 minutes

↓

(5) Resuspend cells in lysis buffer and freeze -70

↓

(6) Thaw at room temperature and shake for 30 minutes for Lysozyme action

↓

(7) Sonicate three times for 30 seconds on ice to disrupt cells and DNA

↓

(8) Centrifuge at 10,000 x g for 30 minutes to clarify

↓

(9) Dialyze against 10mM KP pH=6.0, 24 hrs.

↓

(10) Constitute with 10 time excess Hemin in same buffer with
1 mg/ml Potassium Ferricyanide to keep oxidized.

↓

(11) Cation exchange chromatography (CM-Sepharose)
5 x 85 cm column loaded at pH=6.0 to concentrate
protein at top; pH gradient to 6.8, salt gradient to 50 mM KCl
to wash off GMG4; then 100mM to wash column.

↓

(12) Gel filtration using Sephacryl S-200
7.5 x 80 cm column.

Figure 5. Flow diagram indicating the steps in the isolation and purification of recombinant GMH4. In the case of the perdeuterated recombinant native GMH4 steps 1-3 have been carried out in D_2O solutions. Steps 4-12 have been carried out in H_2O. In H_2O the protein's NHs and OHs will undergo isotope exchange, however after step 12, either soaking the purified protein in D_2O solution or, in the case of preparation for NMR studies, extensive washing with D_2O-buffer solutions in a pressure ultrafiltration cell, will reverse this exchange.

shown in Figure 6 indicates to us that bacterial expression of the GMG4 gene, constitution with protohemin IX, and purification have resulted in a protein with a heme pocket essentially identical to that of the native protein. We further infer from this data and from optical spectroscopy results (not presented) that essential structural similarity exists between native and recombinant GMH4. Interestingly, the minor form of the protein, represented most obviously by one of the minor form heme methyl resonances at ~78.2 ppm, is also present in the recombinant native metGMH4. Since this gene product is the result of a single cDNA sequence, the observation of a similar amount of minor form in the expressed protein strongly supports arguments based upon previous chemical evidence [3,6-8], that this minor form is not a result of the presence of a different (albeit, related) protein in solution. Rather, we believe that this indicates a conformational heterogeneity that is inherently present in otherwise homogeneous preparations of GMH4. The structural source of this minor protein form is currently under investigation, but we speculate at this time that it derives from an earlier folding intermediate.

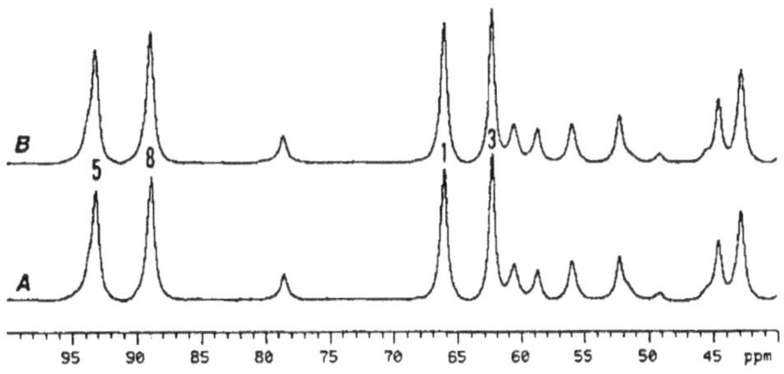

Figure 6. 500 MHz proton 1D spectra showing the high frequency hyperfine shift region of the high-spin ferric monomer hemoglobins: (A) native GMH4; and (B) recombinant GMH4. Both samples were in D_2O-buffer at 20 ^oC, pH' = 6.8. Heme methyl assignments for (A) are labelled.

With confidence that our expression system could produce a recombinant native GMH4 whose spectral characteristics were identical to native GMH4, the perdeuterated protein was expressed, and processed according to the flow diagram given in Figure 5, with the appropriate use of deuterated buffers, solutions and medium, as described in section 2.2 and the Figure 5 caption. The spectral similarities between the native and recombinant perdeuterated proteins are shown in Figure 7, which compares the one-dimensional proton NMR spectra of native GMH4CN (Fig.7A) and DrecGMH4CN (Fig. 7B). It is obvious that the extent of deuteration achieved is significant by realizing that the two spectra are plotted so that the highest intensity peak in the 0-2 ppm region in each spectrum has the same vertical scale. Clearly, the proton hyperfine resonances of DrecGMH4CN are significantly more intense than those of native GMH4CN, indicating the major reduction in intensity of the 10 to -2 ppm region as a result of the deuteration. Using intensity measurements of proton NMR spectra such as shown in Figure 7 we estimate the extent of deuteration to be approximately 90%.

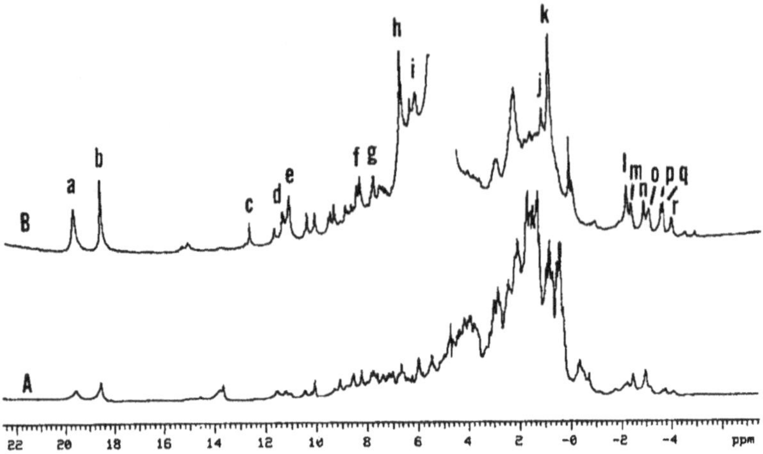

Figure 7. Complete proton one-dimensional spectra taken at 500 MHz of: (A) naturally protonated native GMH4CN; and (B) recombinant perdeuterated GMH4CN. Sample conditions were identical for each protein: 100 mM potassium phosphate/100 mM potassium chloride/20 mM KCN in 99.99% D_2O, pH' = 6.42. The concentrations are (A) 2.5 mM and (B) 0.280 mM. The traces are scaled so that the tallest peak in the 0-2ppm region of each is plotted at equal height. Assignments shown are due to heme protons unless otherwise specified: a) $8CH_3$; b) $3CH_3$; c) H90NpH; d) 4αH; e) 7α1H; f) 6α1H; g) 2αH; h) $5CH_3$; j) 6β2H; k) $1CH_3$; l) 6β1H & 2βHtrans; m) 7β1H; n) 4βHtrans; o) 7β2H; p) 2βHcis; q)4βHcis; r) 6α2H.

3.5. ASSIGNMENTS MADE EASY

The effect of globin deuteration on the efficiency of making heme hyperfine proton resonance assignments is dramatic, as illustrated in Figures 8 and 9. Figure 8 presents MCOSY spectra executed identically for naturally protonated, native GMH4CN (Fig. 8A) and DrecGMH4CN (Fig. 8B). Figure 9 presents phase sensitive NOESY spectra executed identically for naturally protonated, native GMH4CN (Fig. 9A) and DrecGMH4CN (Fig. 9B).

Comparison of the NOESY contour plots offers the most dramatic illustration of how the recombinant perdeuterated protein is advantageous for making heme proton assignments. In the case of naturally protonated, native GMH4CN (Fig. 9A) several cross peak assignments had to be made in severely conjested spectral regions. For example, the heme 5-CH_3, 1-CH_3, 2-vinyl group and all of the heme meso protons. In fact, as reported [1] it was felt that at the completion of assignment efforts with the native, naturally protonated GMH4CN a complete, self-consistent set of heme and proximal histidine assignments had been identified with only two ambiguous assignments. These were the heme 1-CH_3 and δ-meso proton assignments (Fig. 9A).

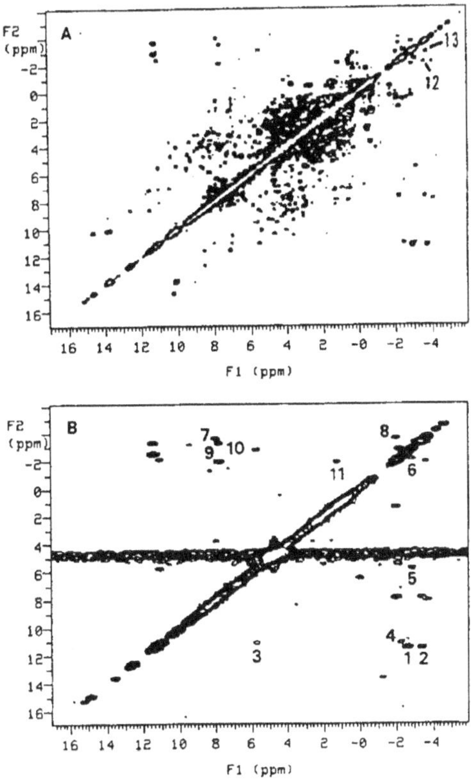

Figure 8. Proton homonuclear magnitude COSY contour plots of data collected at 500 MHz. (A) Symmetrized data for the naturally protonated native GMH4CN taken at 15 °C, pH' 6.8. (B) Unsymmetrized data for recombinant perdeuterated GMH4CN taken at 20 °C, pH 6.8. Both samples were maintained in D_2O-buffer. Cross peak assignments are: (in B): 1) 4α/4βtrans; 2) 4α/4βcis; 3) 7α1/7α2; 4) 7α1/7β2; 5) 7α2/7β2; 6) 7β1/7β2; 7) 6α1/6α2; 8) 6α2/6β1; 9) 2α/2βtrans; 10) 2α/2βcis; 11) 6β1/6β2; (also, shown in A, but seen only in lower contours in B): 12) 2βtrans/2βcis; 13) 4βtrans/4βcis.

Perdeuteration of the globin (DrecGMG4) has the advantage of simplifying the COSY and NOESY spectra of DrecGMH4CN to the point where complete heme proton assignments are easily made. This is obvious in Figure 8 by the elimination of cross peak density due to the diamagnetic envelope in Figure 8B compared to Figure 8A. However, even more dramatic are the results presented in Figure 9. In Figure 9B there is a remarkable simplification of the cross peak pattern compared to Figure 9A. Again, this is primarily due to the reduction or elimination of much of the cross peak intensity due to the diamagnetic spectral region. NOE cross peaks are clearly identified and a careful analysis of the spectrum in Figure 9B shows that one can trace a connectivity

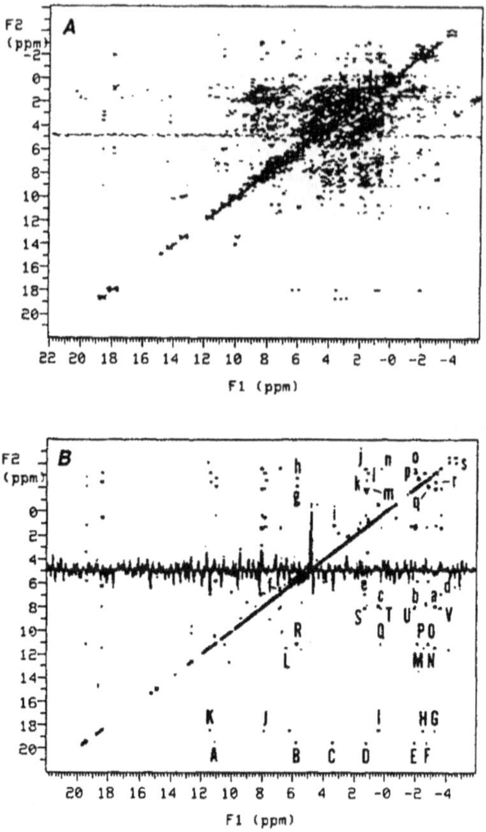

Figure 9. Proton Homonuclear NOESY contour plots of data collected at 500 MHz using a 100 ms mixing time. (A) Unsymmetrized data for the naturally protonated native GMH4CN taken at 30 °C. (B) Unsymmetrized data for recombinant perdeuterated GMH4CN taken at 20 °C. Both samples were maintained in D_2O buffer, pH' = 6.8. Assignments: A) $8CH_3/7\alpha1$; B) $8CH_3/7\alpha2$; C) $8CH_3/\delta$meso; D) $8CH_3/1CH_3$; E) $8CH_3/7\beta1$; F) $8CH_3/7\beta2$; G) $3CH_3/4\beta$cis; H) $3CH_3/4\beta$trans; I) $3CH_3/\alpha$meso; J) $3CH_3/2\alpha$; K) $3CH_3/4\alpha$; L) $4\alpha/\beta$meso; M) $4\alpha/4\beta$trans; N) $4\alpha/4\beta$cis; O) $7\alpha1/7\beta2$; P) $7\alpha1/7\beta1$; Q) $7\alpha1/\gamma$meso; R) $7\alpha1/7\alpha2$; S) $6\alpha1/6\beta2$; T) $6\alpha1/\gamma$meso; U) $6\alpha1/6\beta1$; V) $6\alpha1/6\alpha2$; a) $2\alpha/2\beta$cis; b) $2\alpha/2\beta$trans; c) $2\alpha/\alpha$meso; d) $5CH_3/6\alpha2$: e) $5CH_3/6\beta2$; f) $5CH_3/\beta$meso; g) $7\beta1/7\alpha2$; h) $7\beta2/7\alpha2$; i) $1CH_3/\delta$meso: j) $6\alpha2/6\beta2$; k) $6\beta1/6\beta2$; l) 2βcis$/1CH_3$; m) 2βtrans$/1CH_3$; n) $6\alpha2/\gamma$meso; o) $6\alpha2/6\beta1$; p) 2βcis$/2\beta$trans; q) $7\beta1/7\beta2$; r) 4βcis$/4\beta$trans; s) minor-form 2βcis$/2\beta$trans.

pattern completely around the heme [1,4]. As previously noted it is clear from this comparison that complete, unambiguous heme proton hyperfine shift assignments could be easily made using DrecGMH4CN [4]. Those results have revealed that of the previous assignments made for the naturally protonated, native GMH4CN, only the heme β-meso proton was misassigned [1,4].

Another interesting feature of the DrecGMH4CN spectroscopy is the rapidity with which the complete heme proton assignments could be made. Initial work with the naturally protonated, native GMH4CN required over two years of shared instrument time to assemble a self-consistent set of heme and proximal histidine assignments [1]. By comparison, once a sample of DrecGMH4CN was available, the complete heme proton assignments were obvious in about 72 hours of instrument time with our 280 μM sample.

3.6. ADDITIONAL POTENTIAL USES OF PERDEUTERATED PROTEINS

Although still at an early stage, our work suggests many potential applications where suitably perdeuterated proteins may be beneficial.

i. Among the proteins with prosthetic groups it is most frequently the case that the prosthetic group is integral to the active site of the protein. Since the object of many studies is to understand the chemistry and structure of the active site, prosthetic groups are inherently interesting. Not all prosthetic groups have the capacity of being paramagnetic, as hemes do, yet in combination with a perdeuterated protein, even a reconstitutable diamagnetic prosthetic group may be selectively visualized by proton NMR spectroscopy. Heme proteins, although wide spread in nature, are by no means the only proteins that could be studied in this way. We believe this method to be one of general usefulness.

ii. Highly deuterated proteins have the advantage of reducing the diagonal intensity in two-dimensional proton NMR experiments. This is dramatically shown in Figure 9 and that experiment resulted in the assignment of heme peak f, which lies so close to the diagonal (Fig. 9B) that it was hopelessly obscured in the NOESY spectrum of the fully protonated protein (Fig. 9).

iii. Related to this is the fact that the spectra shown in Figures 8B and 9B were obtained overnight on a 280 μM protein solution. It would have been impossible to have generated a NOESY map showing all of the heme hyperfine shift connectivities on the available equipment, in the same time, if the sample had been the naturally protonated protein. These results illustrate that perdeuteration reduces the sample's inherent total signal intensity, and consequently the dynamic range as regards detection of the broader hyperfine-shifted resonances and, thereby allows higher preamplifier and postamplifier gain levels to be used.

iv. It is obvious to us that perdeuteration of one protein in an interacting, two-protein complex, such as soluble molecular redox complexes, would allow the naturally protonated protein of the inteacting pair to be a selective target for proton NMR spectroscopy. This is a rather inexpensive alternative to more complicated and expensive isotope enrichment/filtered NMR experiments.

v. Finally, it seems likely to us that extensively deuterated proteins may well find use in protein folding studies. We envision experiments like those that employ isotope pulse labelling that, in this context, could be carried out using protons as the pulse label. Our optimism concerning these types of experiments derives from Figure 7, which reveals several individual sharp resonances in the amide proton region (6.5 ppm to 11 ppm). This residual proton intensity is the result of the fact that in order to be as economical as possible our initial purification steps prior to heme constitution ((5)-(9) in Fig. 5) were carried out using H_2O rather than D_2O solutions. Despite the fact that following heme constitution the sample was maintained in D_2O solutions these amide proton intensities were maintained for longer than three weeks. We have concluded that these amide intensities are the result of amide groups undergoing fast isotope exchange in the apoprotein (ie prior to constitution), but whose exchange rates have been significantly slowed upon constitution of the holoprotein.

4. Summary

Three important general conclusions emerge from this work.

First, we have demonstrated that it is possible to create a perdeuterated apoprotein of *Glycera dibranchiata* Monomer Hemoglobin Component IV and to profitably exploit it (as the constituted holoprotein) in proton NMR studies. The results presented here show this to be a method specifically beneficial to making complete heme proton resonance assignments, and demonstrate the potential use of perdueterated proteins for several other types of NMR studies.

Second, the minor form of Monomer Hemoglobin Component IV that has been detected spectroscopically in preparations of the native protein is also present in solutions of the recombinant native protein. This leads us to speculate that there may be an alternative conformation of the protein that approaches the major form conformation in Gibbs Free Energy. The fact that it is present in the expressed protein, which is known to originate from a single gene, leads us to speculate that the minor conformation may be accessed in a folding step prior to insertion of the heme into the globin. Perhaps most important for this work, however, is that the minor form does not compromise our ability to make specific, complete heme and proximal histidine proton resonance assignments in GMH4CN.

Third, the cost of obtaining a recombinant perdeuterated protein is not prohibitive, but success does require that an efficient expression system be used. The sample of DrecGMH4 used in these studies cost about $3600. However, initially our goal has been to determine whether we could carry out this project successfully. We did not concentrate on being economical. Since both the feasibility and usefulness have been demonstrated, we have concentrated on refining our techniques for the most economical production of perdeuterated GMG4. At this point we believe that the same sample as used here could be produced for about one-fourth the original cost.

5. Acknowledgements

Separate parts of the work presented here was supported by different research grants from the National Institutes of Health (GM.47645) and the National Science Foundation (DMB 9018982). The 500 MHz NMR spectrometer was purchased with majority funding from the National Institutes of Health (RR.0631401). We are grateful for this support. We thank Sandra Satterlee for excellent technical assistance in preparing this manuscript.

6. References

1. Alam, S.L. and Satterlee, J.D. (1994) Complete Heme Proton Hyperfine Resonance Assignments of the *Glycera dibranchiata* Component IV Metcyano Monomer Hemoglobin, *Biochemistry* **33**, 4008-4018.
2. Mintorovitch, J., Satterlee, J.D., Pandey, R., Lewis, H., and Smith K.M. (1990) Assignment of Selected Hyperfine Resonances in the Three Cyanide Ligated *Glycera Dibranchiata* Monomer Methemoglobins, *Inorg Chim. Acta* **170**, 157-159.
3. Constantinidis, I., Satterlee, J.D., Pandey, R.K., Leung, H.-K., and Smith, K.M. (1988) Assignments of Selected Hyperfine Proton NMR Resonances in the Met Forms of *Glycera dibranchiata* Monomer Hemoglobins and Comparison with Sperm Whale Metmyoglobin, *Biochemistry* **27**, 3069-3076.
4. Alam, S.L. and Satterlee, J.D. (1994) Unambiguous Heme Proton Hyperfine Resonance Assignments of a Monomeric Hemoglobin from *Glycera dibranchiata* Facilitated with a Completely Deuterated Protein, *J. Am. Chem. Soc.*, in press.
5. Alam, S.L., Dutton, D.P., and Satterlee, J.D. (1994) Expression of Recombinant Monomer Hemoglobins (Component IV) from the Marine Annelid *Glycera dibranchiata:* Evidence for Primary Sequence Positional Regulation of Heme Rotational Disorder, *Biochemistry*, in press.
6. Kandler, R.L. and Satterlee, J.D. (1983) Significant Heterogeneity in the Monomer Fraction of *Glycera dibranchiata* Hemoglobins. Detection, Partial Isolation and Characterization of Several Protein Components, *Comp. Biochem. Physiol.* **75B**, 499-503.
7. Kandler, R.L., Constantinidis, I., and Satterlee, J.D. (1984) Evaluation of the Extent of Heterogeneity in the *Glycera Dibranchiata* Monomer Haemoglobin Fraction by the use of NMR and Ion-Exchange Chromatography, *Biochem. J.* **226**, 131-138.
8. Constantinidis, I. and Satterlee, J.D. (1987) Isoelectric Focusing Purity Criteria and Proton NMR Detectable Spectroscopic Heterogeneity in the Major Isolated Monomer Hemoglobins from *Glycera dibranchiata*, *Biochemistry* **26**, 7779-7786.
9. Alam, S.L., Satterlee, J.D., and Edmonds, C.G. (1994) Complete Amino Acid Sequence of the *Glycera dibranchiata* Monomer Hemoglobin Component IV: Structural Implications, *Journal of Protein Chemistry* **13**, 151-164.
10. Simons, P. and Satterlee, J.D. (1989) cDNA Cloning and Predicted Amino Acid Sequence of *Glycera dibranchiata* Monomer Hemoglobin Component IV, *Biochemistry* **28**, 8525-8530.
11. Kunkel, T.A., Roberts, J.D., and Zakour, R.A. (1987) Rapid and Efficient Site-Specific Mutagenesis without Phenotypic Selection, *Methods in Enzymology* **154**, 367-382.
12. Studier, F.W. and Moffatt, B.A. (1986) Use of Bacteriophage T7 RNA Polymerase to Direct Selective High-level Expression of Cloned Genes, *J. Mol. Biol.* **189**, 113-130.
13. Studier, F.W., Rosenberg, A.H., Dunn, J.J., and Dubendorff, J.W. (1990) Use of T7 RNA Polymerase to Direct Expression of Cloned Genes, *Methods in Enzymology* **185**, 60-89.
14. Rosenberg, A.H., Lade, B.N., Chui, D., Lin, S.-W, Dunn, J.J., and Studier, F.W. (1987) Vectors for selective expression of cloned DNAs by T7 RNA polymerase, *Gene* **56**, 125-135.
15. Inubushi, T., and Becker, E.D. (1983) Efficient Detection of Paramagnetically Shifted NMR Resonances by Optimizing the WEFT Pulse Sequence, *J. Magn. Reson.* **51**, 128-133.
16. Kumar, A., Wagner, G., Ernst, R.R., and Wüthrich, K. (1981) Buildup Rates of the Nuclear Overhauser Effect Measured by Two-Dimensional Proton Magnetic Resonance Spectroscopy: Implications for Studies of Protein Conformation, *J. Am. Chem. Soc.* **103**, 3654-3658.

17. Aue, W.P., Bartholdi, E., and Ernst, R.R. (1976) Two-dimensional spectroscopy. Application to nuclear magnetic resonance, *J. Chem. Phys.* **64**, 2229-2246.
18. Wüthrich, K. (1986) *NMR of Proteins and Nucleic Acids*, Wiley Interscience, New York.
19. States, D.J., Haberkorn, R.A., and Ruben, D.J. (1982) A Two-Dimensional Nuclear Overhauser Experiment with Pure Absorption Phase in Four Quadrants, *J. Magn. Reson.* **48**, 286-292.
20. Arents, G., and Love, W.E. (1989) *Glycera dibranchiata* Hemoglobin Structure and Refinement at 1.5 Å Resolution, *J. Mol. Biol.* **210**, 149-161.
21. Imamura, T., Baldwin, T.O. and Riggs, A. (1972) The Amino Acid Sequence of the Monomeric Hemoglobin Component from the Bloodworm, *Glycera dibranchiata*, *J. Biol. Chem.* **247**, 2785-2797.
22. Mintorovitch, J. and Satterlee, J.D. (1988) Anomalously Slow Cyanide Binding to *Glycera dibranchiata* Monomer Methemoglobin Component II: Implication for the Equilibrium Constant, *Biochemistry* **27**, 8045-8050.
23. Mintorovitch, J., van Pelt, D., and Satterlee, J.D. (1989) Cyanide Binding Rates for the *Glycera dibranchiata* Monomer Methemoglobin Components III & IV: Comparison with Component II Results, *Biochemistry* **28**, 6099-6104.
24. Cooke, R.M., and Wright, P.E. (1987) Structural consequences of heme isomerism in monomeric hemoglobins from *Glycera dibranchiata*, *Biochim. Biophys, Acta* **832**, 357-364.
25. Cooke, R.M., Dalvit, C., Narula, S.S., and Wright, P.E. (1987) NMR studies of the heme pocket conformations of monomeric hemoglobins from *Glycera dibranchiata*, *Eur. J. Biochem.* **166**, 399-408
26. Markley, J.L., Putter, I., and Jardetzky, O. (1968) High-Resolution Nuclear Magnetic Resonance Spectra of Selectively Deuterated Staphylococcal Nuclease, *Science* **161**, 1249-1251.
27. Crespi, H.L., Rosenberg, R.M., Katz, J.J. (1968) Proton Magnetic Resonance of Proteins Fully Deuterated except for H^1-Leucine Side Chains, *Science* **161**, 795-796.
28. Brodin, P., Drakenberg, T., Thulin, E., Forsen, S., and Grundstrom, T. (1989) Selective proton labelling of amino acids in deuterated bovine calbindin D9k. A way to simplify 1H-NMR spectra, *Protein Engineering* **2**, 353-358.
29. Lemaster, D.M. (1988) Protein NMR resonance assignment by isotropic mixing experiments on random fractionally deuterated samples, *FEBS Letters* **233**, 326-330.
30. Lemaster, D.M. (1987) Chiral ß and random fractional deuteration for the determination of protein sidechain conformation by NMR, *FEBS Letters* **223**, 191-196.

REDOX AND SPIN-STATE CONTROL OF THE ACTIVITY OF A DIHEME CYTOCHROME *C* PEROXIDASE - Spectroscopic studies

SUSANA PRAZERES and ISABEL MOURA
Centro de Tecnologia Química Fina e Biotecnologia
Dep. de Química, Faculdade de Ciências e Tecnologia/UNL,
2825 Monte de Caparica, Portugal

RAYMOND GILMOUR and GRAHAM PETTIGREW
Dept. of Preclinical Veterinary Sciences, Royal (Dick) School of Veterinary Studies, University of Edinburgh, Summerhall, Edinburgh EH9 1QH, UK

NATARAJAN RAVI and BOI HANH HUYNH
Department of Physics, Emory University, Atlanta, USA

1. Introduction

Hydrogen peroxide formed in cells, as the result of incomplete reduction of oxygen, can be removed essentially by two ways: by peroxidases in a process of reduction to water or by catalase in a dismutation reaction. The actions of these enzymes are essential to prevent the accumulation of hydrogen peroxide, diminishing the risk of peroxide-induced damage of cell constituents [1].

Recently a periplasmic cytochrome *c* peroxidase was isolated from *Paracoccus denitrificans* (*Pa.d.*) (L.M.D. 52.44) grown under conditions of limiting oxygen as described by Goodhew et al. [2].

The reaction catalyzed by this enzyme is schematically represented in the following equation:

$$2\ \text{Cyt.}\ c\ \text{Fe (II)} + H_2O_2 + 2\ H^+ \dashrightarrow 2\ \text{Cyt.}\ c\ \text{Fe (III)} + 2\ H_2O$$

The *Paracoccus* peroxidase shows a strong preference for the basic mitochondrial cytochrome *c* as electron donor when compared with the acidic cytochrome c_{551} from *Pseudomonas aeruginosa* (*Ps.a.*) but the acidic periplasmic *Paracoccus* cytochrome c_{550} was found to be the best donor in vitro, suggesting it is the physiological donor [2,3].

This peroxidase has some similarities to the one isolated from *Ps. aeruginosa*, which has been extensively studied [4]. Like the *Ps.a.* enzyme [5], the *Paracoccus* peroxidase (molecular mass of 40 kDa) has two *c*-type hemes. The

G.N. La Mar (ed.), Nuclear Magnetic Resonance of Paramagnetic Macromolecules, 141-163.

oxidation-reduction potentials of the two hemes were determined by redox titration and found to be well separated [6]. From this result we can designate the two hemes as high-potential (HP) and low-potential (LP). We propose that the cytochrome *c* peroxidase from *Pa.d.* has two domains. The electron transfer domain contains the HP heme and the peroxidatic domain, where the interaction with hydrogen peroxide occurs, contains the LP heme.

The HP heme acts as the source of one of the two electrons for the reduction of hydrogen peroxide and can be non-physiologically reduced by sodium ascorbate. The LP heme acts as the centre for the peroxidatic reduction and can be reduced by sodium dithionite [6]. The addition of ascorbate to the oxidized enzyme (as it is purified) induces a half-reduced form of the peroxidase, where the HP heme is reduced and the LP heme is oxidized. Only this form is catalytically active in the reduction of hydrogen peroxide and is capable of binding anionic ligands such as cyanide [7].

Figure 1 shows the electron transfer chain in *Pa.d.* Note the localization of cytochrome *c* peroxidase, with the electron transfer and the peroxidatic domains.

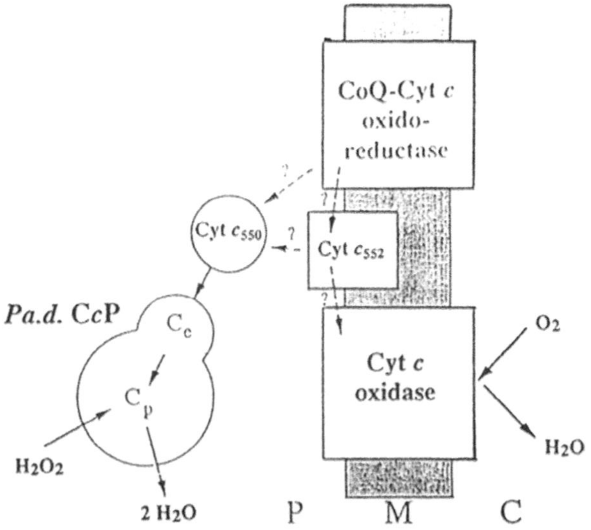

Figure 1. The electron transfer chain of *Paracoccus denitrificans* (adapted from [3]). CoQ-cytochrome *c* oxidoreductase and cytochrome *c* oxidase are located in the cytoplasmic membrane and linked by the membrane-bound cytochrome c_{552}. Cytochrome c_{550} and cytochrome *c* peroxidase (CcP) are located in the periplasmic region. Ce-Electron transfer domain. Cp-Peroxidatic domain. P-Periplasm. M-Cytoplasmic membrane. C-Cytoplasm.

It has been shown that for *Paracoccus denitrificans* cytochrome *c* peroxidase, binding of calcium ions is necessary for its enzymatic activation.

It is important to define the oxidation and spin state of an enzyme, since its activity is greatly dependent on it. Table 1 lists the iron spin states commonly observed in hemeproteins. These spin states are observed in the cytochrome *c* peroxidase from *Paracoccus denitrificans*.

Table 1. Iron spin states usually observed in hemeproteins.

Oxidation states	High-spin	Low-spin
Fe (III) - oxidized	S=5/2	S=1/2
Fe (II) - reduced	S=2	S=0

In general, the high-spin state configuration is observed when a weak ligand (or no ligand) is in the sixth coordination position. When this position is occupied by a ligand having a strong crystal field, the iron is in a low-spin state. The oxidized iron is always observed as a paramagnetic species and the reduced iron as a diamagnetic one.

In this report we present some physical and spectroscopic evidence demonstrating the intrincate interplay between the two heme moieties using redox and spin states as a control mechanism. Binding of calcium ions is also found to be necessary for promoting enzymatic activation.

2. Amino acid sequence of *Pa.d.* cytochrome *c* peroxidase

The amino acid sequence of this cytochrome *c* peroxidase was recently determined by J. Van Beeumen *et al.* Similar to the sequence of *Pseudomonas* enzyme [8,9], the *Pa.* sequence contains two heme attachment sites with the sequence motif Cys-X-Y-Cys-His, typical for *c*-type cytochromes [10].

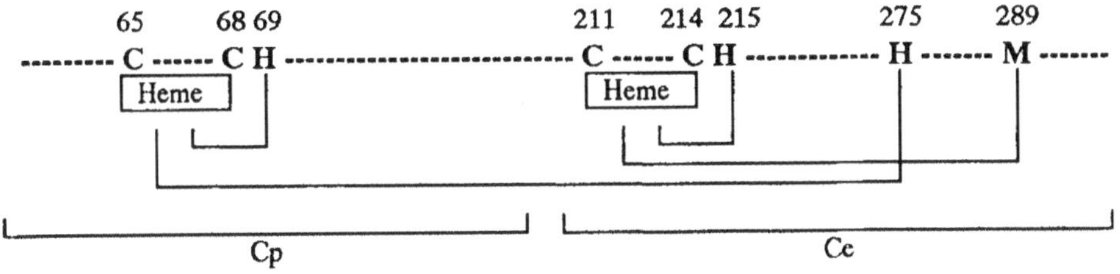

Figure 2. Features of the amino acid sequence of *Paracoccus denitrificans* cytochrome *c* peroxidase.

By analogy to the peroxidase from *Ps. aeruginosa*, the axial ligands are thought to be Met-289 for the HP heme and His-275 for the LP heme. Once again we can distinguish the two domains: the electron transfer domain (Ce) with the HP heme and the peroxidatic domain (Cp) with the LP heme, as shown in Figure 2.

It is important to note that the *Pa.* cytochrome *c* peroxidase sequence contains two similar versions of a calcium binding sequence motif found in other bacterial proteins [11], as shown in Figure 3. Each sequence forms one half of a calcium binding site.

[1]	G	G	X	G	X	D	X	U	X
[2]	**G**	L	**G**	**G**	**V**	**D**	**G**	**L**	**P**
	72								80
	N	**F**	**G**	**G**	**Q**	**D**	**Y**	H	**P**
	219								227

Figure 3. Calcium binding sequence motif found in the amino acid sequence of *Paracoccus denitrificans* cytochrome *c* peroxidase.
[1] Calcium binding sequence motif, where **X** is any residue and **U** is an hydrophobic residue.
[2] *Pa.* cytochrome *c* peroxidase sequence motifs.

The first sequence motif (residues 72-80) is located immediately after the LP heme binding site and only Leu-73 does not match in the calcium binding sequence motif (Gly). The second sequence motif (residues 219-227) is located immediately after the HP heme binding site, but it is less similar to the calcium binding sequence motif than the first. From these observations two conclusions can arise: if these two sequences together constitute a calcium binding site, then the two hemes are very close to each other, with a calcium ion between them. On the other hand, if the first sequence motif represents one half of a calcium binding site, it would suggest that the calcium bound to the protein must be held within a dimer interface.

3. UV/Visible studies of *Pa.d.* cytochrome *c* peroxidase

The UV/Visible spectrum of *Pa.d.* peroxidase (Figure 4) is typical of heme proteins. The oxidized enzyme has a Soret maximum at 408 nm and a weak band at 640 nm. The 640 nm band is characteristic of a high-spin ferric heme [10]. The reduction of the HP heme (half-reduced enzyme) with sodium ascorbate is complete within 2 minutes and results in the disappearance of the 640 nm band. This loss of the band corresponding to a high-spin ferric heme suggests that the oxidized HP heme is in a high-spin state and after reduction changes its spin state to low-spin. A similar behaviour was observed for the *Ps.* peroxidase, by Foote *et al* [12]. The α-band is seen at 556 nm and the β-band at 525 nm.

The reduction of the LP heme with sodium dithionite shifts the Soret band to 418 nm and increases the intensity of the α and β-bands.

Consistent with the HP heme being in a high-spin state, no band at 695 nm (indicative of an axial methionine ligand) was observed. Any small absorption at 695 nm, however, would be difficult to observe given the neighbouring broad 640 nm band.

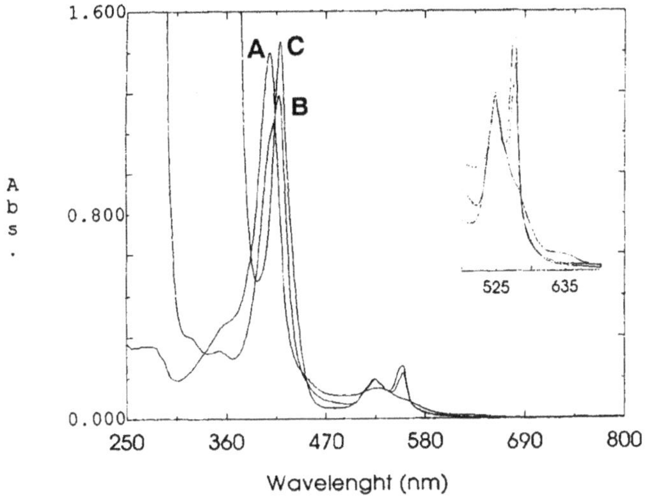

Figure 4. UV/Visible spectra of cytochrome *c* peroxidase from *Paracoccus denitrificans* in 10 mM Hepes buffer pH=7.5.
A-Oxidized enzyme. B-Half-reduced enzyme (1 mM ascorbate-10 μM DAD). C-Reduced enzyme obtained by addition of solid sodium dithionite.

3.1. THE ASSAY FOR CYTOCHROME *C* PEROXIDASE ACTIVITY

The activity of cytochrome *c* peroxidase is determined by following the decrease in the absorbance of the α-band of ferrocytochrome *c* at 550 nm (Figure 5). The reaction can be initiated by two different ways: (1) addition of hydrogen peroxide to a solution containing the reduced cytochrome *c* and the enzyme, and (2) addition of the peroxidase to a solution containing ferrocytochrome *c* and the hydrogen peroxide.

However the two methods can give different results. If the reaction is started with the oxidized enzyme, the lag observed is due to reduction of the enzyme and activation in the cuvette. The lag phase is not present if (a) the enzyme is pre-reduced with ascorbate or (b) hydrogen peroxide is used to initiate the reaction.

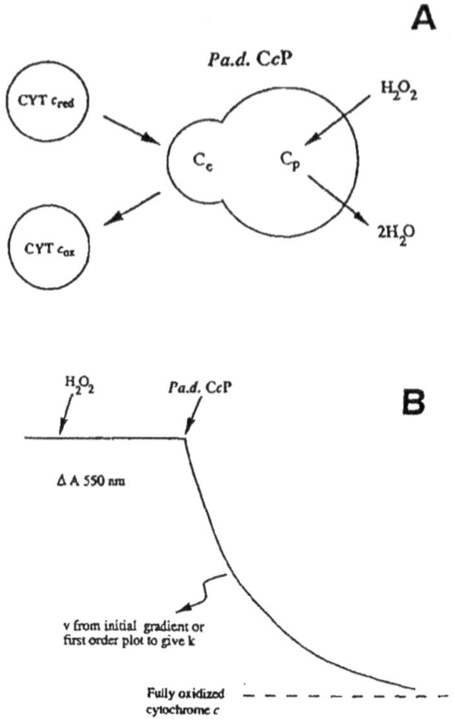

Figure 5. The assay for cytochrome *c* peroxidase activity (adapted from [7]).
A-*Pa.d.* cytochrome *c* peroxidase catalyzes the reduction of hydrogen peroxide in the presence of reduced cytochrome *c*. B-The peroxidase activity is assayed by following the decrease in the absorbance of the α-band (550nm) of the reduced cytochrome *c*.

3.2. THE EFFECT OF CALCIUM IONS ON ACTIVITY AND SPIN STATE OF THE ENZYME

In order to follow the alterations occurring around the two hemes, difference spectroscopy was used.

Figure 6A shows the difference spectrum of the half-reduced peroxidase incubated with ascorbate for 2 minutes minus the spectrum of the oxidized enzyme. The reduction of the HP heme with ascorbate is complete within two minutes and followed by the appearance of some high-spin bands at 380 and 640 nm. Difference spectra recorded at various incubation times versus that of 2 minutes incubation (Figure 6B), show the slow appearance of a high-spin state heme, which reaches an end point after 30 minutes of incubation. The addition of calcium ions induces an increase of the high-spin state.

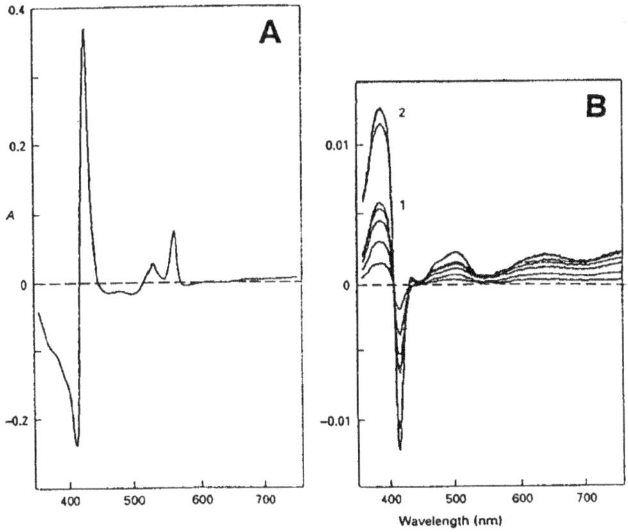

Figure 6. Visible absorption spectra of *Pa.d.* cytochrome *c* peroxidase. Appearance of a high-spin state after ascorbate reduction (reproduced from [6]).
A-Difference spectrum: 2 min ascorbate reduced (1 mM ascorbate-5 μM DAD) minus oxidized form. B-Difference spectra with increasing time after ascorbate reduction. 1) 30 min reduced minus 2 min reduction. 2) After addition of calcium ions (1 mM) minus 2 min reduction.

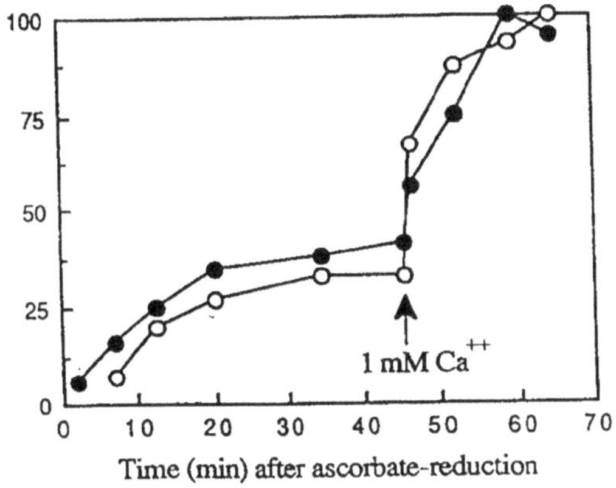

Figure 7. Effect of ascorbate reduction and calcium ions on enzyme activity (adapted from [7]). The high-spin formation was determined by following the increase in absorbance at 380 nm.
(•) % of enzymatic activity. (o) % of high-spin formation.

These results were obtained at pH=6. At pH=7.5 the high-spin formation was only observed after the addition of calcium ions. Probably at pH=6 the calcium ions are more tightly bound to the enzyme, than at pH=7.5.

When the oxidized enzyme is treated with EDTA to remove bounded calcium, no high-spin state is observed after the reduction of the HP heme with ascorbate [6].

The increase of the high-spin state after ascorbate reduction is associated with an increase in the activity of the enzyme, as shown in Figure 7.

The partial activation of the enzyme after ascorbate reduction is probably due to the half-reduced enzyme binding traces of calcium ions from solution. This implies that the half-reduced peroxidase has a higher affinity for Ca^{2+} than the oxidized enzyme. These results clearly show that the LP heme changes its spin-state from low to high-spin after ascorbate reduction, conducting to the enzyme activation.

3.3. THE EFFECT OF DILUTION ON ENZYME ACTIVITY

It was shown by Gilmour *et al* [7] that the *Pa.d.* cytochrome *c* peroxidase loses activity when diluted from a stock solution, prior to the activity assay (Figure 8). This loss in activity increases with increasing dilution.

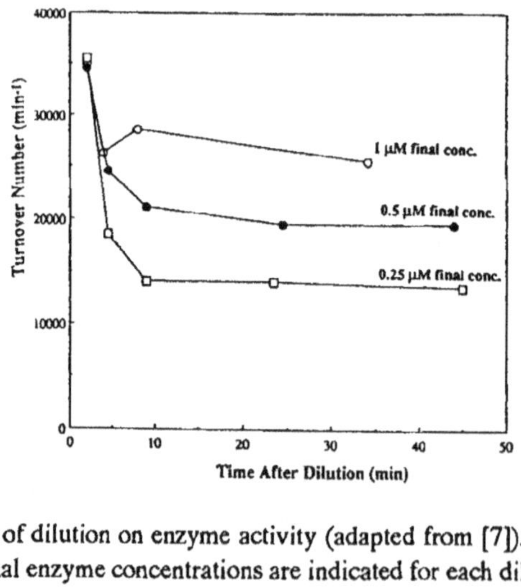

Figure 8. Effect of dilution on enzyme activity (adapted from [7]). Peroxidase stock solution of 100 μM. The final enzyme concentrations are indicated for each dilution.

The same authors proposed that the enzyme exits as a monomer/dimer equilibrium where only the dimer is active. The dilution of the enzyme shifts the equilibrium towards the monomer resulting in a decrease in the activity.

4. NMR studies of *Pa.d.* cytochrome *c* peroxidase

This process of enzyme activation can also be followed by NMR. Thus, the room temperature NMR spectra of the oxidized and half-reduced peroxidase were obtained, in the presence and absence of calcium ions.

Figure 9 shows the low-field region of the NMR spectra of cytochrome *c* peroxidase from *Pa.d.* in the oxidized and half-reduced states. In this region, and because iron in ferric hemes is paramagnetic, protons from the heme methyl groups are detected, shifted from the main diamagnetic envelope region.

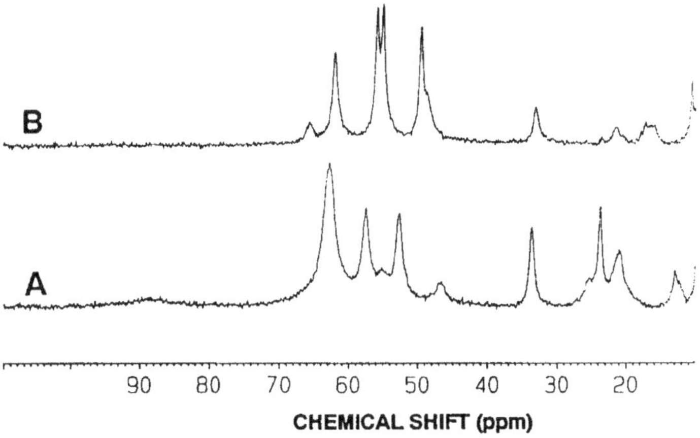

Figure 9. Low-field region of the 300 MHz NMR spectra of cytochrome *c* peroxidase from *Pa.d.* in different oxidation states. Experimental conditions: temperature, 303 K; protein concentration, ~1 mM in 10 mM Mes-Hepes buffer pH* 6.0 (pH*-pH measured without isotopic correction); 5000 scans; 4 K data points.
A-Oxidized state, as isolated. B-Half-reduced state (5 mM ascorbate-5 μM DAD).

In the spectrum of the oxidized enzyme (Figure 9A) two sets of resonances are detected. The resonances between 58 and 52 ppm are assigned to the HP heme which is undergoing a spin-equilibrium [13]. The position of these resonances are not typical for a pure high-spin state cytochrome [14,15]. Moore [16] showed that the chemical shift of the signal arising from the protons of the ε-CH_3 group of the axial methionine can be identified by its correlation with the sum of the chemical shifts of the corresponding four heme methyl resonances. Based in this correlation, the broad resonance around 90 ppm is then assigned to the methyl group of the axial methionine. The two methyl resonances at 33.3 and 23.6 ppm are assigned to the LP heme [13].

In the spectrum of the half-reduced enzyme, achieved by reduction of the HP heme with ascorbate (Figure 9B), the resonances previously observed are replaced by four new heme methyl resonances observable between 64 and 50 ppm, assigned to the LP heme, here in a high-spin state [13]. At -3.1 ppm, a resonance with an intensity of three protons corresponding to the methyl group of the methionine bounded to the reduced iron, is observed (not shown), confirming the coordination of the HP heme as methionine-histidine.

The addition of calcium ions to the half-reduced peroxidase has no effect on the NMR spectrum (not shown).

4.1. EFFECT OF CALCIUM IONS AND EDTA ON THE NMR SPECTRA OF *Pa.d.* CYTOCHROME *c* PEROXIDASE

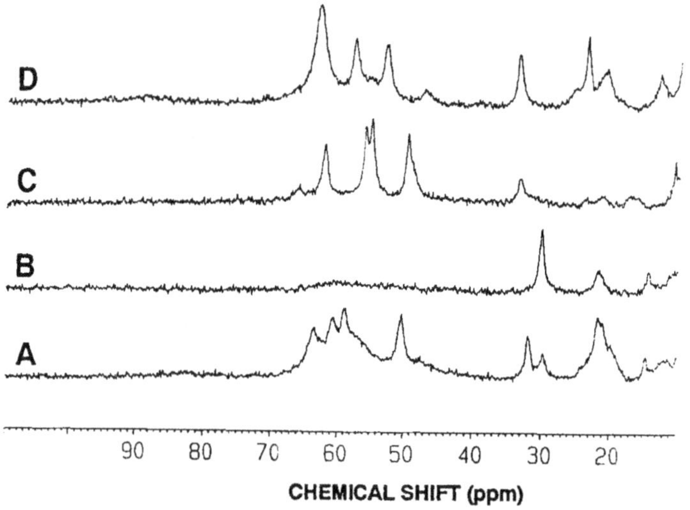

Figure 10. Low-field region of the 300 MHz NMR spectra of cytochrome *c* peroxidase from *Pa.d.*, previously treated with EDTA, in different oxidation states. Experimental conditions: temperature, 303 K; protein concentration, ~0.8 mM in 50 mM Hepes/2 mM EDTA buffer pH* 8.1; 3000 scans; 4 K data points.
A-Oxidized state, EDTA treated. B-Half-reduced state (5 mM ascorbate-5 μM DAD). C-Half-reduced state plus Ca^{2+} (10 mM). D-As C plus ferricyanide (10 mM).

The NMR spectrum of the oxidized peroxidase, previously treated with EDTA to remove bounded calcium (Figure 10A), shows important modifications in the HP heme as well as in the LP heme. At least two forms of the high-spin species are observed: the sharp signals and the broad component extended underneath the sharp peaks, which are in a different position from those of the untreated enzyme.

Some perturbations are also observed in the low-spin LP heme. This result reflects the fact that the calcium binding to the enzyme may affect the global conformation of the enzyme or occurs in a place near the two hemes.

The ascorbate reduction of the peroxidase (Figure 10B) causes the disappearance of the high-spin heme resonances (corresponding to the HP heme, now in a reduced state). Two resonances are observed at 30.2 and 21.8 ppm corresponding to the methyl groups of a low-spin ferric heme.

After the addition of calcium ions to this half-reduced enzyme (Figure 10C), the disappearance of these low-spin heme methyl signals occurs and a set of high-spin heme methyl resonances appears, at the same position as in the half-reduced state of the peroxidase not treated with EDTA. The reoxidation of this enzyme (Figure 10D) results in a spectrum identical to the one of the oxidized peroxidase without EDTA treatment.

In the high-field region spectrum of the half-reduced enzyme (not shown), a resonance assigned to the methyl group of the axial methionine is displayed at -3.7 ppm. After the calcium addition this resonances shifts to -3.1 ppm, the same position as in the spectrum of the half-reduced enzyme without EDTA treatment. This observation indicates that the methyl group of the axial bounded methionine in the HP heme is sensitive to the spin state of the LP heme.

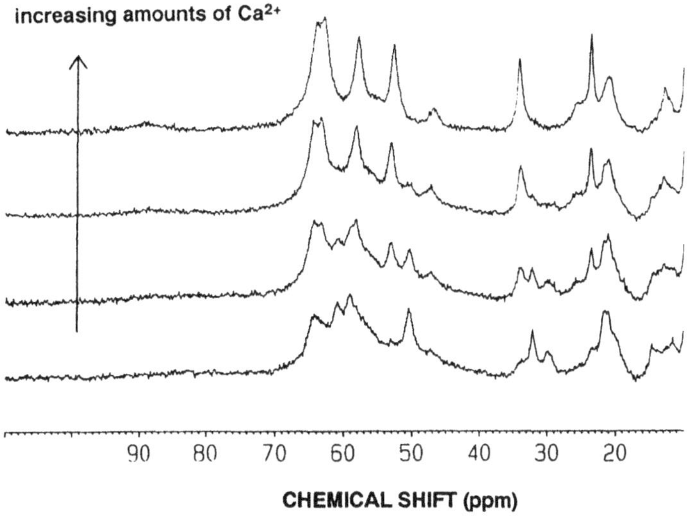

Figure 11. Effect, in the low-field region of the 300 MHz NMR spectra, of calcium addition to the EDTA-treated oxidized cytochrome *c* peroxidase from *Pa.d.* Experimental conditions: temperature, 303 K; 50 mM Hepes buffer pH* 8.6; 3000 scans; 4 K data points. The calcium/enzyme ratios are, from bottom to top, 0, 0.1, 0.3 and 9.2 .

Figure 11 shows the effect of calcium ions addition to the oxidized enzyme, previously treated with EDTA. The bottom spectrum represents the enzyme depleted of calcium (after EDTA treatment) and the top spectrum the enzyme in the presence of an excess of calcium ions (9.2 times excess). This calcium addition causes the appearance of the usual set of resonances observed in the spectrum of the oxidized peroxidase which had not been treated with EDTA.

When small amounts of calcium are added to the half-reduced enzyme, previously treated with EDTA in the oxidized form, a conversion of the low-spin heme signals to a high-spin heme state is observed (Figure 12).

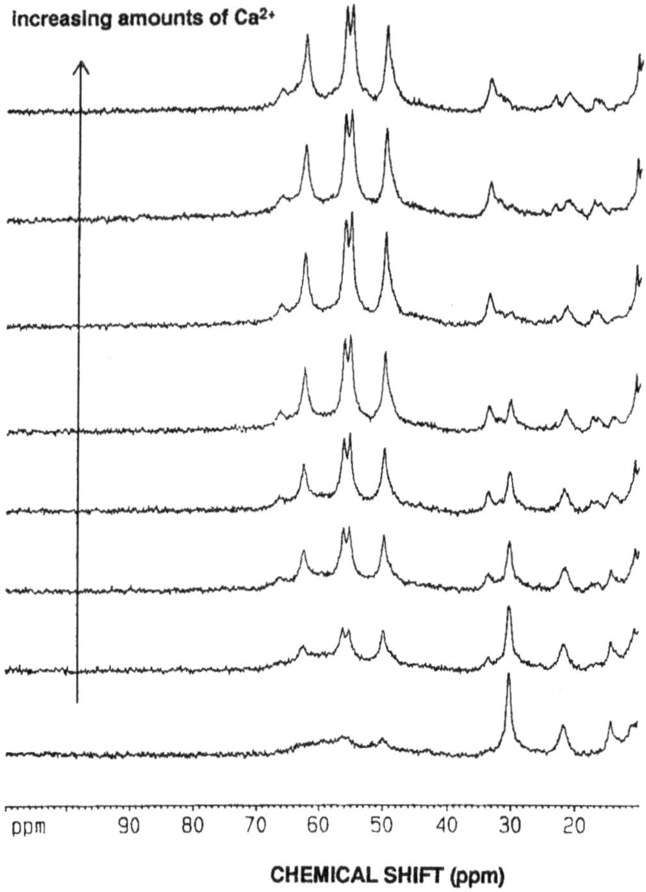

Figure 12. Effect of calcium addition to the half-reduced *Pa.d.* peroxidase, treated with EDTA in the oxidized form, followed by 300 MHz NMR. Experimental conditions: temperature, 303 K; 50 mM Hepes buffer pH* 8.6; 5 mM ascorbate-5 μM DAD; 3000 scans; 4 K data points.

NMR results for the oxidized peroxidase, obtained at room temperature, clearly show the presence of a high-spin HP heme and a low-spin LP heme. The ascorbate reduction of the HP heme switch its spin state from high-spin to low-spin, with the LP heme remaining in a low-spin state. The addition of calcium ions, converts the LP heme from a low-spin to a high-spin state. In this form the enzyme is active and capable of hydrogen peroxide binding. These two processes of reduction and spin state change can be separated in time by the action of EDTA, which removes the calcium ions bound to the enzyme, necessary to promote the spin state change.

In the half-reduced state of the enzyme as isolated, the LP heme is immediately converted to the high-spin state, due to the residual calcium ions already bound to the peroxidase at the protein concentrations required for NMR.

4.2. pK_a VALUES ASSOCIATED WITH THE OXIDIZED AND HALF-REDUCED *Pa.d.* CYTOCHROME *c* PEROXIDASE

The pH dependence of the heme methyl resonances was studied, for the oxidized and half-reduced peroxidase, over a pH* range of 5-10.5. The results are summarised in Table 2.

Table 2. pK_a values associated with two oxidation states of *Pa.d.* peroxidase.

Oxidation state		pK_{a1}	pK_{a2}
Oxidized (as isolated)	HP (high-spin)	5.9 ± 0.2	9.0 ± 0.6
	LP (low-spin)	5.7 ± 0.4	9.2 ± 0.2
Oxidized - Ca^{2+}	HP (high-spin)	6.2 ± 0.4	10.3 ± 0.06
	LP (low-spin)	6.2 ± 0.04	10.5 ± 0.1
Half-reduced + Ca^{2+}	LP (high-spin)	5.9 ± 0.05	9.5 ± 0.1

Two pK_a values were estimated for the oxidized untreated peroxidase for both the HP and LP hemes, and also for the oxidized enzyme depleted of calcium ions. It is important to note that the values found for pK_{a1} and pK_{a2}, in these two conditions of the enzyme, correspond quite well. However, while pK_{a1} value is similar for the two oxidized samples, the value for pK_{a2} is one order of magnitude bigger for the oxidized enzyme in which the calcium was removed.

For the half-reduced peroxidase, again two pK_a values were estimated for the LP heme.

It is interesting to note that the same pK_a are observed for the HP and LP hemes suggesting that the groups that are titrated are close to both hemes.

5. EPR and Mössbauer studies of *Pa.d.* cytochrome *c* peroxidase

A characterisation of the oxidized and half-reduced states of the *Pa.* peroxidase was also performed by EPR and Mössbauer spectroscopies. These techniques are performed at low temperature: EPR at 8 K and Mössbauer between 4.2 and 200 K. For the Mössbauer studies, ^{57}Fe-enriched cells of *Pa.d.* were grown and purified as indicated in [2].

The high-spin/low-spin equilibrium previously described is temperature dependent, and at low temperature almost no high-spin forms are observed.

From the characterisation of *Pa.d.* cytochrome *c* peroxidase by EPR and Mössbauer we can draw several conclusions which include (1) identification of the reduced heme in the half-reduced state of the enzyme, (2) confirmation of the activation mechanism of this enzyme by calcium ions, (3) observation of EPR and Mössbauer spectra of the peroxidase intermediates during the catalytic cycle. These intermediates were obtained by rapid freeze-quench EPR and Mössbauer techniques.

5.1. THE REDUCTION OF THE HP HEME MONITORED BY EPR

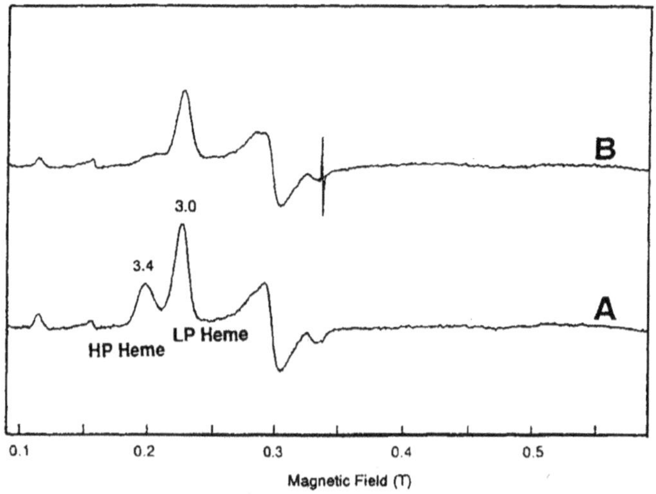

Figure 13. EPR spectra of *Pa.d.* cytochrome *c* peroxidase. Experimental conditions: temperature 8 K; 1 mM sodium phosphate/10 mM NaCl buffer pH 7.0; microwave frequency 9.43 GHz; microwave power 2 mW; modulation amplitude 1 mT; receiver gain $8x10^4$.
A-Oxidized enzyme as isolated. B-Half-reduced enzyme, incubated for 1 min with solid ascorbate.

Figure 13A shows the EPR spectrum of the oxidized (as isolated) peroxidase. Two resonances at g_{max} 3.41 and 3.00 are observed. The presence of these two distinct signals, characteristic for low-spin ferric hemes, reflect different heme environments. A small signal at g=6 represents a high-spin form, arising from the heme undergoing the spin transition.

In the spectrum of the half-reduced state of this peroxidase, obtained after 1 minute of incubation with ascorbate, the features associated with the g_{max}=3.41 signal disappear and the signal at g_{max}=3.00 remains unchanged (Figure 13B).

The disappearance of the signal at g_{max}=3.41, after a short incubation period with the reductant, and the persistence of the features associated with the g_{max}=3.00 signal, indicate the 3.41 signal as the one arising from the HP heme. A similar conclusion is obtained by Mössbauer, as will be shown below.

5.2. EFFECT OF CALCIUM IONS ON THE EPR SPECTRA OF *Pa.d.* CYTOCHROME *c* PEROXIDASE

When the half-reduced peroxidase is incubated with ascorbate for a longer period (60 minutes), new features develop (Figure 14B). Two sharp signals at g_{max}=2.89 (specie I) and g_{max}=2.78 (specie II) appear. The relative proportion of these two species differs from one enzyme preparation to another. This is probably due to the different amounts of calcium ions bound to the enzyme in the different purified preparations that were used.

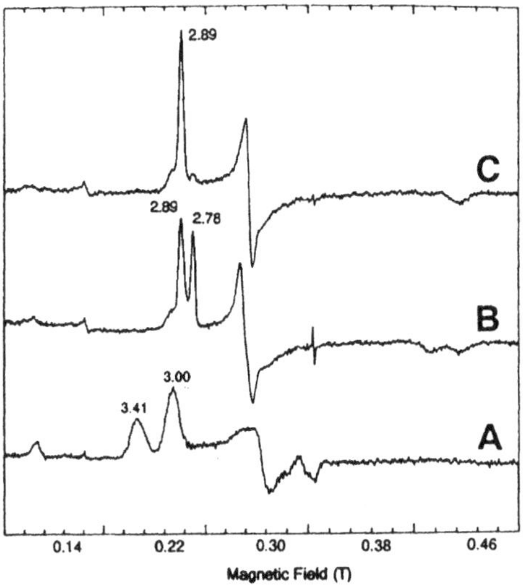

Figure 14. EPR spectra of *Pa.d.* cytochrome *c* peroxidase, in several oxidation states. Experimental conditions: temperature 8 K; 50 mM Hepes buffer pH 8.6; microwave frequency 9.43 GHz; microwave power 2 mW; modulation amplitude 1 mT; receiver gain 8x10^4.
A-Oxidized enzyme (as isolated). B-Half-reduced enzyme (0.4 mM ascorbate), incubated for 60 min with ascorbate. C-Half-reduced enzyme (60 min) + Ca^{2+} (20 mM).

A small signal at g_{max}=3.00 is still noticeable but it only account for 14% and may represent the inactive form of the enzyme (see below).
The addition of calcium ions to the half-reduced peroxidase converts specie II into specie I (Figure 14C).

The signals arising from the half-reduced enzyme are very sharp when compared to the ones from the oxidized peroxidase. This observation is consistent with a modification on the environment around the LP heme induced by the reduction of the HP heme.

The EPR signals obtained at g_{max}=2.89 for the half-reduced enzyme may correspond, at room temperature, to the LP heme signals in a catalytically active high-spin state observed by NMR.

When the oxidized enzyme is treated with EDTA, the correspondent EPR spectrum has some alterations, as shown in Figure 15A. The two signals at g_{max}=3.41 and 3.00 are broader, and the signal at g=6 arising from a high-spin form is also increased.

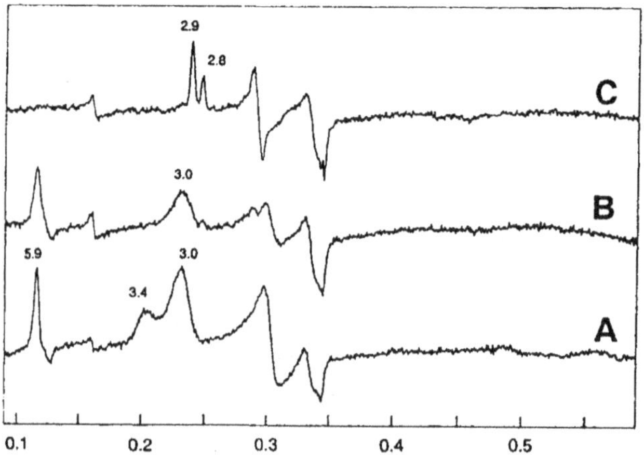

Figure 15. EPR spectra of *Pa.d.* cytochrome *c* peroxidase, previously treated with EDTA in the oxidized state. Experimental conditions: temperature 8 K; 50 mM Hepes buffer pH 8.6; microwave frequency 9.43 GHz; microwave power 2 mW; modulation amplitude 1 mT; receiver gain $8x10^4$.
A-Oxidized enzyme (EDTA treated). B-Half-reduced enzyme (1 mM ascorbate-5 μM DAD), incubated for 60 min with ascorbate. C-Half-reduced enzyme (60 min) + Ca^{2+} (15 mM).

The EPR spectrum of the half-reduced enzyme, depleted of calcium ions by EDTA, after 60 minutes of ascorbate incubation, is again quite different (Figure

15B). The signal at g_{max}=3.41 (and corresponding to the HP heme) has disappeared, but the signal at g_{max}=3.00 (and corresponding to the LP heme) is in the same position. This form may represent the inactive half-reduced enzyme, already seen by NMR at room temperature (the half-reduced low-spin specie, in the absence of calcium ions).

After the addition of calcium ions, specie I (g_{max}=2.89) and specie II (g_{max}=2.78) are observed (Figure 15C). In this sample the total conversion of specie II into specie I was not obtained.

5.3. MÖSSBAUER STUDIES OF *Pa.d.* CYTOCHROME *c* PEROXIDASE IN THE OXIDIZED AND HALF-REDUCED STATES

Mössbauer studies of the oxidized as isolated cytochrome *c* peroxidase reveal again the presence of two distinct low-spin ferric hemes, with different values of isomer shift (δ) and quadrupole splitting (ΔE_Q) (Table 3), reflecting distinct heme environments.

The parameters, ΔE_Q=2.10 mm/s and δ=0.30 mm/s, found for the oxidized HP heme are similar with those reported for several cytochromes with methionine-histidine axial coordination [17,18]. The ΔE_Q value of 2.50 mm/s found for the oxidized LP heme, however, is larger than the usual values for oxidized hemes with bis-histidine coordination [19,20,21,22] and may indicate a different axial ligation for the LP heme in the *Pa.d.* cytochrome *c* peroxidase. The δ value of 0.26 mm/s for the oxidized LP heme is typical for low-spin ferric hemes.

In the half-reduced peroxidase, Mössbauer studies show that the HP heme is reduced and exhibits a quadrupole doublet with ΔE_Q=1.23 mm/s and δ=0.46 mm/s, typical for low-spin ferrous heme with methionine-histidine ligands. Consistent with the EPR results, Mössbauer measurements also show the presence of two low-spin ferric heme species associated with the LP heme in the half-reduced enzyme. The populations of these two species were found to be approximately equal without added calcium ions. In the presence of excess calcium ions, only one specie dominates (~ 90 %).

6. Model of activation proposed for *Pa.d.* cytochrome *c* peroxidase

The combination of these four techniques (UV/Visible, NMR, EPR and Mössbauer) let us to define the oxidation levels and the spin states of this diheme cytochrome *c* peroxidase from *Pa.d.* (Table 3).

Table 3. Evidences for the oxidation levels and the spin states of cytochrome *c* peroxidase from *Pa.d.*

Oxidation state		HP Heme	LP Heme	
Oxidized	UV/Visible	640 nm band(HS)		
	NMR	58-52 ppm(HS) ~90 ppm(Met)	33-23 ppm(LS)	
	EPR	g_{max}=3.41(LS)	g_{max}=3.00(LS)	
	Mössbauer	ΔE_Q=2.10 mm/s δ=0.30 mm/s	ΔE_Q=2.50 mm/s δ=0.26 mm/s	
Half-reduced - Ca^{2+}	UV/Visible	no 640 nm band	--------	
	NMR	-3.7 ppm(Met) (LS)	30-22 ppm(LS)	
	EPR	EPR silent	g_{max}=3.00(LS)	
	Mössbauer	--------	--------	
Half-reduced + Ca^{2+}	UV/Visible	--------	640 nm band(HS)	
	NMR	-3.1 ppm(Met)	64-50 ppm(HS)	
			Comp. I	Comp. II
	EPR	EPR silent	g_{max}=2.89(LS)	g_{max}=2.78(LS)
	Mössbauer	ΔE_Q=1.23 mm/s δ=0.46 mm/s (LS) (Met-His)	ΔE_Q=2.52 mm/s δ=0.31 mm/s (LS)	ΔE_Q=2.37 mm/s δ=0.23 mm/s (LS)

From all the results obtained, it is clear the non-equivalence of the two hemes, and a model for the activation of cytochrome *c* peroxidase is proposed (Figure 16).

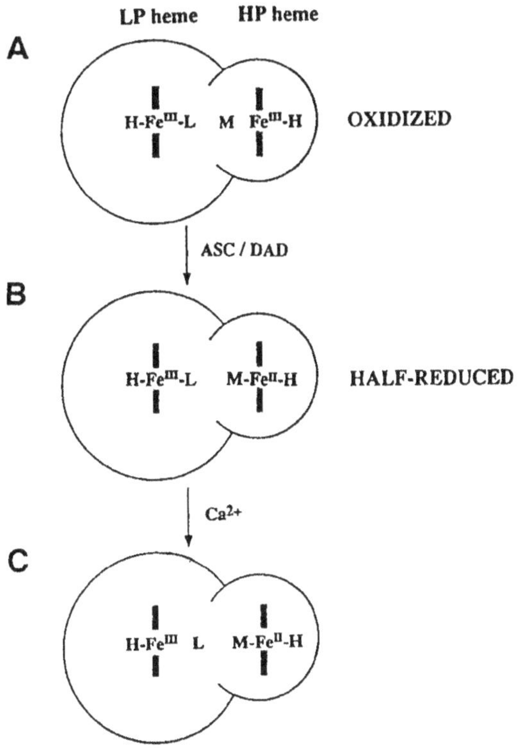

Figure 16. Model of activation proposed for *Paracoccus denitrificans* cytochrome *c* peroxidase (adapted from [7]).
A-Oxidized enzyme. Both hemes are in the oxidized state. The HP heme has methionine-histidine coordination and is undergoing a spin-equilibrium temperature dependent. The LP heme is axially coordinated by bis-histidine or lysine-histidine (by analogy to the *Ps.* enzyme) and is in a low-spin state. B-Half-reduced enzyme in the absence of calcium ions. The HP heme is reduced and becomes low-spin. The LP heme is still oxidized and in a low-spin state. This form cannot bind hydrogen peroxide. C-Half-reduced enzyme in the presence of calcium ions. The reduced HP heme is now in a low-spin state and the LP heme converts into a high-spin state. The enzyme is capable of hydrogen peroxide binding and is enzymatically active.

7. The enzymatic mechanism of *Pa.d.* cytochrome *c* peroxidase

The reaction of active half-reduced *Pa.d.* cytochrome *c* peroxidase with hydrogen peroxide, results in formation of compound I which converts rapidly to compound II (Figure 17). These two intermediates were also seen for the *Ps. aeruginosa* peroxidase (quite similar to the *Pa.d.* enzyme) by EPR [4] and optical techniques [23].

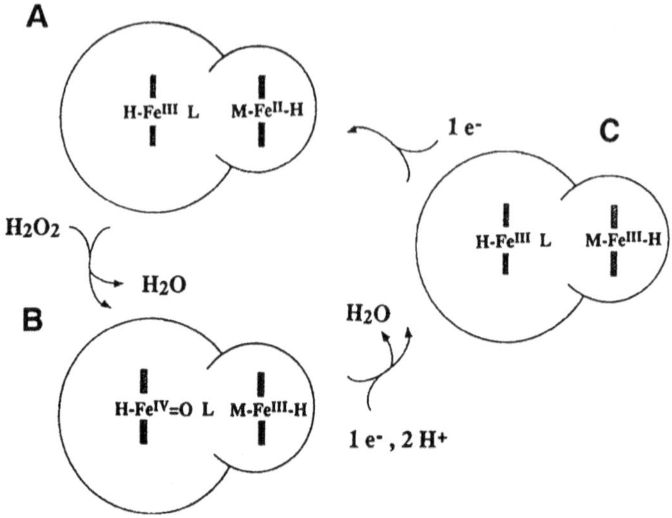

Figure 17. Enzymatic cycle proposed for *Paracoccus denitrificans* cytochrome *c* peroxidase (adapted from [7]).
A-Active half-reduced enzyme. B-Compound I. The LP heme is in the ferryl state and the HP heme is in the ferric state. C- Compound II, formed in a one-electron reduction of compound I. Although both hemes are ferric, this state is not the same as in the oxidized enzyme.

A similar mechanism was proposed to the cytochrome *c* peroxidase from *Ps. aeruginosa* but good experimental evidence is limited.

The preparation of several samples of the two intermediate compounds, by freeze quench kinetics associated with EPR and Mössbauer are planned, to test the validity of the model.

8. Conclusions

At low temperature, EPR and Mössbauer spectra reveal the presence of mainly low-spin species. At room temperature, UV/Visible and NMR spectra reveal the presence of high-spin species. Therefore it is concluded that a temperature dependent spin equilibrium is occurring in the oxidized and half-reduced states of the enzyme. Similar results were obtained for others peroxidases, such as *Pseudomonas aeruginosa* cytochrome *c* peroxidase [12,24,25,26], *Pseudomonas stutzeri* cytochrome *c* peroxidase [27], yeast cytochrome *c* peroxidase [28] and horseradish peroxidase [29]. It is of some interest to note that, in the oxidized enzyme, it is the HP heme the one undergoing the spin transition, while in the half-reduced peroxidase it is the LP heme.

The HP heme (source of electrons for the hydrogen peroxide reduction) is axially coordinated by methionine-histidine. The coordination of the LP heme (centre of the peroxidatic reaction) is not certain but, by analogy to the *Ps. aeruginosa* peroxidase [30] is thought to be bis-histidine or histidine-lysine.

The alterations around the LP heme due to the reduction of the HP heme (shown by the several techniques) are an indication of a heme-heme interaction. Also in cytochrome *c* peroxidase from *Ps. aeruginosa* [12,24,25,26] and *Ps. stutzeri* [27] an interaction between the two hemes was found.

The oxidized enzyme is inaccessible for hydrogen peroxide or cyanide ligation. Only after the reduction of the HP heme (half-reduced peroxidase) does the enzyme become capable of hydrogen peroxide and cyanide binding [7]. This observation suggests that the half-reduced peroxidase has a more open conformation than the oxidized enzyme. A similar observation was found for *Ps. aeruginosa* peroxidase [31].

The present studies reveal that this peroxidase binds calcium ions in a place near the heme sites, necessary to promote the enzyme activation. In contrast for the similar *Ps. aeruginosa* peroxidase there has been no report of a need for calcium ions. However, some eukaryotic peroxidases do contain calcium ions [32,33,34,35,36,37].

It was shown by all these techniques that in *Pa.d.* peroxidase the two processes of reduction and spin state change can be separated in time by removing the calcium ions from the enzyme, by the action of EDTA.

We thank Profs. J.J.G. Moura, J. van Beeumen and Drs. C.F. Goodhew, D. McGinnity, N. Saunder for their contribution to this work.

This work was supported by the JNICT-grant no. STRDA/BIO/359/92, JNICT-Programa Ciencia, a Wellcome Trust Project Grant, British Council, NIH and a Science and Engineering Research Council studentship.

9. References

1. Freeman, B.A. (1984) Biological sites and mechanisms of free radical production, in D. Armstrong *et al* (eds), *Free Radicals in Molecular Biology, Aging, and Disease*, Raven Press, New York, pp. 43-52.
2. Goodhew, C.F., Wilson, I.B.H., Hunter, D.J.B., and Pettigrew, G.W. (1990) The cellular location and specificity of bacterial cytochrome *c* peroxidases, *Biochem. J.* **271**, 707-712.
3. Pettigrew, G.W. (1991) The cytochrome *c* peroxidase of *Paracoccus denitrificans*, *Biochim. Biophys. Acta* **1058**, 25-27.
4. Ellfolk, N., Rönnberg, M., Aäsa, R., Andreasson, L.-E., and Vänngard, T. (1983) Properties and function of the two hemes in *Pseudomonas* cytochrome *c* peroxidase. Evidence for non-equivalence of the hemes, *Biochim. Biophys. Acta* **581**, 325-333.
5. Rönnberg, M., and Ellfolk, N. (1983) Heme-linked properties of *Pseudomonas* cytochrome *c* peroxidase, *Biochim. Biophys. Acta* **743**, 23-30.

6. Gilmour, R., Goodhew, C.F., Pettigrew, G.W., Prazeres, S., Moura, I., and Moura, J.J.G. (1993) Spectroscopic characterization of cytochrome *c* peroxidase from *Paracoccus denitrificans*, *Biochem. J.* **294**, 745-752.
7. Gilmour, R., Goodhew, C.F., Pettigrew, G.W., Prazeres, S., Moura, J.J.G., and Moura, I. (1993) The kinetics of oxidation of cytochrome *c* by *Paracoccus* cytochrome *c* peroxidase, *Biochem. J.* in press.
8. Rönnberg, M., Kalkkinen, N., and Ellfolk, N. (1989) The primary structure of *Pseudomonas* cytochrome *c* peroxidase, *FEBS Letters* **250**, 175-178.
9. Ellfolk, N., Rönnberg, M., and Österlund, K. (1991) Structural and functional features of *Pseudomonas* cytochrome *c* peroxidase, *Biochim. Biophys. Acta* **1080**, 68-77.
10. Moore, G.R., and Pettigrew, G.W. (1990) Cytochromes *c* - Evolutionary, Structural and Physicochemical Aspects, Springer-Verlag.
11. Sprang, S.R. (1993) On a (b-) roll, *TIBS September 1993*, 313-314.
12. Foote, N., Peterson, J., Gadsby, P.M.A., Greenwood, C., and Thomson, A.J. (1985) Redox-linked spin-state change in the di-haem cytochrome c_{551} peroxidase from *Pseudomonas aeruginosa*, *Biochem. J.* **230**, 227-237.
13. Prazeres, S., Moura, I., Moura, J.J.G., Gilmour, R., Goodhew, C.F., and Pettigrew, G.W. (1993) Control of the spin state of the peroxidatic haeme by calcium ions in cytochrome *c* peroxidase from *Paracoccus denitrificans*, *Magn. Reson. Chem.* **31**, S68-S72.
14. La Mar, G.N., Jackson, J.T., Dugad, L.B., Cusanovich, M.A., and Bartsch, R.G. (1990) Proton NMR study of the comparative electronic/magnetic properties and dynamics of the acid <=> alkaline transition in a series of ferricytochromes *c'*, *J. Biol. Chem.* **265**, 16173-16180.
15. Emptage, M.H., Xavier, A.V., Wood, J.M., Alsaadi, B.M., Moore, G.R., Pitt, R.C., Williams, R.J.P., Ambler, R.P., and Bartsch, R.G. (1981) Nuclear Magnetic Resonance studies of *Rhodospirillum rubrum* cytochrome *c'*, *Biochemistry* **20**, 58-64.
16. Moore, G.R. (1985) ^{1}H-NMR studies of the haem and coordinated methionine of Class I and Class II cytochromes *c*, *Biochim. Biophys. Acta* **829**, 425-429.
17. Huynh, B.H., Emptage, M.H., and Münck, E. (1978) Mössbauer study of cytochrome c_2 from *Rhodospirillum rubrum*. Sign of the product $g_xg_yg_z$ of some low spin ferric heme proteins, *Biochim. Biophys. Acta* **534**, 295-306.
18. Dwivedi, A., Toscano Jr, W.A., and Debrunner, P.G. (1979) Mössbauer studies of cytochrome c_{551}. Intrinsic heterogeneity related to g-strain, *Biochim. Biophys. Acta* **576**, 502-508.
19. Lipscomb, J.D., Andersson, K.K., Münck, E., Kent, T.A., and Hooper, A.B. (1982) Resolution of multiple heme centers of hydroxylamine oxidoreductase from *Nitrosomonas*. Mössbauer spectroscopy, *Biochem.* **21**, 3973-3976.
20. Münck, E. (1978) Mössbauer spectroscopy of proteins: electron carriers, in *Methods of Enzymology Vol. LIV-Specialized Techniques*, Academic Press Inc, pp. 346-379.
21. Costa, C., Moura, J.J.G., Moura, I., Liu, M.Y., Peck Jr, H.D., LeGall, J., Wang, Y., and Huynh, B.H. (1990) Hexaheme nitrite reductase from *Desulfovibrio desulfuricans*. Mössbauer and EPR characterization of the heme groups, *J. Biol. Chem.* **265**, 14382-14387.
22. Ravi, N., Moura, I., Costa, C., Teixeira, M., LeGall, J., Moura, J.J.G., and Huynh, B.H. (1992) Mössbauer characterization of the tetraheme cytochrome c_3 from *Desulfovibrio baculatus* (DSM 1743). Spectral deconvolution of the heme components, *Eur. J. Biochem.* **204**, 779-782.
23. Rönnberg, M., Araiso, T., Ellfolk, N., and Dunford, H.B. (1981) The catalytic mechanism of *Pseudomonas* cytochrome *c* peroxidase, *Arch. Biochem. Biophys.* **207**, 197-204.
24. Rönnberg, M., Österlund, K., and Ellfolk, N. (1980) Resonance Raman spectra of *Pseudomonas* cytochrome *c* peroxidase, *Biochim. Biophys. Acta* **626**, 23-30.

25. Foote, N., Peterson, J., Gadsby, P.M.A., Greenwood, C., and Thomson, A.J. (1984) A study of the oxidized form of *Pseudomonas aeruginosa* cytochrome c_{551} peroxidase with the use of magnetic circular dichroism, *Biochem. J.* **223**, 369-378.
26. Ellfolk, N., Rönnberg, M., Aasa, R., Vänngard, T., and Angström, J. (1984) Spin states of and interaction between the hemes of *Pseudomonas* cytochrome *c* peroxidase. Evidence from proton NMR and cyanide binding studies, *Biochim. Biophys. Acta* **791**, 9-14.
27. Villalaín, J., Moura, I., Liu, M.C., Payne, W.J., LeGall, J., Xavier, A.V., and Moura, J.J.G. (1984) NMR and electron-paramagnetic-resonance studies of a dihaem cytochrome from *Pseudomonas stutzeri* (ATCC 11607) (cytochrome *c* peroxidase), *Eur. J. Biochem.* **141**, 305-312.
28. Yonetani, T., and Anni, H. (1987) Yeast cytochrome *c* peroxidase. Coordination and spin states of heme prosthetic group, *J. Biol. Chem.* **262**, 9547-9554.
29. Tamura, N., and Hori, H. (1972) Optical and magnetic measurements of horseradish peroxidase. Electron paramagnetic resonance studies at liquid-hydrogen and -helium temperatures, *Biochim. Biophys. Acta* **284**, 20-29.
30. Aäsa, R., Ellfolk, N., Rönnberg, M., and Vänngard, T. (1981) Electron paramagnetic resonance studies of *Pseudomonas* cytochrome *c* peroxidase, *Biochim. Biophys. Acta* **670**, 170-175.
31. Ellfolk, N., Rönnberg, M., Aasa, R., Andreasson, L.-E., and Vänngard, T. (1984) Anion binding to the resting and half-reduced *Pseudomonas* cytochrome *c* peroxidase, *Biochim. Biophys. Acta* **784**, 62-67.
32. Poulos, T.L., Edwards, S.L., Wariishi, H., and Gold, M.H. (1993) Crystallographic refinement of lignin peroxidase at 2 A^{o}, *J. Biol. Chem.* **268**, 4429-4440.
33. Banci, L., Bertini, I., Bini, T., Tien, M., and Turano, P. (1993) Binding of horseradish, lignin and manganese peroxidases to their respective substrates, *Biochemistry* **32**, 5825-5831.
34. Haschke, R.H., and Friedhoff, J.M. (1978) Calcium-related properties of horseradish peroxidase, *Biochem. Biophys. Res. Commun.* **80**, 1039-1042.
35. Ogawa, S., Shiro, Y., and Morishima, I. (1979) Calcium binding by horseradish peroxidase *c* and the heme environmental structure, *Biochem. Biophys. Res. Commun.* **90**, 674-678.
36. Morishima, I., Kurono, M., and Shiro, Y. (1986) Presence of calcium ion in horseradish peroxidase, *J. Biol. Chem.* **261**, 9391-9399.
37. Shiro, Y., Kurono, M., and Morishima, I. (1986) Presence of endogenous calcium ion and its functional and structural regulation in horseradish peroxidase, *J. Biol. Chem.* **261**, 9382-9390.

METALLOPROTEIN - ENDOR- SPECTROSCOPY

Structure determination of the prosthetic site from randomly oriented specimen and correlations with NMR-spectroscopy

J. HÜTTERMANN, G. P. DÄGES, H. REINHARD AND G. SCHMIDT
Fachrichtung Biophysik und Physikalische Grundlagen der Medizin, Universität des Saarlandes, Klinikum, Bau 76, 66421 Homburg/Saar, Germany

Abstract

The theoretical foundations for the application of high-resolution electron-nuclear-double-resonance spectroscopy to the determination of structural properties of the prosthetic group in randomly oriented metalloproteins are delineated. Applications described involve Superoxide Dismutase as a copper-containing protein and Enoate Reductase as an iron-sulfur protein. Correlations with nuclear magnetic resonance are shown to involve mainly the isotropic coupling for protons as obtained from the paramagnetic shift and the direction of the g-tensor in the molecular framework.

1. Introduction

ENDOR (Electron Nuclear DOuble Resonance)-spectroscopy detects nuclear interactions by a resonant change in the intensity of an EPR (Electron Paramagnetic Resonance) line in a paramagnetic system in which nuclear spins are coupled to the electron spin S. For metalloproteins, S is delivered by the transition metal ion which usually is part or center of the

G.N. La Mar (ed.), Nuclear Magnetic Resonance of Paramagnetic Macromolecules, 165-192.

prosthetic group of the protein. The nuclear spins are those of ligand atoms in the immediate surrounding of the metal ion. Typically, protons (^{1}H, I=1/2) or nitrogens (^{14}N, I=1) are ENDOR-active due to their nuclear spin I and form part of the coordination shell. Other active nuclei can be gained from isotopic substitution. Their interaction with the electron spin is usually small and contributes to the EPR-linewidth but can be retrieved by ENDOR-spectroscopy. The metal ion itself may also have a nuclear spin (e.g. Cu^{++}, I=3/2) in which case its interaction with S often is quite large and thus observable in EPR. The hierarchy of interaction strengths is reflected by a Spin-Hamiltonian $\hat{\mathbf{H}}$ which can be written as:

$$\hat{\mathbf{H}} = \left[\beta \cdot \vec{S} * \mathbf{g} * \vec{H} + \vec{S} * \mathbf{A} * \vec{I} + \vec{I} * \mathbf{P} * \vec{I}\right]_{Metal} \qquad (1)$$
$$+ \sum_{L} \left[\vec{S} * \mathbf{A} * \vec{I} + \vec{I} * \mathbf{P} * \vec{I} + g_n \beta_n * \vec{H} * \vec{I}\right]_{Ligands}$$

The first term in the metal ion part of $\hat{\mathbf{H}}$ is the Zeeman interaction of the electronic spin (operator $\vec{S}$, Bohr magneton β) with the magnetic field $\vec{H}$. Its strength is characterized by the tensorial observable **g**. The next two terms involve metal ion hyperfine interaction (tensor **A**) due to a nuclear spin (operator $\vec{I}$) and quadrupolar interaction (tensor **P**) if the nuclear spin is ≥ 1. The same two terms apply also for the ligand part of $\hat{\mathbf{H}}$ which, in addition, comprises a nuclear Zeeman interaction (nuclear magneton β_n, nuclear g-factor g_n). This form of the Hamiltonian is applicable for metalloproteins with an effective spin S = 1/2.

The task of EPR-ENDOR-spectroscopy of metalloproteins is to describe, as complete as possible, the Hamiltonian $\hat{\mathbf{H}}$ for the system under study by delineating all its tensorial entities. If achieved, one would then be able to construct the metal ion / ligands geometries and to infer on wavefunction properties of the complex. The tensorial nature of the interactions would seem to necessitate single crystals as samples in order to unravel their anisotropy. Fortunately, this severe restriction can be circumvented if the anisotropy of at least one of the metal ion interaction parameters, **g** and **A**, is sufficiently large to spread the EPR resonance condition over a broad range of magnetic field values. In this case, there is a defined, though *a priory* unknown connection between orientations of the paramagnetic complex and a magnetic field value in an EPR spectrum of a randomly oriented sample. While one can determine in such „powder-type“ spectra the principal values of **g** and, if resolved, $\mathbf{A}_{\mathbf{Metal}}$ but not their directions, ENDOR can retrieve the orientation of the ligands in the g-frame and thus provides for the connection between the **g**-tensor and the metal ion / ligand complex.

In this paper we shall delineate the theoretical background of what will be termed orientation-selective ENDOR and its application in the spatial reconstruction of prosthetic groups of metalloproteins from randomly oriented samples. It will be shown that a unique structural determination requires, in addition to ENDOR, information from as many as possible other spectroscopic and structural methods as well as data from suitable model compounds. One technique which has a strong impact on the interpretation of ENDOR data is NMR (Nuclear Magnetic Resonance) spectroscopy of paramagnetic proteins. We shall outline the relation between the two methods and make use of them in the examples of proteins described. For some of the results given, the data are yet to be refined and the paper gives the present status only.

2. Theoretical Background

Solving the Spin-Hamiltonian (1) is equivalent to finding parameters which reproduce, when applied, the experimental data. The usual test of their validity then requires a theoretical simulation which produces a spectrum. Iterative variation of the parameters may eventually yield correspondence with the experimental data. The application of this approach suffers, however, from limitations. For one thing, the number of tensorial entities in (1) which determines the dimension of the Hamiltonian matrix often is very large, depending on the number of ligands. Moreover, for reproducing a spectrum lineshape tensors have to be included also and the number of orientations for which resonance fields and frequencies are to be determined is in the order of 10^4 to 10^5 for systems with large anisotropy. An iterative parameter variation for both EPR and ENDOR interactions in a direct matrix diagonalization of (1) for tensors with arbitrary symmetry is, therefore, beyond the limits of practical application even with the computer power available today. A feasible alternative, which we pursue in our studies, starts with separating the EPR simulation from the ENDOR part and solving them independently under approximations concerning the level of perturbation theory. This approach is justified considering the large difference in interaction energies between the metal ions and the ligand terms in (1). In the ENDOR simulation, each nucleus is treated separately, an approximation which suggests itself since inter-nuclear interactions are negligible compared to the electron-nuclear terms and are not included in (1).

Applying programs developed for fast, interactive parameter variation in both EPR and ENDOR we arrive at parameters which can be used as start values in final simulations using matrix diagonalization.

2.1 EPR-SIMULATION

The static EPR-spectrum is generated basically by finding the values for the magnetic field H_{res} at which, at a given value of the microwave frequency ν, there is a transition between energy levels. At that value, a line is produced by a lineshape function, typically Gaussian or Lorentzian. The powder-type spectrum then is a summation of the field values for all orientations of the paramagnetic axes with respect to the external magnetic field. Applying first-order perturbation theory and neglecting the nuclear quadrupole term, the field H_{res} is given by

$$H_{res} = (h\nu \pm Am) / g\beta \tag{2}$$

(h = Planck constant; m = nuclear spin quantum number).

A and g are the effective values for the tensors **A** and **g** of the metal ions which, in a coordinate system (x, y, z) given by the magnetic field direction can be written as

$$g = \sqrt{g_{xx}^2 \cdot x^2 + g_{yy}^2 \cdot y^2 + g_{zz}^2 \cdot z^2}$$

and

$$A = \frac{1}{g} \cdot \begin{bmatrix} \left(g_{xx} \cdot x \cdot A_{xx} + g_{yy} \cdot y \cdot A_{yx} + g_{zz} \cdot z \cdot A_{zx}\right)^2 \\ +\left(g_{xx} \cdot x \cdot A_{xy} + g_{yy} \cdot y \cdot A_{yy} + g_{zz} \cdot z \cdot A_{zy}\right)^2 \\ +\left(g_{xx} \cdot x \cdot A_{xz} + g_{yy} \cdot y \cdot A_{yz} + g_{zz} \cdot z \cdot A_{zz}\right)^2 \end{bmatrix}^{1/2} \tag{3}$$

The variation of orientations in the reference frame can be done in various ways. Typically, we chose random numbers. This has the advantage that the results of several runs can be added since they involve independent orientations. The result of the simulation is a table of all values of H_{res} with attached orientations. A spectrum is obtained by sorting the H_{res} values in an array of 1024 data points (which is the resolution of our experimental spectra) and folding them with an appropriate lineshape function.

This type of EPR-simulation is the starting point for the fast, interactive ENDOR simulation. Should there be the need for subsequent refinement either on the EPR- or the ENDOR-part we use the direct Hamiltonian diagonalization method described in detail by Kreiter and Hüttermann [1].

2.2 PROTON ENDOR

The response of a single proton in ENDOR derives from the nuclear Hamiltonian which, for a system with an effective electron spin S=1/2, is written as:

$$\hat{\mathbf{H}}_N = \vec{S} * \mathbf{A} * \vec{I} - g_n \beta_n \vec{H} * \vec{I} \tag{4}$$

In the axis system (x, y, z) in which the g-tensor is diagonal the Hamiltonian can be written in components [2, 3].

$$\hat{\mathbf{H}}_N = \sum_{i=1}^{3} \left[\left(\vec{S}_e * \mathbf{A} \right)_i - g_n \beta_n H_0 h_i \right] \vec{I}_i \tag{5}$$

with $\vec{H} = [h_1, h_2, h_3]H_0$ and $\vec{S}_e$ being the vector of the electron spin written in that axis system as

$$\vec{S}_e = \frac{M_s}{g(\theta, \phi)} \cdot \left[g_{xx} h_1, g_{yy} h_2, g_{zz} h_3 \right] \tag{6}$$

where M_S is the spin quantum number and $g(\theta, \phi)$ the value of g at polar angles θ and ϕ with respect to x, y and z. With A_{ij} being the components of **A** and with the selection rules $\Delta M_S = 0$, and $\Delta m_I = \pm 1$ the ENDOR resonance condition is written as:

$$\nu(H,M_s)=\left[\sum_i\left[\frac{M_s}{g(\theta,\phi)}-\left(\sum_j g_j h_j A_{ij}\right)-h_i\nu_0\right]^2\right]^{1/2} \tag{7}$$

ν_0 is the NMR frequency of the proton in the field H_0. The components of the hyperfine tensor comprise the anisotropic (dipolar) and the isotropic part, and can be written in frequency units as:

$$hA_{ij}=-g_n\beta_n\beta_e r^{-3}g_i\left(3r_ir_j-\delta_{ij}\right)+A_{iso}\delta_{ij} \tag{8}$$

In this expression the distance r and the location (r_i, r_j) of the proton with respect to the center of spin-density in the point-dipole approximation, as well as the isotropic coupling are the parameters determining the ENDOR-response (h = Planck-constant, δ_{ij} = Kronecker symbol). In the case of distributed spin-density ρ over various centers, the components for each are weighted and added componentwise. The powder-type ENDOR spectrum is calculated for a given magnetic field by adding the ENDOR-lines of all contributing orientations.

In simulations, the number of orientations used for EPR limits the value of the contributions of orientations of an ENDOR-working point. This may lead to artefacts in the simulated ENDOR-spectrum comparable to those known from insufficient orientations in EPR. An even distribution of contributing orientations can be gained from an analytical expression which holds if **A** and **g** are collinear. In that case, equ. (2) can be reformulated as

$$H_{res}=\frac{h\cdot\nu-\sqrt{g_{xx}^2\cdot A_{xx}^2\cdot x^2+g_{yy}^2\cdot A_{yy}^2\cdot y^2+g_{zz}^2\cdot A_{zz}^2\cdot\left(1-x^2-y^2\right)^2}\cdot m}{\sqrt{g_{xx}^2\cdot x^2+g_{yy}^2\cdot y^2+g_{zz}^2\cdot\left(1-x^2-y^2\right)^2}\cdot\beta} \tag{9}$$

using the normalization condition

$$x^2+y^2+z^2=1$$

Equation (9) can be solved for y^2 to yield

$$y^2 = \frac{-H^2 \cdot g_{xx}^2 \cdot x^2 + H^2 \cdot g_{zz}^2 - H^2 \cdot g_{zz}^2 \cdot x^2 - c_1^2}{H^2 \cdot \left(g_{yy}^2 - g_{zz}^2\right)} + \frac{2 \cdot c_1 \cdot c_2 \cdot \left(\frac{1}{2 \cdot d_1} - d_2 \pm \sqrt{d_2^2 - 4 \cdot d_1 \cdot d_3}\right)}{H^2 \cdot \left(g_{yy}^2 - g_{zz}^2\right)} + \frac{c_c^2 \cdot \left(\frac{1}{2 \cdot d_1} - d_2 \pm \sqrt{d_2^2 - 4 \cdot d_1 \cdot d_3}\right)^2}{H^2 \cdot \left(g_{yy}^2 - g_{zz}^2\right)} \tag{10}$$

with

$$d_1 = c_2^2 \cdot \left(A_{zz}^2 - A_{yy}^2\right) + H^2 \cdot \left(g_{yy}^2 - g_{zz}^2\right)$$

$$d_2 = \left(-2 \cdot c_1 \cdot c_2 \cdot \left(A_{yy}^2 - A_{zz}^2\right)\right)$$

$$d_3 = -A_{xx} \cdot x^2 \cdot H^2 \cdot g_{yy}^2 + A_{xx}^2 \cdot x^2 \cdot H^2 \cdot g_{zz}^2 + A_{yy}^2 \cdot x^2 \cdot H^2 \cdot g_{xx}^2 + A_{yy}^2 \cdot H^2 \cdot g_{zz}^2 - A_{yy}^2 \cdot x^2 \cdot H^2 \cdot g_{zz}^2 - A_{yy}^2 \cdot c_1^2 - A_{zz}^2 \cdot x^2 \cdot H^2 \cdot g_{xx}^2 + A_{zz}^2 \cdot c_1^2 - A_{zz}^2 \cdot H^2 \cdot g_{yy}^2 + A_{zz}^2 \cdot x^2 \cdot H^2 \cdot g_{yy}^2$$

$H = H_{res}$

$$c_1 = \frac{h \cdot \nu}{\beta}$$

$$c_2 = \frac{m}{\beta} \tag{11}$$

Fig. 1 shows an examples of this calculation which is performed in a program denoted „Prometheus“ (**pro**tons in **met**alloproteins: **E**NDOR of **u**noriented **s**ystems). The left window gives, in stereographic projection, the orientations of the four nuclear copper sublevels contributing to the ENDOR response when exiting the EPR spectrum shown in the upper right window at the magnetic field value indicated by the cursor position. The EPR spectrum relates to the CN^- bound derivative of copper-zinc superoxide dismutase to be discussed in more detail below. The bottom right window shows some of

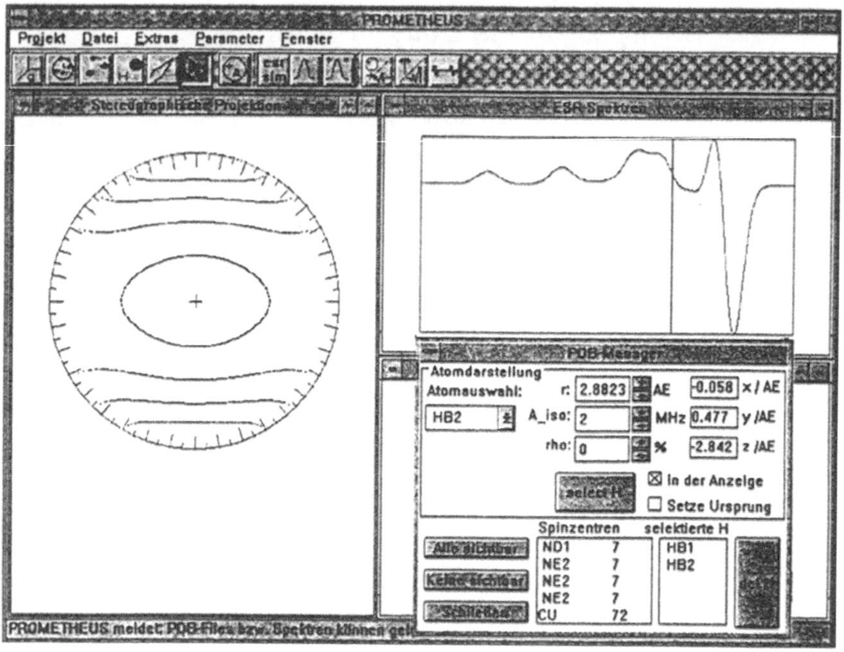

Figure 1. Graphical representation of orientations and nuclear sublevels (left window) excited at a specific magnetic field value in an EPR-spectrum (top right window) and list of variables in ENDOR simulations (bottom right) of protons

the parameters which can be varied in the ENDOR simulation. Five different spin centers and their spin density are seen (copper (72 %) and the four coordinated nitrogens (7 % each)). The protons to be simulated are the two β-protons of the histidine 44 residue, Hβ1 and Hβ2. For Hβ2, the distance, the isotropic coupling and the fractional coordinates are given, which all can be iteratively changed.

An example for the ENDOR simulation capacities of the program is seen in Fig. 2, which displays four windows. The upper left now shows, again in stereographic projection, the positions of all protons in the prosthetic group, two of which (Hβ1 and Hβ2) are selected for simulation and marked by arrows. The EPR spectrum (bottom left) is now for the native superoxide dismutase and shows two field positions at which the spectra of the two protons are simulated simultaneously. The results for the high- and low field positions are seen in the top and bottom right windows, respectively.

The simulation is interactive and gives results nearly in real time. The

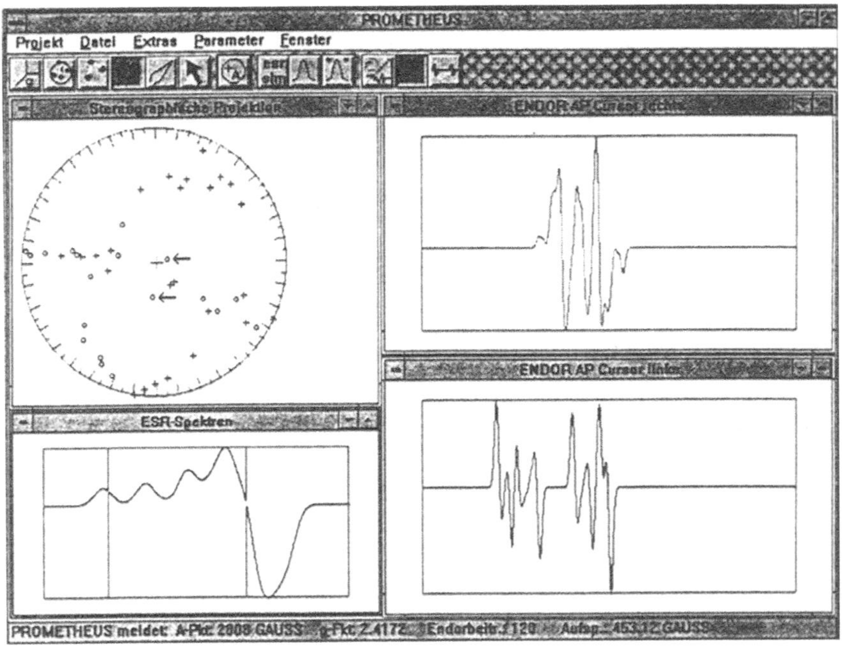

Figure 2. Stereographic projection of protons in SOD together with EPR-spectrum (left windows); simulated proton ENDOR spectra for two protons marked by arrow at magnetic fields (right windows).

proton positions can be changed with respect to the g-axes system. The spectra can be compared on-line with the experimental ones in the same window which are distinguishable by colour codes (not shown here).

2.3. NITROGEN ENDOR

The energy levels E of a system with S =1/2 and I = 1 (^{14}N) are given in first-order perturbation theory as [4]:

$$E = K(M_s)\cdot m_I - (\mathbf{k}^{T} * \mathbf{P} * \mathbf{k})\cdot(1/3\cdot I(I+1) - m_I^2) \tag{12}$$

with

$$(\mathbf{k}^{T} * \mathbf{P} * \mathbf{k}) = \frac{\mathbf{h}^{T} * \mathbf{K}^{T}(M_s) * \mathbf{P} * \mathbf{K}(M_s) * \mathbf{h}}{\mathbf{h}^{T} * \mathbf{K}^{T}(M_s) * \mathbf{K}(M_s) * \mathbf{h}} \tag{13}$$

The first term in (12) represents the hyperfine and the second the quadrupolar interaction with tensor **P**. The latter can be written as $P(M_S)[1/3I\,(I+1) - m_I^2]$ with $P(M_S)$ giving the quadrupolar contribution in the axis system described above for protons. It follows that:

$$P(M_s) = \frac{\mathbf{h}^{T} * \left|\left(\frac{M_s}{g(\theta,\phi)}\right)\mathbf{A} * \mathbf{g} - \nu_N \mathbf{1}\right|^{T} * \mathbf{P} * \left|\left(\frac{M_s}{g(\theta,\phi)}\right)\mathbf{A} * \mathbf{g} - \nu_N \mathbf{1}\right| * \mathbf{h}}{\mathbf{h}^{T} * \left|\left(\frac{M_s}{g(\theta,\phi)}\right)\mathbf{A} * \mathbf{g} - \nu_N \mathbf{1}\right|^{T} * \left|\left(\frac{M_s}{g(\theta,\phi)}\right)\mathbf{A} * \mathbf{g} - \nu_N \mathbf{1}\right| * \mathbf{h}} \tag{14}$$

where the denominator is the expression for the square of $K(M_S)$ used in (12) and where the matrices **1** and **h** represent the unit and the direction cosine matrix, respectively. ν_N is the NMR frequency of the nitrogen nucleus at the respective magnetic field value. Transposed matrices are indicated by ^{T}.

With the selection rules $\Delta M_S = 0$ and $\Delta m_I = \pm 1$ there are four ENDOR lines obtained at frequencies

$$\nu = |K(M_s) \pm P(M_s)| \tag{15}$$

The „powder“ response is again obtained by summing over all orientations contributing to a given magnetic field value.

3. Applications

3.1. SUPEROXIDE DISMUTASE

Superoxide dismutase (SOD) is a dimeric protein with identical subunits each containing a prosthetic group comprising a Cu^{++} and a Zn^{++} ion. The dismutation of $O_2^{-\cdot}$ radicals involves a twofold valence change at the copper ion [5, 6]. The latter is coordinated to four histidine residues (44, 46, 61 and 118 in the enumeration of the bovine isoenzyme) via their nitrogens as depicted in Scheme I.

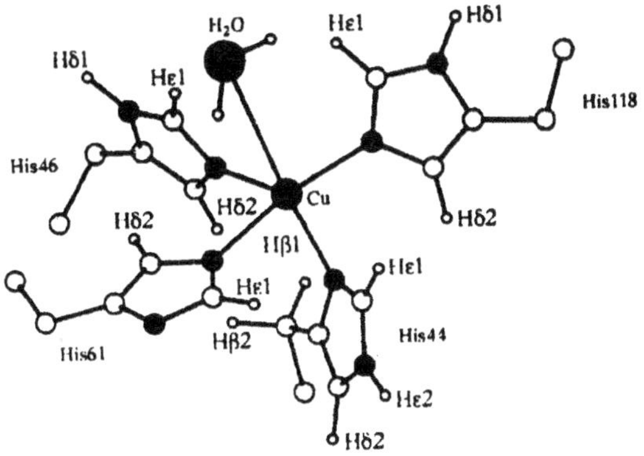

Scheme 1. Schematic representation of the Cu-site in SOD showing coordinated histidine residues and their numbering.

We have recently unravelled the structure of the copper site of the native protein by ENDOR of randomly oriented specimen at low temperatures (5 K) and found, among others, that it is unchanged with respect to the results from X-ray diffraction at room temperature [7, 8]. We now wish to discuss our present understanding of the structural changes involved when inhibitors like CN^- and N_3^- bind to the copper site mimicking the binding of $O_2^{-\cdot}$ but leaving the Cu valence unchanged.

The effect of N_3^- and CN^- addition can be seen first in the EPR-spectra. Fig. 3 shows the experimental tracings together with the orientational gamut of copper nuclear hyperfine sublevels contributing to the spectra which were derived from EPR-simulation. While the native protein showed relatively large rhombicity in both g- (g_{xx} = 2.0257, g_{yy} = 2.1074, g_{zz} = 2.2655) and copper hyperfine interactions (A_{xx} = 3.0, A_{yy} =3.5, A_{zz} = 13.60 mT) [7], the simulation parameters for the N_3^- -derivative (g_{xx} = 2.0344, g_{yy} = 2.0653, g_{zz} = 2.242; A_{xx} = 2.0, A_{yy} = 4.5, A_{zz} = 15.8 mT) reveal less rhombic distortions in the g-tensor and an enhanced A_{zz} hyperfine value. This tendency is continued in the CN^--derivative which shows full axiality in its parameters (g_{xy} = 2.0375, g_z = 2.2104, A_{xy} = 4.17, A_{zz} = 20.87 mT (for the ^{65}Cu isotope)).

Shown in Fig. 3 as capital numbers I, II, ...V are the respective working points in the EPR spectrum at which ENDOR data were taken. The dashed lines give the relation of the working points to the contributing copper hyperfine sublevels and their orientations. The positions are related to those used for the native protein [7].

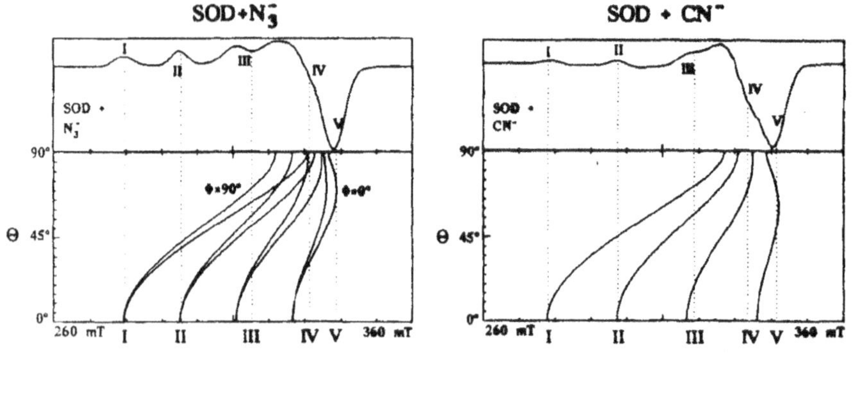

Figure 3. EPR-spectra and correlated orientational variation of copper hyperfine transitions for SOD treated with N_3^- (left) and CN^-, respectively (right). ENDOR working positions I through V are indicated.

Let us consider first the ^{14}N interactions of the coordinated nitrogens. In the native protein we found that there are two groups of interactions which differ sufficiently to be discerned. They have equivalent couplings at field position I but differ at position V [7]. One group gave values identical to those of the model compound copper-tetraimidazole [9] and was assigned as comprising the nitrogens of His 44 and His 46. The other group then contained His 61 and His 118. For the CN^- derivative, previous ENDOR work has shown that one histidine was replaced by CN^-; the remaining interactions were grouped into two classes. One comprised the nitrogen trans to C of CN^- for which the coupling was found to be reduced compared to the native protein. The other contained two equivalent nitrogens cis to that position with an enhanced coupling. This was inferred from measurements at position I only so that a full tensor could not be given [10]. Our experimental results are shown in Fig. 4 for both position I and V together with the respective simulations. At position I we can confirm the previous interpretation [10, 11]; the simulations give about the same values for A_{zz} but slightly different parameters for the quadrupolar interaction. The spectra at position V are interpreted here for the first time revealing a surprisingly large value of the (in-plane) maximum hyperfine coupling for the two cis nitrogens of about 59 MHz.

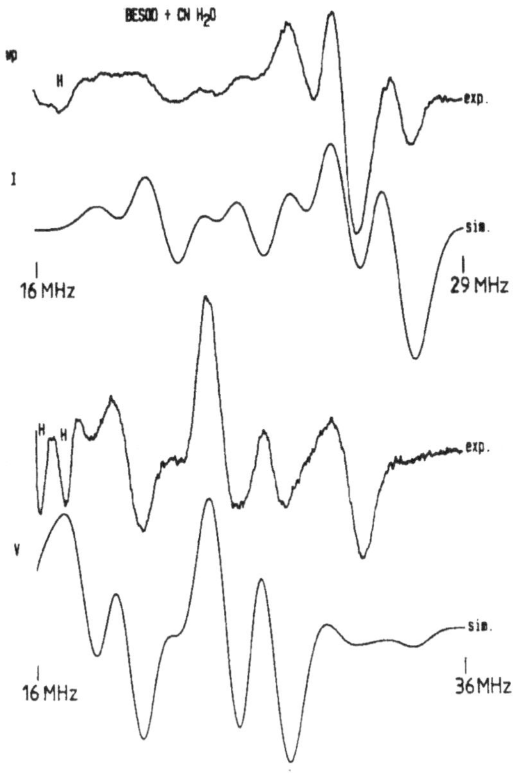

Figure 4. ^{14}N-ENDOR-spectra (exp) and simulations (sim) at working points I (top) and V (bottom) for SOD treated with CN^-

Whereas ENDOR can usually not reveal the number of equivalent nuclei, this can be tested from the intensity relations in the EPR spectrum, if the interaction is resolved. Fig. 5 shows, in an expanded mode, the copper hyperfine EPR-line of field position I in first, and second derivative mode for enhanced resolution. The simulated spectra for both modes use the nitrogen parameters from the ENDOR and two equivalent nuclei for the cis nitrogens. Clearly, using both copper isotopes contributing to the EPR, the agreement with the experimental spectra both in component spacings and intensities is sufficient well to warrant the interpretation, of two large, equivalent and one smaller nitrogen interactions as proposed before [12].

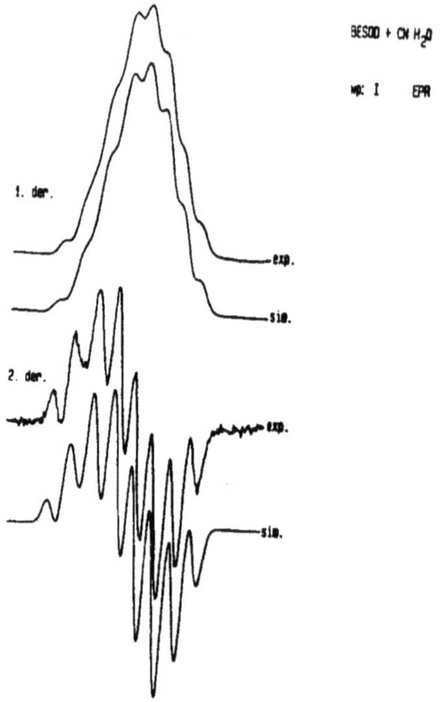

Figure 5. Experimental (exp) and simulated (sim) EPR-line at working position I for SOD treated with CN⁻ using ENDOR parameters for simulation

For the N_3^- derivative the situation proved to be more difficult. Fig. 6 shows the experimental and simulated spectra at the two field positions I and V. As in the native protein, the data require a simulation with one group of nitrogens only at position I, the coupling values being nearly the same in

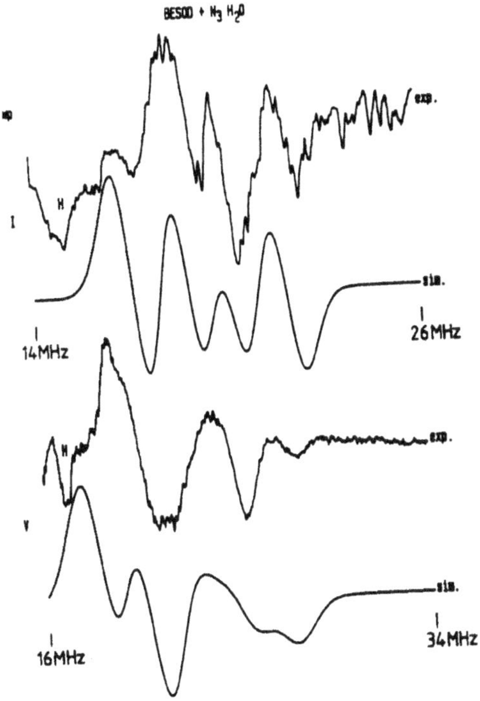

Figure 6. Same as Figure 4 but for SOD treated with N_3^-

both specimen. Field position V, however, needs differentiation into two groups with strongly differing parameters which, moreover, for one group are more rhombically distorted than in the native SOD. We note that it is not possible to relate the ENDOR to the EPR data for the N_3^- derivative since the latter are not sufficiently resolved. Also, in our previous work on this sample we were unable to detect the coupling of the azide ^{14}N interaction [14].We therefore cannot give the number of contributing nuclei. On account of the proton-ENDOR to be discussed below and on the change in EPR it is reasonable to propose that N_3^- binds by replacing one histidine nitrogen as does cyanide but that the binding has smaller effects on structural changes in copper coordination.

The parameters used in simulations for both derivatives are listed in Table 1 which contains, for comparison, also the results for native SOD and previous ENDOR data.

TABLE 1 Compilation of ^{14}N ENDOR parameters for SOD and its derivatives treated with N_3^- and CN^-, respectively

Sample / Field pos.		[a] BESOD		BESOD + N_3^-		[a] BESOD + CN^-		Reference
		A_{zz}	P_{zz}	A_{zz}	P_{zz}	A_{zz}	P_{zz}	
I	(1)	38.7±1	0.86±0.01	37.9±1	1.54±0.02	37.0±1.3	«1.6	(Van Camp et al. 1982)
	(2)					47.8±0.1	1.62±0.01	
	(1)	-	-	-	-	37.2±0.5	-	(Hüttermann et al. 1988)
	(2)					47.8±0.5	1.7±0.1	
	(1)	40.1±0.3	-0.825±0.1	39.0±0.3	-1.50±0.1	37.2±0.3	0.86±0.1	this work
	(2)					48.6±0.3	1.74±0.1	
II		39.0±0.4	1.1±0.1	38.6±0.5	1.4±0.1	-	-	(Hüttermann et al. 1988)
V		A_{xx}	P_{xx}	A_{xx}	P_{xx}	A_{xx}	P_{xx}	this work
	(1)	46.8±0.3	3.45±0.1	43.5±0.3	2.25±0.1	41.4±0.3	-1.72±0.2	
	(2)	50.0±0.8	-1.73±0.2	49.0±0.8	-1.65±0.2	59.0±0.8	-3.08±0.5	
		A_{yy}	P_{yy}	A_{yy}	P_{yy}	A_{yy}	P_{yy}	
	(1)	35.0±0.3	-2.63±0.1	37.5±0.3	-0.75±0.2	37.2±0.3	0.86±0.1	
	(2)	41.9±0.3	2.55±0.1	59.0±0.8	3.15±0.5	48.6±0.3	1.26±0.2	

(1), (2): His61-His118 or His44-His46

a) Adapted from *Reinhard et. al. (1994)*

Turning now to the protons we show in Fig. 7 (top part) for comparison the variation of the outermost coupling values obtained at the five different working positions from the simulations for native SOD. In the range of 1-3 MHz there are numerous lines which make differentiation difficult. However, couplings marked A and G are sufficiently separated at field position I. They belong to the β_2 proton of His 44 (A) and the δ_2 proton of His 46 (G). If we compare, in the bottom part of Fig. 7, the native protein in H_2O and D_2O buffer with the respective data from the N_3^- and the CN^- derivatives at that field position it becomes apparent that the line markedβ (A) in SOD is strongly affected and vanishes whereas the line marked * (G) remains more or less unchanged. This would seem to indicate that both anions replace His 44.

That the β-proton of His 44 is affected by inhibitor binding can also be inferred at working positon IV. Coupling A overlaps with J at that position in native SOD as seen from the top part of Fig. 7. Both in N_3^- and CN^-

treated SOD only J remains. This can be demonstrated convincingly e. g. in Fig. 8 for the CN^- derivative in which the top two experimental spectra are in H_2O and D_2O buffer, respectively. Clearly the labile proton line marked (↓) is fully exchanged in D_2O which was not the case in native SOD due to the overlap with the non-exchangeable His 44 β-proton. Moreover, since the labile proton was assigned to the δ_1 proton of His 118, the results show that this residue is affected only slightly by the binding of CN^- (and N_3^-) to copper. The effect is a small enhancement in coupling value bringing the interaction closer to that of the respective proton in the copper-tetraimidazole model compound [9].

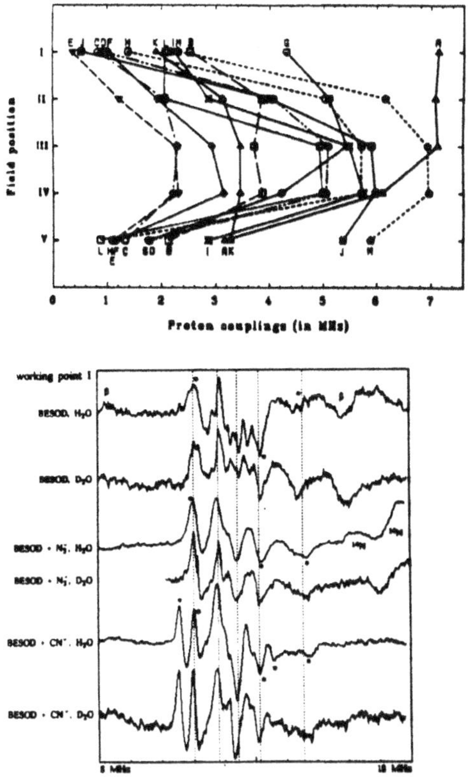

Figure 7. Gamut of outermost ENDOR couplings of various protons in native SOD at different field positions (top) and effect of N_3^- and CN^- treatment on experimental proton ENDOR spectra at working point I in comparison with native SOD (bottom)

The simulations given in the middle part of Fig. 8 show, that the coupling size and the line shape of the δ_1 proton is better reproduced by the coordinates marked „tw“ which is for twisted His 118 rather than for the native configuration. The same goes for another well discernable proton, Hε_1 of His 118. Consider the outermost proton line in the upper tracing of Fig. 8. It is composed of at least two contributions. One is from an ε_1-proton which is seen to be gone in the sample with ε_1-protons specifically deuterated (third trace from top). The other is from a δ_2-proton and lost in the bottom spectrum for the sample in which both classes of protons are deuterated. In the native protein, the protons ε_1 and δ_2 from residue His 118 contributed in a related frequency range (lines L and M in the top panel of Fig. 7) but were separated by about 1 MHz. In the axially symmetrical CN^- derivative, like in the copper-tetraimidazole model compound [9], both protons now are nearly equivalent as expected for an „in-plane“ field position. Again, the simulation using the ε_1-proton as example give a better match with the experimental coupling in the „tw“ configuration. In this, the residue His 118 was twisted by about 60° around the Cu-N bond bringing the imidazole plane in a configuration closer to that of the model compound.

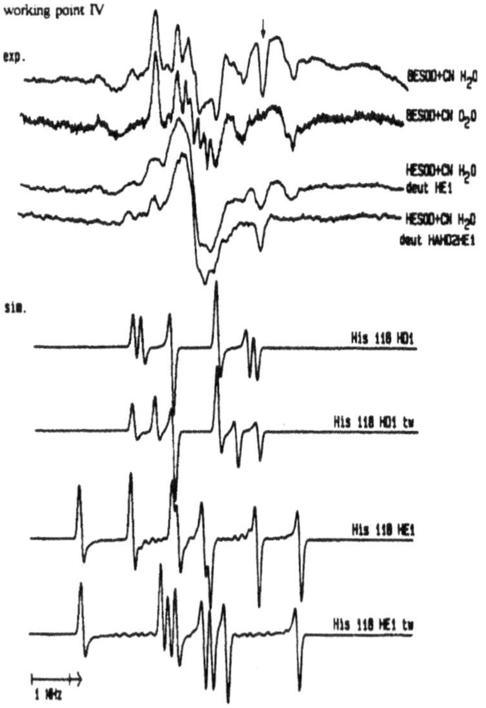

Figure 8. Experimental proton ENDOR spectra at working point IV for different SOD specimen treated with CN^- (top) and simulated spectra for two protons of His118 showing the effect of conformational changes; for details see text

While these data would seem to argue that the anion binding replaces His 44, NMR data strongly argue for the replacement of His 46 [13]. In extensive theoretical simulations of the ENDOR data we have positioned the g-tensor axes directions such as to account for either loss of His 44 and His 46. It was possible to find a good agreement for all discernable protons other than β of His 44 with both assumptions. For example, the simulations of Fig. 8 are performed with the g-axes deriving from loss of His 46 and match well with the His 118 resonances; the same match for these protons, due to the symmetry of the problem, is achieved, however, with the axis system required from loss of His 44. We therefore have to conclude, that if His 46 should be replaced by the anions, the configuration of the new Cu-ion environment must change in a way that the β-protons of His 44 are removed sufficiently far away to give no discernable interaction. This condition is, of course, fullfilled easier if His 44 is replaced. In any way, the anions both bind to the copper ion affecting either His 44 or 46 and leaving the two other residues His 61 and 118 largely unchanged. From the remaining rhombicity in EPR spectra and in the ^{14}N ENDOR parameters one can infer that the total structural changes induced by N_3^- are smaller than those for CN^- but both affect the same histidine residues.

Correlations between NMR spectroscopy and ENDOR rely on the fact that the same nuclear Hamiltonian (4) governs the response of the system under study. A parameter directly related is the isotropic proton coupling A_{iso} of equation (7). For the native protein we have recently compared the values calculated from the paramagnetic shift in NMR [13] with those used in ENDOR simulations and found very good agreement for most of the protons [7]. The CN^- and N_3^- derivatives discussed here still have to await complete assignment of the protons in ENDOR before such a comparison can be made.

Other correlations involve other nuclei if they can be studied by NMR spectroscopy. ENDOR allows e. g. for the simultaneous observation of protons and coordinated nitrogen nuclei. Remote nitrogen interactions in histidines are usually so small that they are better observed in Electron Spin Echo Envelope Modulation (ESEEM) spectroscopy, another high resolution EPR-technique. These nuclei can also be investigated in heteroatom correlation NMR. For native, ^{15}N-substituted SOD and its CN^- and N_3^- derivatives a good match was recently obtained for the isotropic couplings of the remote histidine nitrogens [14,15].

3.2. ENOATE REDUCTASE

This protein catalyzes the stereospecific reduction of 2-enoates at the expense of hydrogen gas. It contains 12 identical subunits with a molecular weight of 73 kDa each. The subunits form trimers, four of which assemble to the functional protein. Besides cofactors, the active site of each subunit contains four iron and four sulfur atoms [16].

We have studied the protein in the dithionite reduced form. The EPR-spectrum is rhombic in the g-tensor (elements 2.013, 1.943, 1.860) with $g_{av} \leq$ 2.0 which is to be expected for systems containing mixed valence iron(II)-iron(III) pairs with a total spin S=1/2 [17]. The signal of the cofactor (flavin radical) is suppressed under these conditions. The experimental EPR-

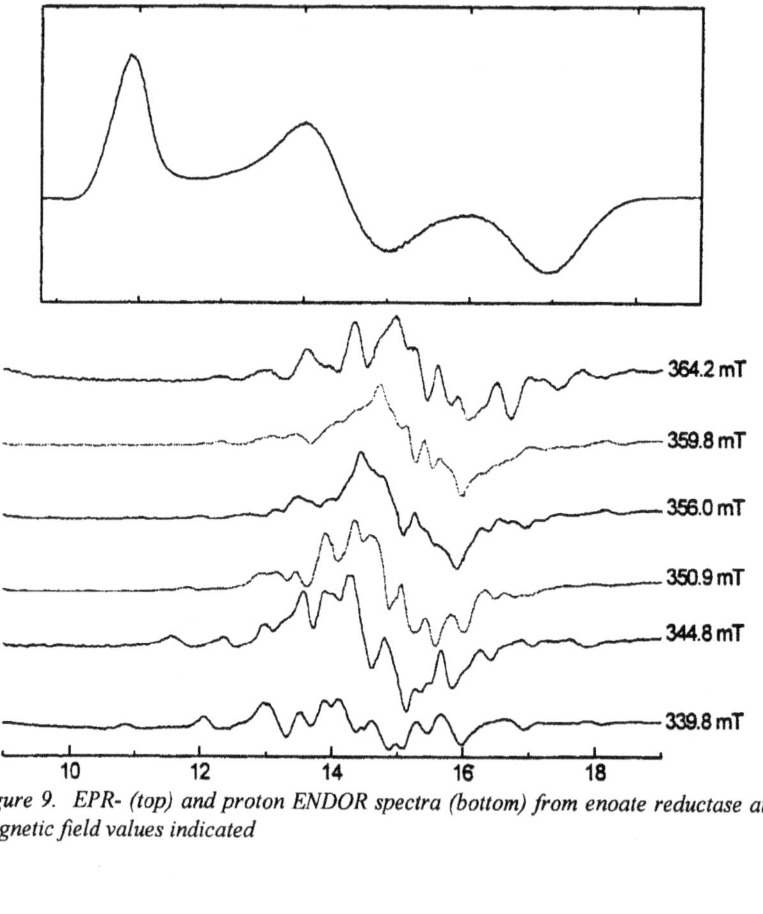

Figure 9. EPR- (top) and proton ENDOR spectra (bottom) from enoate reductase at magnetic field values indicated

spectrum at about 5 K is shown in Fig. 9 together with representative proton ENDOR-spectra in the range 10-20 MHz taken at the magnetic field positions indicated. The questions to be answered by these measurements were whether ENDOR could distinguish between two clusters of the (Fe_2S_2) type and one (Fe_4S_4) cluster and if, what would be the detailed structure in either case.

We started by assuming a reduced (Fe_4S_4) cluster in the formal +1 oxidation state comprising two ferrous irons and a mixed-valence ferric-ferrous pair. There are several organisms for which the X-ray structure of proteins containing (Fe_4S_4) clusters has been solved and which are available in the Protein Data Bank (PDB) files. We have investigated them by means of the „QUANTA"-software (Molecular Simulations Inc) and found that the cubane structures were nearly identical for all but that the positions of the eight cysteine β-protons differed slightly. Also, small positional changes for these protons were observed for proteins at different pH. In order to get an impression about how the simulations of the cysteine protons would reflect the experimental spectra, the cubane of Azotobacter vinelandii (pdb file 1fdb) was used. Its schematic structure is shown in Scheme 2 .

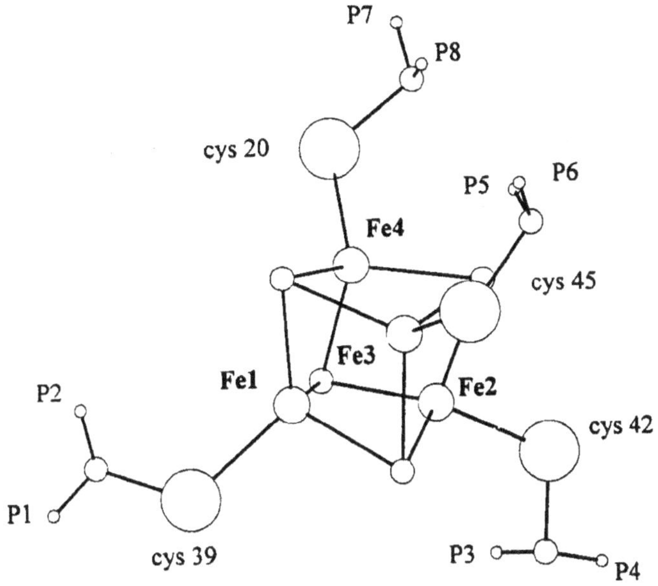

Scheme 2. Schematic representation of the cluster and ist associated cysteine residues as taken from A. vinelandii.

Another set of questions involved the simulation parameters i. e. the spin-densities at the irons, the placement of the mixed valence pair in the cubane, the isotropic proton couplings and the direction of the g-tensor in the cubane framework. Although the most precise data available from EPR- and ENDOR-work on single crystals of a model compound are for a system which models the oxidized state of the cluster and thus pertains e. g. to HIPIP proteins [18,19], we decided to use them as starting values together with the coordinates from A. vinelandii.

As a first parameter, the symmetry of the system was probed using the gamut of outermost proton couplings with magnetic field position as a measure of the influence of the placement of the mixed valence pair in the cubane. The result for all six possibilities is given in Figs. 10a-f. The left part gives the cubane together with the mixed valence pair position, the right shows the variation of the outermost couplings with the magnetic field. In these simulations, the intrinsic g-tensor axes directions were always identical and followed the assignment given in [19] for the oxidized cluster. This

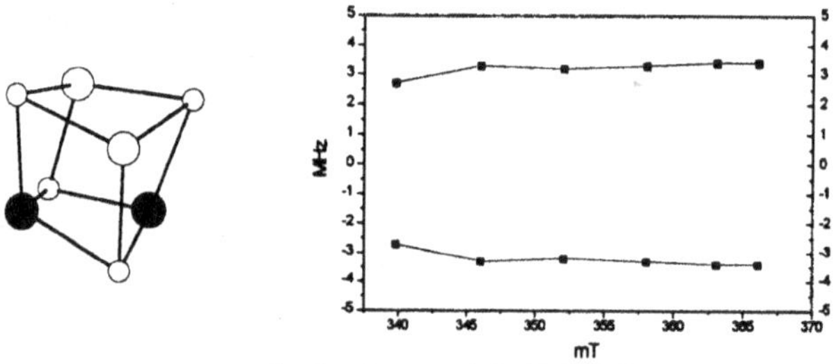

a: mixed valence pair in Fe1-Fe2

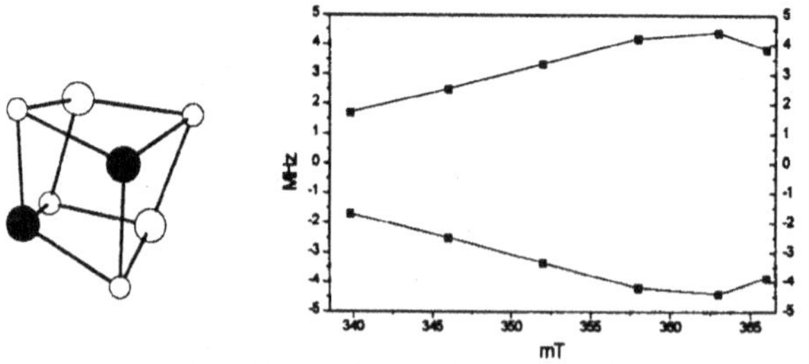

b: mixed valence pair in Fe1-Fe3

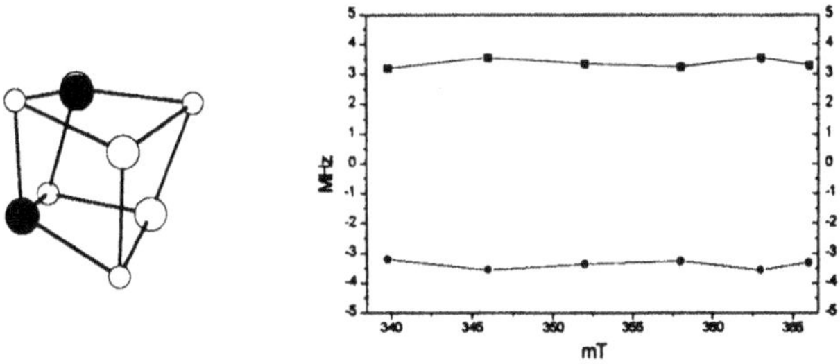

c: mixed valence pair in Fe1-Fe4

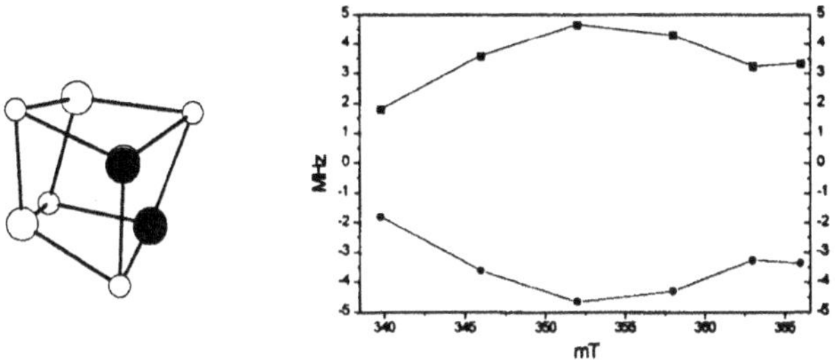

d: mixed valence pair in Fe2-Fe3

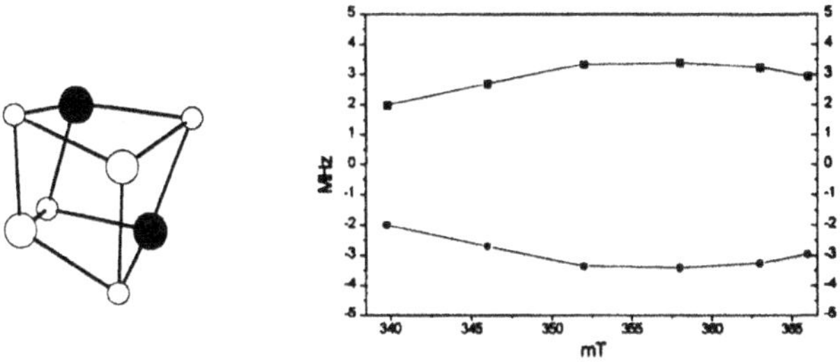

e: mixed valence pair in Fe2-Fe4

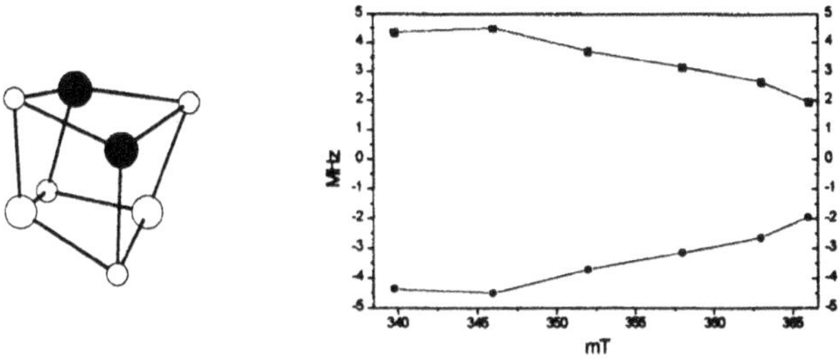

f: mixed valence pair in Fe3-Fe4

Figure 10. Influence of position of mixed valence pair in the cubane on variations of outermost proton couplings with magnetic field positions

means that the maximum g-element is directed along the axis perpendicular to both the mixed valence and the ferrous pair (ferric pair in [19]). The results show that the symmetry of the problem is of a kind which would seem to allow for discerning from the match with experimental gamuts the position of the mixed valence pair.In another set of simulations, the g-tensor axes were rotated in 10° steps around each of the elements directions. Again, very characteristic changes in the gamut of outermost couplings were obtained.

The most promising result was obtained so far with the mixed valence pair in positions Fe1-Fe2 but the g-tensor directions rotated by 100° around the direction of the ferrous pair (g2-axis). The experimental proton line positions centered around ν_0 of the proton NMR-frequency at each magnetic field together with the simulated values at 13 out of the 15 experimental field positions measured are given in Fig. 11. Marked in black are those protons which contribute most to the outermost couplings discussed before. The proton numbering is identical to that shown in Scheme 2.

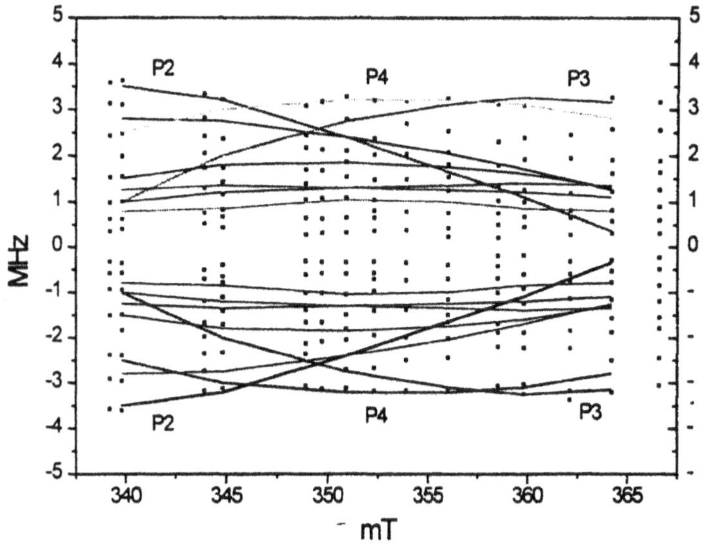

Figure 11. ENDOR measurements (dots) and ENDOR simulations (lines for the different protons).

The comparison shows that the match with the outermost line positions is very convincing. However, some of the ENDOR lines in the center part still remain unexplained. This problem will be attacked in the near future by refining the isotropic couplings for the individual protons. Nevertheless, from the extensive set of simulations we can infer that the basic placement of the mixed valence pair and the g-tensor direction may be considered as fixed either on Fe1-Fe2 or Fe1-Fe4. The simulation parameters are shown in Table 2. For the iron spin densities we used recent Mössbauer data [20], for the cysteine sulfurs we retained the values given in [19].

TABLE 2 Coordinates of cluster atoms, isotropic couplings (taken from [19]), spin densities and g-tensor elements and directions used in proton ENDOR simulations.

	x	y	z	a-iso
P1	-3.66	-2.38	0.61	+2.5 (+3.63)**
P2*	-3.30	-1.20	-0.66	+1.86
P3*	4.54	-2.87	1.17	+1.6
P4*	6.08	-2.21	1.75	+3.4 (+2.60)**
P5	2.58	4.79	-1.21	-1.8
P6	1.95	6.28	-0.46	-1.95
P7	2.77	1.69	-6.23	-2.0
P8	2.51	2.74	-4.83	-1.04

* These protons produce the maximal splittings in the ENDOR simulations.
** Values in brackets from [19].

	x	y	z	ρ
Fe1	0	0	0	+141%
Fe2	2.65	0	0	+141%
Fe3	1.24	2.21	0	-66%
Fe4	1.36	0.70	-2.22	-53%
S_{cys}39	-1.69	-1.13	1.11	3.7%
S_{cys}42	4.78	-0.56	0.63	3.9%
S_{cys}45	1.45	4.34	0.84	-5.6%
S_{cys}20	1.54	0.59	-4.52	-4.8%

	x	y	z
g1=2.013	-0.985	-0.144	0.098
g2=1.943	0	0.562	0.826
g3=1.86	-0.174	0.814	-0.553

We can conclude that the iron-sulfur clusters is of the (Fe_4S_4) type since, among others, the number of protons would be doubled in two (Fe_2S_2) clusters which is nearly incompatible with our results. Further, the spin densities on the irons and the cysteine sulfurs in the oxidized model compounds and in the reduced complex are very much alike. However, the g-tensor directions are changed. We do not know whether this latter finding is due to specific properties of the cluster in Enoate reductase or if this is a general difference for the reduced clusters. One can hypothesize that the former case might apply since in a most recent EPR-study of the reduced state in a model complex the intrinsic g-tensor directions are found to be unchanged [21].

Concerning correlations with NMR we note that no data exist from Enoate reductase so far. Probably, it cannot be studied due to its very large molecular weight. Nevertheless, some comparisons can be made since the isotropic proton ENDOR couplings listed in Table 2 match closely with those of the model compound described in [19]. Only two values had to be reduced (for protons P1 and P4) so far but still lie in a range which should be acceptable for comparison. Since the model compound data, in turn, reflect the situation of an oxidized (Fe_4S_4) cluster one can look for correlations with e. g. NMR data from high-potential iron proteins. In a recent paper, the paramagnetic shifts of all eight cysteine β-protons for the HIPIP protein of E. halophila are given [22]. Calculating from these the isotropic couplings we arrive at values which correspond closely to those of the oxidized model compound and, therefore, to those used in Enoate reductase.

It follows that the spin density distribution in the oxidized and the reduced cluster as seen by the cysteine protons are very much alike. This conclusion is also reflected by the respective Mössbauer data [17]. It is further supported by the results obtained in an ENDOR study of ^{57}Fe and ^{1}H interactions from the oxidized HIPIP protein of E. halophila which is presently carried out in our laboratory.

4. Acknowledgements

Work from the authors' laboratory was funded through grants from the Deutsche Forschungsgemeinschaft and the European Community. We gratefully acknowledge the collaborations with the groups of I. Bertini (University of Florence, Italy) on SOD and of J. J. G. Moura (University of Lisbon, Portugal) on Enoate Reductase.

5. References

1. Kreiter, A. and Hüttermann, J. (1991) Simultaneous EPR and ENDOR Powder- Spectra Synthesis by Direct Hamiltonian Diagonalization, *Journal of Magnetic Resonance* **93,** 12-26.
2. Hurst, G. C., Henderson ,T. A. and Kreilick, R. W. (1985) Angle-Selected ENDOR Spectroscopy. 1. Theoretical Interpretation of ENDOR Shifts from Randomly Orientated Transition-Metal Complexes, *J. Am. Chem. Soc.* **107**, 7294-7299.
3. Däges, G. P. and Hüttermann, J. (1992) ESR and ENDOR of pentacoordinated cobalt(II)porphyrins, *J. Phys. Chem.* **96**, 4787-4794.
4. Iwasaki, M. (1974) Second-Order Perturbation Treatment of the General Spin Hamiltonian in an Arbitrary Coordinate System, *Journal of Magnetic Resonance* **16**, 417-423.
5. McCord, J. M. and Fridovich, I. (1969) Superoxide Dismutase: An Enzymatic Function for Erythrocuprein (Hemocuprein), *J. Biol. Chem.* **244**, 1581-1588.
6. Fridovich, I. (1986) Biological Effects of the Superoxide Radical,*Archives of Biochemical Biophysics* **247**, 1-11.
7. Reinhard, H., Kappl, R., Hüttermann, J. and Viezzoli, M. S. (1994) ENDOR of Superoxide Dismutase: Structure Determination of the Copper Site from Randomly Oriented Specimen, *J. Phys. Chem.* **in press.**
8. Tainer, J. A., Getzoff, E. D., Richardson, J. S. and Richardson, D. C. (1983) Structure and mechanism of copper, zink superoxide dismutase, *Nature* **306**, 284-287.
9. Scholl, H. J. and Hüttermann, J. (1992) ESR and ENDOR of Cu(II) Complexes with Nitrogen Donors: Probing Parameters for Prosthetic Group Modeling of Superoxide Dismutase, *J. Phys. Chem.* **96**, 9684-9691.
10. Van Camp, H. L,.Sands, R. H and.Fee, J. A. (1982) An Examination of the Cyanide Derivative Bovine Superoxide Dismutase with Electron-Nuclear Double Resonance, *Biochimica et Biophysica Acta* **704**, 75-89.
11. Hüttermann, J., Kappl, R., Banci, L. and Bertini, I. (1988) An ENDOR study of human and bovine erythrocytesuperoxide dismutase: 1H and 14N interactions,*Biochimica et Biophysica Acta* **956**, 173-188.
12. Rotilio, G., Morpurgo, L., Giovagnoli, C. and Calabrese, L. (1972),*Biochemistry* **11**, 2187-2192.
13. Banci, L., Bertini, I., Luchinat, C., Piccioli, M., Scozzafava, A. and Turano, P. (1989) 1H NOE Studies on Dicopper(II) Dicobalt(II)Superoxide Dismutase, *Inorganic Chemistry* **28**, 4651-4656.
14. Dikanov, S., Felli, I, Viezzoli, M. S., Spoyalov, A. and Hüttermann, J. (1994) X-Band ESEEM-spectroscopy of ^{15}N substituted native and inhibitor- bound superoxide dismutase. Hyperfine couplings with remote nitrogen of histidine ligands, *FEBS letters* **345**, 55-60.
15. Bertini, I., Banci, L., Picciolo, M. and Luchinat, C. (1990) Spectroscopic Studies on Cu2Zn2SOD: A Continuous Advancement of Investigation Tools,*Coordination Chemistry Reviews* **100**, 67-103.
16. Simon, H. (1991) Enoate Reductase, in F. Müller (ed), *Chemistry and Biochemistry of Flavoenzymes* CRC-Press, pp. 318-328.

17. Trautwein, A. X., Bill E., Bominaar, E. L. and Winkler, H. (1991) Iron-Containing Proteins and Related Analogs-Complementary Mössbauer, EPR and Magnetic Susceptibility Studies, *Structure and Bonding* **78**, 1-95.
18. Rius, G. and Lamotte, B. (1989) Single-Crystal ENDOR Study of a ^{57}Fe-Enriched Iron-Sulfur $[Fe_4S_4]^{3+}$ Cluster, *J. Am. Chem. Soc.* **111**, 2464-2469.
19. Mouesca, J.-M., Rius, G. and Lamotte, B. (1993) Single Crystal Proton ENDOR Studies of the $[Fe4S4]^{3+}$ Cluster: Determination of the Spin Population Distribution and Proposal of a Model To Interpret the ^{1}H NMR Paramagnetic Shifts in High-Potential Ferredoxins, *J. A. Chem. Soc.* **115,** 4714-4731.
20. Moura, I., Lisbon, personal comunication.
21. Gloux, J., Gloux, P., Lamotte, B., Mouesca, J.-M. and Rius, G. (1994) The Different $[Fe_4S_4]^{3+}$ and $[Fe_4S_4]^{+}$ Species Created by γ Irradiated in Single Chrystals of the $(Et_4N)_2[Fe_4S_4(SBenz)_4]$ Model Compound: Their EPR Description and Their Biological Significance, *J. Am. Chem. Soc.* **116**, 1953-1961.
22. Banci, L., Bertini, I., Capozzi, F., Carloni, P., Ciurli, S., Lucinat, C. and Piccioli, M. (1993) The Iron-Sulfur Cluster in the Oxidized High-Potential Iron Protein from Ectothiorhodosspira halophila, *J. Am. Chem. Soc.* **115**, 3431-3440.

NMR STUDIES OF NONHEME IRON PROTEINS

ZHIGANG WANG AND LAWRENCE QUE, JR.
Department of Chemistry, University of Minnesota
Minneapolis, MN 55455
U.S.A.

1. Introduction

The application of NMR spectroscopy to paramagnetic iron proteins lacking a heme prosthetic group or an iron sulfur cluster has lagged behind their more chromophoric counterparts. This situation is a result of their generally broader lines which impede the facile interpretation of these features. Furthermore, lacking a common prosthetic group, these proteins exhibit greater structural diversity, and all the ligands need to be identified. Our interest in the structure and function of nonheme iron proteins has led us to explore the utility of NMR to identify amino acid residues that are coordinated to the active site metal centers of these proteins. Our identification strategy consists of a) isotropic shift comparisons with appropriate model complexes to establish the general characteristics expected when a particular residue is coordinated to an iron center, and b) confirmation of the assignment by solvent exchangeability, bond connectivity, or through space correlations. We demonstrate these strategies on two proteins of known structure, iron superoxide dismutase and hemerythrin, and on two proteins of unknown structure, isopenicillin N synthase and uteroferrin, the purple acid phosphatase from porcine uterus. Our studies have been limited to forms of the proteins that contain a high spin Fe(II) center, as such centers generally have the fast electronic spin lattice relaxation times that give rise to NMR features sharp enough for investigation [1]. In some cases, the Fe(II) center has been substituted with Co(II) which has even more favorable relaxation properties [1].

Among amino acid residues, the most likely ligands to be found in the active sites of nonheme iron proteins are imidazole, carboxylate, phenolate, and thiolate. The latter two give rise to intense ligand-to-metal charge transfer transitions for the iron(III) form of the protein, which may be probed by visible and Raman spectroscopies [2]. However imidazole and carboxylate are ligands for which there are no other convenient spectroscopic probes besides NMR. Scheme I summarizes data collected on model

G.N. La Mar (ed.), Nuclear Magnetic Resonance of Paramagnetic Macromolecules, 193-211.

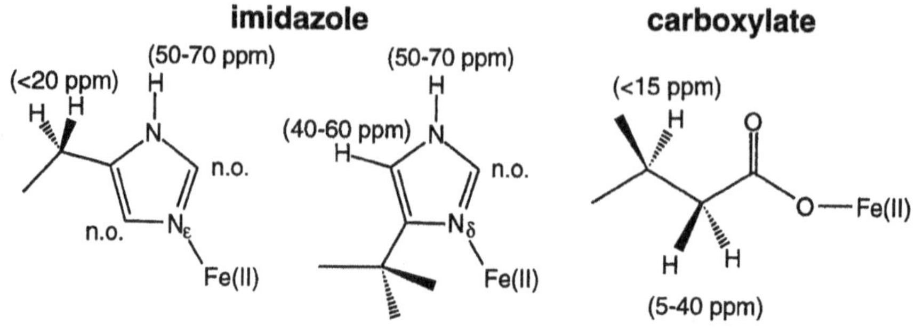

Scheme I

complexes that give an idea of where various protons on these residues may be found if they are coordinated to a high spin iron(II) center [3-9].

Preliminary assignment of NMR features in a protein may be made using these isotropic shift comparisons. T_1 values for these resonances are helpful for establishing relative distances of the various protons from the metal center, but an independent observation is usually required for the corroboration of the initial assignment. The imidazole data presented in Scheme I derive from various synthetic imidazole complexes [3-7]. Although the protons on carbons adjacent to the coordinated nitrogen are usually too broad to be observed, the N-H proton can readily be observed and is assigned by demonstrating its disappearance when the protein is placed in D_2O solvent. The C_δ-H proton is observable only when the histidine is coordinated via N_δ and may be confirmed by the presence of NOE when the N-H proton is irradiated. The data on carboxylate shifts in Scheme I come from studies of heterobimetallic complexes wherein carboxylates bridge between an Fe(II) and a diamagnetic metal center [8,9]. The CH_2 protons adjacent to the carboxylate function are diastereotopic and exhibit shifts that depend on the cosine of the angle defined by the C-H bond and the p_π orbital on the carboxylate carbon [10]. The C-H protons β to the carboxylate function exhibit smaller contact shifts, as expected for a σ delocalization mechanism. These paramagnetic shifts are further modulated by the dipolar effects of the metal center. The carboxylate protons do not resonate in a region unique enough to allow for unequivocal assignment, so an NOE experiment demonstrating the presence of an ABX or and ABMNX system and/or COSY data is needed to establish the presence of such a residue. These strategies are illustrated in the following examples.

2. Mononuclear Iron(II) Enzymes

2.1. IRON SUPEROXIDE DISMUTASE

Superoxide dismutase catalyzes the disproportionation of superoxide into O_2 and H_2O_2 [11]. The CuZn enzyme from erythrocytes has been extensively studied with NMR by Bertini and the Florence group [12]; indeed many of the strategies used for applying NMR to paramagnetic metalloproteins have been developed using the CuCo derivative of this enzyme [13]. The iron enzyme (FeSOD) is a dimeric protein of 32 kDa and often found in prokaryotes. Crystal structures of FeSOD from two sources are available [14,15]; these show that the Fe center is coordinated to the polypeptide chain via three His and one Asp residue. The ligands are arranged in a trigonal pyramid around the iron with 2 His and the Asp comprising an approximate trigonal plane (Figure 1 inset).

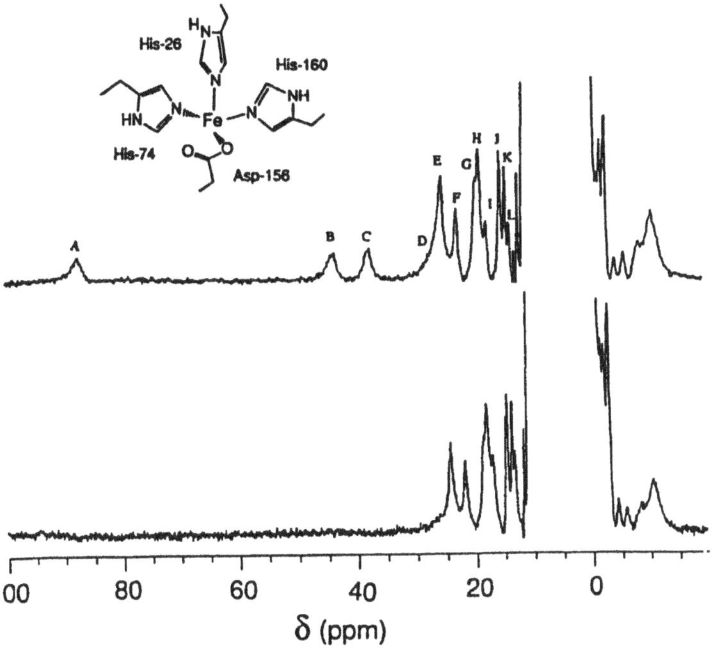

Figure 1. Proton NMR spectra of Fe(II)SOD from *E. coli* at 300 MHz and 27 °C in 50 mM phosphate in H_2O buffer (top, pH 7.4) and D_2O buffer (bottom, pD 7.4). The insert is a schematic drawing of the active site of the enzyme from *P. ovalis* based on the X-ray crystallographic studies. In the *E. coli* enzyme, the ligands His-26, His-74, His-160, and Asp-156 should be replaced with corresponding ligands His-26, His-75, His-162, and Asp-158 based on the sequence alignment. Reprinted with permission from the American Chemical Society.

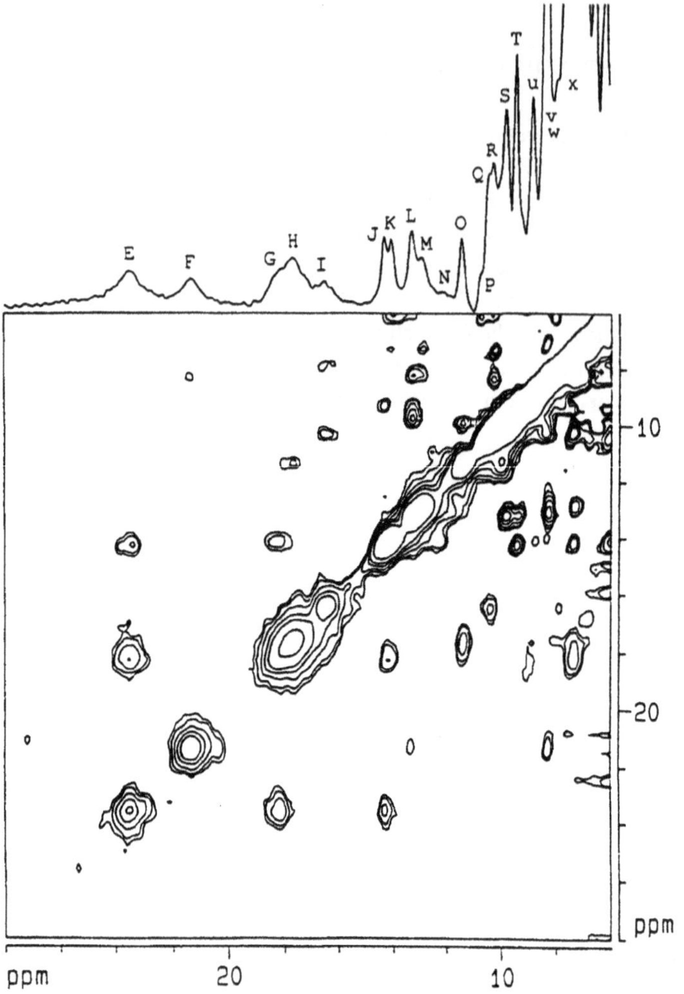

Figure 2. Proton NOESY spectrum of Fe(II)SOD from *E. coli* at 360.13 MHz at 30 °C in 50 mM phosphate in D_2O buffer at pD 7.4. This spectrum is obtained in the TPPI phase sensitive mode with a mixing time of 15 ms. A 5% random variation of the mixing time was used to eliminate zero-quantum coherence in the NOESY experiments. Persaturation was used on the solvent resonance during both relaxation delay and mixing time. A 10%-shifted Gaussian weighting function was applied in both dimensions prior to Fourier transformation, followed by a fifth-order polynomial baseline correction. The top trace is a resolution-enhanced 1D spectrum with the labeling of the signals. Reprinted with permission from the American Chemical Society.

The ^1H NMR spectrum of the Fe(II) enzyme from *E. coli* shown in Figure 1 consists of relatively sharp features found over a 100-ppm range [16]. Three resonances labeled A, B, and C disappear when the protein is dissolved in D_2O. These are associated with the N-H protons of the three coordinated His residues. Since there are no solvent-nonexchangeable resonances in the same region, where the C_δ-H protons of His residues N_δ-coordinated to Fe(II) centers would be expected [6,7], the three His residues must be coordinated via N_ε, as found for the oxidized form of the enzyme by x-ray crystallography [14].

The remaining features in the paramagnetically shifted region of the ^1H NMR spectrum must arise from the coordinated Asp-158 residue or other residues in close proximity to the Fe(II) center. In light of the recent successful application of 2D NMR techniques to paramagnetic metalloproteins [17], a NOESY experiment was attempted to identify NMR signals arising from this residue [14]. Figure 2 shows a phase sensitive (TPPI) NOESY spectrum of Fe(II)SOD in the region of < 30 ppm, which reveals many cross signals. A set of cross peaks relating signals E, G, and J is observed in the 14 to 24 ppm region, suggesting that these three signals comprise an ABX system. The cross signal between E and G is more intense than those between E and J and between G and J, indicating that signals E and G are from a geminal proton pair. This ABX spin system is further confirmed by 1D NOE and COSY results [14]. Based on the chemical shifts and T_1 values of signals E, G, and J, they are assigned to the $-C_\beta H_2$-$C_\alpha H$< moiety of the coordinated Asp 158 ligand with E and G arising from the $-C_\beta H_2$ geminal proton pair and J from the $-C_\alpha H$ proton.

2.2. ISOPENICILLIN N SYNTHASE

Isopenicillin N synthase (IPNS) catalyzes the four-electron oxidative ring closure reactions of δ-(L-α-aminoadipoyl)-L-cysteinyl-D-valine (ACV) to form isopenicillin N (Scheme II), the precursor of other penicillins and cephalosporins, in β-lactam-antibiotic-producing micro-organisms [18]. This enzyme has a MW of 38.4 kDa and contains a single high-spin nonheme Fe(II) center in its active site. The mechanism of

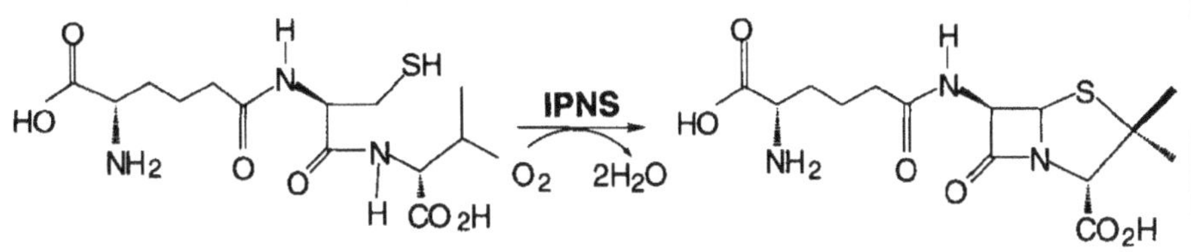

Scheme II

IPNS catalysis has been extensively studied using many different substrate analogues [19], but the role of the iron(II) center is not firmly established. As a first step towards this goal, the coordination environment of the Fe(II) active site must be determined. Exogenous ligands such as the substrate thiolate, solvent, and NO have been shown to bind to the Fe(II) center on the basis of EPR, Mössbauer, and electronic spectral studies [20], while the endogenous ligands have been identified by NMR [21,22].

The ^1H NMR spectrum of Fe(II)-IPNS exhibits isotropically shifted signals at 65 (~3H), 42 (~2H), and 24 (~2H) ppm, of which only the 42 ppm feature remains in D_2O buffer [21,22]. The three solvent exchangeable resonances at 65 ppm are assigned to the N-H portons of His coordinated to the Fe(II) center based on their chemical shifts [3-7]. This suggests that there are three His ligands coordinated to the Fe(II) center. The nonexchangeable resonance at 42 ppm may be associated with coordinated carboxylate ligands because their chemical shifts fall in the chemical shift region for $-C_\beta H_2$ (or $-C_\gamma H_2$) of Asp (or Glu) coordinated to Fe(II) [8,9]. Anaerobic addition of ACV to Fe(II)IPNS perturbs its ^1H NMR spectrum by splitting the overlapped solvent-exchangeable resonances into two, one at 66 ppm corresponding to 2 protons and the other at 56 ppm corresponding to one proton. The nonexchangeable peak at 45 ppm is also split into two signals. Thus ACV binding does not appear to displace any of the endogenous ligands .

The broadness of the NMR resonances of the Fe(II)IPNS-ACV complex prevents us from performing any NOE experiments on the nonexchangeable -CH resonances. However, the substitution of Co(II) for Fe(II) gives rise to sharper isotropically shifted proton resonances [22]. There is a solvent-exchangeable peak at 78 ppm corresponding to 3 protons, which falls in the chemical shift range for imidazole NH protons of coordinated His residues in Co(II) proteins with similar geometry, such as those in Co(II)-substituted concanavalin A [23]. There is also a solvent-nonexchangeable signal at 31.7 ppm, comparable to the 44 and 33 ppm signals associated with the $-C_\beta H_2$ protons of the Asp residue coordinated to the Co(II) site of Cu(I)Co(II)SOD [24]. A significant NOE correlation between the 31.7 ppm signal and a signal at 21 ppm is observed, suggesting that they belong to the $-CH_2$ protons of a ligand [22].

Addition of ACV to Co(II)IPNS engenders a new set of NMR signals due to the formation of the Co(II)IPNS-ACV complex. There are three solvent-exchangeable signals at 94, 82, and 79 ppm, suggesting that there are still three His ligands coordinated to the metal ion [22]. The 86 ppm nonexchangeable signal is likely to arise from one of the $-C_\beta H_2$ protons of the Cys moiety of ACV, which has been demonstrated to directly coordinate to the active site in IPNS [20,21,25,26]. 1D NOE studies show that the signals at 38.6, 24.3 and 14.6 ppm comprise an ABX system (Figure 3). The observed isotropic shifts and the stronger NOE response between the 38.6 ppm and 24.3 ppm signals suggest that they are from the geminal proton pair of the

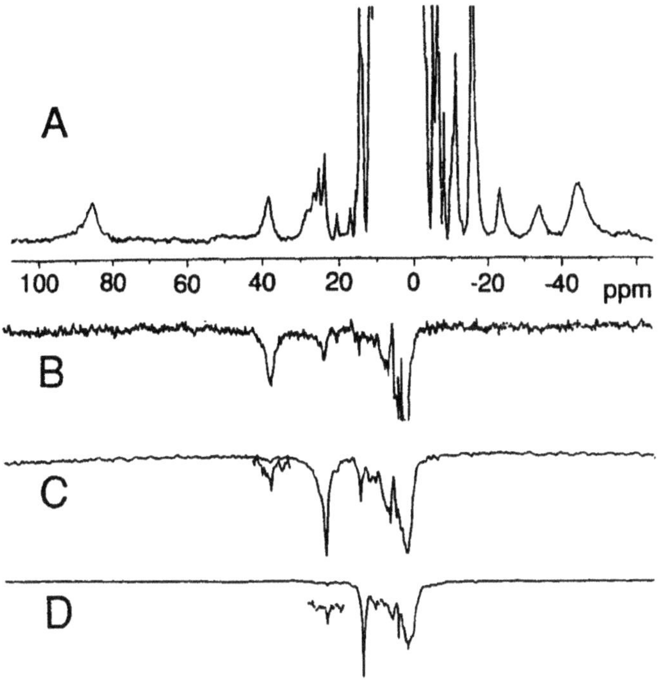

Figure 3. Proton NMR spectrum (A) and NOE difference spectra (B-D) of Co(II)IPNS-ACV complex at 300 MHz at 14 °C in 10 mM MOPS buffer at pD 7.1. The NOE difference spectra were obtained by subtraction of the reference FID's from the FID's in which the resonances at 38.6 ppm (B), 24.3 ppm (C), and 14.6 ppm (D) were saturated by a decoupler pulse for 80 ms, respectively. Reprinted with permission from the American Chemical Society.

$C_{\beta}H_2$-$C_{\alpha}H$< moiety in an Asp ligand. This confirms that there is a carboxylate ligand coordinated to the Co(II) center in the Co(II)IPNS-ACV complex.

On the basis of our NMR studies, the metal binding site of IPNS thus consists of three His ligands and one carboxylate ligand, probably Asp; the binding of ACV to the enzyme does not displace any of these ligands. These conclusions are supported by our recent EXAFS studies on Fe(II)IPNS and its ACV complex [27]. A picture of the active site that has evolved from the various spectroscopic studies is illustrated in Scheme III and looks remarkably similar to that of FeSOD. Together with the recent crystal structure of soybean lipoxygenase [28,29] showing the same ligand set, these three proteins may form the nucleus of a subset of nonheme iron proteins having 3-His-1-carboxylate active sites.

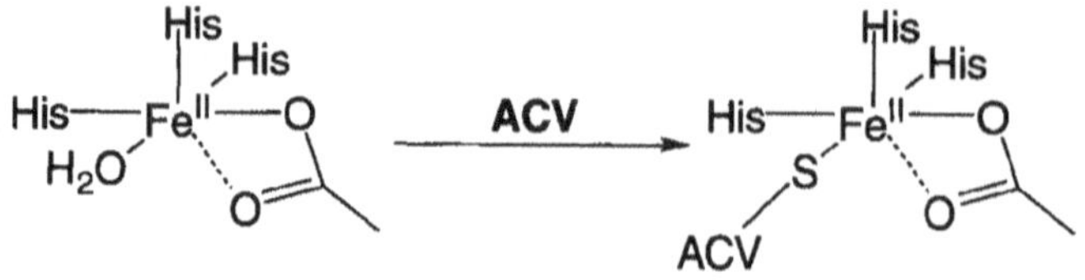

Scheme III

3. Dinuclear Enzymes

3.1. HEMERYTHRIN AND MYOHEMERYTHRIN

Hemerythrin (Hr) and myohemerythrin (MHr) are the reversible dioxygen-binding proteins respectively isolated from the blood and muscle of a variety of marine invertebrates including sipunculids, annelids, priapulids, and brachiopods [30]. Hr is most often found as an octamer, but dimeric, trimeric, and tetrameric forms are also found. Each 13.5 kDa subunit of Hr contains a dinuclear iron center serving a function similar to that of hemoglobin in mammalian systems. MHr is monomeric and plays a role similar to that of myoglobin in mammalian systems. Hrs and MHrs are the best characterized of the subclass of oxo-bridged dinuclear iron proteins [30].

All Hrs, including MHrs, are typically isolated in their met Fe(III)Fe(III) forms designated as metHr or metMHr. The met forms can be reduced to the deoxy or Fe(II)Fe(II) states with dithionite [31], which is capable of binding O_2. X-ray crystallographic structures of deoxyHr, metHr, metHrN_3, and metMHrN_3 have all been solved to high resolution [32-34]. Together with the results obtained from spectroscopic studies, it is now clear that the two high-spin iron centers are bridged by two endogenous carboxylates (Asp 106 and Glu 58 for Hr, and Asp 111 and Glu 58 for MHr) together with an additional oxo (all met forms) or hydroxo (deoxy forms) bridge. The coordination environments of the iron atoms are completed by five terminal His ligands with three on one iron and two on the other iron. Exogenous ligands, such as N_3^- and F^-, can bind to the iron atom with only two His ligands [30].

Paramagnetically shifted NMR spectroscopy has provided some insight into the coordination chemistry of the dinuclear iron centers of Hr and MHr, especially their deoxy forms [35-37]. The proton NMR spectrum of deoxyHr from *P. gouldii* shows three broad solvent exchangeable peaks at 43.2, 46.3 and 63.2 ppm which are assigned to the five His N$_\delta$-H protons (Figure 4B) [35,36]. All the nonexchangeable signals appear within the 20 to -10 ppm region with very poor resolution. The wide range of His N-H chemical shifts observed suggests a significant dipolar component to the

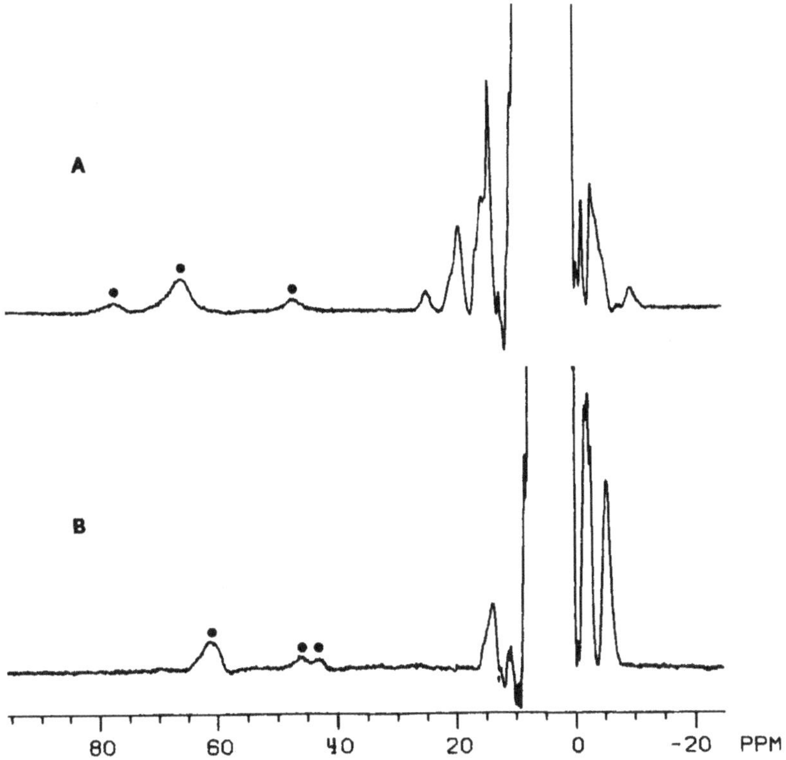

Figure 4. Proton NMR spectra of *P. gouldii* deoxyHr-N_3 complex (A) and deoxyHr (B) at 300 MHz at 45 °C in 50 mM phosphate buffer at pH 7.5. The signals marked by black dots are solvent-exchangeable resonances that disappear in D_2O buffer. Reprinted with permission from the American Chemical Society.

isotropic shift and the large shifts observed indicate the absence of a strong antiferromagnetic interaction. This is consistent with the existence of an hydroxo bridge between the two irons. Upon addition of azide, the proton NMR spectrum of deoxyHr undergoes significant changes as shown in Figure 4A. Apparently, one of the solvent exchangeable signals shifts to 78.3 ppm, with the other two remaining roughly unchanged. The nonexchangeable resonances near 20 ppm are also perturbed by azide bindng. These changes confirm that the addition of some anions to deoxyHr will alter the coordination environment of the diiron centers. Unfortunately, the molecular weight of Hr is too big (108 kDa) and the proton NMR signals observed for deoxyHr are very broad. Indeed the protons due to the bridging Asp and Glu residues cannot be identified.

The significantly smaller size of deoxyMHr (13.9 kDa) provides a better opportunity for carrying out NMR studies. As shown in Figure 5, the ^{1}H NMR spectra of the deoxyMHr from *T. zostericola* exhibit a number of relatively sharp paramagnetically shifted signals ranging from 70 ppm to -20 ppm [37]. For the sake of clarity, we have labeled the signals with letters, one letter for one proton. The chemical shifts of signals a to e (65.9, 61.1, and 49.5 ppm) are typical of the His N-H protons coordinated to Fe(II) centers [3-7]. Indeed, we find that signals a, b, c, and d disappear upon solvent deuteration. This confirms that signals a, b, c, and d are from the N_δ-H protions of four of the five His ligands. Interestingly, signal e also disappears but only after storing the sample in D_2O buffer at 4 °C for about a month. Therefore, signal e accounts for the N_δ-H proton of the fifth His ligand. The slow water exchange rate of the N_δ-H proton corresponding to signal e may be due to hydrogen bonding with another amino acid residue. This is consistent with the x-ray crystallographic results of deoxyHr [33] and metMHrN$_3$ [34], which shows a hydrogen bond between the N_δ-H proton of His 54 with a carboxylate oxygen of Glu 24 [34]. Aside from the loss of signals a to e, the ^{1}H NMR spectrum of deoxyMHr in D_2O buffer is almost identical to that in H_2O buffer outside the diamagnetic region. As the ^{1}H NMR signals of C_δ-H and C_ε-H protons of the His ligands are too broad to be seen due to their proximity to the Fe(II)Fe(II) core, many of the non-exchangeable paramagnetically shifted signals, with chemical shifts ranging from 36.1 ppm to -17.5 ppm, must arise from the bridging Glu and Asp ligands based on their chemical shifts [8,9].

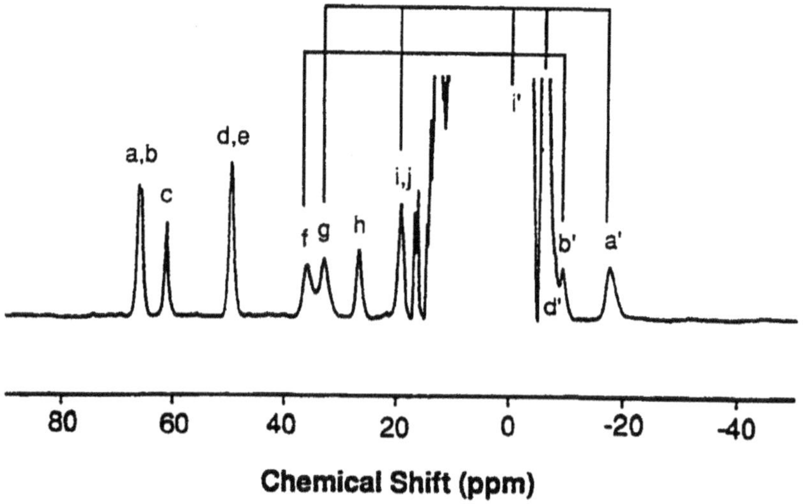

Figure 5. Proton NMR spectra of *T. zostericola* deoxyMHr at 300 MHz at 25 °C in 50 mM Tris-Cl in H_2O buffer at pH 8.5). Resonances with NOE correlations to each other are linked by solid lines.

The relative sharp features found for deoxyMHr make it possible to perform more sophisticated NMR studies. 1D NOE experiments have been performed on deoxyMHr in D_2O buffer [37]. When signal f at 36.1 ppm is saturated, significant NOEs are observed on signal i at 19.4 ppm and signal b' at -8.7 ppm. When signal b' is irradiated, NOEs are observed on signals f and i. Signals f, i and b' thus comprise three-spin system. On the other hand, when signal g at 33.0 ppm is irradiated, significant NOEs are observed on signals j at 19.4 ppm, signal i' at -0.9 ppm, and signal a' at -17.5 ppm. When signal a' is saturated, significant NOEs are observed on signals g, j, i' and signal d'. These signals comprise a five-spin system where the proton corresponding to signal d' is in close proximity to the proton corresponding to signal a' but distant from the proton corresponding to signal g. We mark the NOE connectivitives of the nonexchangeable signals of deoxyMHr in Figure 5.

Considering that there are two bridging carboxylates, Asp and Glu, coordinated to the Fe(II)Fe(II) core in deoxyMHr, there should be paramagnetically shifted ABX (from the $-C_\beta H_2-C_\alpha H<$ of Asp) and ABMNX (from the $-C_\gamma H_2-C_\beta H_2-C_\alpha H<$ of Glu) systems present. For the Asp $-C_\beta H_2-C_\alpha H<$ moiety, NOE responses between the two geminal β protons, and between each of the geminal protons and the α protons should be observed. The NOE connections among signals f, i, and b' are observed and associated with the bridging Asp. Signals f and i correspond to the two β geminal

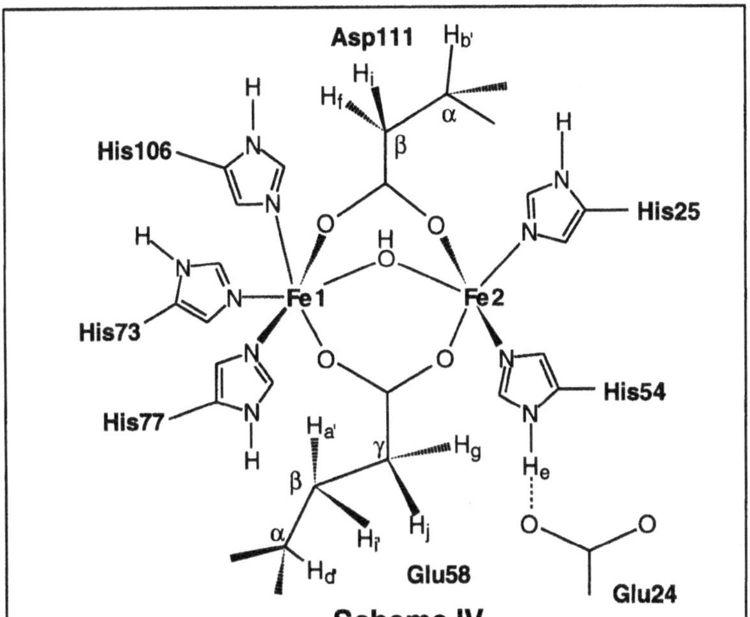

Scheme IV

protons since they exhibit a relative strong NOE effect and signal b' is from the α proton. For the Glu $-C_{\gamma}H_2-C_{\beta}H_2-C_{\alpha}H<$ moiety, NOE connections should exist between the two γ geminal protons, between the two β geminal protons, between each of the γ protons to each of the two β protons, and between each of the two β protons and the α proton. The NOE connectivities observed among signals g, j, i', a' and d' likely arise from the binding Glu. Signals g and j are from the two γ geminal protons, signals i' and a' are from the two β geminal protons, and signal d' are from the α proton. This assignment is consistent with the chemical shift data of carboxylate ligands coordinated to Fe(II) obtained from model compounds [8,9]. Our tentative assignments of the Asp and Glu ligands are illustrated in Scheme IV.

3.2. PURPLE ACID PHOSPHATASES

Purple acid phosphatases (PAPs), a class of enzymes isolated from a wide variety of animal and plant sources, catalyze the hydrolysis of certain phosphate esters, including nucleotide di- and triphosphates and aryl phosphates *in vitro*. The most thoroughly studied PAPs are the mammalian enzymes from porcine uterus (also called uteroferrin, Uf) and bovine spleen (BSPAP) [38,39]. Both proteins consist of a single polypeptide chain with a molecular weight of 35 kDa. One important structural feature of the mammalian PAPs is the Fe(III)Fe(II) center found in the catalytically active form of the enzyme [38-41]. The purple color associated with PAPs results from a tyrosinate-to-Fe(III) charge-transfer transition, as demonstrated by resonance Raman spectroscopy [40-41]; thus a tyrosinate ligand is bound to the Fe(III) site. Since there is no X-ray crystallographic structure available for PAPs, paramagnetically shifted ^1H NMR spectroscopy has extensively used to identify the ligands coordinated to the dinuclear iron centers of these proteins in our laboratory [5,10,42]. We have taken advantage of the favorable electronic relaxation of the antiferromagnetically coupled Fe(III)Fe(II) centers in order to observe well-resolved hyperfine-shifted NMR signals.

Figure 6A shows the ^1H NMR spectra of reduced Uf in H_2O. Seven well-resolved ^1H NMR features representing nine protons (labeled a to i) appear in the downfield region. Signals b and e disappear upon switching to D_2O buffer, suggesting that two His ligands are coordinated to the diiron center. Based on the chemical shifts of the solvent-exchangeable N-H signals, we assign signal b at 88.4 ppm to the N-H proton of the His coordinated to Fe(III) [5,7], and signal e at 44.0 ppm to the one coordinated to Fe(II) [3-7]. The binding of inhibitor tungstate causes small changes of the Uf NMR spectrum (Figure 6B). The two solvent-exchangeable signals (signal b at 88.3 ppm and e at 54.ppm) become better resolved from their neighbors and allow NOE experiments to be carried out.

When signal b is saturated, a significant NOE is observed on signal i (Figure 7A), suggesting that H_b and H_i are close to each other. So H_i is either the C_{δ}-H on an

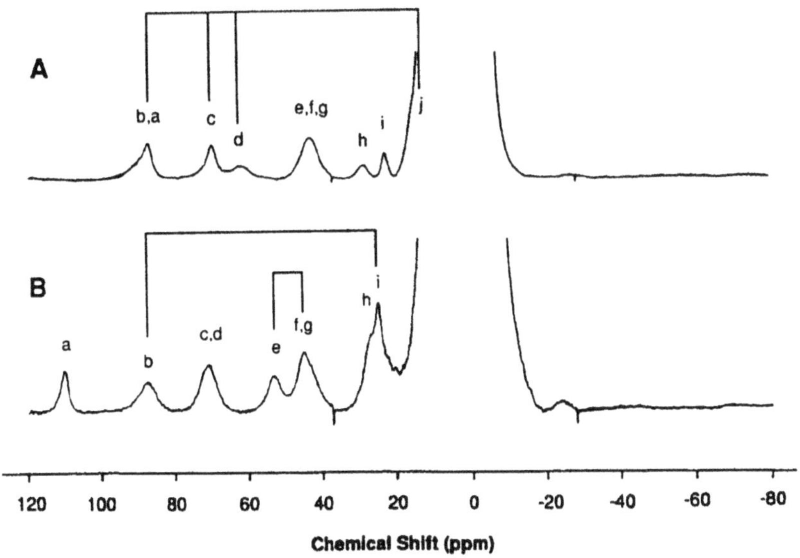

Figure 6. Proton NMR spectra of reduced Uf (A) and reduced Uf-tungstate complex (B) at 300 MHz at 30 °C in 100 mM acetate buffer at pH 4.9. Resonances with NOE correlations to each other are linked by solid lines. Adapted with permission from the American Chemical Society.

N_δ-coordinated His or one of the two geminal protons on the ß-CH_2 group of a N_ε-coordinated His. However the chemical shift of H_i is inconsistent with that expected for the C_δ-H on an N_δ-coordinated His (~70 ppm) to an Fe(III) site , thus favoring the latter assignment. This implies that this His residue is N_ε-coordinated to the Fe(III) site. The signal due to the other ß-CH_2 proton could not be identified by irradiation of H_i and is presumably buried among the bulk protein resonances. When signal e is saturated, a significant NOE on signal g is observed (Figure 7B), indicating that H_e and H_g are in close proximity of each other. Since H_g exhibits an isotropic shift of 44.2 ppm, which is that expected for the C_δ-H on an N_δ-coordinated His to an Fe(II) site, we conclude that the His on Fe(II) is N_δ-coordinated, with signals e and g corresponding to the N_ε-H and C_δ-H protons, respectively.

Irradiation of the solvent-nonexchangeable signals in the Uf spectra allows the resonances of the Fe(III)-coordinated Tyr residue to be identified. When signal a is saturated, a significant NOE is observed on signal j and weaker but observable NOEs are also found for signals c and d (Figure 6). These results suggest that H_a and H_j are in close proximity to each other with H_c and H_d being somewhat more distant. When

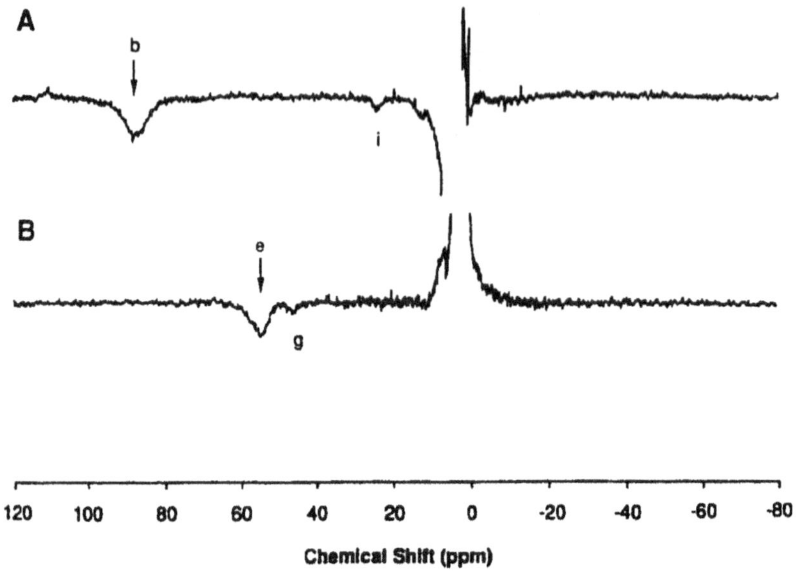

Figure 7. NOE difference spectra of reduced Uf-tungstate complex at 300 MHz at 30 °C in 100 mM acetate buffer at pH 4.9. The NOE difference spectra were obtained by subtraction of the reference FID's from the FID's in which the decoupler pulse was set at signal b (A) and signal e (B) for 50 ms. Reprinted with permission from the American Chemical Society.

either signal c or d is saturated, NOEs on signal a and j are observed. These observations support the assignment of the NMR signals to protons on the Fe(III)-coordinated tyrosine residue, made on the basis of chemical shifts [5,10]; signals a and j are associated with the β-CH_2 protons while signals c and d arise from the Tyr C_δ-H protons. Thus, all the downfield features except f and h have been assigned. Irradiation of the latter two signals does not reveal any discernible NOEs. Parallel studies on BSPAP reveal the same results, indicating that the active sites of the two enzymes are very similar [42]. At this point, three terminal ligands to the Fe(III)Fe(II) center have been identified: a Tyr and an N_ε-coordinated His residue to the Fe(III) ion and an N_δ–coordinated His residue to the Fe(II) ion. Both bridging and terminal carboxylate ligands have been proposed to complete the coordination environment of the diiron center of PAPs, but there is no NMR evidence to support this speculation. Signal f and h have the right chemical shifts for the carboxylate proton signals, but cannot be unequivocally assigned.

Substitution of the Fe(II) in Uf with Co(II), affords catalytically active Fe(III)Co(II)Uf [43]. Because of the very favorable electronic relaxation properties of

Co(II), many sharp isotropically shifted NMR signals are observed for Fe(III)Co(II)Uf, which have relaxation times (T_1's) 2 to 4 times longer than those of the native enzyme [43]. The somewhat unsymmetric peak at 99 ppm is solvent exchangeable and actually consists of one sharper signal and one broader signal, which are assigned to the

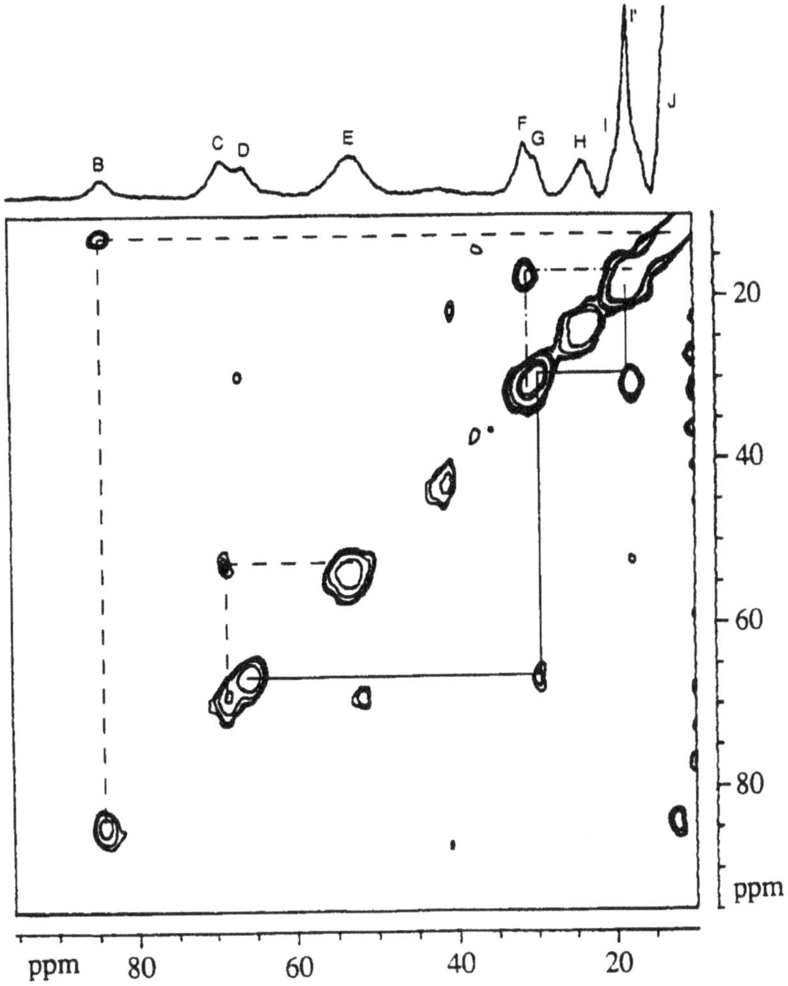

Figure 8. Proton NOESY spectrum of Fe(III)Co(II)Uf at 360.13 MHz at 40 °C in 100 mM acetate in D_2O buffer at pH 5.3. This spectrum is obtained in the TPPI phase sensitive mode with a mixing time of 15 ms. A 5%-shifted Gaussian weighting function was applied in both dimensions prior to Fourier transformation, followed by baseline correction. The top trace is a resolution-enhanced 1D spectrum with the labeling of the signals. Reprinted with permission from the American Chemical Society.

imidazole N-H protons of the His ligands coordinated to Co(II) and Fe(III), respectively. This confirms our early conclusion that there are two His ligands to the diiron center of Uf. The increased relaxation times of the isotropically shifted signals in the 90 to 10 ppm region in the Fe(III)Co(II)Uf spectrum allows a NOESY experiment to be carried out (Figure 8) and a number of cross peaks are evident. The strong cross peaks observed between signals B (81 ppm) and J (10.8 ppm), and signals C (68 ppm) and E (54 ppm) and weak 1D NOE correlations between signal B and signal C or E indicate that these arise to the Tyr ligand on Fe(III). In addition, there are remarkably clear cross signals between signal D (64 ppm) and G (28.5 ppm) and between signals G and I (17.7 ppm) which indicate that they belong to an ABX system. Such a pattern is consistent with the $-C_{\beta}H_2-C_{\alpha}H<$ moiety of a coordinated Asp residue. The chemical shifts and T_1 values for signals D, G, and I are indeed consistent with those assigned to the Asp residue bound to Co(II) center of CuCoSOD [24]. We thus assign signals D, G, and I to an Asp or Glu residue that chelated to the Co(II) site of Fe(III)Co(II)Uf. Considering the similar coordination properties of Co(II) and Fe(II), it is reasonable to extend conclusion to the Fe(III)Fe(II) case. Taking all the information obtained from our NMR studies and the data from our recent EXAFS studies [44], we have proposed the model shown in Scheme V for the dinuclear iron center of Uf and BSPAP.

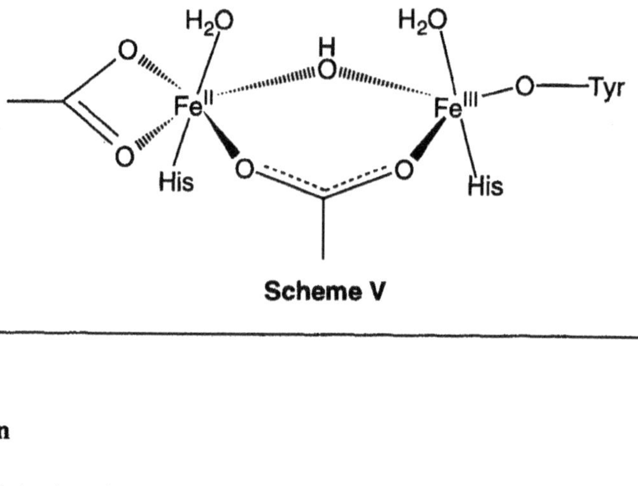

Scheme V

4. Conclusion

In summary, it is clear that modern NMR techniques are applicable for the investigation of nonheme iron proteins and can provide useful information about the metal binding residues. The success of such experiments depends on one's ability to manipulate the electronic spin lattice relaxation time of the paramagnetic metal center so as to obtain signals with T_1 values of 10 ms or longer.

5. References

1. Banci, L., Bertini, I., Luchinat, C. (1991) *Nuclear and Electronic Relaxation*, VCH, Weinheim, Germany.
2. Spiro, T.G. (1988) *Biological Applications of Raman Spectroscopy, Vol. 3, Resonance Raman Spectra of Heme and Metalloproteins*, John Wiley & Sons, New York, NY.
3. Goff, H. and La Mar, G.N. (1977) High spin ferrous porphyrin complexes as models for deoxymyoglobin and hemoglobin. A proton nuclear magnetic resonance study, *J. Am. Chem. Soc.* **99**, 6599-6606.
4. Pillai, R.P., Lenkinski, R.E., Sakai, T.T., Geckle, J.M., Krishna, N.R., and Glickson, J.D. (1980) Proton NMR studies of iron(II)-bleomycin: Assignment of resonances by saturation transfer experiments, *Biochem. Biophys. Res. Commun.* **96**, 341-349.
5. Lauffer, R.B., Antanaitis, B.C., Aisen, P., and Que, L., Jr. (1983) ^{1}H NMR studies of porcine uteroferrin. magnetic interactions and active site structure, *J. Biol. Chem.* **258**, 14212-14218.
6. Balch, A.L., Chan, Y.-W., La Mar, G.N., Latos-Grazynski, L., and Renner, M.W. (1985) Proton and deuterium nuclear magnetic resonance studies on iron(II) complexes of N-substituted porphyrins, *Inorg. Chem.* **24**, 1437-1443.
7. Wu, F.-J. and Kurtz, D.M., Jr. (1989) ^{1}H NMR spectroscopy of imidazole ligands in paramagnetic ferric and ferrous complexes and clusters. Relevance to non-heme proteins, *J. Am. Chem. Soc.* **111**, 6563-6572.
8. Borovik, A.S., Hendrich, M.P., Holman, T.R., Münck, E.; Papaefthymiou, V., and Que, L., Jr. (1990) Models for diferrous forms of iron-oxo proteins. Structure and properties of $[Fe_2BPMP(O_2CR)_2]BPh_4$ complexes, *J. Am. Chem. Soc.* **112**, 6031-6038.
9. Wang, Z., Holman, T.R., and Que, L., Jr. (1993) Two-dimensional ^{1}H NMR studies of paramagnetic bimetallic mixed-metal complexes, *Magn. Reson. Chem.* **31**, S78-S84.
10. Scarrow, R.C., Pyrz, J.W., and Que, L., Jr. (1990) NMR studies of the dinuclear iron site in reduced uteroferrin and its oxoanion complexes, *J. Am. Chem. Soc.* **112**, 657-665.
11. Oberley, L.W. (1982) *Superoxide Dismutase*, CRC Press, Boca Raton, FL.
12. Bertini, I., Turano, P., and Vila, A. J. (1993) Nuclear magnetic resonance of paramagnetic metalloproteins, *Chem. Rev.* **93**, 2833-2832.
13. Bertini, I. and Luchinat, C. (1986) *NMR of Paramagnetic Molecules in Biological Systems*, Benjamin/Cummings, Menlo Park, CA.
14. Stoddard, B.L., Howell, P.L., Ringe, D., and Petsko, G.A. (1990) The 2.1 Å resolution structure of iron superoxide dismutase from *Pseudomonas ovalis*, *Biochemistry* **29**, 8885-8893.
15. Carlioz, A., Ludwig, M. L., Stallings, W.C., Fee, J.A., Steinman, H.M., and Touati, D. (1988) Iron superoxide dismutase. Nucleotide sequences of the gene from *Escherichia coli* K12 and correlations with crystal structures, *J. Biol. Chem.* **263**, 1555-1562.
16. Ming, L.-J., Lynch, J.B., Holz, R.C., and Que, L., Jr. (1994) One- and two-dimensional ^{1}H NMR studies of the active site of iron(II) superoxide dismutase from *Escherichia coli*, *Inorg. Chem.* **33**, 83-87.

17. La Mar, G. N. and de Ropp, J.S. (1992) NMR methodology for paramagnetic proteins, in L.J. Berliner and J. Reuben (eds), *NMR of Paramagnetic Molecules*, Plenum Press, New York, NY, pp. 1-78.

18. Baldwin, J.E. (1989) Recent advances in the biosynthesis of penicillins and cephalosporins, in P.H. Bentley and R. Southgate (eds), *Recent Advances in the Chemistry of β-Lactam Antibiotics*, Royal Socierty of Chemistry, London, pp. 1-22.

19. Baldwin, J.E. and Bradley, M. (1990) Isopenicillin N synthase: mechanistic studies, *Chem. Rev.* **90**, 1079-1088.

20. Chen, V.J., Frolik, C. A., Orville, A. M., Harpel, M.R., Lipscomb, J.D., Surerus, K.K., and Münck, E. (1989) Spectroscopic studies of isopenicillin N synthase, *J. Biol. Chem.* **264**, 21677-21681.

21. Ming, L.-J., Que, L., Jr. Kriauciunas, A., Frolik, C.A, and Chen, V.J. (1990) Coordination chemistry of the metal binding site of isopenicillin N synthase, *Inorg. Chem.* **29**, 1111-1112.

22. Ming, L.-J., Que, L., Jr. Kriauciunas, A., Frolik, C.A, and Chen, V.J. (1991) NMR studies of the active site of isopencillin N synthase, a non-heme iron(II) enzyme, *Biochemistry* **30**, 11653-11659.

23. Bertini, I., Viezzoli, M.S., Luchinat, C., Stafford, E., Cardin, A.D., Behnke, W.D., Bhattacharyya, L., and Brewer, C. F. (1987) Circular dichroism and ^{1}H NMR studies of Co(II)- and Ni(II)-substituted concanavalin A and the lentil and pea lectins, *J. Biol. Chem.* **262**, 16985-16994.

24. Bertini, I., Luchinat, C., Piccioli, M., Vicens Oliver, M., and Viezzoli, M.S. (1991) H NMR investigation of reduced copper-cobalt superoxide dismutase, *Eur. Biophys. J.* **20**, 269-.279.

25. Scott, R.A., Wang, S., Eidsness, M.K., Kriauciunas, A., Frolik, C.A., and Chen, V.J. (1992) X-ray absorption spectroscopic studies of the high-spin iron(II) active site of isopenicillin N synthase: Evidence for Fe-S inteaction in the enzyme-substrates complex, *Biochemistry* **31**, 4596-4601.

26. Orville, A.M., Chen, V.J., Kriauciunas, A., Harpel, M.R., Fox, B.G., Münck, E., and Lipscomb, J.D. (1992) Thiolate ligation of the active site Fe(II) of isopenicillin N synthase derives from sbustrate rather than endogenous cysteines: Spectroscopic studies of site-specific Cys->Ser mutated enzymes, *Biochemistry* **31**, 4602-4612.

27. Randall, C.R., Zang, Y., True, A.E., Que, L., Jr., Charnock, J. M., Garner, C.D., Fujishima, Y., Schofield, C. J., and Baldwin, J.E. (1993) X-ray absorption studies of the ferrous active site of isopenicillin N synthase and realted model complexes, *Biochemistry* **32**, 6664-6673.

28. Boyington, J.C., Gaffney, B.J., and Amzel, L.M. (1993) The three-dimensional structure of an arachidonic acid 1,5-lipoxygenase, *Science* **260**, 1482-1486.

29. Minor, W., Steczko, J., Bolin, J.T., Otwinowski, Z., and Axelrod, B. (1993) Crystallographic determination of the active site iron and its ligands in soybean lipoxygenase L-1, *Biochemistry* **32**, 6320-6323.

30. Kurtz, D.M., Jr. (1992) Molecular structure/function relationships of hemerythrins, *Adv. Comp. Environ. Physiol.* **13** (Blood and Tissue Oxygen Carriers), 151-171.

31. Wilkins, R.G. and Harrington, P.C. (1983) The chemistry of hemerythrin, *Adv. Inorg. Biochem.* **5**, 51-85.

32. Stenkamp, R.E., Sieker, L.C., and Jensen, L.H. (1991) Structures of met and azidomet hemerythrin at 1.66 Å resolution, *J. Mol. Biol.* **220**, 723-737.

33. Holmes, M.A., Trong, I.L., Turley, S., Sieker, L.C., and Stenkamp, R.E. (1991) Structures of deoxy

and oxy hemerythrin at 2.0 Å resolution, *J. Mol. Biol* **218**, 583-593.

34. Sheriff, S., Hendrickson, W.A., and Smith J.L. (1987) Structure of myohemerythrin in the azidomet state at 1.7/1.3 Å resolution, *J. Mol. Biol.* **197**, 273-296.
35. Nocek, J.M., Kurtz, D.M., Jr., Sage, J.T., Debrunner, P.G., Maroney, M.J., and Que, L., Jr. (1985) Nitric oxide adduct of the binuclear iron center in deoxyhemerythrin from *Phascolopsis gouldii*. Analogue of a putative intermediate in the oxygenation reaction, *J. Am. Chem. Soc.* **107**, 3382-3384.
36. Maroney, M.J., Kurtz, D.M., Jr., Nocek, J.M., Pearce, L.L., and Que, L., Jr. (1986) ^{1}H NMR probes of the binuclear iron clusters in hemerythrin, *J. Am. Chem. Soc.* **108**, 6871-6879.
37. Wang, Z. (1994) Spectroscopic studies of oxo-bridged dinuclear iron-containing proteins and related model complexes, *Ph.D. Thesis*, University of Minnesota.
38. Doi, K., Antanaitis, B.C., and Aisen, P. (1988) The Binuclear iron centers of uteroferrin and the purple acid phosphatases, *Struct. Bonding (Berlin)* **70**, 1-26.
39. Vincent, J.B. and Averill, B.A. (1990) An enzyme with a double identity: purple acid phasphatase and tartrate-resistant acid phosphatase, *FASEB J.* **4**, 3009-3014.
40. Antanaitis, B.C., Strekas, T., and Aisen, P. (1982) Characterization of pink and purple uteroferrin by resonance Raman and CD spectroscopy, *J. Biol. Chem.* **257**, 3766-3770.
41. Averill, B.A., Davis, J.C., Burman, S., Zirino, T., Sanders-Loehr, J., Loehr, T.M., Sage, J.T., and Debrunner, P.G. (1987) Spectroscopic and magnetic studies of the purple acid phosphatase from bovine spleen, *J. Am. Chem. Soc.* **109**, 3760-3767.
42. Wang, Z., Ming, L.-J., Que, L., Jr., Vincent, J.B., Crowder, M.W., and Averill, B.A. (1992) ^{1}H NMR and NOE studies of the purple acid phosphatases from porcine uterus and bovine spleen, *Biochemistry* **31**, 5263-5268.
43. Holz, R.C., Que, L., Jr., and Ming, L.-J. (1992) NOESY studies on the Fe(III)Co(II) active site of the purple acid phosphatase uteroferrin, *J. Am. Chem. Soc.* **114**, 4434-4436.
44. True, A.E., Scarrow, R.C., Randall, C. R., Holz, R.C., and Que, L., Jr. (1993) EXAFS studies of uteroferrin and its anion complexes *J. Am. Chem. Soc.* **115**, 4246-4255.

Acknowledgments. This work has been supported by National Institutes of Health grant GM-33162 and National Science Foundation grants MCB-9104669 and MCB-9405723. We thank Drs. Randall B. Lauffer, Michael J. Maroney, Robert C. Scarrow, Li-June Ming, and Richard C. Holz for their efforts in obtaining NMR spectra of nonheme iron proteins while in the Que group. We are grateful to the following for providing the proteins for our studies: Dr. John B. Lynch (FeSOD), Dr. Victor J. Chen (IPNS), Dr. Donald M. Kurtz, Jr. (Hr), Dr. Walter B. Ellis (MHr) and Dr. Bruce A. Averill (BSPAP).

COBALT SUBSTITUTED PROTEINS

A. DONAIRE, J. SALGADO, H.R. JIMENEZ and J.M. MORATAL
Department of Inorganic Chemistry, University of Valencia, C/ Dr. Moliner 50, 46100 Burjassot (Valencia), Spain.

Abstract. Cobalt(II) has been extensively used as a spectroscopic probe in many proteins, mainly replacing zinc, but also substituting iron, manganese and copper ions. The relatively short electronic relaxation times of high spin cobalt(II) makes this ion suitable as a paramagnetic probe for Nuclear Magnetic Resonance spectroscopy. A survey of the NMR studies performed in cobalt substituted proteins is shown. In the zinc enzymes Carboxypeptidase A, Carbonic Anhydrase and Superoxide Dismutase the implications of these studies on their catalytic mechanisms are commented. Finally, a further insight in the research of the blue copper protein Azurin by applying NMR to its cobalt derivative is also reported.

Keywords: Nuclear Magnetic Resonance, cobalt, spectroscopic probes, metallo-substituted proteins, Carboxypeptidase, Carbonic Anhydrase, Superoxide Dismutase, Azurin.

1. Cobalt(II) as a spectroscopic probe for NMR

Cobalt (II) has been extensively used as a spectroscopic probe substituting mainly zinc, but also iron, manganese and copper ions [1]. Its spectroscopic features make this ion suitable not only for Nuclear Magnetic Resonance [2,3], but also for UV-Vis, EPR, CD and MCD spectroscopies and for magnetic susceptibility measurements [4].

A paramagnetic metal ion is an adequate spectroscopic probe in NMR when its unpaired electrons relax fast. In fact, there is a qualitative statement, derived from the Solomon equation, that can be enunciated as follows: the shorter the electronic relaxation times of the metal ions, the sharper the NMR signals of the nuclei coupled to the unpaired electrons [2,3]. Cobalt(II) in proteins has usually short electron relaxation times and, consequently, is a good spectroscopic probe for NMR.

In the majority of the metalloproteins substituted by cobalt, this ion is found in the high spin state. The lowest energy spectroscopic terms for a high spin (S=3/2) d^7 ion in a octahedral, trigonal bipyramidal and tetrahedral environment are represented in Figure 1. Octahedral high spin cobalt(II) presents a $^4T_{1g}$ ground state. This triplet converts into three energy levels that give six Kramer's doublets if spin-orbit coupling is accounted for. Because of this, electron relaxation in this kind of systems is relatively fast (in the order of 10^{-12} s), and, consequently, ^1H NMR signals are narrow enough to be observed.

High spin cobalt(II) in five coordinated complexes also presents fast electronic relaxation times. For example, for a trigonal bipyramidal geometry, the $^4A_2$

G.N. La Mar (ed.), Nuclear Magnetic Resonance of Paramagnetic Macromolecules, 213-244.

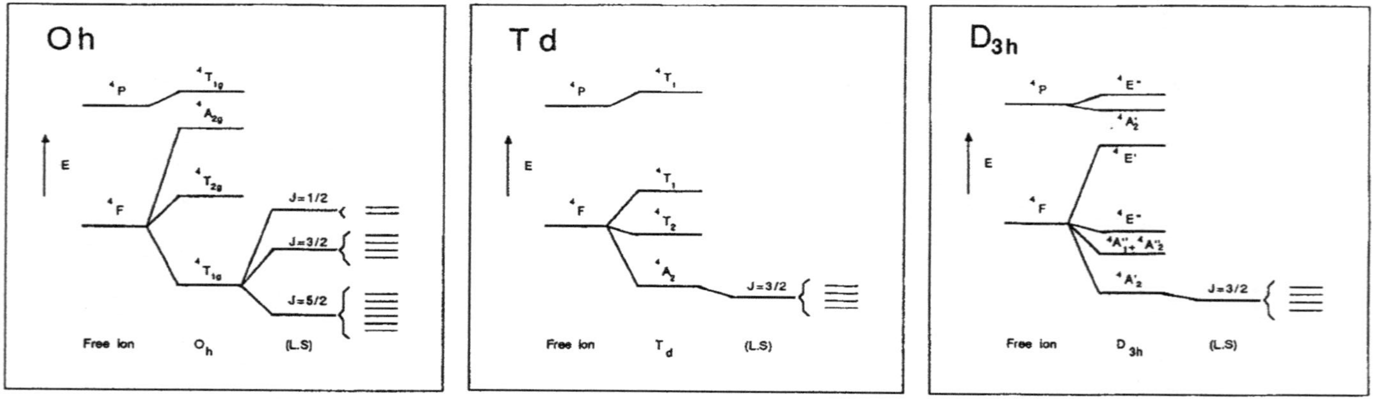

Figure 1. Splitting and spin-orbit coupling of ion terms of high spin cobalt(II) in octaedral (O_h), Tetrahedral (T_d) and trigonal bipyramidal (D_{3h}) geometries. The scaling of the terms is arbitrary.

ground state is not degenerated and does not split, but the existence of low energy levels populated at room temperature makes the electron relaxation easy. NMR signals are also usually not very broad and they can be detected even when the electron relaxation is slower than in octahedral complexes. The other pentacoordinated geometries have also similar electronic relaxation patterns.

Finally, high spin cobalt(II) in a tetrahedral geometry is also suitable for ^{1}H NMR spectroscopy. In four coordinated complexes, the difference in energy between the first excited level and the non-degenerated $^{4}A_2$ ground state is higher than in five coordinated complexes. Therefore, four coordinated cobalt(II) has shorter electronic relaxation times (about 10^{-11} s). However, even in this case, ^{1}H NMR lines are sharp enough to be detected.

To sum up, the electronic relaxation rates in high spin cobalt(II) complexes increase with the coordination number. Similarly, the relaxation times of protons coupled to the unpaired electrons are found to follow the same order. Thus, in these systems, signals are sharper and better resolved in the order: $O_h > D_{3h} > T_d$.

A comment must be made on the magnetic anisotropy of these systems. This arises from the existence of populations of different electronic states. Therefore, the magnetic anisotropy in Co(II) complexes follows the order: $O_h > D_{3h} > T_d$. Likewise, when magnetic anisotropy is increased, the dipolar effects on protons near the metal ion are also increased, and so, a large number of signals with pseudo-contact contribution to the hyperfine shift are observed.

2. Zinc substituted by Cobalt

Zinc is, after iron, the most abundant element in biological systems among the transition and group II elements. It has been found to act mainly either with a specific catalytic function (as in hydrolytic enzymes), or with a structural role [5].

Zn(II) ion is diamagnetic and its only isotope with nuclear magnetic moment, ^{67}Zn, is not detectable by NMR at usual protein concentrations. On the other hand, hydrolytic zinc enzymes have a medium-size molecular weight (usually larger than 30 kDa) that, up till now, makes these proteins unsuitable for standard multidimensional NMR spectroscopy. Because of this, zinc substitution is so far the only way to study specifically the residues of the active site in solution, where the enzyme is active [6]. Zn substitution by different paramagnetic metal ions has been extensively used [1,4,7]. Among them, Co(II) is the only one which restores partially or totally the enzymatic activity of all the cobalt-substituted zinc-enzymes studied up till now. A covalent radius similar to that of Zn facilitating the adoption of similar coordination geometries (mainly tetrahedral) similar to those of the zinc in the native enzymes, and a not very different way of interaction with typical zinc-enzymes ligands make cobalt(II) the best paramagnetic metal ion for zinc(II) substitution, Figure 2. Adding to this the excellent spectroscopic characteristics of this transition element for several techniques (see above) it is easy to understand why Co(II) has been chosen as the first spectroscopic probe for zinc enzymes.

From the study of the cobalt substituted derivatives through ^{1}H NMR, important conclusions on the mechanism of the enzymatic reactions of zinc enzymes have been achieved. Some examples of these conclusions are set out below.

Zn-proteins:

- Carboxypeptidase, CPD
- Carbonic Anhydrase, CA
- Superoxide dismutase, SOD
- Alkaline phosphatase, AP
- Liver Alcohol Dehydrogenase, LADH
- Zinc finger peptide CP-1

Copper proteins:

- Azurin
- Stellacyanin

Iron proteins:

- Rubredoxin
- Uterferrin
- Isopenicillin N synthase
- Conalbumin

Manganese proteins:

- Lectins

Metal-storage proteins:

- Metallothioenins

Figure 2. Cobalt(II) substituted proteins studied by NMR. (From ref. 1)

2.1. CARBOXYPEPTIDASE A

Carboxypeptidase A is a Zn enzyme that catalyzes the hydrolysis of polipeptides in which the carboxylate-terminal aminoacid has a hydrophobic or an aliphatic residue. Zn ion is essential for enzymatic activity [8]. The substitution of zinc(II) by cobalt(II) leads to the formation of a derivative that is even more active (215%) than the native enzyme [6]. Zinc(II) and cobalt(II) Carboxypeptidase A (CoCPD, hereafter) have been characterized by X-Ray crystallography at 1.54 Å [9] and 1.7 Å resolution [10], respectively. The cobalt ion, as the zinc in the native enzyme [9], is coordinated to two imidazole groups from His-69 and His-196, to a bidentate carboxylate group from

Glu-72 and to a water molecule (Figure 3). At physiological pH values, this water molecule is hydrogen bound to Glu-270. CoCPD UV-Vis and ^1H NMR spectra indicate that the pentacoordination is kept in solution [11,12].

CPD has two critical deprotonations that affect its catalytic properties (Scheme 1). The EH_2 form is inactive [13] and shows high affinity for substrates and inhibitors [14]. The first deprotonation (with a pKa of 6.1 for the native enzyme) leads to the formation of the active form of the enzyme (EH). This deprotonation has been identified with the ionization of Glu-270 [15]. When this residue is modified [16], the enzyme is inactive. The second deprotonation (with a pKa value around 9) gives an inactive form of the enzyme (E). This form shows much less affinity for substrates and inhibitors [14].

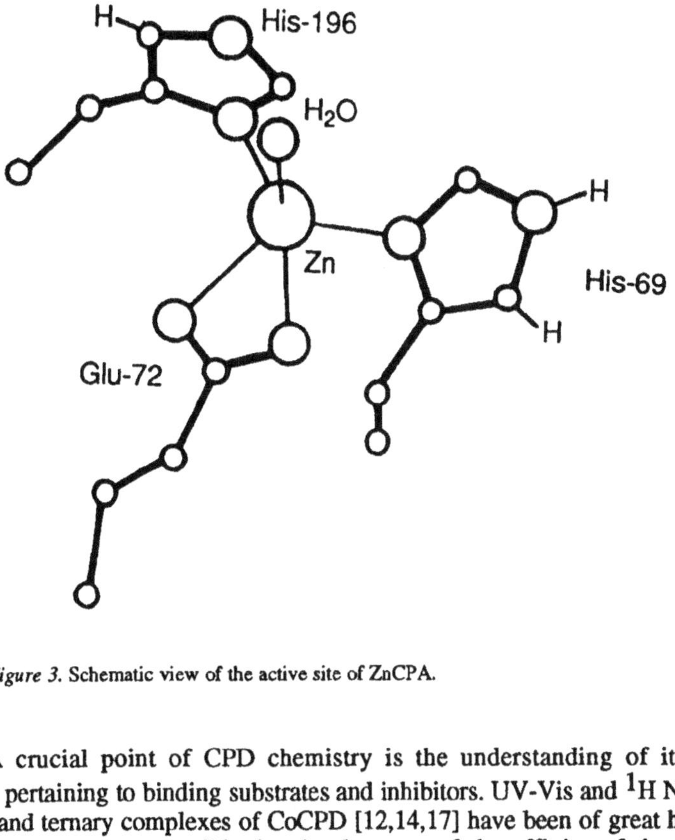

Figure 3. Schematic view of the active site of ZnCPA.

A crucial point of CPD chemistry is the understanding of its acid-base properties pertaining to binding substrates and inhibitors. UV-Vis and ^1H NMR studies of binary and ternary complexes of CoCPD [12,14,17] have been of great help in order to clarify the group responsible for the decrease of the affinity of the enzyme for inhibitors at high pH.

Two different binding sites for inhibitors, analogous to substrates, have been identified in CPD. The first one, called S'_1, is placed near the metal ion. Inhibitors such as D-Phenylalanine (D-Phe), L-Phenylalanine (L-Phe), Acetate (Ac), phenylacetate and β-phenylpropionate tightly bind to this binding site as it has been observed by

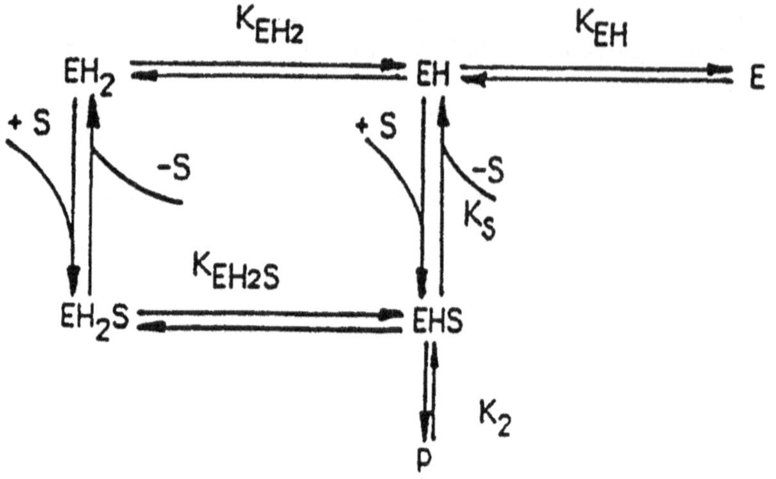

Scheme 1

^{13}C NMR linewidth measurements [14,18,19] and ^1H NMR spectra [12]. On the other hand, X-Ray crystallographic data on the ZnCPD/DPhe complex and with several inhibitors analogous to substrates have been performed. They show that carboxylate groups of these inhibitors specifically bind to the enzyme through the interaction with the positively charged Arg-145 guanidinium group at the S'_1 site (Figure 4A,B) [20-22]. The other binding site (S_1) is the metal ion itself. The binding of small anions (halides and pseudohalides) to the zinc depends on the occupation of the S'_1 site by pseudosubstrates of peptide hydrolysis. When these pseudosubstrates are placed in the S'_1 site, the Glu-270-metal-water hydrogen bond network is disrupted allowing the binding of small inorganic inhibitors to the metal ion (Figure 4C). Interestingly, it has been observed by ^{31}P NMR with the cobalt and nickel derivatives that phosphate and pyrophosphate bind directly to the metal without the need of a pseudosubstrate in the S'_1 site [23,24]. In this case these two anions bind to the enzyme replacing the water molecule and keeping the hydrogen bond network between the metal and the Glu-270 (Scheme 2).

CoCPD ^1H NMR spectrum at pH 6.0 (Figure 5A) shows three well resolved isotropically shifted signals in the downfield region at 63 (a), 52 (c) and 45(d) ppm and a broader signal at 56 ppm(b). Signals a, c, and d have been assigned to the NH and δH protons of His-69 and to the δH proton of His-196, respectively. When pH is increased, signals shift slightly upfield, showing a pK_a value of 8.8 (Table 1), coincident with the pK_{a2} of the enzyme. This pK_a has also been obtained by UV-Vis spectroscopy [17].

From the ^1H NMR titration with the pH of the binary adducts CoCPD/L-Phe (Figure 5B), CoCPD/DPhe and CoCPD/Ac pK_a values of 7.7, 7.8 and 9.2 were obtained (Table 1). Both L-Phe and D-Phe have an amino group that interacts with the

Glu-270-H_2O-Co system (Figure 4B) and they show a synergetic effect for the binding of small anionic inhibitors (halides and pseudohalides) to the enzyme.

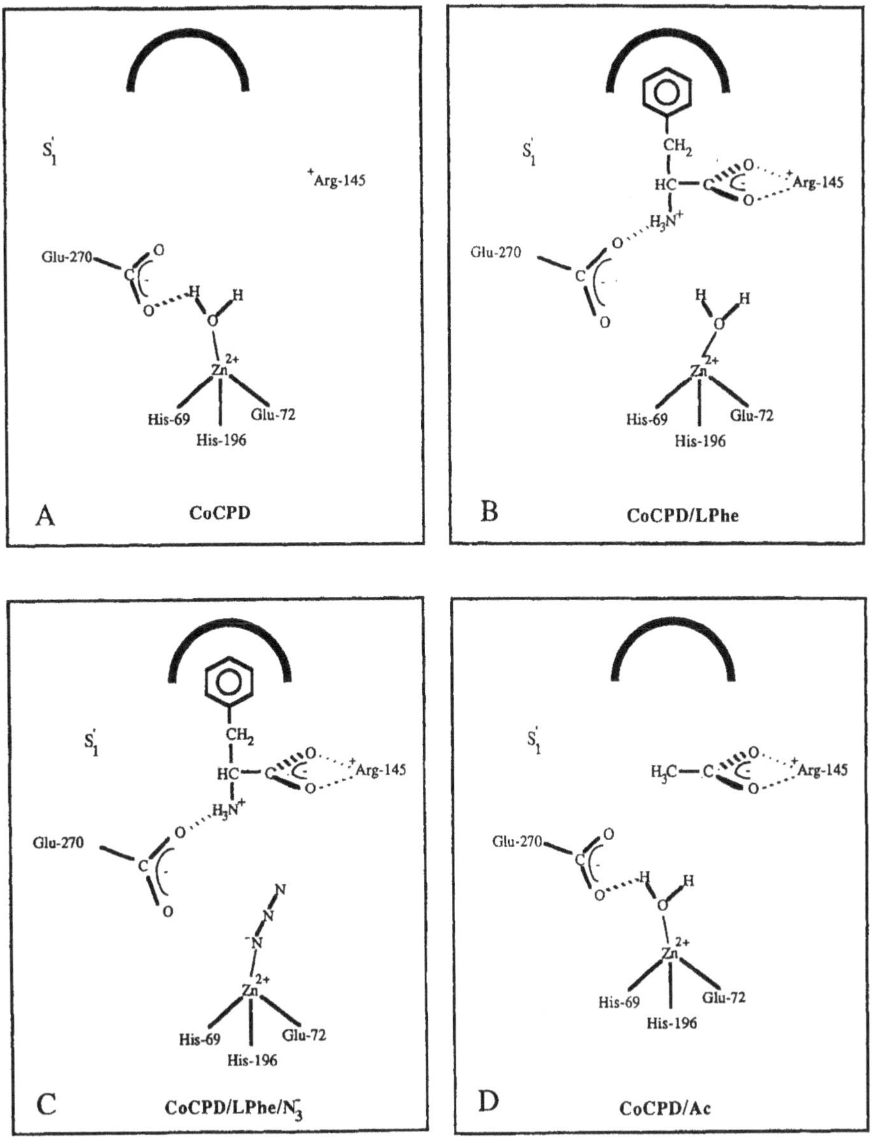

Figure 4. Binding mode of inhibitors to carboxypeptidase A.

In the ternary complexes CoCPD/LPhe/N_3^- and Co/DPhe/DPhe both azide and the second molecule of DPhe are directly bound to the metal ion. In both cases, the water molecule bound to the metal is displaced (Figure 4C). The UV-vis and the ^{1}H NMR (Figure 5C) spectra indicate that the ternary complex with azide is tetracoordinated, while

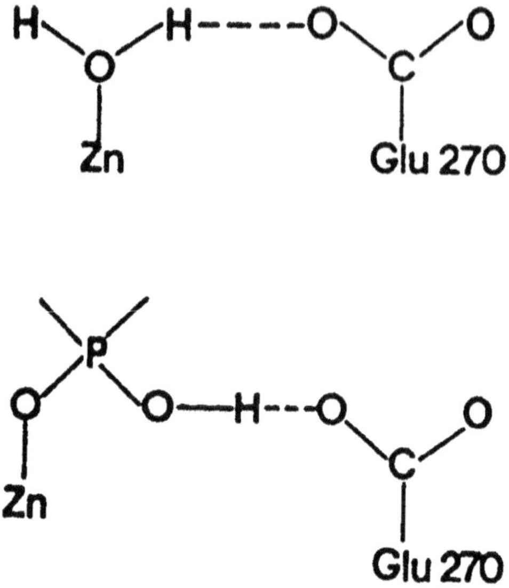

Sheme 2

TABLE 1. Protonation constants of CoCPD with differents adducts in H_2O 1 M NaCl / 20 mM Mes at 25 °C.

Adduct	pK_a
CoCPD	8.8 ± 0.1
CoCPD/LPhe (L-Phe 100 mM)	7.7 ± 0.1
CoCPD/DPhe (D-Phe 2.5 mM)	7.8 ± 0.1
CoCPD/Ac (Ac 10 mM)	9.2 ± 0.2
CoCPD/LPhe/N_3^- (L-Phe 100mM, N_3^- 100 mM)	9.7 ± 0.2
CoCPD/DPhe/DPhe (DPhe 100 mM)	No changes until pH values of 10.7

the DPhe complex is hexacoordinated. ^{1}H NMR and UV-vis titrations with the pH for these adducts give an apparent pK value of 9.7 for the azide complex at a concentration 0.1 M in azide. This pK changes with the concentration of azide, indicating that the observed effect comes from the competence between the azide and the hydroxide anions for the same binding site. In the DPhe ternary complex, the ^{1}H NMR spectrum of this adduct is not altered until pH 10.8, due to the high affinity of DPhe for the second (metal) binding site.

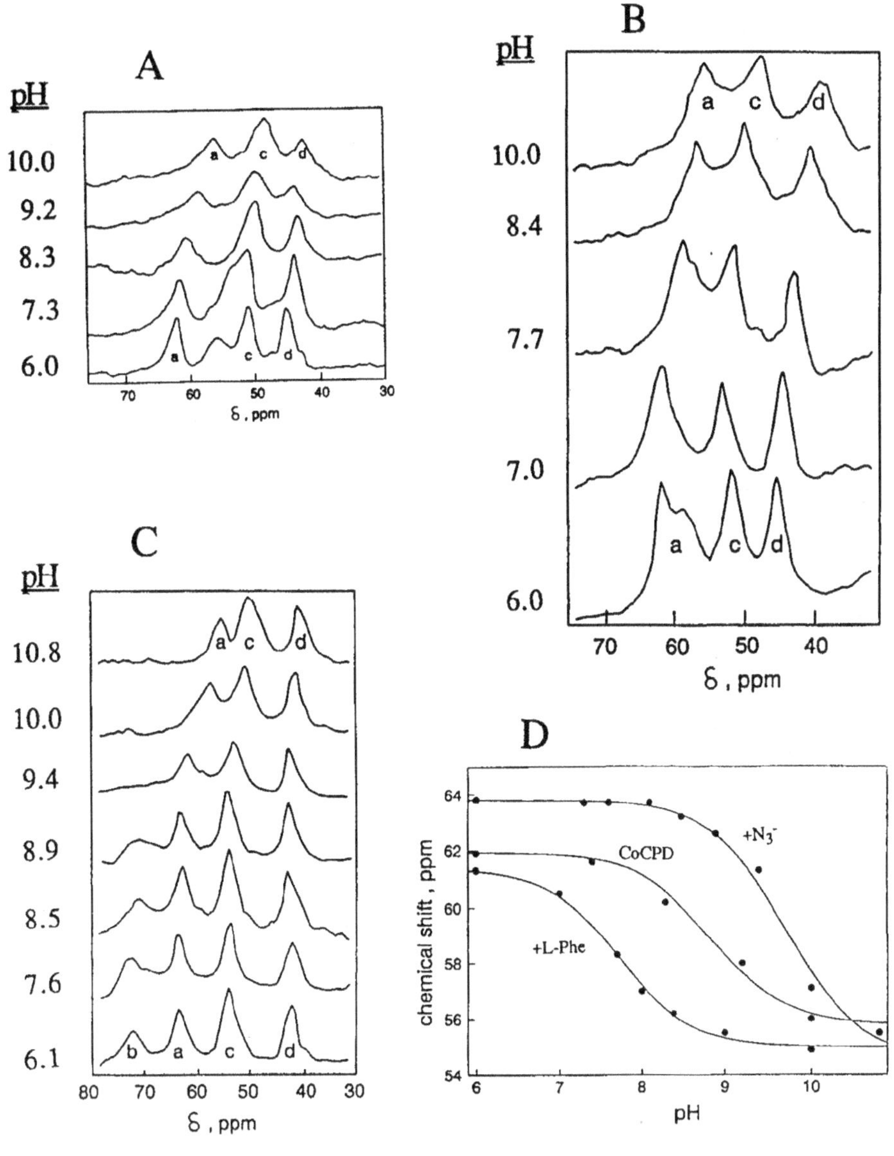

Figure 5. pH dependence of ^{1}H NMR spectra of CoCPD (1 mM) in 1 M NaCl/20 mM Mes (A); CoCPD (1mM) and L-Phe (100 mM) (B); CoCPD (1mM), L-Phe (100 mM) and N_3^- (100 mM) (C); dependence of the ^{1}H NMR signal a on pH for CoCPD (middle) and the cobalt enzyme in the presence of L-Phe (100 mM) (bottom) and of L-Phe (100 mM) plus N_3^- (100 mM) (top), the lines are theoretical pH titration curves using the parameters for the pK_a (D). (Reprinted from ref. 17).

Therefore, in the binary complexes of CoCPD in which the Glu-270-H_2O-Co system is perturbed by the binding of the inhibitor in the S_1 site (CoCPD/LPhe and CoCPD/DPhe) the pKa of the enzyme is modified from 8.8 to 7.7. This higher acidity is consistent with the disruption of the hydrogen bond between the water molecule and the Glu-270. The small acetate anion, which lacks the amino group, does not significantly modify the pKa of the enzyme because it does not affect the interaction between the coordinated water molecule and Glu-270 (Figure 4D). In the ternary complexes CoCPD/LPhe/N_3^- and CoCPD/DPhe/DPhe, in which the second inhibitor binds directly to the metal ion displacing the coordinated water molecule, no pK_a was observed. All these facts are consistent with the assignment of the zinc bound water molecule as the group responsible for the second pK_a of the enzyme [25]. This has been recently confirmed by X-ray absorption fine structure (XAFS) studies on the native enzyme in solution [26].

2.2. CARBONIC ANHYDRASE

Carbonic Anhydrase catalyzes the interconversion between carbon dioxide and bicarbonate anion at physiological conditions [27-31]:

$$CO_2 + H_2O \Leftrightarrow HCO_3^- + H^+$$

Recently, many crystallographic studies on CA from several isoenzymes [32,33], metallo-derivatives [34,35] and mutants [36-38], with different inhibitors [39-46] and even with the substrate bicarbonate [35,37] have been performed. In the native and cobalt substituted enzymes at pH 8 the metal ion is coordinated to three imidazole groups from His-94, -96 and -119 and to one solvent molecule (numbering are refers to the human isoenzyme II) (Figure 6). The enzyme molecule has two different domains, one of them dominated by hydrophobic residues and the other one by charged aminoacids. Two ionizable groups have been proposed to regulate the catalysis of the enzyme as it has been observed by the spectroscopic titration of the cobalt derivative with the pH [47]. The first one is the coordinated water molecule while the second is the Nδ imidazole of His-64. The deprotonation of the coordinated water molecule provides the active form of CA.

The study of CA metallo-derivatives in presence of their two main groups of inhibitors, sulfonamides and inorganic anions, has shed light to many of the questions about the catalytic mechanism of the enzyme. It has been proposed that the coordination number of the metal ion is crucial for the catalysis [48]. Indeed, among the different metallosubstituted carbonic anhydrases, only the cobalt one shows significant activity, and this has been related with the fact that only Co and Zn Carbonic Anhydrases are tetracoordinated at pH 8 [48]. Likewise, UV-Vis and ^1H NMR spectra indicate that the cobalt enzyme is tetracoordinated in solution at high pH, while the nickel derivative is octahedral [49].

The factors that govern the coordination number in CoCA and in its adducts with inhibitors in solution have been extensively studied by UV-Vis and ^1H NMR spectroscopies [50-53].

The ^1H NMR spectrum of bovine CA at pH 6 (Figure 7) shows three downfield shifted signals that have been assigned to the protons of the imidazole rings of the three coordinated histidines (Table 2). Two criteria have been established for determining the coordination number in cobalt CA by ^1H NMR spectroscopy [50]. The first one is based on the T_1 of the hyperfine shifted signals. Adducts with T_1 values lower than 5 ms are considered essentially tetrahedral, while those with T_1s higher than 15 ms are considered pentacoordinated. Intermediate relaxation times of the hyperfine shifted signals of the cobalt derivative are indicative of equilibrium between four and five coordinated species. The second criterion refers to the spread of the signals corresponding to protons which isotropic shifts have only pseudo-contact contribution (Figure 7). Magnetic anisotropy in cobalt five coordinated complexes is larger than that in cobalt tetrahedral ones. This makes the former species to show a larger set of isotropically shifted signals in the -20, 20 ppm region. Therefore, from these criteria, it can be deduced that the nitrate, acetate and thiocyanate CoCA adducts are pentacoordinated, while the CNO adduct is tetracoordinated. Analogously, CoCA at pH 6 and its complexes with azide and perchlorate are pentacoordinated. These two criteria are nicely consistent with the visible spectra. The CoCA UV-Vis at pH 6 has a molar extinction coefficient of 300 M^{-1} cm^{-1}, which is associated with an equilibrium between four and five coordinated cobalt(II) complexes. Azide, perchlorate and sulfate adducts have similar absorbances in the visible region. On the other hand, thiocyanate forms an adduct with CoCA in which cobalt remains pentacoordinated as it can be deduced from the UV-Vis spectrum (ε=100 M^{-1} cm^{-1}).

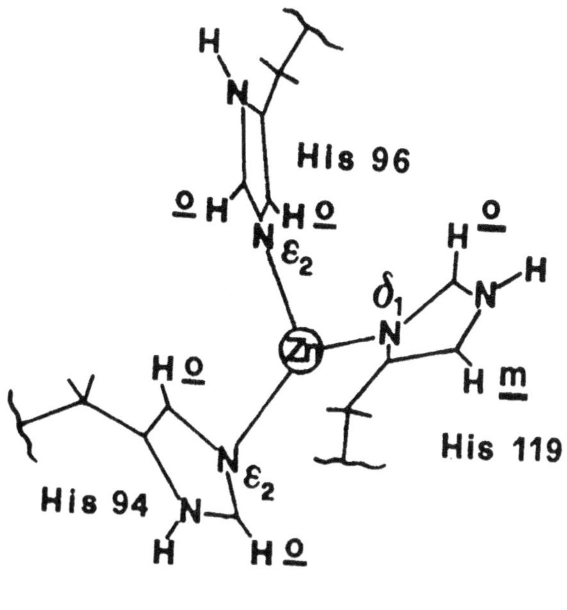

Figure 6. Schematic view of the active site of ZnBCA.

From X-Ray crystallographic data on the native enzyme with thiocyanate, it is known that this anion binds directly to the metal ion [54], being its soft sulfur atom directed towards the hydrophobic patch of the cavity, where it interacts with Val-143,

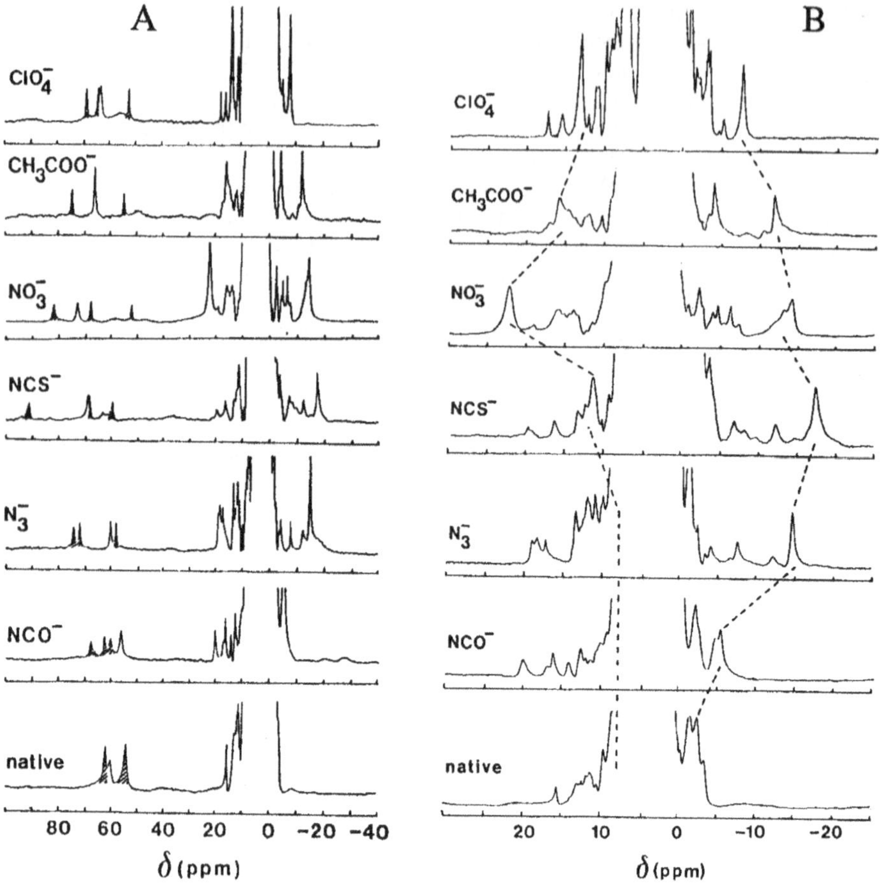

Figure 7. ^{1}H NMR spectra (200 MHz) at 300 K of some anion adducts of cobalt(II)-substituted carbonic anhydrase. The dashed signals disappear when the spectra are recorder in D_2O (A); Expanded "quasi-diamagnetic region" (B). (Reprinted from ref. 50).

Leu-198 and Trp-209. The pentacoordination is completed by a water molecule. Analogously, from studies with the cobalt derivative in solution it was proposed that cyanate would bind directly to the metal ion, being its hard oxygen atom interacting with charged residues (as the NH of Thr-199) at the hydrophilic site of the enzyme. The intermediate hard/soft pattern of the azide nitrogen atom would explain the equilibrium between tetra and penta-coordinated species observed with its CoCA adduct. The versatility of CoCA for both tetra and penta-coordination was taken as a factor that facilitate the interconversion between bicarbonate and carbon dioxide during the catalysis. A mechanism of action of CA was proposed from this study [51] (Scheme 3).

TABLE 2. 200 MHz ^1H NMR Shifts (ppm) and T_1 values (ms) for Some Anion Derivatives of Bovine Cobalt(II)-Substituted Carbonic Anhydrase[a,b].

Signal	no anion	$+CH_3COO^-$	$+NO_3^-$	$+ClO_4^-$	$+NCS^-$	$+NCO^-$	$+N_3^-$
His NH	62.1 (7.3)	66.0 (20.3)	6.2 (17)	69.0 (9.5)	68.7 (18)	62.1 (3.1)	57.8 (6.2)
His-119 m-H	60.3 (7.3)	66.0 (20.3)	70.1 (12)	63.4 (5.7)	69.0 (18)	55.5 (3.6)	59.9 (7.8)
His-119 NH	54.2 (8.1)	75.0 (19.5)	78.7 (12)	64.3 (5.4)	91.2 (14)	59.8 (2.5)	74.1 (5.5)
His NH	54.2 (8.1)	55.0 (22.5)	51.1 (27)	52.4 (14.6)	59.5 (25)	67.3 (2.2)	71.7 (5.4)
downfield CH_3	<6	15.7 (24)	20.1 (19)	13.3 (10)	11.4	<10	<10
upfield CH_3	≃ -3	-11.8 (40)	-13.1 (28)	-7.7 (20)	-17.3 (44)	-5.6 (7)	-14.9 (17)

[a] Only the meta-like signals for the histidine rings are reported.
[b] From ref. 51.

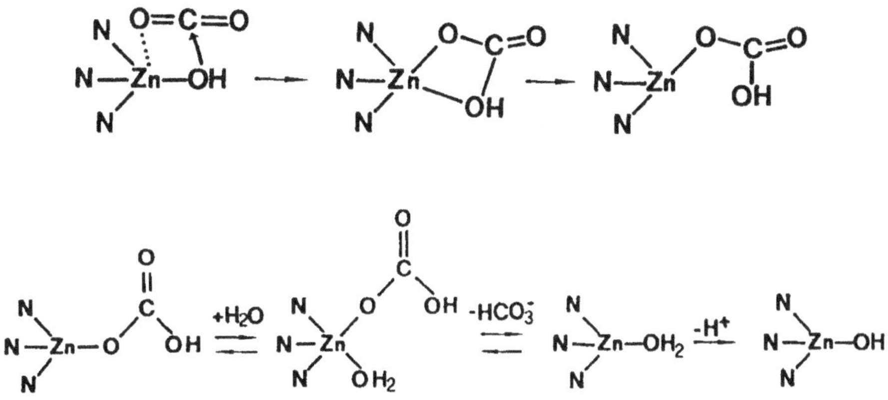

Scheme 3

Recent X-ray crystallographic data on the adducts of these anions with the native enzyme are only partially in agreement with the spectroscopic data from the cobalt derivatives [39-46]. In solid state, nitrate and cyanate anions are not directly bound to the metal ion in the zinc enzyme. Discrepancies probably arise from the different affinity of cobalt and zinc ions towards the donor atoms and also perhaps from the high concentration of ammonium sulfate necessary to crystallize the protein. A study of the cobalt derivative with the substrate bicarbonate has also been carried out [35] and it has been found that the cobalt ion has an octahedral distorted geometry. However, even when a new mechanism based on an octahedral intermediate has been proposed, the main data available from solution conditions, where the enzyme is active, are in agreement with a direct coordination to the metal ion of all the above mentioned inhibitors. Moreover, NMR studies on the ^{67}Zn enriched enzyme indicate that CNO^- binds in solution to the metal ion [55], as suggests the results obtained with the cobalt derivative.

Recently, from the ^1H NMR assignments of signals with pseudocontact contribution, the magnetic susceptibility anisotropy in the nitrate, perchlorate and thiocyanate adducts has been obtained [56,57]. In the case of the nitrate adduct, it has been also calculated by applying the angular overlap model (AOM) of the ligand field [58]. From these studies, a weak bond between the inhibitor and the cobalt ion has been proposed to account for the general spectroscopic properties in all these five-coordinated adducts. Wether these data are transferable to the native enzyme or not and its implications in the catalytic mechanism are questions still not answered.

Cobalt Carbonic Anhydrase have been also studied to obtain the electronic structure of cobalt and the magnetic susceptibility tensor that determines the magnetic anisotropy. Thus, from the hyperfine shifts of protons belonging to non-coordinated residues, the orientation of the magnetic axes of the magnetic susceptibility tensor in the $CoCA/NO_3^-$, $CoCA/ClO_4^-$ and $CoCA/SCN^-$ adducts has been achieved [56,57]. The values of the calculated magnetic susceptibility anisotropies for the choice of the z axis close to the Co-N(His119) bond have been found to be $\Delta\chi ax = -6.4 \times 10^{-8}\ m^3\ mol^{-1}$ and $\Delta\chi eq = 2.4 \times 10^{-8}\ m^3\ mol^{-1}$ for the perchlorate adduct and $\Delta\chi ax = -9.5 \times 10^{-8}\ m^3\ mol^{-1}$ and $\Delta\chi eq = 5.2 \times 10^{-8}\ m^3\ mol^{-1}$ for the nitrate adduct. In both cases, the large values of $\Delta\chi eq$ are indicative of a sizeable rhombicity of the χ tensor, being larger in the nitrate adduct. This is completely consistent with the UV-Vis spectra of the adducts and with the two criteria for the determination of the coordination number based on the ^1H NMR spectra discussed above. The computed values for the magnetic susceptibility anisotropies obtained from AOM calculations [58] are in agreement with those obtained from the ^1H NMR assignments. Once the orientation and the values of the magnetic susceptibility tensor is found, new assignments of the non-coordinated residues can be estimated.

2.3. SUPEROXIDE DISMUTASE

Cu, Zn superoxide dismutases (SOD) are dimeric enzymes built by two identical subunits, each one containing one copper(II) and one zinc(II) ion [59-62]. The copper ion is coordinated to four histidine residues: His-46, His-48, His-63 and His-120 (numbering refers to the human enzyme), in a distorted square-planar geometry (Figure 8) [63]. A water molecule occupies the fifth coordination position, with a Cu-O distance of about 2.8 Å [64]. The ligands of the tetrahedral zinc are His-63, His-71, His-80 and Asp-83. Therefore, His-63 bridges the copper and the zinc ions.

Metal derivatives of the enzyme, by substitution in one or in the two metal binding sites, have been obtained and extensively studied by NMR [65-68]. The Cu_2Co_2SOD derivative in the oxidized state is specially interesting because of the antiferromagnetic coupling which occurs between Cu(II) and Co(II) [69]. In fact, in this system, the electronic relaxation times of Cu(II) are shortened from about 2×10^{-9} s (typical of Cu(II) complexes) to values around 10^{-11} s (typical of Co(II) tetrahedral

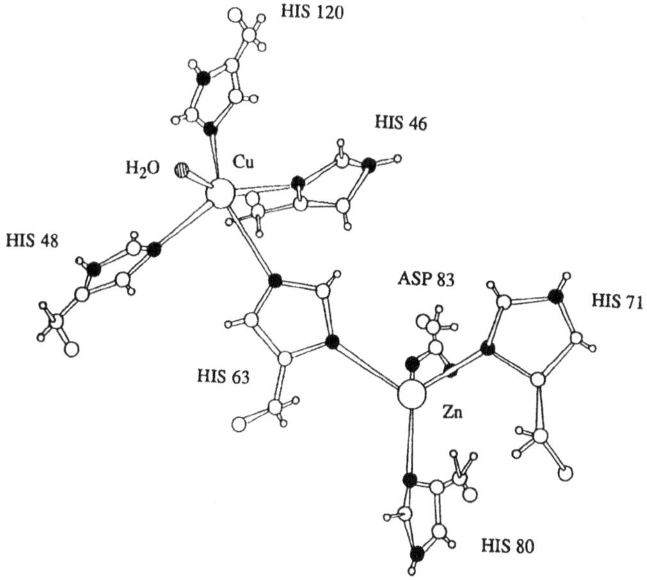

Figure 8. Schematic drawing of the active site of bovine Cu_2Zn_2SOD in the native form.

complexes) [70]. This has allowed the detection of hyperfine shifted signals belonging to protons of the histidines bound to the copper ion, that, otherwise, would not be observed. Moreover, due to the smaller S value of copper(II) with regard to the cobalt(II) ion, protons belonging to residues coordinated to the former have even longer relaxation rates than those of the cobalt histidine ligands and, consequently, appear better resolved.

The ^1H NMR spectrum of this derivative (Figure 9A) displays a set of hyperfine shifted signals that correspond to the protons of the ligands of both metal ions [65]. The characterization by NMR of this metallo substituted enzyme is probably one of the best examples of how the development of the application of this technique to paramagnetic systems has been carried out. The first attempt to assign these signals was based on the magnetic field dependence of the signal linewidths and on their nuclear longitudinal relaxation times by applying the Solomon equation [65]. Due to the existence of strong ligand centered dipolar effects on the imidazole rings of the histidine ligands, this approach led to incorrect metal-proton distances. Subsequently, the unequivocal assignment of most of the hyperfine shifted signals with contact contribution was carried out through the pattern of the observed NOEs with the help of the crystallographic X-Ray data [66-68]. Lately, the successful application of 2D NMR techniques in paramagnetic systems[71,72], together with the use of selectively deuteriated derivatives [73,74] have permitted the assignment of all the resolved hyperfine shifted signals, as well as some dipolar shifted ones that appear in the diamagnetic envelope (Table 3).

All these studies have taken to the better comprehension of the structure and function of the enzyme active site in solution. At this point, from the studies with chemically modified residues [75,76] and, recently, with several SOD mutants prepared by site-directed mutagenesis [77-82] the exact catalytic function of several residues have been discovered. In the most accepted among the proposed catalytic mechanisms of the enzyme

$$Cu^{2+} + O_2^{-} \Leftrightarrow Cu^{+} + O_2$$

$$Cu^{+} + O_2^{-} + 2H^{+} \Leftrightarrow Cu^{2+} + H_2O_2$$

the rate-limiting step for the overall enzymatic reaction is the binding of the substrate to the active cavity. This cavity is charged due to positive residues like Lys-122, Lys-136 and Arg-143, and also negative groups as the carboxylates of Glu-132 and Glu-133.

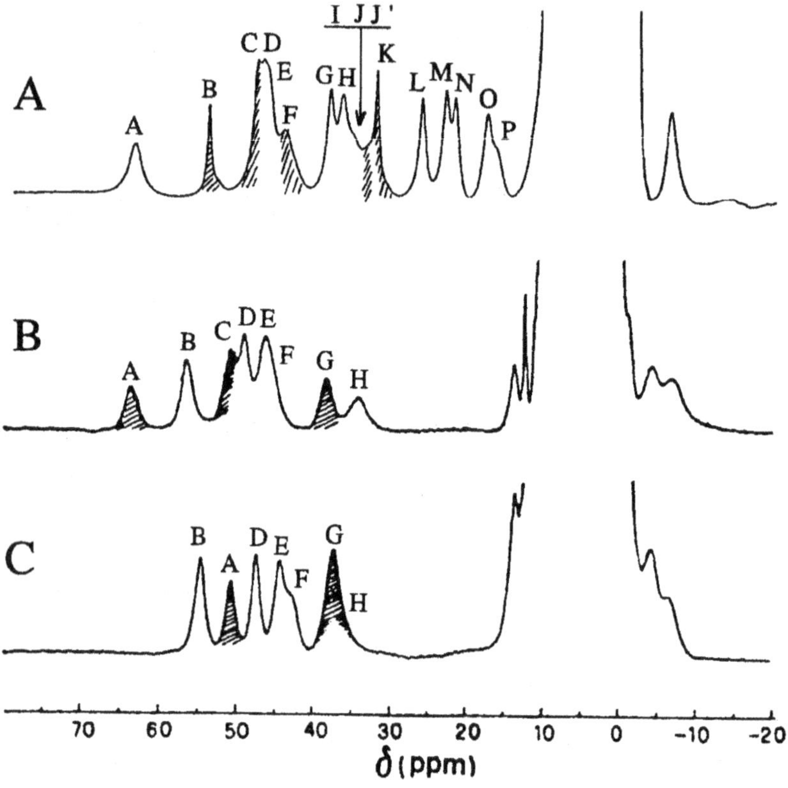

Figure 9. ^1H NMR spectra (200 MHz) in H_2O at 300 K: (A) $Cu^{II}{}_2Co_2SOD$ at pH 7.5; (B) $Cu^{I}{}_2Co_2SOD$ at pH 5.5; (C) E_2Co_2SOD at pH 5.5. The shaded signals indicate protons that are exchangeable in a D_2O solution. (Part A, reprinted from ref. 66. Part B and C, reprinted from ref. 74).

Electrostatic interactions between the superoxide anion and these charged groups determine the way of binding of the substrate to the enzyme. The studies with the pH on the cobalt derivatives of the native enzyme and the Lys-136 mutants (where Lys is substituted by Arg, Ala or Gln) have taken to discard this residue, and to propose the copper semicoordinated water molecule, as the responsible group for the pH dependent properties of the enzyme at high pH [80]. ^{1}H NMR studies show that the T137-R Cu_2Co_2SOD mutant has about 20 times more affinity for azide, which is considered as a good probe of the superoxide anion, than the wild type enzyme [82]. The increase in the affinity of the enzyme for azide is consistent with an increment in the positive charge in the cavity. The fact that the activity is significantly reduced by this mutation has led to propose that the change of Thr137 by Arg alters the active-site channel, disrupting the hydrogen bond network between Glu-132, Glu-133 and Lys-136 [82]. Therefore, ^{1}H NMR spectroscopy in the Cu_2Co_2SOD derivative and the application of this technique on its mutants has been showed to be a very powerful technique to understand the roles of every specific residues in the catalytic mechanism.

TABLE 3. ^{1}H NMR Chemical shifts, T_1 values, and assignments of the hyperfine-shifted signals in Cu_2Co_2SOD belonging to the protons of coordinated residues (From ref. 72).

Signal	δ(ppm)	T_1 (ms)	assignment
A	66.2	1.5	Hδ2 His 63
B	56.5	7.8	Hδ1 His 120
C	50.3	4.2	Hε2 His 46
D	49.4	3.8	Hδ2 His 71
E	48.8	4.6	Hδ2 His 80
F	46.7	2.1	Hε2 His 80 (His 71)
G	40.6	3.5	Hδ2 His 46
H	39.0	1.8	Hε1 His 120
I	37.4	1.7	(Hβ1) (Hβ2) Asp 83
J'	35.6	1.7	(Hβ2) (Hβ1) Asp 83
J	35.4		Hε2 His 71 (His 80)
K	34.5	8.0	Hδ1 His 48
L	28.4	4.3	Hδ2 His 48
M	25.3	2.7	Hε1 His 46
N	24.1	2.9	Hδ2 His 120
O	19.6	1.9	Hε1 His 48
P	18.7	1.6	Hβ1 His 46
Q	- 6.2	2.4	Hβ2 His 71
R	- 6.2	2.4	Hβ2 His 46
a'	14.05		Hε2 His 43
b'	12.85		Hδ1 His 43
c'	12.30		Hβ2 His 48
d'	11.21		Hβ2 His 120
e'	10.29		Hε1 Trp 30
f'	8.63		Hε1 His 43
g'	7.38		Hζ1 Trp 30
h'	7.06		Hδ1 Trp 30
i'	6.40		Hβ2 His 48
k'	3.13		Hβ1 His 120
l'	1.23		β-CH_3 Ala 140
m'	0.53		γ1-CH_3 Val 118
n'	0.28		Hγ1 Arg 143
o'	- 1.51		γ2-CH_3 Val 118

The reduced derivative ($Cu^{I}_2Co_2SOD$) has also been studied by ^1H NMR spectroscopy [83,84]. Its spectrum (Figure 9B) displays three exchangeable protons. Since in this derivative the only paramagnetic probe is the cobalt(II) ion, this has confirmed the protonation of His-63 and, consequently, the disruption of the bridge between the two metal ions in the reduced state [83]. As in the case of pentacoordinated derivatives of Carbonic Anhydrase, the orientation of the axis of the magnetic susceptibility tensor has been tentatively determined from the assignments of pseudo-contact shifted signals[84]. The ^1H NMR spectrum of the E_2Co_2SOD derivative (where E represents an empty site) (Figure 9C) only shows two NH protons, corresponding to the Hδ2 and Hε2 of the histidines 71 and 80, respectively [65]. This indicates that the extraction of the copper ion makes the NH of the bridged histidine enter fast exchange conditions with the solvent.

The cobalt(II) has also been introduced in the copper site, and the disubstituted Co_2Co_2SOD has been obtained. The ^1H NMR spectrum of this derivative displays a set of well resolved hyperfine shifted signals belonging to the residues coordinated to both cobalt ions [85-87]. Protons of cobalt coordinated residues in the copper site show narrower signals than those of the cobalt ligands in the zinc site, in agreement with a pentacoordinated coordination for the former one. When phosphate binds to the enzyme[86], signals make broader. This indicative the existence of antiferromagnetic coupling between the two metal ions, that is reduced by entrance of the phosphate inhibitor which breaks the His-63 bridge.

2.4. OTHERS ZINC BY COBALT SUBSTITUTED PROTEINS

2.4.1. *Alkaline phosphatase*

Alkaline phosphatase (AP) is a dimer, in which each subunit contains two zinc ions in the so called sites A and B and one magnesium ion in the site C. Metal ions are coordinated to three histidines in sites A, with a four coordination in site A and a five coordination in site B [88].

Derivatives in which sites A and B have been substituted by cobalt and/or copper ions and with or without the third position occupied by a magnesium ion have been obtained. Such derivatives have been studied by UV-Vis, ^1H NMR and water proton nuclear magnetic relaxation dispersion (NMRD) spectroscopies [89,90]. The NMRD measurements show that the electronic relaxation times of copper ion are shortened down from 4×10^{-9} s for the Cu_2E_2AP (E = empty site) enzyme to around 10^{-11} s for the Cu_2Co_2AP. This is indicative of magnetic coupling between the two metal ions. In agreement with this, the ^1H NMR spectrum of the $Cu^{II}_2Co^{II}_2AP$ at pH 6 has been reported and the signals belonging to the copper ligands have been detected [90]. Sites A and B are 3.9 Å apart, a distance small enough to allow dipolar magnetic coupling between the two metal ions. In the Cu_2Cu_2AP derivative a large magnetic coupling was observed and it was proposed that a hydroxo or alkoxo group could bridge the two ions[91]. However, in the X-Ray structure of the native enzyme (Zn_2Zn_2AP) with phosphate is the phosphate group which is found as a bridge. Finally, the occupation of site C by magnesium produces minimal changes in the ^1H NMR spectrum of the cobalt derivative.

2.4.2. *Liver Alcohol dehydrogenase (LADH)*
Liver Alcohol dehydrogenase is a dimer of 80 kDa molecular weight. Every dimer contains two non-equivalent zinc ions, one of them with a catalytic role and the other with a structural function [4]. Cobalt(II) have been substituted in both sites. The non-catalytic zinc is tetrahedral with four cysteine sulfur ligands [92]. Its substitution by cobalt confirms that there is a CoS_4 cromophore[93]. The catalytic zinc is also four-coordinated but, in this case, the ligands are a histidine, two cysteines and a water molecule. The electronic and ^1H NMR spectra are indicative of tetra-coordination for cobalt in this site [94]. A set of very broad isotropically shifted signals are observed in the downfield region, all of them with T_1 values lower than 1 ms. The spectrum changes with the pH, in agreement with the existence of two acidic groups in the active cleft. From these variations a pK_a of 9 was obtained and assigned to the NHδ of the coordinated histidine [4]. This is coincident with the pK_a for the dissociation rate constant of NAD^+ in the native enzyme. The interaction of LADH with NAD^+ and NADH have been also studied by ^1H NMR on the cobalt derivative. The observed changes in the spectra would indicate the same coordination for the cobalt when these cofactors bind to the enzyme.

2.4.3. *Zn finger peptide CP-1*
Zinc fingers are small peptides (25-30 amino acids) that bind one Zn(II) ion by molecule. In the zinc finger peptide CP-1, the metal ion is bound to two histidine and two cysteine residues in a distorted tetrahedral geometry. The substitution of zinc by cobalt and subsequent study by NMR has allowed the determination of the axial and equatorial contributions to the magnetic susceptibility tensor [95].

3. Cobalt in copper proteins

Blue or type 1 copper proteins (some times also called cupredoxins) are small single copper electron transfer carriers which play important roles in the redox chains of plants and bacteria [96,97]. In their oxidized form, they are characterized by the presence of a broad intense charge transfer band around 600 nm and small hyperfine coupling constants in the $g_{||}$ region of the EPR spectrum [96,97]. The structural features which account for such spectroscopic features are now very well known since it has been solved the X-Ray crystal structure of many of these proteins. It consists of a 96-159 polipeptide chain folded to form a β-sandwich structure, and in the case of azurins a short segment of α-helix [98]. The type-1 site is defined by three strongly coordinated equatorial ligands: the Sγ of a cysteine and the Nδ of two histidines. Completing the coordination sphere there is a weakly ligated methionine, although azurins has as a fifth ligand the carbonyl group of a glycin [96,97].

Despite the very good spectroscopic characteristics of copper(II) for UV-vis and EPR spectroscopies, it is in general not so adequate for NMR studies. Cu^{2+} has no zero field splitting, small magnetic anisotropy and, generally, far-off excited states. Therefore, electron relaxation times (≈ $2x10^{-9}$ s) are so long that the line broadening effect on protons surrounding the metal is too large to allow their detection [1]. This fact has limited the NMR study of these proteins to the reduced diamagnetic Cu(I) species.

It is generally accepted that the blue copper proteins govern the coordination of the metal due to the rigidity of the metal site as a consequence of the abundance of hydrogen bonds [99]. This idea means that different transition metals would accommodate in a very similar way in the metal site, as it has been confirmed through the structural study of the Zn derivative of *Pseudomonas aeruginosa* azurin. In this Zn-azurin the overall coordination sphere is kept and there is only a slight approach between the Gly45 axial ligand and the metal, being now the Met121 not coordinated [100].

Cobalt (II) has been used as a substitute of copper in some of the blue copper proteins, trying to get chemical and structural information about the metal site [101-109]. The early studies of McMillin et al. on cobalt (II) substituted azurin, plastocyanin and stellacyanin allowed the assignment of the intense visible bands found in these metalloproteins as S-(Cys)----M(II) (M=Cu or Co) charge transfer bands [104, 105]. This approach is still being used in the investigation of some blue copper proteins whose metal centers are not very well known [107]. More information can be obtained from these cobalt substituted proteins by NMR [108,109], as it is set out below.

3.1. AZURIN

Azurin is one of the best known blue copper proteins which has been used extensively as a model in the investigation of biological electron transfer reactions [96,97,110]. The copper ion is strongly bound to the Sγ of Cys-112 and to the Nδ of both His-46 and His-117, and weakly ligated by the Sδ of Met-121 and the carbonyl oxygen of Gly-45, resulting in a distorted trigonal-bipyramidal geometry, Figure 10, [98]. The crystal structure of azurin from different sources and at different conditions (different pHs and redox states) has been solved [98-100,111] and the structure-function relationships have been extensively studied by site directed mutagenesis [112-116]. Now it seems to be clear that the protein uses a hydrophobic patch, situated in the surface of the molecule to interact with other proteins in the electron transfer reaction [98,115,116]. The coordinated His117, placed in the center of the hydrophobic patch, would constitute the way for the transit of electrons from (or to) the copper center [98].

Hill et al. performed the first ^{1}H NMR experiments on Co(II) azurin [117]. Although they hardly could observe the signals form the coordinated residues these experiments gave important information about the characteristics of the metal site and the acid-base properties of the protein. More recently, we have started a deeper characterization of Co(II) *Pseudomonas aeruginosa* azurin [108]. The spectrum of this derivative (Figure 11) shows 15 isotropically shifted signals with T_1s ranging from < 1 to 9 ms (Table 4). Among the broadest signals, those at 284 and 231 ppm are particularly characteristic. A similar pair of signals has been found in Ni(II)-azurin and they have been assigned to the two $C\beta H_2$ protons of the Cys112 coordinated residue [118]. Signals a and b, not present in D_2O, correspond to the NεH protons of the two histidines bound to the metal ion (His46 and His117). Close to the diamagnetic region of the spectrum there are hyperfine shifted signals which would correspond to non-coordinated residues that are close to the cobalt ion. They would have only pseudocontact contribution to their isotropic shifts indicating the existence of magnetic anisotropy in the cobalt coordination center.

The application of 2D NMR to the study of this spectrum was one of the first examples of the use of such sophisticated techniques on a paramagnetic non iron

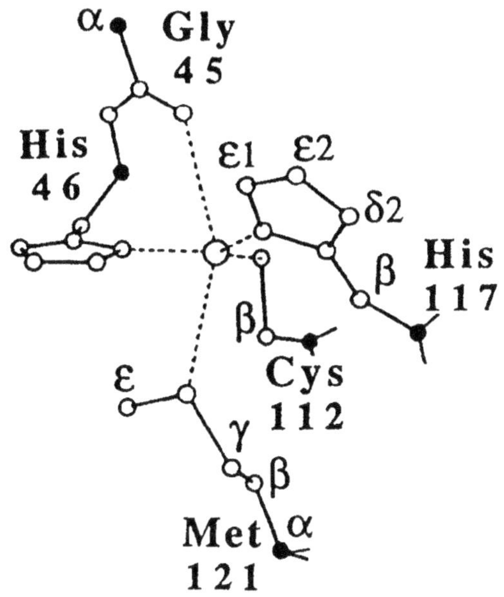

Figure 10. The metal center of copper(II) azurin.

metalloprotein [108]. It allowed the identification of the rest of the hyperfine signals. The NOESY spectrum of Co(II)-azurin in water solvent (Figure 12) permit the

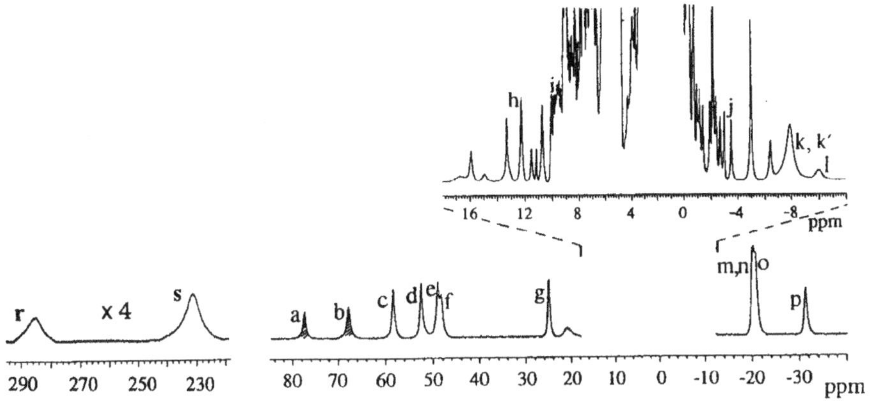

Figure 11. 1D ^1H NMR spectrum (400 MHz, 50mM CH_3COONH_4, 25 °C) of Co(II)-azurin in H_2O solvent at pH 4.4. Expanded is shown the close to the diamagnetic region of the same spectrum. Shaded signals disappear in D_2O.

assignment of signals c and d, connected through NOESY cross peaks with signals b and a respectively, as the CδH protons of the two coordinated histidines. Many other NOE connectivities are observed from which one can distinguish three sets of signals corresponding to the coordinated residues Met121 and Gly45 and to Met13.

TABLE 4. Assignment of the most significant signals of the ^{1}H NMR spectrum of *Pseudomonas aeruginosa* Co(II) azurin (37 °C, pH=5.0).

Signal	δ(ppm)	T_1 (ms)	$\Delta\nu_{1/2}$(Hz)	assignment
r	284.16	< 1.0	1600	Cys-112 $C\beta H_2$
s	231.17	< 1.0	1650	
p	-29.35	2.7	205	Gly-45 Cα2
e	47.37	3.7	255	Gly-45 Cα1
l	-9.10	4.3	230	Gly-45 NH
d	50.64	6.8	195	His-46 CδH
a	75.02	2.9	215	His-46 NεH
c	56.65	8.7	180	His-117 CδH
b	65.68	3.9	250	His-117 NεH
j	-2.87	29.8	35	Met-121 CαH
m	-18.69			Met-121 CβH
n	-19.02	3.9		Met-121 CβH
o	-19.19			Met-121 Cγ1H
f	45.64	4.2	120	Met-121 Cγ2H
k'	-7.20	3.0		Met-121 $C\varepsilon H_3$

As occurs in the native protein, [98,119-121] as well as in the Ni(II) metalloderivative, [103] the ^{1}H NMR spectrum of cobalt(II)-azurin changes with the pH in slow exchange conditions (Figure 13). This effect has been interpreted as being due to the local conformational change originated by the ionization equilibrium of His35 [103, 108]. More significant, however, can be the disappearance of signal b as the pH increases, suggesting that the corresponding NεH proton enters fast exchange conditions with bulk water. This fact is used to specifically assign signal b to the NεH proton of His117, since it is much more accessible to the solvent than the other coordinated histidine, His46 [98]. The increase of the exchange rate of the His117 NεH proton can be relevant as far as the electron transfer mechanism is concerned, since it has been suggested that this proton would play an important role in the transit of electrons from or to the copper center through the imidazole ring of the coordinated His117 [98].

The isotropic shifts of Co(II)-*Alcaligenes denitrificans* azurin are practically identical to those of the corresponding *Pseudomonas aeruginosa* protein, as expected for a very similar coordination environment in the two azurins. On the other hand, the ^{1}H NMR study of the cobalt(II) derivative of the M121Q mutant of *Alcaligenes denitrificans* azurin confirms the assignments made on the wild type protein [122]. These experiments demonstrate the usefulness of cobalt(II) as a powerful probe in the structural investigation of the blue copper proteins by NMR. The use of this approach can be extended to the characterization of the metallic site of other copper proteins which structure is currently unknown.

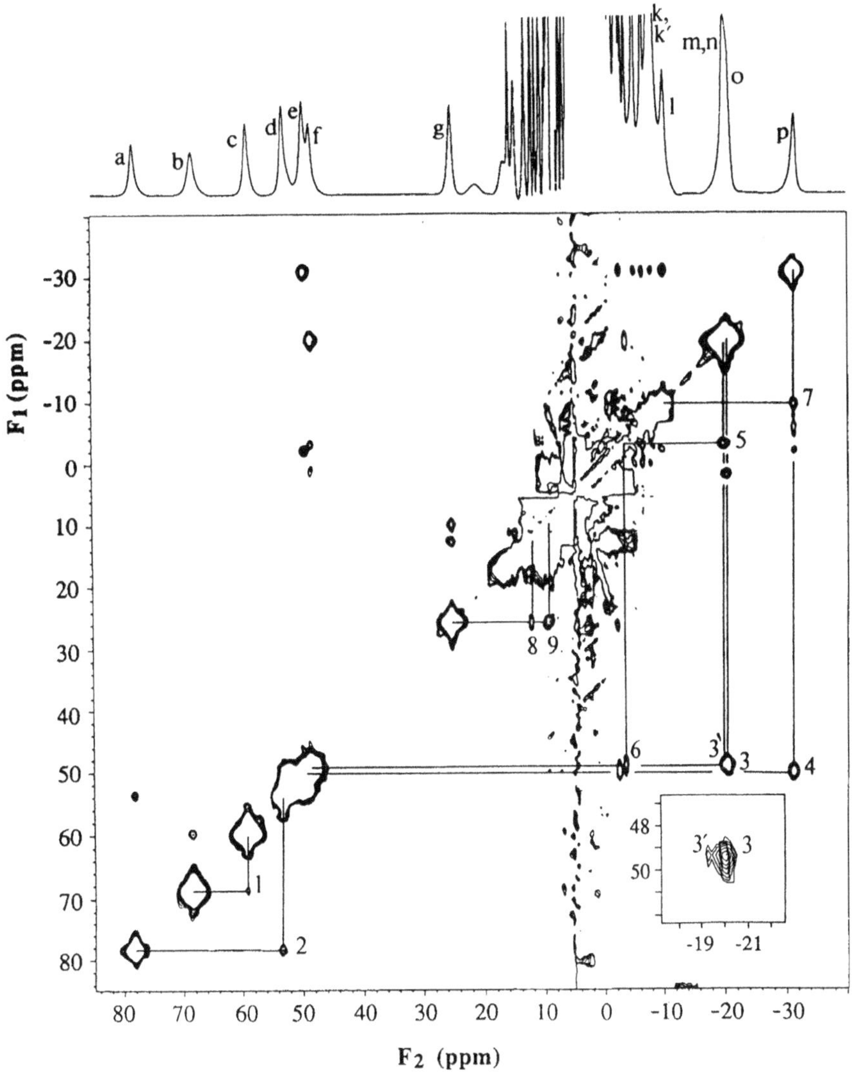

Figure 12. Phase-sensitive (TPPI) ^{1}H NOESY spectrum obtained at 400 MHz on a 4 mM sample of Co(II)-azurin in water (pH 4.5, 25 °C). This map was collected with a 6-ms mixing time, 256 *t*1 values (4096 scans each) over a 70-kHz bandwidth using 1 K data points in the *F*2 dimension. A 80°-shifted sine-bell-squared weighting function was applied in both dimensions. The inset shows cross peaks 3 and 3' of the same spectra processed to find more resolution.

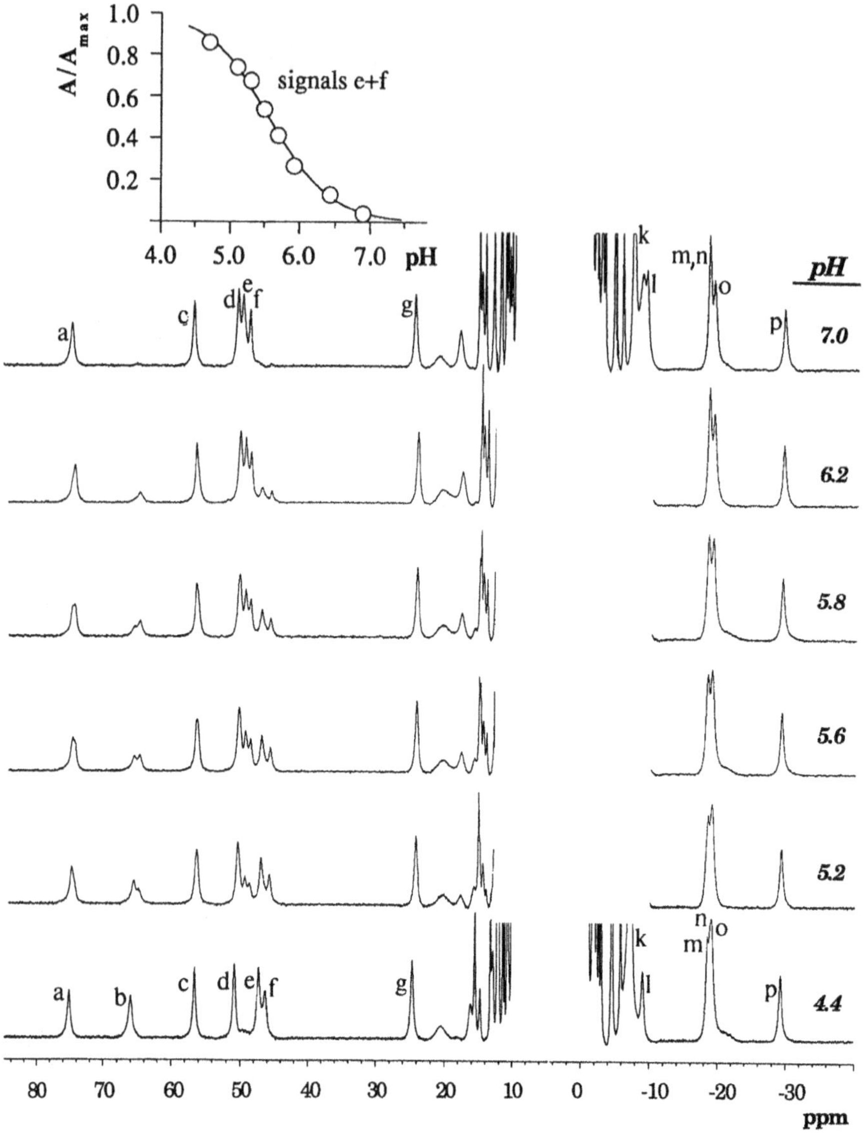

Figure 13. The 400 MHz ^{1}H NMR spectra of Co(II) azurin at different pHs (50 mM CH_3COONH_4, H_2O solvent, 37°C). The relative area of the best resolved signals (e, f group) as a function of pH is also represented in the upper left corner.

3.2. STELLACYANIN

Stellacyanin has a molecular weight of around 20,000 although 40% of it is carbohydrate. It has the lowest reduction potential (184 mV) in the family of the blue copper proteins, and although it was one of the earliest known members of this family its structure is still unknown. The three residues providing the essential ligands to conform the type 1 site are His46, His92 and Cys87 [96]. More controversial is the nature of the axial ligand since there is not methionine in stellacyanin [123].

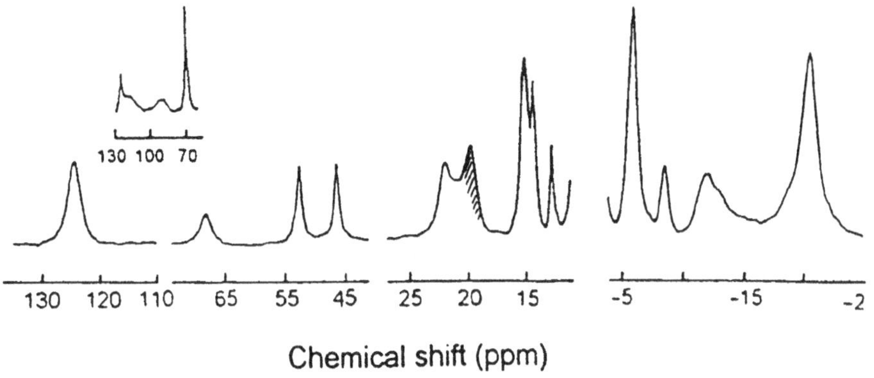

Figure 14. ^{1}H NMR spectrum (270 MHz, unbuffered H_2O, 30°C) of cobalt(II)-stellacyanin from *Rhus vernicifera* at pH 7.3. Shaded signal disappear in D_2O. (Reprinted from ref. 109).

An NMR study on the cobalt derivative of this protein has been performed by S. Dahlin et al. [109]. The ^{1}H NMR spectra of this protein (Figure 14) shows significant differences with regard to the spectra of the cobalt azurin. An attempt to assign this spectrum was done trying to get structural information about the metal site. Assignments were done on the basis of the proton-metal distances calculated from the relaxation rates by the Solomon equation [124]. This information is used by the authors to build a three-dimensional model of the metal site where they suggest the Sγ sulfur of Cys59, which constitutes a disulfide bridge together with Cys93, as being the fourth ligand. These results are not in agreement with many other studies like sequence homology [125], structure prediction [126] and Met121 by Gln substitution in azurin [113], which suggest that the axial ligand in Stellacyanin is either the amide or the carbonyl group of Gln97. A thorough assignment of the cobalt(II)-stellacyanin spectrum using 1D NOE and NOESY techniques could help to clarify this point.

4. Other cobalt substituted proteins

4.1. IRON AND MANGANESE METALLOPROTEINS

Iron by cobalt substitutions have been employed in several systems such as Rubredoxin[127], uteroferrin[128], the isopenicillin N synthase [129] and ovotransferrin [130]. This substitution makes the ^1H NMR signals usually sharper than those corresponding to the native iron proteins, due to the lower electronic relaxation rates of the cobalt(II). Rubredoxin is an iron-sulfur protein that has only one iron ion tetracoordinated to four cysteines residues. The ^1H NMR spectrum of the cobalt substituted rubredoxin shows signals from -50 up to 270 ppm. Some of them are assigned to the β-CH_2 coordinated cysteines[127]. Uteroferrin is a 35 kDa purple acid phosphatase with a $Fe^{III}Fe^{II}$ dinuclear active site. In the $Fe^{III}Co^{II}$ derivative successful 1D NOE and NOESY experiments have been performed despite the short relaxation times of the ^1H NMR signals. From these experiments, a carboxylate group has been proposed as a ligand of the Co(II) [128]. The same conclusion has been obtained from 1D NOE experiments on the cobalt derivative of the isopenicillin N synthase [129]. The cobalt substituted ovotransferrin ^1H NMR spectrum [130] and that of its oxalate complex [131] have been obtained and some signals have been tentatively assigned on the basis of the T_1 values and the observed solvent exchangeable protons. Neither 1D nor 2D NOE experiments have been performed on this derivative.

Finally, cobalt has also replaced manganese ions in lectins. Only one ^1H NMR study on this derivative have been carried out without any specific assignment [132].

4.2. METALLOTHIONEINS

Metallothioneins are low molecular weight (6-7 kDa) proteins rich in metals and cysteine residues. They contain two clusters of the type M_4S_{11} and M_3S_9 (M=zinc or cadmium). ^1H NMR spectrum of the Co_7 rabbit liver metallothionein displays a large set of well resolved signals from -50 to 300 ppm, arising essentially from the four-metal clusters [133]. Magnetic susceptibility measurements have been achieved in this derivative showing a strong coupling between the cobalt ions, specially in the four-metal cluster. Recently, NOESY and 1D NOE experiments on the Co_7 derivative have allowed the pairwise assignment of the isotropically shifted signals of the $C\beta H_2$ groups of the metal-coordinated cysteines [134]. Moreover, from the temperature dependence of these signals, a tentative specific assignment of these groups have been proposed.

5. Acknowledgments

We thank the CICYT (Ministerio de Educación y Ciencia, Spain) financial support of this work (Proyecto No. PB 91-0639).

6. References

1. Bertini, I., Turano, P. and Vila, A. (1993) Nuclear Magnetic Resonance of Paramangetic Metalloproteins, *Chem. Rev.* **93**, 2833-2932 and references therein.
2. La Mar, G.N., Horrocks, W.D. and Holm, R.H. (1973) *NMR of Paramagnetic Molecules*, Academic Press, New York.
3. Bertini, I. and Luchinat, C. (1986) *NMR of Paramagnetic Molecules in Biological Systems*, The Benjamin/Cummings Publishing Company, Menlo Park.
4. Bertini, I. and Luchinat, C. (1984) High Spin cobalt(II) as a probe for the investigation of metalloproteins, in G.L. Eichorn and L.G. Marzilli (eds), *Advances in Inorganic Biochemistry*, Elsevier, New York, pp. 72-111.
5. Vallee, B.L. (1986) A Synopsis of Zinc Biology and Pathology, in I. Bertini, C. Luchinat, W. Maret and M. Zeppezauer (eds.), *Zinc Enzymes*, Birkhäuser, Boston, pp. 1-15.
6. Bertini, I and Luchinat, C.L. (1986) Metal Substitution as a Tool for the Investigation of Zinc Proteins, in I. Bertini, C. Luchinat, W. Maret, M. Zeppezauer (eds.), *Zinc Enzymes*, Birkhäuser, Boston, pp. 27-48.
7. Bertini, I. (1982) The Coordination Properties of the Active Site of Zinc Enzymes, in I. Bertini, R.S. Drago, and C. Luchinat (eds.), *The Coordination Chemistry of Metalloenzymes*, Reidel Publishing Company, Dordrecth, pp. 1-18.
8. Auld, D.S. and Vallee, B.L. (1987) Carboxypeptidase A. in Hewberger, K. Brockjehurst (eds.), *Hydrolitic enzymes*, Elsevier Sci. Publ. Co., New York, pp. 201-253.
9. Rees, D.C. and Lipscomb, W.N. (1983) Crystallographic studies on apocarboxypeptidase A and the complex with glycil-L-tyrosine, *Proc. Nat. Acad. Sci. U.S.A.* **80**, 7151-7154.
10. Hardman, K.D. and Lipscomb, W.N. (1984) Structures of nickel(II) and cobalt(II) Carboxypeptidase A, *J. Am. Chem. Soc.* **106**, 463-464.
11. Latt, S.A. and Valle, B.L. (1971) Spectral properties of cobalt carboxypeptidase. Effects of substrates and inhibitors, *Biochemstry* **10**, 4263-4270.
12. Bertini, I., Luchinat, C., Messori,L., Monnanni, R., Auld, D.S. and Riordan, J.F. (1988) ^{1}H NMR Spectroscopic characterization of binary and ternary complexes of cobalt(II) carboxypeptidase A with inhibitors, *Biochemistry* **27**, 8318-8325.
13. Auld, D.S. and Vallee, B.L. (1970) Kinetics of carboxypeptidase A. The pH dependence of tripeptide hydrolysis catalyzed by zinc, cobalt, and manganese enzymes, *Biochemistry* **9**, 4352-4359.
14. Bertini, I., Luchinat, C., Monnanni, R., Moratal, J.M., Donaire, A. and Auld, D.S. (1990) Azide and Chloride Binding to Carboxypeptidase A in the Presence of L-Phenylalanine, *J. Inorg. Biochem.* **39**, 9-16.
15. Banci, L., Bertini, I. and La Penna, G. (1993) A Molecular Dynamics Study of Carboxypeptidase A: Effect of Protonation of Glu 270, *Inorg. Chem.* **32**, 2207-2211.
16. Geoghegan, K.F., Galdes, K.F., Martinelli, R.A., Holmquist, B., Auld, D.S. and Vallee, B.L. (1983) Cryospectroscopy of intermediates in the mechanism of carboxypeptidase A, *Biochemistry* **22**, 2255-2262
17. Auld, D.S., Bertini, I. Donaire, A., Messori, L. and Moratal, J.M. (1992) pH-Dependent Properties of Cobalt(II) Carboxypeptidase A-Inhibitor Complexes, *Biochemistry* **31**, 3840-3846.
18. Bertini, I., Monnanni, R., Pellacani, G.C., Sola, M., Vallee, B.L. and Auld, D.S. (1988) ^{13}C NMR Studies of carboxylate inhibitor binding to cobalt(II) Carboxypeptidase A, *J. Inorg. Biochem.* **32**, 13-20.
19. Luchinat, C., Monnanni, R., Roelens, S., Vallee, B.L. and Auld, D.S. (1988) ^{13}C NMR Studies of D- and L-phenylalanine binding to cobalt(II) Carboxypeptidase A, *J. Inorg. Biochem.* **32**, 1-6.
20. Christianson, D.W., Mangani, S., Shoham, G. and Lipscomb, W.N. (1989) Binding of D-phenylalanine and D-tyrosine to carboxypeptidase A, *J. Biol. Chem.* **264**, 12849-12853.
21. Christianson, D.W. and Lipscomb, W.N. (1987) Carboxypeptidase A: Novel Enzyme-Substrate-Product Complex, *J. Am. Chem. Soc.* **109**, 5536-5538.
22. Kim, H. and Lipscomb, W.N. (1991) Comparison of the structures of three carboxypeptidase A-phosphonate complexes determined by X-ray crystallography, *Biochemestry* **30**, 8171-8180.

23. Bertini, I., Donaire, A., Messori, L. and Moratal, J.M. (1990) Interaction of Phosphate and Pyrophosphate with Cobalt(II) Carboxypeptidase, *Inorg. Chem.* **29**, 202-205.

24. Moratal, J.M., Donaire, A., Castells, J., Jiménez, H.R., Salgado, J. and Hillerns, F. (1992) Spectroscopic Studies of the Interaction of Nickel(II) Carboxypeptidase with Phosphate and Pyrophosphate, *J. Chem. Soc. Dalton Trans.* 713-717.

25. Donaire, A. (1990) Propiedades Acido-Base de la Carboxipeptidasa A: Asignación del Grupo Responsable del pKa_2 del Enzima, in Valencia University. Doctoral Thesis, *Estudio Químico y Estructural de la Carboxipeptidasa A.*, pp 221-262.

26. Zhang, K. and Auld, D.S. (1993) XAFS Studies of Carboxypeptidase A: Detection of a Structural Alteration in the Zinc Coordination Sphere Coupled to the Catalytically Important Alkaline pKa, *Biochemistry* **32**, 13844-13851.

27. Lipscomb, W.N. (1983) Structure and catalysis of enzymes, *Annu. Rev. Biochem.* **52**, 17-34.

28. Lindskog, S., Engberg, P., Forsman, C., Ibrahim, S.A., Jonsson, B.-H., Simonsson, I. and Tibell, L. (1984) Kinetics and mechanism of carbonic anhydrase isoenzymes, *Ann. N. Y. Acad. Sci.* **429**, 61-75.

29. Silverman, D.N. and Lindskog, S. (1988) The catalytic mechanism of carbonic anhydrase: Implications of a rate-limiting protolysis of water, *Acc. Chem. Res.* **21**, 30-36.

30. Aresta, M. and Schloss, J.V. (1990) *Enzymatic and Model Carboxylation and Reduction Reactions for Carbon Dioxide Utilization*, Kluwer Academic Publishers, Dordrecht.

31. Christianson, D.W. (1991) Structural biology of zinc *Adv. Protein Chem.* **42**, 281-355.

32. Eriksson, E.A. and Liljas, A. (1991) X-ray crystallographic studies of carbonic anhydrase isoenzymes I, II and III, in S.J. Dodgson, E.E.Tashian, G. Gros and N.D. Carter (eds.), *The carbonic anhydrases*, Plenum Press, New York, pp. 33-48.

33. Hakansson, K., Carlsson, M., Svensson, L.A. and Liljas, A. (1992) Structure of Native and Apo Carbonic Anhydrase II and Structure of Some of its Anion-Ligand Complexes, *J. Mol. Biol.* **227**, 1192-1204.

34. Hakansson, K., Wehnert, A. and Liljas, A. (1994) X-ray Analysis of Metal-Substituted Human Carbonic Anhydrase II Derivatives, *Acta Cryst.* **D50**, 93-100.

35. Hakansson, K. and Wehnert, A. (1992), Structure of Cobalt Carbonic Anhydrase Complexed with Bicarbonate, *J. Mol. Biol.* **228**, 1212-1218.

36. Hakansson, K., Wehnert, A. and Liljas, A. (1994) Wild-Type and E106Q Mutant Carbonic Anhydrase Complexed with Acetate, *Acta Cryst.* **D50**, 101-104.

37. Xue, Y., Vidgre, J., Svensson, L.A., Liljas, A., Jonsson, B.H. and Lindskog, S. (1993) Crystallographic analysis of Thr-200 → His human carbonic anhydrase II and its complex with the substrate HCO_3^-, *Proteins* **15**, 80-87.

38. Xue, Y., Liljas, A., Jonsson, B.H. and Lindskog, S. (1993) Structural Analysis of the Zinc Hydroxide-Thr-199-Glu-106 Hydrogen-Bond Network in Human Carbonic Anhydrase II, *Proteins*, **17**, 93-106.

39. Merz, K.M.Jr., Murcko, M.A. and Kollman P.A. (1991) Inhibition of carbonic anhydrase, *J. Am. Chem. Soc.* **113**, 4484-4490.

40. Lindahl, M., Liljas, A., Habash, J., Harrop, S. and Helliwell, J.R. (1992) The Sensitivity of the Synchrotron Laue Method to Small Structural Changes: Binding Studies of Human Carbonic Anhydrase II (HCAII), *Acta Cryst.* **B48**, 281-285.

41. Mangani, S. and Hakansson, K. (1992) Crystallographic studies of the binding of protonated and unprotonated inhibitors to carbonic anhydrase, *Eur. J. Biochem.* **210**, 867-871.

42. Jönsson, B., Hakansson, K. and Liljas, A. (1993) The structure of human carbonic anhydrase II in complex with bromide and azide, *FEBS Lett.* **322**, 186-190.

43. Lindahl, M., Svensson, L.A. and Liljas, A. (1993) Metal poison inhibition of carbonic anhydrase, *Proteins* **15**, 177-182.

44. Mangani, S. and Liljas, A. (1993) Crystal structure of the complex between human carbonic anhydrase II and the aromatic inhibitor 1,2,4-triazole, *J. Mol. Biol.* **232**, 9-14.

45. Nair, S. K. and Christianson, D.S. (1993) Crystallographic studies of azide binding to human carbonic anhydrase II, *Eur. J. Biochem.* **213**, 507-515.

46. Vidgren, J., Svensson, L.A. and Liljas, A. (1993) Refined structure of the aminobenzolamide complex of human carbonic anhudrase II at 1.9 Å and sulphonamide modelling of bovine carbonic anhydrase III, *Int. J. Biol. Macromol.* **15**, 97-100.

47. Bertini, I. and Luchinat, C. (1983) Cobalt(II) as a probe of the structure and function of carbonic anhydrase, *Acc. Chem. Res.* **16**, 272-279.

48. Liljas, A., Hakansson, K., Jonsson, B.H. and Xue, Y. (1994) Inhibition and catalysis of carbonic anhydrase. Recent crystallographic analyses, *Eur. J. Biochem.* **219**, 1-10.

49. Moratal, J.M., Martínez, M.J., Donaire, A., Castells, J., Salgado, J. and Jiménez, H.R. (1991) Spectroscopic Studies of Nickel(II) Carbonic Anhydrase and its Adducts with Inorganic Anions, *J. Chem. Soc. Dalton Trans.* 3393-3399.

50. Banci, L., Bertini, I., Luchinat, C., Donaire, A., Martínez, M.J. and Moratal, J.M. (1990) The Factors Governing the Coordination Number in the Anion Derivatives of Carbonic Anhydrase, *Comments Inorg. Chem.* **9**, 245-261.

51. Banci, L., Bertini, I., Luchinat, C. and Moratal, J.M. (1990) The mechanism of Action of Carbonic Anhydrase, in *Enzymatic and Model Carboxylation and Reduction Reactions for Carbon Dioxide Utilization*, Kluwer Academic Publishers, Dordrecht, pp. 181-197.

52. Moratal, J.M., Donaire, A., Salgado, J. and Martínez, M.J. (1990) Interaction of sulphate and Chloride with Cobalt(II)-Carbonic Anhydrase, *J. Inorg. Biochem.* **40**, 245-252.

53 Moratal,.J.M., Martínez, M.J., Donaire, A. and Aznar, L. (1992) Thermodynamic Parameters of the Interaction between Co(II)Bovine Carbonic Anhydrase and Anionic Inhibitors, *J. Inorg.Biochem.* **44**, 65-71.

54. Eriksson, A.E., Kylsten, P.M., Jones, T.A. and Liljas, A. (1989) Crystallographic studies of inhibitor binding sites in human carbonic anhydrase II: a pentacoordinated binding of the SCN^- ion to the zinc at high pH, *Proteins*, **4**, 283-293.

55. Bertini, I., Luchinat, C., Pierattelli, R. and Vila, A.J. (1992) A multinuclear Ligand NMR Investigation of Cyanide, Cyanate and Thiocyanate binding to Zinc and Cobalt Carbonic Anhydrase, *Inorg. Chem.* **31**, 3975-3979.

56. Banci, L., Dugad, L.B., La Mar, G.N., Keating, K.A., Luchinat, C. and Pierattelli, R. (1992) ^1H Nuclear magnetic resonance investigation of cobalt(II) substituted carbonic anhydrase, *Biophys. J.* **63**, 530-543.

57. Bertini, I., Jonsson, B.H., Luchinat, C., Pierattelli, R. and Vila, A.J. (1994) Strategies of signal assignments in paramagnetic metalloproteins. An NMR investigation of the thiocyanato adduct of the cobalt(II) substituted human carbonic anhydrase II, *J. Magn. Res. B.*, in press.

58. Bencini, A., Bertini, I. and Bini, T. (1993) Angular Overlap Interpretation of the Spectromagnetic Properties of the Nitrate Derivative of Cobalt(II)-Substituted Carbonic Anhydrase, *Inorg.Chem.* **32**, 3312-3315.

59. Valentine, J.S. and Pantoliano, M.W. (1981) Protein-metal ion interactions in cuprozinc protein (superoxide dismutase), in T.G. Spiro (ed.), Wiley, New York, pp. 291-358.

60. Fridovich, I. (1974) Superoxide dismutases, *Adv. Enzymol. Relat. Areas Mol. Biol.* **41**, 35-97.

61. Fridovich, I. (1986) Superoxide dismutases, *Adv. Enzymol. Relat. Areas Mol. Biol.* **58**, 61-97.

62. Keller, G.A., Warner, T.G., Steimer, K.S. and Hallewell, R.A. (1991) Copper-zinc superoxide dismutase is a peroxisomal enzyme in human fibroblasts and hepatoma cells, *Proc. Natl. Acad. Sci., U.S.A.* **88**, 7381-7385.

63. Parge, H. E., Hallewell, R.A. and Tainer, J.A. (1992) Atomic structures of wild-type and thermostable mutant recombinant human copper-zinc superoxide dismutase, *Proc. Natl. Acad. Sci. U.S.A.* **89**, 6109-6113.

64. Tainer, J.A., Getzoff, E.D., Richardson, J.S. and Richardson, D.C. (1983) Structure and mechanism of copper, zinc superoxide dismutase, *Nature* **306**, 284-287.

65. Bertini, I., Lanini, G., Luchinat, C., Messori, L., Monnanni, R. and Scozzafava, A. (1985) Investigation of Cu_2Co_2SOD and its anion derivatives. ^1H NMR and electronic spectra, *J. Am. Chem. Soc.* **107**, 4391-4396.

66. Banci, L., Bertini, I., Luchinat, C., Piccioli, M., Scozzafava, A. and Turano, P. (1989) ^1H NOE studies on dicopper(II) dicobal(II) superoxide dismutase, *Inorg. Chem.* **28**, 4650-4656.

67. Paci, M., Desider, A., Sette, M., Falconi, M. and Rotilio, G. (1990) Mapping the copper lignads of Cu,Zn superoxide dismutase by nuclear Ovehauser enhancement of the isotropically shifted ^1H NMR lines of the Cu, Co derivative, *FEBS Lett.* **261**, 231-236.

68. Banci, L., Bertini, I., Luchinat, C. and Piccioli, M. (1990) Transient versus steady state NOE in paramagnetic molecules. Cu_2Co_2SOD as an example, *FEBS Lett.* **272**, 175-180.

69. Morgenstern-Badarau, I., Cocco, D., Desideri, A., Rotilio, G., Jordanov, J. and Dupre, N., (1986) Magnetic susceptibility studies of the native cupro-zinc superoxide dismutase and its cobalt-substituted derivatives. Antiferromagnetic coupling in the imidazolate-bridged copper(II)-cobalt(II) pair, *J. Am. Chem. Soc.* **108**, 300-302.

70. Banci, L., Bertini, I., Luchinat, C. and Scozzafava, A. (1987) Nuclear relaxation in the magnetic coupled system Cu_2Co_2SOD. Histidine-44 is detached upon anion binding, *J. Am. Chem. Soc.* **109**, 2328-2334.

71. Sette, M., Paci, M., Desideri, A. and Rotilio, G. (1993) Two-dimensional NMR assignment of hyperfine-shifted resonances of very fast relaxing metal binding sites of proteins by NOE spectroscopy. The case of Cu,Co superoxide dismutase, *Eur. J. Biochem.* **213**, 391-397.

72. Banci, L., Bertini, I., Luchinat, C., Piccioli, M and Scozzafava, A. (1993) 1D versus 2D ^{1}H NMR experiments in Dicopper, Dicobalt Superoxide Dismutase: A Further Mapping of the Active Site, *Gazz. Chim. Ital.* **123**, 95-100.

73. Bertini, I., Piccioli, M, Scozzafava, A. and Viezzoli, M.S. (1993) Copper-Cobalt Superoxide Dismutase: a Re-examination of the ^{1}H NMR Spectrum Through a Novel Selectivity Deuteriated Derivative, *Magn. Res. Chem.* **31**, S17-S22.

74. Banci, L., Bertini, I., Luchinat, C. and Viezzoli, M.S. (1990) A comment on the ^{1}H NMR spectra of cobalt(II)-substituted superoxide dismutases with histidines deuteriated in the ε1-position, *Inorg. Chem.* **29**, 1438-1440.

75. Cocco, D., Rossi, L., Barra, D., Bossa, F., Rotilio, G. (1982) Carbamoylation of copper-zinc-superoxide dismutase by cyanate. Role of lysines in the enzyme action *FEBS Lett.* **150**, 303-306.

76. Cudd, A. and Fridovich, I. (1982) Electrostatic interactions in the reaction mechanism of bovine erythrocyte superoxide dismutase, *J. Biol. Chem.* **257**, 11443-11447.

77. Beyer, W.F.Jr., Fridovich, I., Mullenbach, G.T. and Hallewell, R.A., (1987) Examination of the role of arginine-143 in the human copper and zinc superoxide dismutase by site-specific mutagenesis, *J. Biol. Chem.* **262**, 11182-11187.

78. Banci, L., Bertini, I., Luchinat, C. and Hallewell, R.A. (1988) Investigation of superoxide dismutase Lys-143, Ile-143 and Glu-143 mutants: Cu_2Co_2SOD derivatives, *J. Am. Chem. Soc.* **110**, 3629-3633.

79. Azab, H.A., Banci, L., Borsari, M., Luchinat, C., Sola, M. and Viezzoli, M.S. (1992) Redox Chemistry of Superoxide Dismutase. Cyclic Voltammetry of Wild-Type Enzyme and Mutants on Functionally Relevant Residues, *Inorg. Chem.* **31**, 4649-4655.

80. Banci, L., Bertini, I., Luchinat, C., Viezzoli, M.S. (1993) pH-Dependent Properties of SOD Studied through Mutants on Lys-136, *Inorg. Chem.* **32**, 1403-1406.

81. Bertini, I., Hiromi, K., Hirose, J., Sola, M. and Viezzoli, M.S. (1993) Electron Transfer between Copper and Zinc Superoxide Dismutase and Hexacyanoferrate(II), *Inorg. Chem.* **32**, 1106-1110.

82. Banci, L., Bertini, I., Bauer, D., Hallewell, R.A. and Viezzoli, M.S. (1993) Investigation of a New Cu,Zn Superoxide Dismutase Mutant: The Thr --Arg 137 Derivative, *Biochemistry* **32**, 4348-4388.

83. Moss, T.H. and Fee, J.A. (1975) Magnetic properties of cobalt substituted bovine superoxide dismutase derivatives, *Biochem. Biophys. Res. Commun.* **66**, 799-808.

84. Bertini, I., Luchinat, C., Piccioli, M., Vicens Oliver, M. and Viezzoli, M.S. (1991) *Eur. J. Biophys.* **20**, 269.

85. Banci, L., Bertini, I., Luchinat, C., Monnanni, R. and Scozzafava, A. (1986) A spectroscopic investigation of Co_2Zn_2- and Co_2Co_2-superoxide dismutase, *Gazz. Chim. Ital.* **116**, 51-54.

86. Banci, L., Bertini, I., Luchinat, C., Monnanni, R. and Scozzafava, A. (1987) Characterization of the cobalt(II)-substituted superoxide dismutase-phosphate systems, *Inorg. Chem.* **26**, 153-156.

87. Ming, L.J. and Valentine, J.S. (1990) NMR studies of cobalt(II)-substituted derivatives of bovine copper-zinc superoxide dismutase. Effects of pH, phosphate, and metal migration, *J. Am. Chem. Soc.* **112**, 4256-4264.

88. Kim, E.E. and Wyckoff, H.W. (1991) Reaction mechanism of alkaline phosphatase based on crystal structures. Two-metal ion catalysis, *J. Mol. Biol.* **218**, 449-464.

89. Coleman, J.E. and Gettins, P. (1986) Multinuclear NMR Probes of Structure and Mechanism in Zinc Enzymes, in I. Bertini, C. Luchinat, W. Maret and M. Zeppezauer (eds.), *Zinc Enzymes*, Birkhäuser, Boston, pp. 77-100.

90. Banci, L., Bertini, I., Luchinat, C., Viezzoli, M.S. and Wang, Y. (1988) Characterization of Cu_2Co_2- and Co_2Co_2-alkaline phosphatase complexes at acidic pH, *Inorg. Chem.* **27**, 1442-1446.

91. Bertini, I., Luchinat, C., Scozzafava, A., Maldotti A. and Traverso, O. (1983) Investigation of the copper-magnesium-alkaline phosphatase system, *Inorg. Chim. Acta* **78**, 19-22.

92. Colonna-Cesari, F., Perahia, D., Karplus, M., Eklund, H., Branden, C.I. and Tapia, O. (1986) Interdomain motion in liver alcohol dehydrogenase. Structural and energetic analysis of the hinge bending mode, *J. Biol. Chem.* **261**, 15273-15278.

93. Maret, W., Andersson, I. Dietrich, H., Schneider-Bernloehr, H., Einarsson, R. and Zeppezauer, M. (1979) Site-specific substituted cobalt(II) horse liver alcohol dehydrogenases. Preparation and characterization in solution, crystalline and immobilized state, *Eur. J. Biochem.* **98**, 501-512.

94. Bertini, I., Garber, M., Lanini, G., Luchinat, C., Maret, W., Rawer, S. and Zeppezauer, M. (1984) ^{1}H NMR Investigation of the active site of cobalt(II)-substituted liver alcohol dehydrogenase, *J. Am. chem. Soc.* **106**, 1826-1830.

95. Harper, L.V., Amann, B.T., Vinson, V.K. and Berg, J.M. (1993) NMR Studies of a Cobalt-Substituted Zinc Finger Peptide, *J. Am. Chem. Soc.* **115**, 2577-2580.

96. Sykes, A.G. (1991) in A.G. Sykes (ed), *Advances in Inorganic Chemistry*; Academic Press Inc, New York, Vol. 26, pp. 377-408.

97. Chapman, S.K.(1991) in R.W. Hay, J.R. Dilworth and K.B. Nolan (eds), *Perspectives on Bioinorganic Chemistry*; Jai Press Ltd, London, Vol 1, pp. 95-140.

98. Nar, H., Messerschmidt, A., Huber, R., van de Kamp, M. and Canters, G.W. (1991) Crystal structure analysis of oxidized *Pseudomonas aeruginosa* azurin at pH 5.5 and pH 9.0, *J. Mol. Biol.* **221**, 765-772.

99. Shepard, W.E.B., Anderson, B.F., Lewandoski, D.A., Norris, G.E. and Baker, E.N. (1990) Copper coordination geometry in azurin undergoes minimal changes on reduction of copper(II) to copper(I), *J. Am. Chem. Soc.* **112**, 7817-7819.

100. Nar, H., Huber, R., Messerschmidt, A., Filippon, A.C., Barth, M., Jaquinod, M., van de Kamp, M. and Canters, G.W. (1992) Characterization and crystal structure of zinc azurin, a by-product of heterologous expression in *Escherichia coli* of *Pseudomonas aeruginosa* azurin, *Eur. J. Biochem.* **205**, 1123-1129.

101. Tennent, D.L. and McMillin, D.R. (1979) A detailed analysis of the charge-transfer bands of a blue copper protein. Studies of the nickel(II), manganese(II) and cobalt(II) derivatives of azurin, *J. Am. Chem. Soc.* **101**, 2307-2311.

102. Suzuki, S., Sakurai, T., Shidara, S. and Iwasaki, H. (1989) Spectroscopic Characterization of Cobalt(II)-Substituted Achromobacter Pseudoazurin: Similarity of the Metal Center in Co(II)-Pseudoazurin to those in Co(II)-Plastocyanin and Co(II)-Plantacyanin, *Inorg. Chem.* **28**, 802-804.

103. Moratal, J.M., Salgado, J., Donaire, A., Jiménez, H.R. and Castells, J. (1993) 1D- and 2D-NMR studies of the pH effects on the metal-site geometry in nickel(II)-azurin from *Pseudomonas aeruginosa*, *J. Chem. Soc., Chem. Commun.* 110-112.

104. McMillin, D.R., Holwerda, R.A. and Gray, H.B. (1974) Preparation and spectroscopic studies of cobalt(II)-Stellacyanin, *Proc. Nat. Acad. Sci. USA* **71**, 1339-1341.

105. McMillin, D.R., Rosenberg, R.C. and Gray, H.B. (1974) Preparation and spectroscopic studies of cobalt(II) derivatives of blue copper proteins, *Proc. Nat. Acad. Sci. USA* **71**, 4760-4762.

106. Di Bilio, A.J., Chang, T.K., Malmström, B.G., Gray, H.B., Karlsson, B.G., Nordling, M., Pascher, T. and Lundberg, L.G. (1992) Electronic absorption spectra of M(II)(Met121X) azurins (M=Co, Ni, Cu; X=Leu, Gly, Asp, Glu): charge-transfer energies and reduction potentials, *Inorg. Chim. Acta* **198-200**, 145-148.

107. Strong, C., Harrison, S.L. and Zeger, W. (1994) Preparation and characterization of cobalt (II)-substituted rusticyanin, *Inorg. Chem.* **33**, 606-608.

108. Moratal, J.M., Salgado, J., Donaire, A., Jiménez, H.R. and Castells, J. (1993) COSY and NOESY characterization of cobalt(II)-substituted azurin from *Pseudomonas aeruginosa*, *Inorg. Chem.* **32**, 3587-3588.

109. Dahlin, S., Reinhammar, B. and Ångström, J. (1989) Proton-Metal distance determination in cobalt(II) stellacyanin by ^{1}H Nuclear Magnetic Resonance relaxation measurements including Curie-Spin effects:A proposed structure of the metal-binding region, *Biochemistry* **28**, 7224-7233.

110. Canters, G.W. and van de Kamp, M. (1992) Protein-mediated electron transfer, *Current Opinion in Structural Biology* **2**, 859-869.

111. Baker, E.N. (1988) Structure of azurin from *Alcaligenes denitrificans*. Refinement at 1.8 Å resolution and comparison of the two crystallographically independent molecules, *J. Mol. Biol.* **203**, 1071-1095.

112. Nar, H., Messerschmidt, A., Huber, R., van de Kamp, M. and Canters, G.W. (1991) X-ray crystal structure of the two site-specific mutants His35Gln and His35Leu of azurin from *Pseudomonas aeruginosa*, *J. Mol. Biol.* **218**, 4427-447.

113. Romero, A., Hoitink, C.W.G., Nar, H., Huber, R., Messerschmidt, A. and Canters, G.W. (1993) X-ray analysis and spectroscopic characterization of M121Q azurin. A copper site model for stellacyanin, *J. Mol. Biol.* **229**, 1007-1021.

114. Canters, G.W. and Gilardi, G. (1993) Engineering type I copper sites in proteins, *FEBS Lett.* **1**, 39-48.

115. van de Kamp, M., Floris, R., Hali, F.C. and Canters, G.W. (1990) Site-directed mutagenesis reveals trhat the hydrophobic patch of azurin mediates electron transfer, *J. Am. Chem. Soc.* **112**, 907-908.

116. van de Kamp, M., Silvestrini, M.C., Brunori, M., van Beeumen, J., Hali, F.C. and Canters, G.W. (1990) Involment of the hydrophobic patch of azurin in the electron-transfer reactions with cytochrome c_{551} and nitrite reductase, *Eur. J. Biochem.* **194**, 109-118.

117. Hill, H.A.O., Smith, B.E. and Storm, C.B. (1976) The proton magnetic resonance spectra of a cobalt(II) azurin, *Biochem. Biophys. Res. Commun.* **70**, 783-790.

118. Moratal, J.M., Salgado, J., Donaire, A., Jiménez, H.R., Castells, J. and Martinez-Ferrer, M.J. (1993) ^{1}H 2D-NMR characterization of Ni(II)-substituted azurin from *Pseudomonas aeruginosa*, *Magn. Res. Chem.* **31**, S41-S46.

119. Adman, E.T., Canters, G.W., Hill, H.A.O. and Kitchen, N.A. (1982) The effect of pH and temperature on the structure of the active site of azurin from *Pseudomonas aeruginosa*, *FEBS Lett.* **143**, 287-292.

120. Corin, A.F., Bersohn, R. and Cole, P.E. (1983) pH dependence of the reduction-oxidation reaction of azurin with cytochrome c_{551}: Role of histidine-35 of azurin in electron transfer, *Biochemistry* **22**, 2032-2038.

121. Groeneveld, C.M. and Canters, G.W. (1988) NMR study of structure and electron transfer mechanism of *Pseudomonas aeruginosa* azurin, *J. Biol. Chem.* **263**, 167-173.

122. Moratal, J.M. et al., unpublished results.

123. Bergman, C., Gandvik, E.K., Nyman, P.O. and Strid, L. (1977) The amino acid sequence of stellacyanin from the lacquer tree, *Biophys. Res. Commun.* **77**, 1052-1059.

124. Solomon, I. (1955) Relaxation processes in a system of two spins, *Phys. Rev.* **99**, 559-565.

125. Murata, M., Begg, G.S., Lambron, F., Leslie, B., Simpson, R.J., Freeman, H.C. and Morgan, F.J. (1982) Amino acid sequence of a basic blue protein from cucumber seedings, *Proc. Nat. Acad. Sci. USA* **79**, 6434-6437.

126. Fields, B.A., Guss, J.M. and Freeman, H.C. (1991) Three-dimensional model of stellacyanin, a blue copper-protein, *J. Mol. Biol.* **222**, 1053-1065.

127. Moura, I., Teixeira, M., LeGall, J. and Moura, J.J.G. (1991) Spectroscopic studies of cobalt and nickel substituted rubredoxin and desulforedoxin, *J. Inorg. Biochem.* **44**, 127-139.

128. Holz, R.C., Que, L.Jr. and Ming, L.J. (1992) NOESY Studies on the Fe(III)Co(II) Active Site of the Purple Acid Phosphatase Uteroferrin, *J. Am. Chem. Soc.* **114**, 4434-4436.

129. Ming, L.J., Que, L.Jr., Kriauciunas, A., Frolik, C.A. and Chen, V.J. (1991) NMR studies of the active site of isopenicillin N synthase, a non-heme iron(II) enzyme, *Biochemistry* **30**, 11653-11659.

130. Bertini, I., Luchinat, C., Messori, L. and Scozzafava, A. (1984) Cobalt(II) as an NMR probe for the investigation of the coordination sites of conalbumin, *Eur. J. Biochem.* **141**, 375-378.

131. Bertini, I., Luchinat, C., Messori, L., Monnanni, R. and Scozzafava, A. (1986) The metal-binding properties of ovotransferrin. An investigation of cobalt(II) derivatives, *J. Biol. Chem.* **261**, 1139-1146.

132. Bertini, I., Viezzoli, M.S., Luchinat, C., Stafford, E., Cardin, A.D., Behnke, W.D., Bhattacharyya, L. and Brewer, C.F. (1987) Circular dichroism and proton NMR studies of cobalt(2+)- and nickel(2+)-substituted concanavalin A and the lentil and pea lectins, *J. Biol. Chem.* **262**, 16985-16994.

133. Bertini, I., Luchinat, C., Messori, L. and Vasak, M. (1989) Proton NMR Studies of the Cobalt(II)-Metallothionein System, *J. Am. Chem. Soc.* **111**, 7296-7300.

134. Bertini, I., Luchinat, C., Messori, L. and Vasak, M. (1993) A two-dimensional NMR study of Co(II)7 rabbit liver metallothionein, *Eur. J. Biochem.* **211**, 235-240.

PARAMAGNETIC LANTHANIDE(III) IONS AS NMR PROBES FOR BIOMOLECULAR STRUCTURE AND FUNCTION

LI-JUNE MING
Department of Chemistry
and Institute for Biomolecular Science
University of South Florida
Tampa, Florida 33620-5250

Abstract: Paramagnetic lanthanide(III) ions (Ln^{3+}) have been used successfully as spectroscopic probes for the study of chemical, physical and physiological properties of many Ca^{2+}-dependent biological systems in the past several years. Some paramagnetic Ln^{3+} (e.g. Pr^{3+}, Eu^{3+}, and Yb^{3+}) possessing very short electronic relaxation times are able to exhibit relatively sharp isotropically shifted ^1H NMR features attributable to the protons in the close proximity of the metal. Thus, the use of such Ln^{3+} ions as substitutes for Ca^{2+} in proteins raises the possibility of detailed NMR study of the Ca^{2+} binding environment in proteins. These paramagnetic Ln^{3+} ions have also been used as NMR probes for the study of the structural and functional roles of several metal-dependent antibiotics, such as the anthracyclines. We report in this contribution the use of two dimensional NMR techniques for the studies of paramagnetic Ln^{3+}-substituted Ca^{2+} proteins in terms of the configuration of the metal-binding sites and the interactions with ligands, and for the study of metallo-antibiotics.

1. Introduction

Paramagnetic transition metal ions have been used successfully as spectroscopic probes for the study of metal binding sites in many metalloproteins [1,2]. When a paramagnetic metal ion with fast electronic relaxation rates is used as a probe, ^1H NMR spectroscopy becomes a valuable tool for such study via the detection and assignment of isotropically shifted signals outside of the normal 0-13 ppm region attributable to protons close to the metal ion [2]. Information about the coordinated ligands and the interactions with substrates analogues and the medium can thus be obtained. For example: A coordinated His can be identified via the detection of the isotropically shifted solvent exchangeable imidazole NH signal and its correlation with a ring CH signal, and a coordinated carboxylate-containing residue can also be identified by showing the isotropically shifted $(C_\gamma H_2)$-$C_\beta H_2$-$C_\alpha H$ spin patterns, which can be differentiated from the spectral features of Cys and Met showing different chemical shifts and relaxation times. The metal ions $Fe^{2+/3+}$, Co^{2+}, and Ni^{2+} have been used extensively as probes for structural and functional studies of many metalloenzymes by means of NMR techniques. Particularly, Co^{2+}-substituted derivatives of Zn enzymes can always show high activity, thus can serve as good models for better understanding of the structure and mechanism of the native enzymes [2]. The contributions in this volume by Lawrence Que (on nonheme and

G.N. La Mar (ed.), Nuclear Magnetic Resonance of Paramagnetic Macromolecules, 245-264.

non-FeS Fe proteins) and by J. M. Moratal and Antonio Donaire (on Co^{2+}- and Ni^{2+}-substituted proteins) provide a general view on the use of NMR techniques for the study of the coordinated ligands in several paramagnetic metalloproteins.

In addition to the widely used paramagnetic transition metal ion NMR probes, paramagnetic lanthanide(III) ions (Ln^{3+}) with short electronic relaxation times (~10^{-13} s) have long been used as shift reagents to simplify the assignment of complicated ^{1}H NMR spectra [3] and for the study of biomolecules.[4,5] However, the use of these Ln^{3+} (as well as other paramagnetic transition metal ions) as NMR probes for the study of the structure and mechanism of proteins has been hampered due to the lack of appropriate techniques for conclusive signal assignment. The use of 2D NMR techniques for the study of paramagnetic metal sites in metalloproteins has been impeded in the past several years owing to the fast nuclear relaxation rates of the isotropically shifted signals, although these techniques have been extensively used for structural studies of diamagnetic proteins [6]. In the past three years, such techniques have been applied successfully in our [7] and other laboratories [8] for the study of the metal sites in several paramagnetic metalloproteins.

Dipolar shift is the predominant isotropic shift mechanism in paramagnetic Ln^{3+} complexes, as opposed to the presence of large contribution of contact shift mechanism in paramagnetic transition metal complexes (although dipolar shift is also an important shift mechanism in Fe-heme systems as discussed in several other chapters in this volume, such as the ones by Gerd La Mar and James Satterlee), because the valence f orbitals are highly shielded and cannot participate in direct chemical bonding [3]. The dipolar shift of a nucleus depends on the relative position of the nucleus with respect to the metal as shown below [2,3]:

$$\frac{\Delta \nu^{dip}}{\nu} = -D\left(\frac{3\cos^2\theta - 1}{r^3}\right) - D'\left(\frac{\sin^2\theta\cos 2\Omega}{r^3}\right) \qquad (1)$$

where $D = (1/3N)[(\chi_z - \frac{1}{2}(\chi_x + \chi_y)]$ and $D' = (1/2N)(\chi_x - \chi_y)$, r is the nucleus-metal distance, θ is the geometric angle of $\vec{r}$ and the Z axis, and Ω is the angle of the X axis and the projection of $\vec{r}$ on the XY plane. The configuration of the metal site is obtainable when the principal components of the magnetic susceptibility tensor (χ's) are known, although which are not always available in practice [3]. Moreover, the isotropically shifted signals which are buried in the diamagnetic envelope cannot be easily resolved and assigned; and the signals with similar isotropic shifts and relaxation times may also render signal assignment difficult solely based on eq 1. Although proton-metal distances can be estimated by the Solomon equation and Curie relaxation [2], the structure of the metal site cannot be uniquely defined unless the spatial and bond correlations of the nuclei with respect to each other can be precisely determined. One advantage of utilizing paramagnetic Ln^{3+} ions over transition metal ions as probes is that their causing large dipolar shift of nearby nuclei allows the study of the whole metal environment, but not restricted to only the coordinated ligands as in most cases of the latter.

Proton NOE techniques have been used successfully for the assignment of isotropically shifted resonances in the past several years [8]. The determination of NOE between a pair of nuclei (i and j) in a paramagnetic species can give the internuclear distance (r_{ij}) as shown in eq 2 [8,9],

$$\mathrm{NOE}(i) = (\sigma_{ij}/\rho_i)\,[1 - \exp(-\rho_i t)] \qquad (2)$$

where $\sigma_{ij} = -\hbar^2\gamma^4\tau_c/10r_{ij}^6$ is the cross relaxation with γ the gyromagnetic ratio and τ_c the rotational correlation time, ρ_i is the intrinsic relaxation rate of i, and t is saturation duration time. A NOESY [9,10] experiment can also be described by a slight modification of eq 2 to include ρ_j [11]. Coherence transfer NMR techniques have also been useful for signal assignment in paramagnetic species when cross signals in COSY (as functions of $\sin(\pi J_{ab}t_1)\exp(-t_1/T_2)$ with J_{ab} the scalar coupling constant, T_2 the spin-spin relaxation time, and t_1 the evolution time for the second dimension) [10] due to scalarly coupled nuclei could be observed. These techniques have recently been applied for the study of paramagnetic metalloproteins [7,8] (which are also discussed in several contributions of this volume such as the theory and technique-oriented ones by Bertini and by Luchinat), although the observation of relaxation-based coherence transfer in a COSY experiment becomes possible when Curie relaxation is present (to a less extent for a small S = 1/2 metalloprotein at elevated temperatures) [12]. A precise assignment of the isotropically shifted features, along with the geometric-dependent dipolar shift, can provide detailed structural information about the paramagnetic metal-binding environment in metalloproteins. In this contribution, the application of 1D and 2D NMR techniques for structural and mechanistic studies of several metallo-biomolecular systems in our laboratory using paramagnetic Ln^{3+} as probes will be summarized and discussed.

2. Calcium Proteins

The versatility of Ca^{2+} in biological processes has been well documented, including its participation in regulation, signaling, and catalysis [13]. Some of these processes require the participation of protein molecules, such as phospholipases, calmodulin, and some proteases and nucleases, whose stability and proper action rely on the binding of Ca^{2+} to the protein molecules. In contrast to the much better biochemical studies of most Ca^{2+} proteins, their spectroscopic studies have been greatly hampered due to the inertness of the Ca^{2+} toward most spectroscopies. ^{47}Ca NMR (I = 7/2) has been one of the very few cases that has provided detailed Ca^{2+}-binding properties of proteins, although the low natural abundance and the quadrupole moment of this nucleus prohibit its being commonly used as an NMR probe [14]. Nevertheless, an I = 7/2 nucleus under extreme narrowing conditions would give reasonably sharp central m = 1/2 → -1/2 transition (as a function of $\chi^2/\nu_o^2\tau_c$) [15] on a spectrometer of higher magnetic field, which is now readily available commercially. Important molecular and metal binding properties of Ca^{2+} proteins and their interactions with ligands (a substrate analogue or an antagonist) have also been provided by the use of other metal ion probes, such as ^{113}Cd as an NMR probe [14,16].

Owing to the similar binding properties (essentially ionic binding preferentially to oxygen-rich sites) and radii (0.93-1.10 Å vs. 1.06 Å for 7-coordination sphere) of Ln^{3+} ions to those of Ca^{2+} and their rich optical and magnetic properties, these ions have become the most widely used probes for the study of the physical and molecular properties of Ca^{2+} proteins [17]; particularly, Eu^{3+} and Tb^{3+} ion have been widely used as luminescent probes [18]. In some cases, such as α-amylase [19] and calmodulin [20], the Ln^{3+}-substituted derivatives still afford high activity and can serve as good models for better understanding of the structural and mechanistic properties of the native proteins. In other cases, Ln^{3+}-substitution affords derivatives with a "build-in inhibitor", the Ln^{3+} ion, which binds substrates tightly to form enzyme-substrate complexes without proceeding to the subsequent catalysis, such as in staphylococcal nuclease [21]; thus affording a different view from an enzyme-inhibitor complex. In the past three years, we have been applying both 1D and 2D NMR techniques for the study of several Ca^{2+} proteins, including α-lactalbumin, parvalbumin, invertebrate sarcoplasmic Ca^{2+} binding proteins, and calmodulin, using paramagnetic Ln^{3+} ions as probes. Ln^{3+}-substitution has allowed us to look directly into the metal binding environment of these Ca^{2+} proteins by the use of the versatile NMR techniques, and has greatly enhanced our understanding of these proteins on the molecular basis.

2.1 PARVALBUMIN

Parvalbumin (PV) is a group of small acidic Ca^{2+}-binding proteins (M_r ~ 11-12 kDa) possessing similar Ca^{2+}-binding environments as in calmodulin and troponin C, and has been considered the prototype of the EF-hand Ca^{2+} protein family [13,22]. Each PV molecule binds two Ca^{2+} ions in its helix-loop-helix moieties, i.e. the CD and the EF loop, with distorted pentagonal bipyramidal coordination geometry [23]. This protein can be found in the muscles of all vertebrates, and has been proposed to play an important role in muscle relaxation [22]. Since the crystal structure of parvalbumin has been determined [23], a better understanding of the spectrum-structural correlation of the protein can be achieved if its spectral features can be precisely assigned, which would provide further insights into the correlation of structure and function of this protein via spectroscopic studies. The crystal structures of a Yb^{3+} and a Tb^{3+}-substituted carp parvalbumin have also been determined by X-ray crystallography that showed a great similarity to the native Ca^{2+} protein [24]. Thus a Ln^{3+}-substituted derivative can serve as a structural deputy of the native protein that can be studied by means of optical and magnetic resonance techniques.

Several factors in the Ln^{3+}-PV systems may facilitate significantly the understanding of the NMR properties of paramagnetic Ln^{3+} ions in proteins, and are thus considered the model systems in our studies: (i) The electronic relaxation rates of several paramagnetic Ln^{3+} ions (e.g. Pr^{3+}, Eu^{3+}, and Yb^{3+}) are fast that give rise to relatively slow relaxing isotropically shifted features which can be detected and studied more easily. (ii) The small size of PV (affording smaller rotational correlation time, τ_r) decreases the contribution of Curie relaxation (CR) proportionally [2]. (i.e. $T_{2CR}^{-1} \propto B_0^2 f(\tau_r) S^2(S+1)^2/r^6$) (iii) Some lanthanides have small spin-orbital J values (Eu^{3+} has a $J = 0$ ground state and low-lying $J = 1$ and $J = 2$ excited states, and Yb^{3+} has a ground state of $J = 7/2$) which may further decrease the contribution of CR to line broadening. (iv) The small CR contribution allows experiments to be performed at higher magnetic fields (B_0) to provide better spectral resolution and signal sensitivity, and would also decrease the relaxation-based coherence transfer [12]. (v) Fish PV can be purified in large quantities, that allows us to optimize and perform an NMR experiment of a concentrated paramagnetic derivative in a short period of time. (vi) Different isotypes of PV can be isolated that allows the study of minor perturbation of the structure of the metal binding sites (due to natural-occuring mutation of one or a few amino acids in or near the metal binding sites) using the sensitive paramagnetic Ln^{3+} probes.

The apparent binding affinity of Yb^{3+} to carp PV is larger in the EF site than in the CD site resulting in a sequential Yb^{3+} binding [25]. A Yb^{3+}-substituted derivative of pike PV (component III, pI = 5.0) with the Yb^{3+} bound in the EF site, CaYbPV (M,M'-PV with M in the CD site and M' in the EF site), can be prepared by the addition of 1 equiv Yb^{3+} to native pike PV in Tris buffer at pH 7.6 which exhibits many isotropically shifted ^1H NMR resonances in a wide spectral range attributable to the protons in the proximity of the Yb^{3+} ion (Figure 1), similar to the spectra of the spectra of other fish and rabbit CaYbPV derivatives [5,26]. Owing to the smaller affinity of Yb^{3+} binding to the CD site, the derivative YbYbPV cannot be fully formed when <10 equiv Yb^{3+} was introduced to native PV.

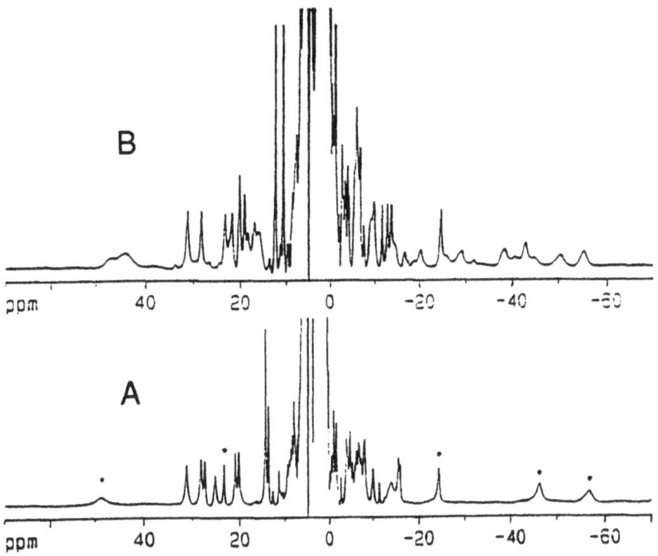

Figure 1. Proton NMR spectra (300.13 MHz, 303 K) of Yb^{3+}-substituted derivatives of pike PV: (A) CaYbPV and (B) YbYbPV in Tris buffer at pH 7.6. Several solvent exchangeable peptidyl NH signals can be identified (marked in A). Some of the slowly relaxing signals are suppressed by the short recycle times.

Many cross signals in a NOESY spectrum originated from the isotropically shifted features of the pike CaYbPV derivative can be detected with a mixing time of 60 ms attributable to pairs of protons close to each other in the vicinity of the Yb^{3+} in the EF site, such as the $C_\beta H_2 C_\alpha H$ moiety of an Asp residue, the geminal pair of Gly, and the very complicated spin systems of Leu and Ile residues (Figure 2). The downfield portion of the spectrum is shown in Figure 3 that reveals the "finger print" of the two residues Leu86 and Ile97 [7e]. The observation of cross relaxations associated with the methyl signals at 13.5 and 14.1 ppm provides a clue for the assignment of the Leu and the Ile residue due to their unique spin patterns, i.e., the signals at 13.5, 10.5, 9.6, 8.1, 6.5, and 6.0 due to the Leu residue ($C_\delta H_3$, $C_\alpha H$, $C_\gamma H$, $C_\delta H_3$, $C_\beta H$, and $C_\beta H'$, respectively) and the signals at 21.1, 20.2, 14.1, 7.9, and 0.6 ppm due to the Ile residue ($C_\gamma H$, $C_\gamma H'$, $C_\delta H_3$, $C_\beta H$, and $C_\gamma H_3$, respectively). This assignment has also further confirmed by means of a bond correlated COSY and TOCSY experiment (Figure 2B). A comparison of the spectrum of CaYbPV with that of YbYbPV also provides a clue for signal assignment (Figure 1). A few isotropically shifted signals in CaYbPV are not significantly changed when the Ca^{2+} in CaYbPV is replaced by Yb^{3+}, suggesting that they are due to the residues farther away from the CD site. For example: while both Leu86 and Ile97 located between the CD and EF sites are dramatically perturbed, the not-significantly-perturbed three spin system at 31.3, 28.3, and 6.9 ppm as revealed by NOESY can be assigned to the coordinated Asp92 located far distant from the CD site.

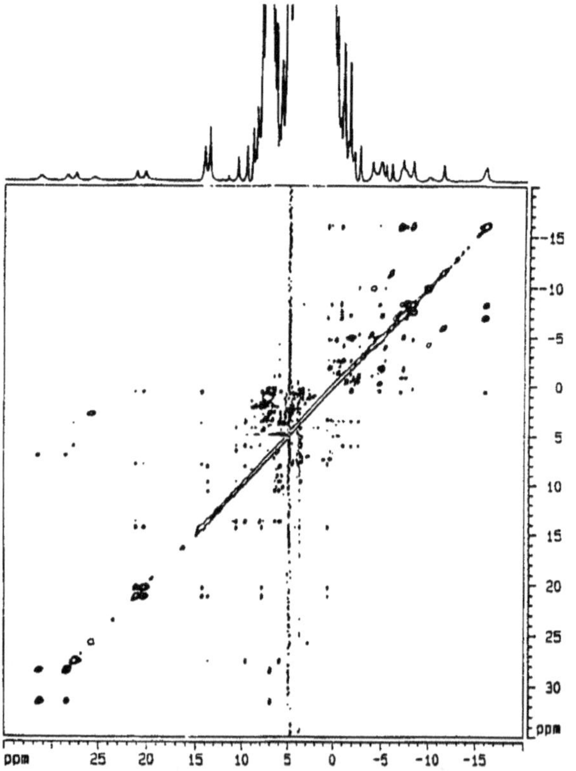

Figure 2. A 60-ms mixing NOESY spectrum of pike CaYbPV in Tris D_2O buffer at pD 7.6.

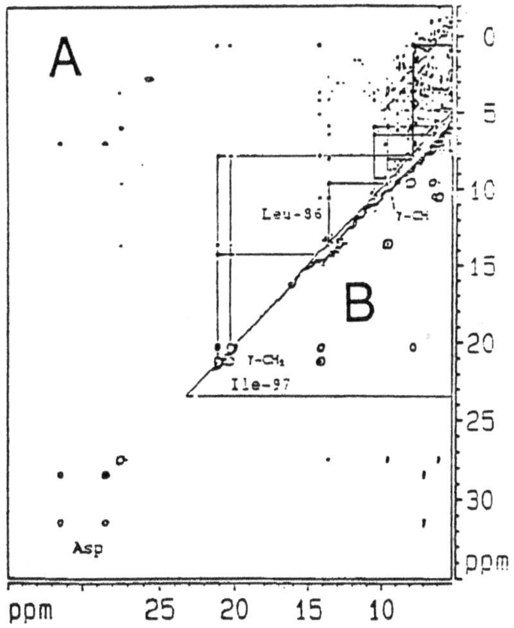

Figure 3. Downfield region of (A) NOESY and (B) COSY spectra of pike CaYbPV which shows clearly the finger prints of the two residues Leu86 and Ile 97.

2.2 BOVINE α-LACTALBUMIN [7d]

The biochemical significance of α-lactalbumin (LA) is due to its participation in lactose synthesis in the mammary gland in conjunction with the enzyme glucosyltransferase [27]. LA has also been widely used as a model for better understanding of the thermodynamic and structural properties (e.g. folding and unfolding) and metal binding properties of proteins [27]. This protein has been shown to bind tightly a Ca^{2+} ion via three Asp residues (Asp82, Asp 87, and Asp88), two peptidyl C=O (Lys79 and Asp84), and two water molecules in a distorted pentagonal-bipyramidal coordination sphere according to the crystal structures of the human and the baboon proteins [28]. Several other divalent metal ions have also been reported to bind LA, and to modulate the catalytic properties of this protein in lactose synthesis [29]. It is worth mentioning that the binding of some divalent metal ions, such as Zn^{2+}, to LA changes the spectroscopic properties of the protein into those of a more "apo-like" configuration, even though the Ca^{2+} is still bound in the protein molecule. Such a change is accompanied by a surprising promotion of its activation in lactose synthesis. Ln^{3+} ions have also been reported to compete with Ca^{2+} for binding to the protein, presumably retaining a similar metal binding configuration as in the native form [29], suggesting that Ln^{3+} ions can also serve as probes for the study of the structural and functional properties of this protein.

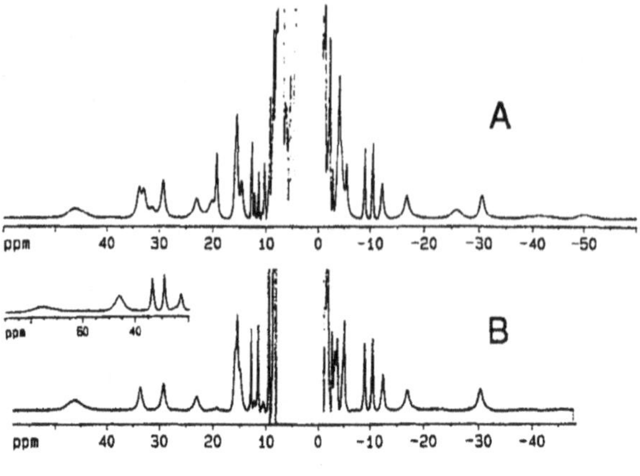

Figure 4. Proton NMR spectra (360.13 MHz) of Yb^{3+}-substituted bovine LA at pH 6.9 and 303 K in (A) H_2O and (B) D_2O.

Both Yb^{3+} and Pr^{3+} bind bovine LA affording derivatives which show ^1H NMR spectra in a large spectral range. The Yb^{3+} derivative exhibits isotropically shifted ^1H NMR features in a spectral range of > 100 ppm (Figure 4). We have also noticed that the signals widths of most of the isotropically shifted signals are much larger than those of CaYbPV of similar molecular weight (14 kD vs. 12 kD). An NOE build-up experiment according to eq (2) has revealed that the YbLA derivative of mM concentration has an apparent molecular weight of ~30 kD (to give a rotational correlation time of ~12 ns), suggesting the formation of a dimeric coagulate under the experimental conditions [7d]. Despite the broadness of the isotropically shifted features, cross signals in both NOESY and COSY spectra have been detected (Figure 5 [7d]). Three ABX spin systems can be clearly identified in this spectrum, i.e., a, b, and d; c, e, and g; and x,y, and zz (the latter revealed by the use of a 8-ms mixing time due to the short T_1 of zz at 77 ppm). The much higher intensity of the cross signals 1 and 5 relative to the rest signals in the upfield region of the NOESY spectrum with a mixing time of 35 ms (Figure 4) suggests that signals a, b and c, g are due to proton pairs in the metal site very close to each other, likely the geminal $C_\beta H_2$ pairs of the coordinated Asp residues. While signals d and e that show cross relaxations with proton pairs (a, b) and (c, g), respectively, (giving cross signals (2, 3) and (4, 6) of lower intensity than that of 1 and 5) are due to protons farther distant from the geminal pairs, likely the $C_\alpha H$ protons of the coordinated Asp residues. The shorter relaxation times and larger line widths of the proton pairs a, b and c, g relative to d and e also reflect such a proposal that they are closer to the metal, being due to the $C_\beta H_2$ protons of the coordinated Asp residues. The identification of the signals z, u, and s as an AMX spin system with integrations of 2, 2, and 1, respectively, suggest that they might be attributable to a fast rotating Phe, likely due to the ring protons of Phe80 located in the metal binding loop. A better resolved NOESY spectrum in the region of 10 to -5 ppm has also been achieved by the use of narrower spectral width and longer mixing and recycle times (spectrum not shown here).

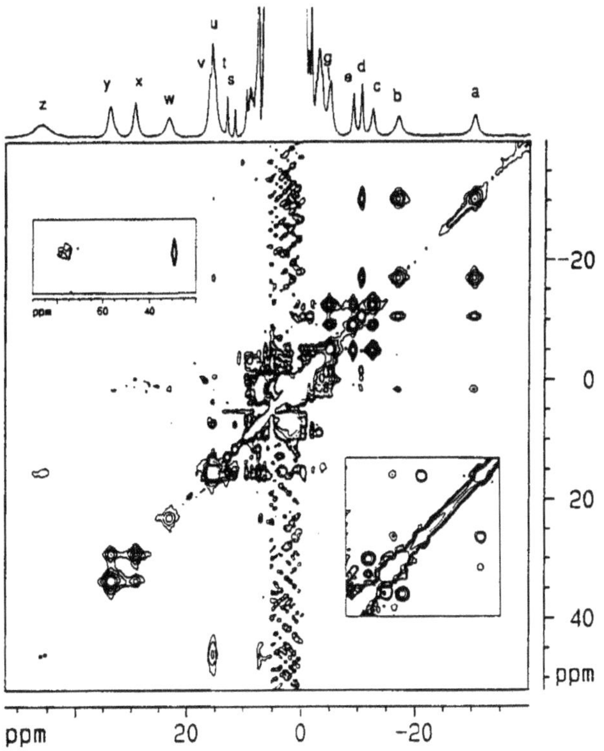

Figure 5. NOESY (with a mixing time of 35 ms; and 8 ms, inset in the downfield region for the signal zz) and upfield region COSY (inset) spectra of YbLA (in D_2O at "pH" = 6.9).

2.3 CALMODULIN

Calmodulin (CaM) is a wildly distributed EF-hand Ca^{2+}-binding protein in eukaryotic cells which participates in many significant biological processes, including Ca^{2+}-dependent activation of several enzymes such as nucleotide phosphodiesterase, kinases, and adenylate cylase [30]. CaM also represents an EF-hand protein of increasing Ca^{2+}-binding complicity, with 4 Ca^{2+} ions bound in each molecule. The binding of Ca^{2+} to apo-CaM promotes a conformational change of the protein molecule that initiates the significant biochemical activity of CaM. Ln^{3+} derivatives of CaM show significant bioactivity [20], thus can serve as good models for spectroscopic studies of the structure and function of CaM.

Bovine brain CaM can be easily purified via ammonium sulfate precipitation, heat treatment, and a hydrophobic column [31]. Yb^{3+} has been introduced to both the native and apo protein. Slightly different Yb^{3+} binding patterns have been noticed which may be due to the presence of a competition of Yb^{3+} with Ca^{2+} for the metal binding site in the native form. Such a difference has also been observed on Yb^{3+} binding to PV. Our preliminary results show different ^{1}H NMR spectra with 1 to 4 equiv Yb^{3+} added to apo-CaM, suggestive of a possible sequential Yb^{3+} binding. The binding of the antipsychotic drug trifluoperazine (TFP) to Yb^{3+}-substituted CaM has also been studied in our laboratory. A broadening of the signals is observed with the addition of TFP which may be due to fast exchange of the free and CaM-bound forms of TFP, with a sharpening of the signals observed when the protein is saturated with TFP (> 2 equiv). The far shifted features are not affected as significantly as some signals near the diamagnetic range indicating that the drug is bound to CaM in a remote site from the metal. In a previous study of Cd^{2+} binding to CaM, a cooperative tight binding of the metal to two sites followed by a weaker binding to the other two sites has been observed [32]. The influence of the binding of TFP on the metal binding sites of CaM has also been noticed in the previous study of the Cd_4 derivative. We have discussed briefly here that paramagnetic Ln^{3+} can be useful probes for the study of protein-ligand interaction as well. Further assignment of the isotropically shifted signals, especially the less shifted ones which are due to protons farther away from the metal that may interact directly with the ligand, may provide further structural information of the CaM-ligand complex.

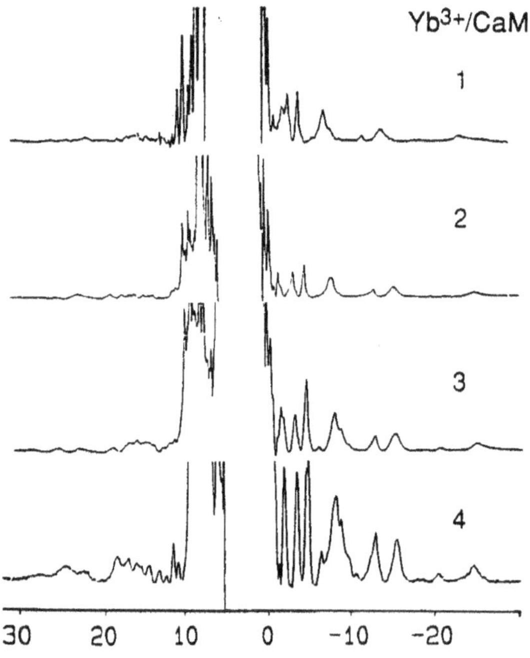

Figure 6. Proton NMR spectra of apo CaM at 360.13 MHz and 298 K with increasing amount of Yb^{3+} in Tris buffer at pD 7.6.

3. Metal-Dependent Antibiotics

The discovery of several antibiotics with antitumor activities opened a new era of cancer chemotherapy [33]. Functional studies of a few antitumor antibiotics revealed a direct participation of metal ions in the drug-DNA interactions and the subsequent DNA scission [34], such as bleomycin [35], streptonigrin [36,37], anthracyclines [38,39], and aureolic acids [40]. These observations suggest the importance of metal ions in maintaining the proper action of these antibiotics. Nevertheless, the structure of the metal complexes and the functional roles of the metal ions were not fully defined in the previous studies. Because the primary target of these antibiotics is DNA, a precise determination of the structure of metal-drug-DNA ternary complexes is ultimately important which will provide the mechanistic and molecular basis for a better understanding of the action of these antitumor antibiotics. Structural and mechanistic studies of the metal complexes of these antibiotics and their interactions with nucleic acids can also provide clues for future structure-based design [41] of metal complexes with specific DNA recognition as tools for the study of DNA structure. We have been studying several Ln^{3+} complexes of a few antibiotics by the use of NMR techniques, and is discussed in this section.

In recent years, the use of metal complexes as tools for the studies of DNA structure has attracted a great attention [42], such as: (a) the use of Fe^{II}EDTA for DNA footprinting; (b) the use of Cu^{I}(1,10-phenanthroline)$_2$ as an "artificial nuclease" for the study of DNA structure; (c) the use of metal-1,10-phenanthroline complexes for chiro-recognition of DNA; and (d) the promotion of nucleic acid hydrolysis by metal complexes, including Ln^{3+} complexes [43]. Extensive use of metal complexes as tools for such studies relies on a thorough understanding of metal complex-DNA interactions and a precise determination of the structures of the complexes and their DNA complexes.

3.1 THE ANTHRACYCLINE ANTICANCER ANTIBIOTICS [44]

The anthracycline antibiotics isolated from several *Streptomyces* species have currently been widely used as chemotherapeutic agents with high efficacy in the treatment of human cancers despite their severe side effects, including dose-dependent cardiomyopthy [45]. These antibiotics consist of a 4-ring anthraquinone chromophore and an amino sugar substituent at the 7-position as shown in Figure 7.

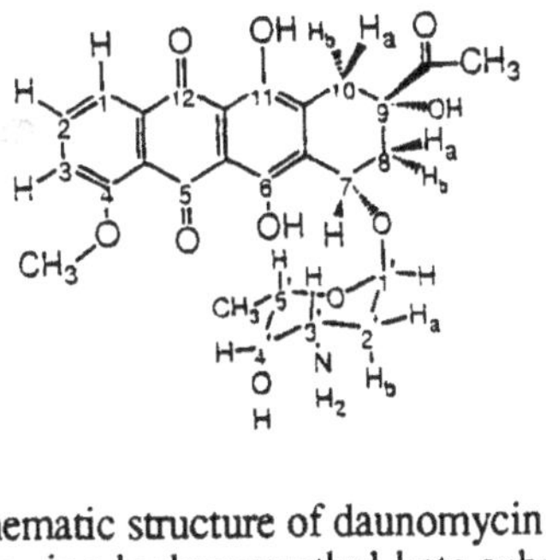

Figure 7. Schematic structure of daunomycin (or adriamycin when the 9-position is a hydroxymethyl-keto substituent).

The biological activity of these antibiotics is largely due to their DNA binding capability (via intercalation of the 4-ring moiety) and the redox activity of the quinone functional group, which engender significant perturbation of nucleic acid metabolism and electron transfer reactions in the cells and cause oxidative damage of the membrane of cells and organelles [46]. Several reports have suggested the participation of metal ions (such as Mg^{2+}, Ca^{2+}, and transition metals) in the action of these antibiotics and subsequent damage of biomolecules (e.g. DNA and phospholipids) [47]. Iron has further been postulated to take part in the cardiotoxicity of these antibiotics via free radical damage of the inner membrane of cardiac mitochondria [48]. However, the metal binding properties and the structure of the metal complexes were still not clearly defined. Although Ln^{3+} ions bind ligands with much less covalency than does Fe^{3+}, they have been used successfully as substitutes for the Fe in the O-rich binding sites in proteins, such as transferrin [49]. We discuss here the study of a prototypic anthracycline antibiotic, daunomycin, and the determination of the structure of a Yb^{3+}-daunomycin complex by the use of optical and 2D ^{1}H NMR techniques (EXSY and COSY), which provides us with the tools for future studies of the interaction of metal-anthracycline complexes with biomolecules.

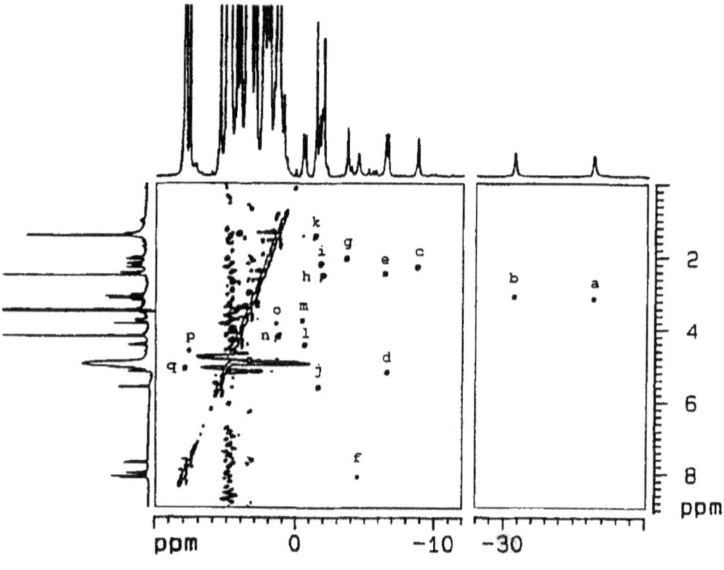

Figure 8. Proton EXSY spectrum (Bruker AMX360 at 360.13 MHz and 298 K) of the 1:1 Yb^{3+}-daunomycin complex and free drug (Yb^{3+}:drug = 1:3) in methanol-d_4. A standard phase sensitive NOESY pulse sequence with a mixing time of 30 ms and 1024 (f2) x 512 (f1) data points were applied for data acquisition, and bandwidths of 20 (f2) x 7.2 (f1) kHz were used to provide a better resolution along the f1 dimension. A 45°-shifted sine-squared apodization function was applied to both dimensions prior to Fourier transformation.

The introduction of Yb^{3+} to a methanol solution of daunomycin•HCl engenders the formation of a 1:1 complex on the basis of a Job plot with absorptions at 498 (sh), 532, and 570 nm (ε_{570} = 15.2 mM^{-1} cm^{-1}), despite that Ln^{3+} ions can form easily 8-coordinate complexes. The ^{1}H NMR spectrum of the complex in methanol-d_4 exhibits 13 clearly detected isotropically shifted signals in the upfield region (Figure 8, top trace). Owing to the presence of chemical exchange between the free drug and its complexed form in the solution, a full assignment of the isotropically shifted features of the 1:1 complex can be achieved by the use of 2D magnetization transfer techniques (EXSY) because of the easy assignment of the free drug. All the 17 non-exchangeable protons of the complex can be correlated to the metal-free antibiotic and unambiguously assigned as follows: Signals f, p, and q arise from the aromatic protons 1, 3, and 2, respectively; a and b are due to the geminal protons at position 10; c, d, and e are due to the protons at 7 and 8 positions, respectively; h, k, and n are associated with the 14, 5', and 4 methyl protons, respectively; j, (i, g), m, o, and l are associated with the sugar protons 1' to 5', respectively. The much shorter relaxation times (22.6 and 16.5 ms, respectively) of the signals at -30.8 and -36.4 ppm (which have been assigned to the geminal $C_{10}H_2$ protons) also suggest that the metal is coordinated to the aglycon rings via one of the β-ketophenolate moiety in positions 11 and 12 close to the $C_{10}H_2$ protons. The much longer relaxation times of the sugar protons (185-335 ms) are indicative of that the Yb^{3+} does not interact with the amino sugar to a great extent, as opposed to an early study of a Pd^{2+}-anthracycline complex where the metal has been proposed to bind with the sugar from a second complex [50].

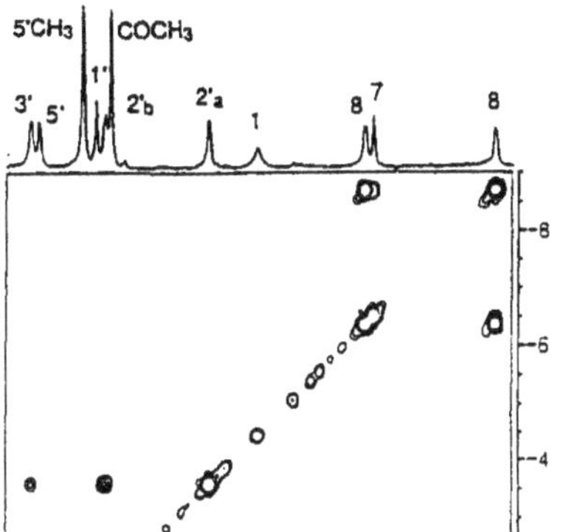

Figure 9. A partial plot of the COSY spectrum (at 360.13 MHz and 298 K) of the 1:1 Yb^{3+}-daunomycin complex in methanol-d_4. A bandwidth of 11.6 kHz was applied to both dimensions with 1024 (f2) x 512 (f1) data points which gave an acquisition time of 44 ms. A 0°-shifted sine-squared window function was applied to both dimensions prior to Fourier transformation.

The sharpness of the isotropically shifted features also allows the use of coherence transfer techniques for signal assignment, and for the elucidation of the configuration of vicinal proton pairs which correlates with their scalar coupling constants. Figure 9 shows the magnitude-mode ^1H COSY spectrum of the isotropically shifted features of the 1:1 Yb^{3+}-daunomycin complex. The $C_{10}H_2$ protons (-36.2 and -30.6 ppm) and the C_8H_2-C_7H spin system (-6.3 and -8.7 ppm and -6.5 ppm) can be recognized clearly where a vicinal C_8H-C_7H pair shows weak cross signals as shoulders. The bond correlation of the sugar moiety can also be observed, where the connectivities of $C_2'H_2$-$C_3'H$ and H-C_5'-CH_3 can be established. The crystal structure of daunomycin shows that the C_3'-NH_2 and the C_5'-CH_3 groups are in the equatorial positions, leaving the $C_3'H$ and $C_5'H$ protons in the axial positions and the $C_1'H$ and $C_4'H$ protons in the equatorial positions [51]. On the basis of the intensity of the coherence-transfer cross signals, one of the $C_2'H_2$ signals at -1.75 ppm, which shows an intense cross signal with the vicinal axial $C_3'H$ proton at -0.4 ppm, can be further assigned to the axial $C_2'H$ proton. The vicinal proton pairs with anti configuration afford large coupling constants, thus more intense cross signals. That the cross signals of the $C_3'H$-$C_4'H$ and $C_4'H$-$C_5'H$ pairs were not detected could be attributable to the smaller coupling constants between these axial-equatorial proton pairs. We have shown here by means of NMR that the configuration of the sugar moiety in the metal-daunomycin complex in solution is similar to that of the crystal structure of the free drug [51].

Metal ions have previously been reported to facilitate the reduction of quinones to form semiquinones by lowering their electrode potential via the formation of more stable metal-semiquinone complexes with an extra charge on the semiquinones.[52,53] Thus, the influence of metal ions in the action of anthracycline antibiotics toward the damage of biomolecules can be partially due to a fine-tuning of the redox potential of these antibiotics via metal ion binding, which may change their dioxygen-activation and free radical-generation capability. Owing to the presence of large amount of metal ions (such as the mM-range Mg^{2+}) in the body, the use of paramagnetic Ln^{3+} ions (as alkaline earth metal substitutes) along with the use of transition metal ions as spectroscopic probes would provide further insights into the mechanism of in vivo action of these antibiotics.

3.2 OTHER METAL-DEPENDENT ANTIBIOTICS

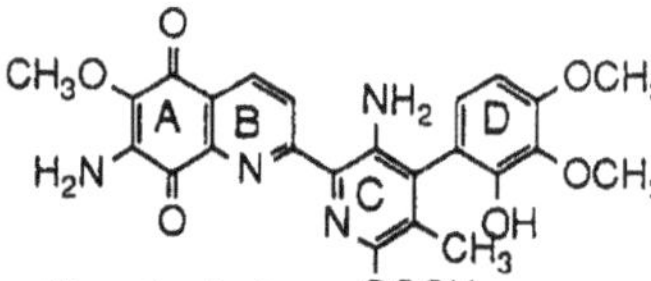

Streptonigrin

Streptonigrin (SN) produced by *Streptomyces flocculus* has been shown to be a potent antitumor antibiotics, being active against lymphoma, melanoma, and breast, cervix, and some other cancers, regardless of its severe side effect [54]. The metal ions Fe^{2+} and Cu^+ have been shown to enhance the antitumor and bactericidal activities of the drug [36,37]. Although the mechanism of DNA cleavage by SN has been studied in some detail, the role that metal ions play in facilitating such process has never been clearly defined, as opposed to the much better understood role of Fe^{2+} in bleomycin. Other than transition metal ions, we have also found that SN binds Ln^{3+} ions and exhibits well resolved isotropically shifted features.

Chromomycin A_3 ($ChrA_3$), mithramycin A (MitA), and olivomycin are glycosylated antitumor antibiotics of the aureolic acid group isolated from streptomyces

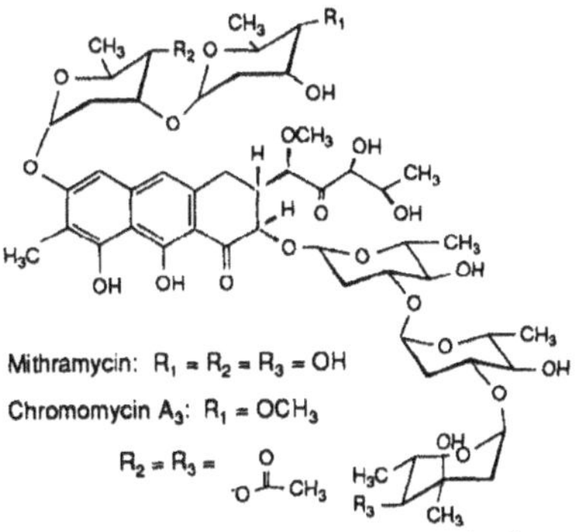

species which display a metal-dependent preference for GC-rich DNA sequences in the minor groove [55], although a duplex of $(AGGATCCT)_2$ has also been reported to bind 2 $ChrA_3$ in the presence of metal ions [56]. These antibiotics have also be observed to bind metal ions in the absence of DNA via their β-keto phenolate moiety [57], reminiscent of the similar metal binding site in anthracyclines. In addition to Mg^{2+}, several divalent metal ions have been reported to facilitate the binding of these drugs to a DNA duplex [40]. These previous studies also concluded a radius-specific metal binding with only those divalent metal ions smaller than 0.85 Å can be incorporated for DNA binding [40]. However, we have observed that a trivalent Yb^{3+} of 0.93 Å can also bind $ChrA_3$ and exhibits many isotropically shifted signals. Such a result is not surprising after we have learned that Ln^{3+} can bind well to the anthracyclines via a same β-keto phenolate moiety. The capability of Ln^{3+} ions binding to the antibiotics SN and the aureolic acids suggests that these metal ions can be spectroscopic probes as well for mechanistic and structural studies of a variety of different metal-dependent antibiotics.

4. Future Perspectives

Ln^{3+} ions have long been used as probes for the study of biomolecular structure and function, such as in the study of the structure of Ln^{3+}-nucleotide complexes and compared with other M^{n+}-nucleotide complexes, and the use of Ln^{3+}-ATP complexes as inhibitors in enzymological studies of kinases [58]. However, further investigation by means of NMR techniques has been impeded in the past several years owing to the presence of the paramagnetism of the several spectroscopically active Ln^{3+} ions. With the recent development of the multi-dimensional NMR techniques for the study of paramagnetic species, paramagnetic Ln^{3+} ions can be further used as a probe for the study of metalloproteins and metallo-biomolecules to provide detailed physical properties of these molecules, and to gain further insights into their biochemical and physiological functions.

5. Acknowledgments

The author is grateful for the funding provided by the University of South Florida (USF). This work has also been partially supported by the USF Research and Creative Scholarship Grant Program (on antibiotics), and the Petroleum Research Fund administered by the American Chemical Society (on Ln^{3+} NMR probes; PRF26483-G3). The contributions from X. David Wei and Hongzhu Yang and other coworkers in the laboratory to this work are also acknowledged.

6. References

1. Williams, R. J. P. (1978) Enzyme action: Views derived from metalloenzyme studies, *Chem. Britain* **14**, 25. (b) Hughes, M. N. (1981) *The Inorganic Chemistry of Biological Processes*; 2nd Ed.; Wiley, New York.
2. (a) Bertini, I. and Luchinat, C. (1986) *NMR of Paramagnetic Molecules in Biological Systems*; Benjamin/Cumming, Menlo Park, CA. (b) Berliner, L. J. and Reuben, J., eds. (1993) *NMR of Paramagnetic Molecules*; Plenum, New York.
3. (a) Morrill, T. C., Ed. (1986) *Lanthanide Shift Reagents in Stereochemical Analysis*; VCH, NY. (b) La Mar, G. N., Horrocks, W. DeW., Jr., and Holm, R. H. eds, (1973) *NMR of Paramagnetic Molecules*, Chapters 12 & 13; Academic, NY.
4. Lenkinski, R. E. (1984) Lanthanide complexes of peptides and proteins, *Biol. Magn. Reson.* **6**, 23-71.
5. Lee, L. and Sykes, B. D. (1980) High resolution NMR, *Adv. Inorg. Biochem.* **2**, 183-210.
6. Wüthrich, K. (1986) *NMR of Proteins and Nucleic Acids* , Wiley, NY.
7. (a) Holz, R. C., Que, L., Jr., and Ming, L.-J. (1992) NOESY Studies on the Fe(III)Co(II) Active Site of the Purple Acid Phosphatase Uteroferrin, *J. Am. Chem. Soc.* **114**, 4434-4436. (b) Bertini, I., Luchinat, C., Ming, L.-J., Piccioli, M., Sola, M., and Valentine, J. S. (1992) Two-dimensional ^{1}H NMR studies of the paramagnetic metalloenzyme copper-nickel superoxide dismutase, *Inorg. Chem.* **31**, 4433-4435. (c) Ming, L.-J., Lynch, J. B., Holz, R. C., and Que, L., Jr. (1994) One- and two-dimensional ^{1}H NMR studies of the active Site of iron(II) superoxide dismutase from *Escherichia coli*, *Inorg. Chem.* **33**, 83-87. (d) Ming, L.-J. (1993) Two-dimensional ^{1}H NMR studies of Ca(II)-binding site in proteins using paramagnetic lanthanides(III) as probes and Yb(III)-substituted bovine α-lactalbumin as an example, *Magn. Reson. Chem.* **31**, S104-S109. (e) Ming, L.-J. (1993) Two-Dimensional ^{1}H NMR Study of Paramagnetic Lanthanide(III)-Substituted Ca(II) Proteins, *J. Inorg. Biochem.* **51**, 99.
8. For general reviews, see (a) La Mar, G. N. and de Ropp, J. S. (1993) NMR Methodology for Paramagnetic Proteins, In Berliner, L. J.; Reuben, J., Eds.; *NMR of Paramagnetic Molecules*; Plenum, New York.. (b) Bertini, I., Turano, P., Vila, A. J. (1994) Nuclear Magnetic Resonance of Paramagnetic Metalloproteins, *Chem. Rev.* **93**, 2833-2932.
9. (a) Noggle, J. H.and Schirmer, R. E. (1971) *The Nuclear Overhauser Effect*; Academic, NY. (b) Neuhaus, D. and Williamson, M. P. (1989) *The Nuclear Overhauser Effect in Structural and Conformational Analysis*; VCH, NY.
10. Ernst, R. R., Bodenhausen, G., and Wokaun, A. (1987) *Principles of Nuclear Magnetic Resonance in One and Two Dimensions*; Oxford.
11. Sette, M., de Ropp, J. S., Hernandez, G., and La Mar, G. N. (1993) Determination of interproton distances from NOESY spectra in the active site of paramagnetic metalloproteins: Cyanide-inhibited horseradish peroxidase, *J. Am. Chem. Soc.* **115**, 5237-5245.
12. Bertini, I, Luchinat, C., and Tarchi, D. (1993) Are true scalar proton-proton connectivities ever measured in COSY spectra of paramagnetic macromolecules? *Chem. Phys. Lett.* **203**, 445.
13. (a) Marmé, D., ed. (1985) *Calcium and Cell Physiology*; Springer-Verlag, NY. (b) Sigel, H., Ed. (1984) *Calcium and Its Role in Biology;* Dekker, NY. (c) Heizmann, C. W., Ed. (1991) *Novel Calcium-Binding Proteins*, Springer-Verlag, NY.

14. Vogel, H. and Forsén, S. (1987) NMR studies of calcium-binding proteins, *Biol. Magn. Reson.* **7**, 249.
15. (a) Westlund, P. O.and Wennerström, H. (1982) *J. Magn. Reson.* **50**, 451-466. (b) Aramini, J. M. and Vogel, H. J. (1994) A scandium-45 NMR study of ovotransferrin and its half-molecules, *J. Am. Chem. Soc.* **116**, 1988-1993.
16. Summers, M. F. (1988) ^{113}Cd NMR spectroscopy of coordination compounds and proteins, *Coord. Chem. Rev.* **86**, 43-134.
17. (a) Evans, C. H. (1990) *Biochemistry of the Lanthanides*; Plenum, NY. (b) Bünzli, J.-C. G. and Choppin, G. R. (1989) *Lanthanide Probes in Life, Chemical and Earth Sciences*; Elsevier, NY.
18. (a) Horrocks, W. DeW., Jr. (1982) Lanthanide ion probes of biomolecular structure, *Adv. Inorg. Biochem.* **4**, 201-260. (b) O'Hara, P. B. (1987) Lanthanide ions as luminescent probes of biomolecular structure, *Photochem. Photobiol.* **46**, 1067-1070.
19. Smolka, G. E., Birnbaum, E. R., and Darnall, D. W. (1971) Rare earth metal ions as substitutes for the calcium ion in *Bacillus subtilis* α-amylase, *Biochemistry* **10**, 4556-4561.
20. Chao, S.-H., Suzuki, Y., Zysk, J. R., and Cheung, W. Y. (1984) Activation of calmodulin by various metal cations as a function of ionic radius, *Molec. Pharmacol.* **26**, 75-82.
21. (a) Furie, B., Eastkake, A., Schechter, A. N., and Anfinsen. C. B. (1973) The interaction of the lanthanide ions with staphylococcal nuclease, *J. Biol. Chem.* **248**, 5821-5825.
22. (a) Wnuk, W., Cox, J. A., and Stein, E. A. (1982) in *Calcium and Cell Function*; Cheung, W. Y., Ed; Vol. II, Academic, NY. (b) Gerday, C. (1988) in *Calcium and Calcium Binding Proteins*; Gerday, C. Ed; Springer, NY.
23. (a) Swain, A. L., Kretsinger, R. H., and Amma, E. L. (1989) Restrained least squares refinement of native (calcium) and cadmium-substituted carp parvalbumin using x-ray crystallographic data at 1.6-Å resolution, *J. Biol. Chem.* **264**, 16620. (b) Kumar, V. D., Lee, L., and Edwards, B. F. P. (1990) Refined crystal structure of calcium-liganded carp parvalbumin 4.25 at 1.5-Å resolution, *Biochemistry* **29**, 1404. (c) Declercq, J.-P., Tinant, B., Parello, J., Etienne, G., and Huber, R. (1988) Crystal structure determination and refinement of pike 4.10 parvalbumin (Minor component from *Esox lucius*), *J. Mol. Biol.***202**, 349. (d) Declercq, J.-P., Tinant, B., Parello, J., Rambaud, J. (1991) Ionic interactions with parvalbumins, *J. Mol. Biol.* **220**, 1017. (e) Roquet, F., Declercq, J.-P., Tinant, B., Rambaud, J., and Parello, J. (1992) Crystal structure of the unique parvalbumin component from muscle of the leopard shark (*Triakis semifasciata*), *J. Mol. Biol.* **223**, 705.
24. (a) Kumar, V. D., Lee, L., and Edwards, B. F. P. (1991) Refined crystal structure of ytterbium-substituted carp parvalbumin 4.25 at 1.5Å, and its comparison with the native and cadmium-substituted structures, *FEBS Lett.* **283**, 311. (b) Wéry, J. P., Dideberg, O., Charlier, P., and Gerdy, C. (1985) Crystallization and structure at 3.2 Å resolution of a terbium parvalbumin, *FEBS Lett.* **182**, 103-106. (c) Sowadski, J., Cornick, G., and Kretsinger, R. H. (1978) Terbium replacement of calcium in parvalbumin, *J. Mol. Biol.* **124**, 123-132.
25. McNemar, C. W. and Horrocks, W. DeW., Jr. (1990) Europium(III) ion luminescence as a structural probe of parvalbumin isotypes, *Biochim. Biophys. Acta* **1040**, 229 and references therein.
26. Capozzi, F., Cremonini, M. A., Luchinat, C., Sola, M. (1993) Assignment of

pseudo-contact-shifted ^{1}H NMR resonances in the EF site of Yb^{3+}-substituted rabbit parvalbumin through a combination of 2D techniques and magnetic susceptibility tensor determination, *Magn. Reson. Chem.* **31**, S118-S127.
27. McKenzie, H. A. and White, Jr., F. H. (1991) Lysozyme and α-lactalbumin: structure, function, and interrelationship, *Adv. Protein Chem.* **41**, 174.
28. (a) Acharya, K. R., Stuart, D. I., Walker, N. P. C., Lewis, M., and Phillips, D. C. (1989) Refined structure of baboon α-lactalbumin at 1.7 Å resolution, *J. Mol. Biol.* **208**, 99-127. (b) Harata, K., Muraki, M. (1992) X-ray structural evidence for a local helix-loop transition in α-lactalbumin, *J. Biol. Chem.* **267**, 1419. (c) Acharya, K. R., Ren, J., Stuart, D. I., Phillips, D. C., and Fenna, R. E. (1991) Crystal structure of human α-lactalbumin at 1.7 Å resolution, *J. Mol. Biol.* **221**, 571-581.
29. M. J. Kronman (1989) Metal-ion binding and the molecular conformational properties of α-lactalbumin, *Critical Rev. Biochem. Mol. Biol.* **24**, 565.
30. Cox, J. A., Comte, M., Malnoë, A., Burger, D, and Stein, E. A. (1984) Mode of action of the regulatory protein calmodulin, *Metal Ions Biol. Syst.* **17**, 215.
31. Gopalakrishna, R., Anderson, W. (1982) Ca^{2+}-induced hydrophobic site on calmodulin: application for purification of calmodulin by phenyl-sepharose affinity chromatography, *Biochem. Biophys. Res. Commun.* **104**, 830-836.
32. Forsén S., Thulin, E., Drakenberg, T., Krebs, J., and Seamon, K. (1980) A ^{113}Cd NMR study of calmodulin and its interaction with calcium, magnesium and trifluoperazine, *FEBS Lett.* **117**, 189-194.
33. Remers, W. A. (1979) *The Chemistry of Antitumor Antibiotics*, Wiley, New York.
34. (a) Sigel, H.; Ed. (1980) *Metal Ions in Biological Systems*, Vol 11; Dekker, New York. (b) Sigel, H.; ed. (1985) *Metal Ions in Biological Systems*, Vol 19; Dekker, New York.
35. Hecht, S. M. (1986) The chemistry of activated bleomycin, *Acc. Chem. Res.* **19**, 383-391.
36. (a) Cone, R., Hasan, S. K., Lown, J. W., and Morgan, A. R. (1976) The mechanism of the degradation of DNA by streptonigrin, *Can. J. Biochem.* **54**, 219-223. (b) Lown, J. W. and Sim, S.-K. (1976) Studies related to antitumor antibiotics. Part VII. Synthesis of streptonigrin analogues and their single strand scission of DNA, *Can. J. Chem.* **54**, 2563-2572. (c) White, J. R. and Yeowell, H. N. (1982) Iron enhances the bactericidal action of streptonigrin, *Biochem. Biophys. Res. Commun.* **106**, 407-411. (d) Yeowell, H. N. and White, J. R. (1982) Iron requirement in the bactericidal mechanism of streptonigrin, *Antimicro. Agen. Chemother.* **22**, 961-968. (e) Sugiura, Y., Kuwahara, J. and Suzuki, T. (1984) DNA interaction and nucleotide sequence cleavage of copper-streptonigrin, *Biochim. Biophys. Acta* **782**, 254-261. (f) Cohen, M. S., Chai, Y., Britigan, B. E., McKenna, W., Adams, J., Svendsen, T., Bean, K., Hassett, D. J., and Sparling, P. F. (1987) Role of extracellular iron in the action of the quinone antibiotic streptonigrin: Mechanism of killing and resistance of *Neisseria gonorrhoeae*, *Antimicro. Agen. Chemother.* **31**, 1507-1513.
37. (a) Rao, K. V. (1979) Interaction of streptonigrin with metals and with DNA, *J. Pharm. Sci.* **68**, 853-856. (b) White, J. R. (1977) Streptonigrin-transition metal complex: Binding to DNA and biological activity, *Biochem. Biophys. Res. Commun.* **77**, 387-391.
38. Martin, R. B. (1985) Tetracyclines and daunorubicin, In Ref. 36b; Chapter 2.
39. Kiraly, R. and Martin, R. B. (1982) Metal binding to daunorubicin and quinizarin, *Inorg. Chim. Acta* **67**, 13-18.

40. (a) Nayak, R., Sirsi, M., and Podder, S. K. (1973) Role of magnesium ion on the interaction between chromomycin A_3 and deoxyribonucleic acid, *FEBS Lett.* **30**, 157-162. (b) Banvill, D. L., Keniry, M. A., Kam, M., and Shafer, R. H. (1990) NMR studies of the interaction of chromomycin A_3 with small DNA duplexes. Binding to GC-containing sequences, *Biochemistry* **29**, 6521. (c) Itzhaki, L., Weinberger, S., Livnah, N., and Berman, E. (1990) A unique binding cavity for divalent cations in the DNA-metal-chromomycin A_3 complex, *Biopolymers* **29**, 481-489.
41. (a) Kuntz, I. D. (1992) Structure-based strategies for drug design and discovery, *Science* **257**, 1078-1082. (b) Erickson, J. W. and Fesik, S. W. (1992) Macromolecular X-ray crystallography and NMR as tools for structure-based drug design, *Ann. Rep. Med. Chem.* **27**, 271-289
42. (a) For a general review, see: Tullius, T. D.; ed. (1989) *Metal-DNA Chemistry*, ACS, Washington DC. (b) Sigman, D. S. (1990) Chemical nucleases, *Biochemistry* **29**, 9097-9105. (c) Sigman, D. S., Bruce, T. W., Mazumder, A., and Sutton, C. L. (1993) Targeted chemical nucleases, *Acc. Chem. Res.* **26**, 98.
43. Morrow, J. R. (1993) Artificial ribonucleases, In *Models in Inorganic Chemistry*, Eichhorn, G. L. and Marzilli, L. G., Eds.; Prentice Hall, Englewood Cliffs, NJ.
44. Ming, L.-J. and Wei, X. (1994) Ytterbium (III) Complex of Daunomycin, a Model Metal Complex of Anthracycline Antibiotics, *Inorg. Chem.* **33**, in press.
45. Weiss, R. B., Sarosy, G., Clagett-Carr, K., Russo, M., and Leyland-Jones, B. *Cancer Chemother. Pharmacol.* **18**, 185-197.
46. Lown, J. W. (1993) *Chem. Soc. Rev.* **22**, 165-176.
47. (a) Ref 40. (b) Phillips, D. R. and Carlyle, G. A. (1981) The effect of physiological levels of divalent metal ions on the interaction of daunomycin with DNA: evidence of a ternary daunomycin-Cu^{2+}-DNA complex, *Biochem. Pharmacol.* **30**, 2021-2024. (c) Spinelli, M. and Dabrowiak, J. C. (1982) Interaction of copper(II) ions with the daunomycin-calf thymus deoxyribonucleic acid complex, *Biochemistry* **21**, 5862-5870. (d) Mariam, Y. H. and Glover, G. P. (1986) Degradation of DNA by metalloanthracyclines: requirement for metal ions, *Biochem. Biophys. Res. Commun.* **136**, 1-7. (e) Eliot, H., Gianni, L., and Myers, C. (1984) Oxidative destruction of DNA by the adriamycin-iron complex, *Biochemistry* **23**, 928-936. (f) Beraldo, H., Garnier-Suillerot, A., Tosi, L., and Lavelle, F. (1985) Iron(III)-adriamycin and iron(III)-daunomycin complexes: physicochemical characteristics, interaction with DNA, and antitumor activity, *Biochemistry* **24**, 284-289. (g) Cullinane, C. and Phillips, D. R. (1990) Induction of stable transcriptional blockage sites by adriamycin: GpC specificity of apparent adriamycin-DNA adducts and dependence on iron(III) ions, *Biochemistry* **29**, 5638-5646.
48. (a) Paradies, G. and Ruggiero, F. M. (1983) The effect of doxorubicin on the transport of pyruvate in rat-heart mitochondria, *Biochem. Biophys. Res. Commun.* **156**, 1302-1307. (b) Marcillat, O., Zhang, Y., and Davies, K. J. (1989) Oxidative and non-oxidative mechanisms in the inactivation of cardiac mitochodrial electron transport chain components by doxorubicin, *Biochem. J.* **259**, 181-189. (e) Fu, L. X., Waagstein, G., and Hjalmarson, Å. (1990) A new insight into adriamycin-induced cardiotoxicity, *Intl. J. Cardiol.* **29**, 15-20. (h) Myers, C. E., Gianni, L., Simone, C. B., Klecker, R. and Green, R. (1982) Oxidative destruction of erythrocyte ghost membranes catalyzed by the doxorubicin-iron complex, *Biochemistry* **21**, 1707-1713.
49. Welch, S. (1992) *Transferrin: The Iron Carrier*; CRC, Boca Raton; Chapter 5.

50. Fiallo, M. M. L. and Garnier-Suillerot, A. (1986) Metal anthracycline complexes as a new class of anthracycline derivatives. Pd(II)-adriamycin and Pd(II)-daunorubicin complexes: physicochemical characteristics and antitumor activity, *Biochemistry* **25**, 924-930.
51. Courseille, C.; Buseta, B.; Geoffre, S.; Hospital, M. (1979) Complex daunomycin-butanol, *Acta Cryst.* **B35**, 764-767.
52. (a) F. Müller (1983) and (b) D. E. Edmondson and G. Tollin (1983) in *Radicals in Biochemistry,* Springer, NY.
53. Scott, S. L., Bakac, A., and Espenson, J. H. (1992) Preparation and reactivity of a cationic dichromium-semiquinone complex, *J. Am. Chem. Soc.* **114**, 4605-4610, and references therein.
54. Hajdu, J. (1985) in Ref. 34b.
55. (a) Van Dyke, M. W. and Dervan, P. B. (1983) Chromomycin, mithramycin, and olivomycin binding sites on heterogeneous deoxyribonucleic acid. Footprinting with (methidiumpropyl-EDTA)iron(II), *Biochemistry* **22**, 2373-2377. (b) Fox, K. R. and Howarth, N. R. (1985) Investigations into the sequence-selective binding of mithramycin and related ligands to DNA, *Nucleic Acids Res.* **13**, 8695-8714. (c) Stankus, A., Goodisman, J., and Dabrowiak, J. C. (1992) Quantitative footprinting analysis of the chromomycin A_3-DNA interaction, *Biochemistry* **31**, 9310-9318.
56. Gao, X. and Patel, D. J. (1990) Chromomycin dimer-DNA oligomer complexes. Sequence selectivity and divalent cation specificity, *Biochemistry* **29**, 10940-10956.
57. (a) Aich, P. and Dasgupta, D. (1990) Role of Mg^{2+} in the mithramycin-DNA interaction: evidence for two types of mitramycin-Mg^{2+} complex, *Biochem. Biophys. Res. Commun.* **173**, 689-696. (b) Silva, D. J. and Kahne, D. E. (1993) Studies of the 2:1 chromomycin A_3-Mg^{2+} complex in methanol: role of carbohydrates in complex formation, *J. Am. Chem. Soc.* **115**, 7962-7970. (c) Silva, D. J., Goodnow, R., Jr., Kahne, D. E. (1993) The sugar in chromomycin A_3 stabilize the Mg^{2+}-dimer complex, *Biochemistry* **32**, 463-471.
58. (a) Tanswell, P., Thornton, J. M., Korda, A. V., and Williams, R. J. P. (1975) Quantitative determination of the configuration of ATP in aqueous solution using the lanthanide cations as nuclear magnetic resonance probes, *Eur. J. Biochem.* **57**, 135-145. (b) Morrison, J. F. and Cleland, W. W. (1983) Lanthanide-adenosine 5'-triphosphate complexes: determination of their dissociation constants and mechanism of action as inhibitors of yeast hexokinase, *Biochemistry*, **22**, 5507-5513.

Chemical Functions of Single and Double NH---S Hydrogen Bond in Iron-Sulfur Metalloproteins; Model Ligands with Cys-containing Peptide and Simple Acylaminobenzenethiolate

Akira NAKAMURA
Coordination Chemistry Laboratories, Institute for Molecular Science, Okazaki 444 Japan

Norikazu UEYAMA
Department of Macromolecular Science, Faculty of Science, Osaka University, Osaka 560 Japan

1. Introduction

Metalloenzymes and metalloproteins functioning in biological redox systems contain various transition metal ions surrounded by many Cys thiolate ligands. In the active center of electron transfer metalloproteins having the thiolate ligand, the existence of NH---S hydrogen bond has been proposed by the X-ray crystallographic analysis of some of metalloproteins, e.g. rubredoxins [1], plant-type ferredoxins [2], bacterial ferredoxins [3] and blue-copper proteins [4] as shown in Fig. 1.

The redox potential of model complexes having simple thiolato ligands is remarkably different from the redox potential of the corresponding native metalloproteins. Many of the reported model complexes are unstable in air in organic solvent or in aqueous solution. Thus, chemical properties of these models are not reproducing the important functions of native metalloproteins.

Hydrophobic environments are well known to influence the polar interactions inside of the proteins. In order to clarify chemical functions of these interactions, we have synthesized a series of peptide model complexes containing an invariant sequence around the [1Fe] core in rubredoxin [5-7], the [2Fe-2S] core in plant-type ferredoxin[8] or the [4Fe-4S] core in ferredoxin [9]. The peptide model complexes are found to have a preferable conformation to readily form the NH---S hydrogen bond. In the case of simple alkanethiolate complexes, the redox potential in a low-dielectric-constant organic solvents shifted to the negative side, whereas the peptide model complexes forming a NH---S hydrogen bond exhibit a positive-shifted redox potential in an organic solvent. For example, we have studied the relation between the NH---S hydrogen bond and redox potential at the FeS_4 core in the active center of rubredoxin. Specific hydrogen bonds are formed in a preferable structure of the peptide ligand and supported in a low dielectric constant solvent. Furthermore, we also studied the model complexes of peptide-mimetic simple thiolate ligands, e.g. 2-acylaminobenzenethiolate, having a structure suitable for the intramolecular NH---S hydrogen bond. To get information on

G.N. La Mar (ed.), Nuclear Magnetic Resonance of Paramagnetic Macromolecules, 265-279.

the nature of the NH---S hydrogen bond, we also synthesized the peptide-mimetic complexes containing various metal ions, e.g. Mo(V) [10], Mo(IV), Co(II) [11] and Cu(I) [12], found in biologically important metalloenzymes and metalloproteins.

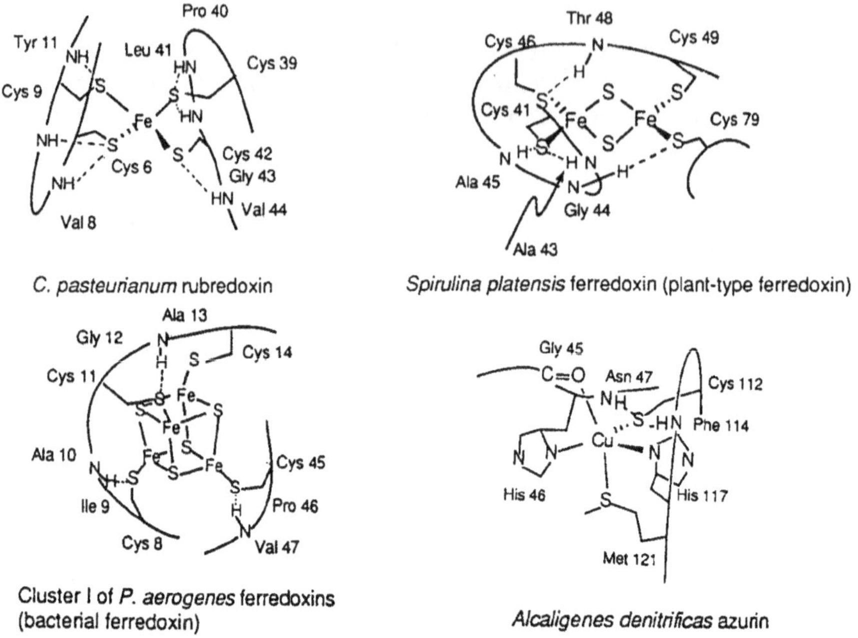

Fig. 1. The proposed NH---S hydrogen bonds in the active sites of rubredoxin, plant-type [2Fe-2S] ferredoxin, bacterial [4Fe-4S] ferredoxin and azurin.

2. Shift of Redox Potential through NH---S Hydrogen Bond Formation in Peptide Model Complexes

Rubredoxin acting as an electron transfer mediator in biological systems has a Fe ion surrounded by four cysteine thiolates of two invariant Cys-X-Y-Cys (X, Y = amino acid residues) fragments [13,14]. We have synthesized the rubredoxin model complexes having a chelating tetrapeptide ligand, $[Fe^{II}(Z\text{-}\mathbf{cys}\text{-}X\text{-}Y\text{-}\mathbf{cys}\text{-}OMe)_2]^{2-}$ (X-Y = Ala-Ala, Pro-Leu, Thr-Val) to construct spectral and electrochemical models having a characteristic properties of native rubredoxin [5]. Furthermore, model complexes containing a series of Cys-X-Y-Cys-Gly-A (A = Val, Phe) hexapeptide fragment existing within 3 Å of the Fe ion were synthesized [6,7,15,16].

The ^{2}H-NMR spectral data of a deuterated amide N^{2}H-peptide complex, $[Fe^{II}(Z\text{-}\mathbf{cys}\text{-}Pro\text{-}Leu\text{-}\mathbf{cys}\text{-}Gly\text{-}Val\text{-}OMe)_2]^{2-}$, show the presence of two N^{2}H--S hydrogen bonds in the chelating Cys-Pro-Leu-Cys moiety and one such hydrogen bond by the Cys-Gly-Val-OMe fragment in 1,2-dimethoxyethane (DME) [7]. Figure 2 shows the ^{2}H-NMR spectra of the amide deuterated complex, $[Fe^{II}(Z\text{-}\mathbf{cys}\text{-}Pro\text{-}Leu\text{-}\mathbf{cys}\text{-}Gly\text{-}Val\text{-}OMe)_2]^{2-}$, in acetonitrile at room temperature. The contact shifted amide N^{2}H signals are observed at 50 ppm, 20 ppm and -10 ppm. The shift mainly due to a Fermi contact clearly indicates the presence of the NH---S hydrogen bond.

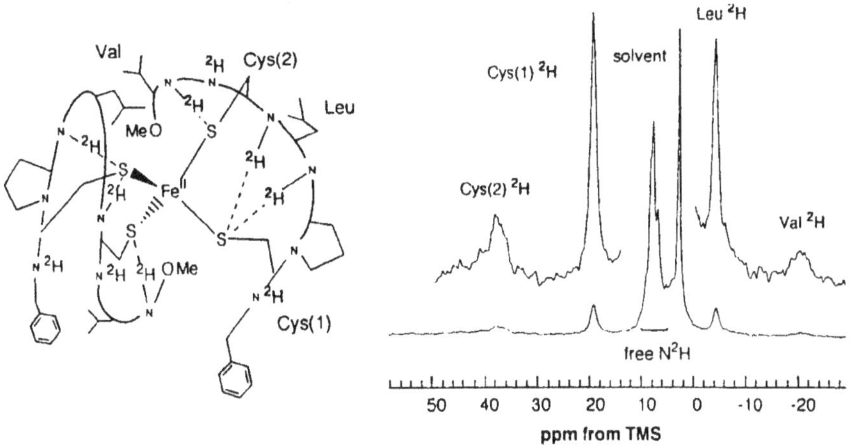

Fig. 2. ^{2}H NMR spectrum of amide deuterated [Fe^{II}(Z-**cys**-Pro-Leu-**cys**-Gly-Val-OMe$)_2]^{2-}$ complex in acetonitrile.

The Cys-Gly-Val-OMe unit provides a turn structure to give the specific NH---S hydrogen bond in a low-dielectric-constant solvent. Instead of the Val residue, *p*-substituted anilide was introduced in the tripeptide fragment as Cys-Gly-NHC_6H_4-*p*-X (X = OMe, H, F, CN) to investigate the distal electronic effect through the NH---S hydrogen bond [6]. The absorption maxima and CD extrema of the anilide-containing peptide Fe(II) complexes in acetonitrile resemble to the reported maxima for native rubredoxin in an aqueous solution (Table 1). Especially, a blue-shifted Fe-S LMCT absorption band is observed at 312 nm just as that of native rubredoxin. These peptide model studies indicate that the blue-shift is ascribed to the electronic perturbation to the Fe-S bond by the NH---S hydrogen bonding.

TABLE 1. Solvent dependence of the redox potentials of Fe(II) peptide complexes in CH_3CN and DME at room temperature.

Complexes	σ_p	Redox potential, V vs SCE CH_3CN	DME
[Fe^{II}(Z-**cys**-Pro-Leu-**cys**-OMe$)_2]^{2-}$		-0.54	-0.59
[Fe^{II}(Z-**cys**-Pro-Leu-**cys**-Gly-Val-OMe$)_2]^{2-}$		-0.46	-0.35
[Fe^{II}(Z-**cys**-Pro-Leu-**cys**-Gly-NHC_6H_4-*p*-OMe$)_2]^{2-}$	-0.27	-0.36	-0.33
[Fe^{II}(Z-**cys**-Pro-Leu-**cys**-Gly-$NHC_6H_5)_2]^{2-}$	0	-0.38	-0.33
[Fe^{II}(Z-**cys**-Pro-Leu-**cys**-Gly-NHC_6H_4-*p*-F$)_2]^{2-}$	0.06	-0.39	-0.31
[Fe^{II}(Z-**cys**-Pro-Leu-**cys**-Gly-NHC_6H_4-*p*-CN$)_2]^{2-}$	0.66	-0.30	-0.24

The electrochemical properties of these peptide model complexes were studied using cyclic voltammograms. The results are listed in Table I. The positive-shifted redox potential at -0.35 V vs SCE for $[Fe^{II}(Z\text{-}\mathbf{cys}\text{-Pro-Leu-}\mathbf{cys}\text{-Gly-Val-OMe})_2]^{2-}$ in acetonitrile is close to that (-0.06 V vs NHE and -0.31 V vs SCE) reported for *Clostridium pasteurianum* rubredoxin [13]. The complexes having various kinds of *p*-substituted (OMe, H, F, CN) anilide derivatives exhibit the redox potentials obeying to a Hammett rule with the σ_p values. The results indicates that the electronic effect of *p*-substituents extends to Fe(II) ion through the benzene ring and the NH---S hydrogen bond. The ^2H NMR spectra of $[Fe^{II}(Z\text{-}\mathbf{cys}\text{-Pro-Leu-}\mathbf{cys}\text{-Gly-NHC}_6H_4\text{-}p\text{-F})_2]^{2-}$ in acetonitrile show contact-shifted signals at 50 ppm, 20 ppm and -10 ppm at -30°C similar to those of the above hexapeptide model complex. The ^{19}F NMR spectra of $[Fe^{II}(Z\text{-}\mathbf{cys}\text{-Pro-Leu-}\mathbf{cys}\text{-Gly-NHC}_6H_4\text{-}p\text{-F})_2]^{2-}$ show signals at -122.6 ppm and -122.9 ppm (free liagnd ^{19}F signal at 199.9 ppm) due to the presence of two isomers (δ and λ) that formed by the two types of coordinations with the two chelating Cys-X-Y-Cys fragments [15, 17]. The corresponding *m*-fluoro derivative, $[Fe^{II}(Z\text{-}\mathbf{cys}\text{-Pro-Leu-}\mathbf{cys}\text{-Gly-NHC}_6H_4\text{-}m\text{-F})_2]^{2-}$, exhibits the corresponding ^{19}F signals at -103.5 ppm and -108.3 ppm (free ligand ^{19}F signal at -114.0 ppm). Each signal for the p-^{19}F and m-^{19}F in the two complexes shows the opposite shift by paramagnetic Fe(II). Thus, the unpaired electron delocalizes from Fe(II) ion through the NH---S hydrogen bond and then the antibonding MO on the benzene ring and reaches the ^{19}F atom.

In conclusion, the active site of electron transfer metalloproteins is significantly influenced by electrostatic interactions including the NH---S hydrogen bond that is supported by hydrophobic environments. Concurrently, the NH---S hydrogen bond remarkably shifts the redox potential to proceed biologically important redox reactions under mild conditions.

3. Cooperation of an Aromatic Ring with the NH---S Hydrogen Bond to Stabilize the Peptide Model Complexes in Aqueous Micellar Solution

Recently studies have been extensively carried out on the function of an aromatic ring around the active site of metalloproteins [18]. Three functions have been proposed for the metalloproteins. One is the formation of hydrophobic environments to support electrostatic interactions by the aromatic rings around the metal center. Second is the function as an electron transfer mediator by taking a specific orientation among several aromatic and heterocyclic rings. Third is a direct $d\pi-p\pi$ interaction between metal ion and an aromatic ring [19].

Novel peptide thiolate ligands having an isolated aromatic ring without $\pi-\pi$ conjugation between the ring and the amide group were designed and allowed to form Fe(II) complexes. Thus, $(NEt_4)_2[Fe^{II}(Z\text{-}\mathbf{cys}\text{-Pro-Leu-}\mathbf{cys}\text{-Gly-NHCH}_2C_6H_4\text{-}p\text{-F})_2]$ and $(NEt_4)_2[Fe^{II}(Z\text{-}\mathbf{cys}\text{-Pro-Leu-}\mathbf{cys}\text{-Gly-NHCH}_2CH_2C_6H_4\text{-}p\text{-F})_2]$ were synthesized. Similarly, $[Fe^{II}(Z\text{-}\mathbf{cys}\text{-Pro-Leu-}\mathbf{cys}\text{-Gly-Phe-OMe})_2]^{2-}$ containing a peptide fragment containing a phenyl group found in the active site of rubredoxin was also prepared. The ^2H-NMR results indicate the presence of the hydrogen bonds in the same way as that of the above-mentioned hexapeptide model complexes.

Table 2 lists the absorption maxima and CD extrema of these complexes in acetonitrile. The introduction of the aromatic ring in the peptide complexes results in appearance of a new LMCT band at 330 ~ 332 nm. This band corresponds to the absorption maximum at 333 nm reported for native reduced rubredoxin. When the peptide ligand has an aromatic ring in a suitable place, remarkably large $\Delta\varepsilon$ CD extrema for these peptide complexes are observed. The complex, [Fe^{II}(Z-**cys**-Pro-Leu-**cys**-Gly-Phe-OMe)$_2$]$^{2-}$, exhibits a large $\Delta\varepsilon$ (-27) similar to the value ($\Delta\varepsilon$ = -36) of reduced rubredoxin. Most of synthetic peptide model complexes have shown only small $\Delta\varepsilon$ extrema that have been considered to be ascribed to the non-rigidity of the peptide conformation around the metal center. However, the aromatic ring cooperates with the NH---S hydrogen bond to give a rigid structure even in the short peptide ligand. The formation of NH---S hydrogen bonds is also supported by the contact-shifted N^{2}H NMR signals at 40.5, 35.7, 18.8, -2.4 and -4.6 ppm in acetonitrile at 30°C. The temperature dependence of chemical shifts of these signals is subject to a Curie-Weiss law. This behavior indicates that the distant amide NH forms a NH---S hydrogen bond with Cys thiolate. The adjacent aromatic ring on the Phe residue is thought to interact directly with the hydrogen bonded sulfur atom.

TABLE 2. UV-visible and CD spectral data for Fe(II)-peptide complexes in acetonitrile and rubredoxin in aqueous solution.

Complexes	UV-vis[a]		CD[b]		
[Fe^{II}(Z-**cys**-Gly-Val-OMe)$_4$]$^{2-}$	316 (5300)		320 (-3.5)		341 (1.2)
[Fe^{II}(Z-**cys**-Pro-Leu-**cys**-OMe)$_2$]$^{2-}$	314 (5900)		320 (-6.3)		340 (3.3)
[Fe^{II}(Z-**cys**-Pro-Leu-**cys**-Gly-Val-OMe)$_2$]$^{2-}$	312 (4500)		309 (-5.3)		338 (2.8)
[Fe^{II}(Z-**cys**-Pro-Leu-**cys**-Gly-NHC$_6$H$_4$-*p*-OMe)$_2$]$^{2-}$	311 (4250)	331 (3900)	318 (-7.7)		340 (4.2)
[Fe^{II}(Z-**cys**-Pro-Leu-**cys**-Gly-NHC$_6$H$_5$)$_2$]$^{2-}$	312 (5400)	330 (4400)	307 (-10)	323 (-10)	339 (5.8)
[Fe^{II}(Z-**cys**-Pro-Leu-**cys**-Gly-NHC$_6$H$_4$-*p*-F)$_2$]$^{2-}$	312 (5900)	332 (4800)	308 (-10)	322 (-9.6)	339 (6.9)
[Fe^{II}(Z-**cys**-Pro-Leu-**cys**-Gly-NHC$_6$H$_4$-*p*-CN)$_2$]$^{2-}$	312 (5400)	330 (4600)	311 (-9.9)	323 (-10)	340 (5.8)
[Fe^{II}(Z-**cys**-Pro-Leu-**cys**-Gly-NHCH$_2$C$_6$H$_4$-*p*-F)$_2$]$^{2-}$	312 (6780)	331 (6030)	315 (-12.0)		335 (6.0)
[Fe^{II}(Z-**cys**-Pro-Leu-**cys**-Gly-NHCH$_2$CH$_2$ C$_6$H$_4$-*p*-F)$_2$]$^{2-}$	312 (7400)	330 (6200)	316 (-19.2)		335 (9.0)
[Fe^{II}(Z-**cys**-Pro-Leu-**cys**-Gly-Phe-OMe)$_2$]$^{2-}$	312 (8240)	332 (6750)	316 (-27.0)		334 (16.0)
Reduced rubredoxin[c]	312 (10900)	333 (6000)	314 (-36)		334 (18)

[a] In nm (ε, $M^{-1}cm^{-1}$). [b] In nm ($\Delta\varepsilon$, $M^{-1}cm^{-1}$). [c] Eaton and Lovenberg (1973).

The complexes having no-π-conjugation of phenyl rings with the amide, (NEt$_4$)$_2$[Fe^{II}(Z-**cys**-Pro-Leu-**cys**-Gly-NHCH$_2$C$_6$H$_4$-*p*-F)$_2$] and (NEt$_4$)$_2$[Fe^{II}(Z-**cys**-Pro-Leu-**cys**-Gly-NHCH$_2$CH$_2$C$_6$H$_4$-*p*-F)$_2$] exhibit slightly contact shifted ^{19}F signals due to a dipolar contact which does not obey the Curie-Weiss law [16]. The sulfur atom probably prefers a tetrahedral structure having two types of interactions with the amide NH group and the aromatic ring as shown in Fig. 3.

Although solutions of alkanethiolato Fe(II) complexes are unstable in air, a 10 % Triton X-100 aqueous micellar solution of (NEt$_4$)$_2$[Fe^{II}(Z-**cys**-Pro-Leu-**cys**-Gly-Phe-OMe)$_2$] is found stable to air and remains unchanged upon standing at room temperature. The solution gives a quasi-reversible redox couple at -0.26 V vs SCE under argon. The aromatic ring at the Phe residue cooperating with the NH---S hydrogen bond completely covers the sulfur atom to prevent hydrolysis by water and oxidation by air.

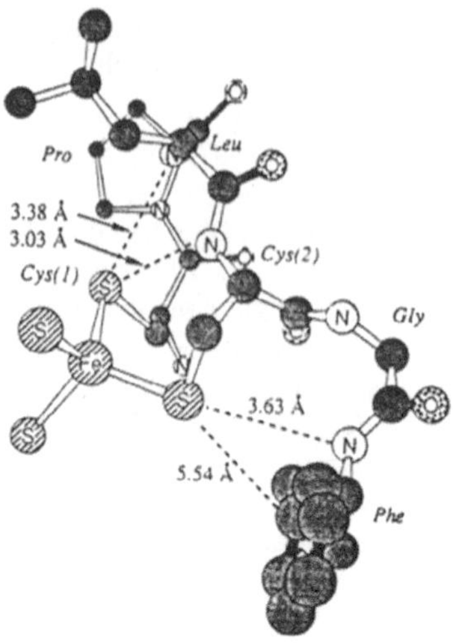

Fig. 3. Schematic structure of $(NEt_4)_2[Fe^{II}(\text{Z-}\mathbf{cys}\text{-Pro-Leu-}\mathbf{cys}\text{-Gly-Phe-OMe})_2]$ obtained by BIOGRAF energy-minimum calculations.

4. Intermolecular Association Between Peptide Rubredoxin Model Complex and Small Additive Molecule

Recently ^1H NMR and X-ray analyses of the active site of metalloproteins and metalloenzymes have demonstrated the change of their solution structure upon association with relevant substrates or other proteins [20, 21]. The structure variation is considered to be associated with unique control of the reactivity. We studied the intermolecular association between a peptide model complex and an artificially designed additive as models of the metalloprotein-protein interaction.

Although a redox partner for native rubredoxin is still unknown, it is easily predicted that the interaction region is located near the Cys-X-Y-Cys chelating chains because of the exposure of these parts to outside. The conformation at these parts thus determines the extent of binding to the redox partner. Previously, we demonstrated that the identity of amino acid residues in the Cys-X-Y-Cys chelating parts significantly contributes to the regulation of the redox potential by the NH---S hydrogen bonding through the unique conformation.

In order to clarify further important roles of the peptide ligand in the electron transfer through rubredoxin, intermolecular association of a rubredoxin peptide model Fe(II) complex with a variety of multiamide additives was studied. The molecular design of various multiamide additives was carried out considering possible orientation of the amide dipoles and typical ones are shown in Fig. 4. These additives are expected to realize the intermolecular association through NH---O=C hydrogen bonds in low-dielectric-constant organic solvent. The formation of the NH---S hydrogen bond is also expected with metal thiolate partners.

The addition of PaDaPy to the peptide complex in 1:2 molar ratio results in the

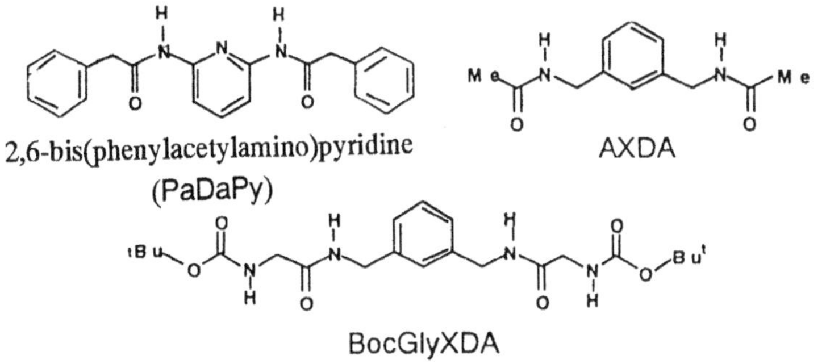

Fig. 4. Multiamide additives designed as redox partners for rubredoxin peptide model Fe(II) complexes.

large shift of the ^{1}H NMR signals of [Fe^{II}(Z-**cys**-Pro-Leu-**cys**-Gly-Val-OMe)$_2$]$^{2-}$ in acetonitrile-d_3 due to formation of the NH---O=C hydrogen bonding through the corresponding change of the peptide ligand conformation. Thus specific intermolecular association was also detected by the NOE measurements in the (1:2) complex. The ^{1}H NMR titration indicates a specific association between PaDaPy NH and peptide CO hydrogen bond. The NOESY spectra also show the presence of NOE between PaDaPy NH and Leu *i*Pr groups.

When PaDaPy was added to a solution of the peptide complex, the redox potential in acetonitrile is found at -0.34 V vs SCE. The value is shifted by 130 mV from the value without PaDaPy. The observed shift is caused by the selective

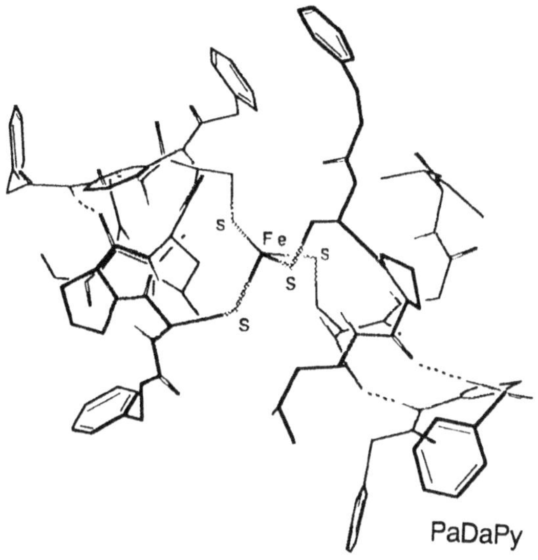

Fig. 5. A proposed structure of the intermolecular association between [Fe^{II}(Z-**cys**-Pro-Leu-**cys**-Gly-Val-OMe)$_2$]$^{2-}$ and PaDaPy.

association of PaDaPy to the peptide ligand since the use of 2,6-diaminopyridine and other amide additives does not induce such a large change which is significant for the biological electron transfer chains.

Most of the electron transfer metalloproteins work in a range of several dozens of mV unit. On the basis of the results of the NOESY spectra, a structure of the intermolecular association complex was speculated using a molecular dynamic simulation. The energy minimum calculations indicate a preferable association structure with two binding sites of the peptide Pro C=O and Leu C=O groups to each of the two amide NH groups of PaDaPy. The calculated structure is shown in Fig. 5. It is interesting that the association structure coincides with that obtained from the NOESY results.

In the ^{2}H NMR study, a remarkable shift of amide N^{2}H NMR signals of the peptide ligand was observed upon addition of PaDaPy. The results indicate that the association changes the strength of the NH---S hydrogen bonds in the peptide ligand. The association also contributes to the positive shift of redox potential. Thus, the intermolecular association by a rigid multiamide molecule induces a conformational change of the peptide ligand to regulate the electrochemical properties even in a small peptide model complexes.

5. Intramolecularly Single and Double NH---S Hydrogen Bonded Model Complexes

In order to study a more detailed aspects on the role of the NH---S hydrogen bond, novel metal thiolate complexes intramolecular singly or doubly hydrogen bonded ligands were designed as shown in Fig. 6. Mononuclear Fe(II) and Co(II) complexes having the above thiolate ligands, e.g. $(NEt_4)_2[Fe^{II}(2\text{-}t\text{-}BuCONHC_6H_4)_4]$ and $(NEt_4)_2[Co^{II}(2\text{-}t\text{-}BuCONHC_6H_4)_4]$, were synthesized as simple peptide-mimetic models of rubredoxin [11]. The presence of intermolecular NH---S hydrogen bonds between thiolate and amide cation in the solid state has been reported for $(Me_3NCH_2CONH_2)_2[Co^{II}(SPh)_4]$ [22].

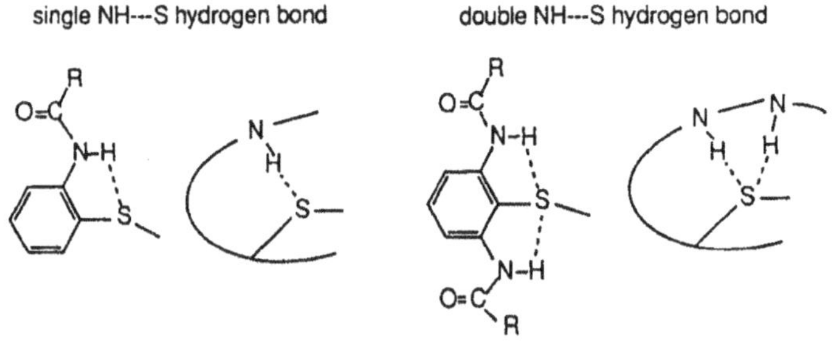

Fig. 6. Design of intramolecularly singly and doubly NH---S hydrogen bonded arenethiolate ligands.

The crystal structures of $(NEt_4)_2[Fe^{II}(2\text{-}t\text{-}BuCONHC_6H_4)_4]$ and $(NEt_4)_2[Co^{II}(2\text{-}t\text{-}BuCONHC_6H_4)_4]$ show a *Td* geometry for the MS_4 cores with the unique location of each NH group on the sulfur atoms as shown in Fig. 7.

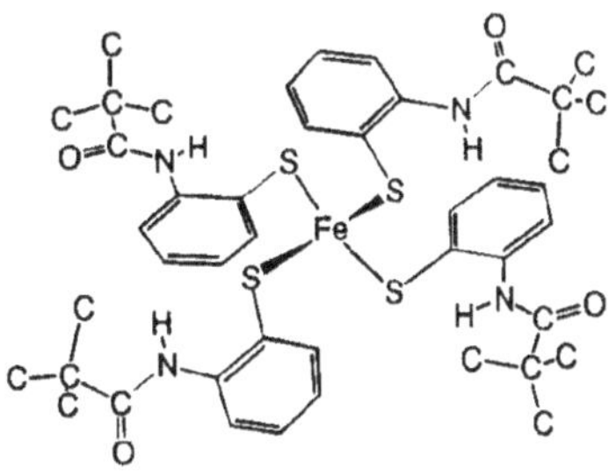

Fig. 7. Structure of $[Fe^{II}(2\text{-}t\text{-}BuCONHC_6H_4)_4]^{2-}$.

Each of the bond distances of Fe-S (2.329 Å) and Co-S (2.296 Å) in these complexes is shorter than that of the corresponding $[Fe^{II}(SPh)_4]^{2-}$ (2.353 Å) [23] and $[Co^{II}(SPh)_4]^{2-}$ (2.328 Å) [24] cases. The observed shortening at the metal-sulfur bonds can be explained as follows. The singly-occupied orbitals (SOMO) in these complexes are of mostly d-character and are antibonding combination of metal d and sulfur p orbitals. As the NH---S hydrogen bonding occurs to the sulfur p-orbital, the electron-density of the SOMO will be decreased depending on the strength of the hydrogen bond.

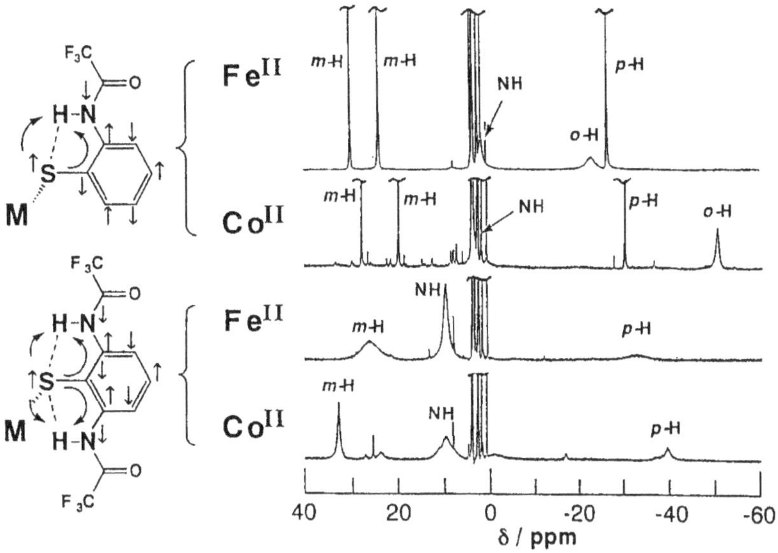

Fig. 8. ^1H NMR spectra of $(NMe_4)_2[M^{II}(2\text{-}CF_3CONHC_6H_4)_4]$ (M = Fe(II), Co(II)) and $(NMe_4)_2[M^{II}\{2,6\text{-}(CF_3CONH)_2C_6H_3\}_4]$ in acetonitrile-d_3 at 30°C.

The ^1H-NMR spectra of various Fe(II) and Co(II) complexes in CD_3CN are shown in Fig. 8. The isotropic contact shift of each proton signal is negative for *o*- and *p*-protons and positive for *m*-protons in comparison with those of the corresponding disulfide. The contact shift of the amide NH signal at *o*-position in $(NEt_4)_2[M^{II}(2\text{-}t\text{-}BuCONHC_6H_4)_4]$ (M = Fe(II), Co(II)) is positive whereas that at *p*-position in $(NEt_4)_2[Fe^{II}(4\text{-}t\text{-}BuCONHC_6H_4)_4]$ is negative. When the NH---S hydrogen bond is absent in $[Fe^{II}(2\text{-}t\text{-}BuCONHC_6H_4)_4]^{2-}$, the isotropic contact shift of the NH signal is expected to be positive due to a Fermi contact mechanism since the spin density is distributed on the conjugated benzene and amide groups in opposite sign. Actually, $(NEt_4)_2[Co^{II}$(*o*-methylbenzenethiolato)$_4]$ shows a methyl proton signals at $(\Delta H/Ho)^{iso}$ = 24.5 ppm in acetonitrile-d_3 at 30°C. The observed negative value for $[Fe^{II}(2\text{-}t\text{-}BuCONHC_6H_4)_4]^{2-}$ indicates that the amide NH proton at *o*-position is directly influenced by a Fermi contact from the sulfur atom. The results also support the presence of NH---S hydrogen bond.

The IR spectrum of $(NEt_4)_2[Co^{II}(2\text{-}t\text{-}BuCONHC_6H_4)_4]$ in the solid state shows an amide NH band at 3281 cm^{-1} and a CO band at 1667 cm^{-1}. The corresponding disulfide, bis(2-*t*-$BuCONHC_6H_4$) disulfide, in the solid state gives a free NH band at 3389 cm^{-1} and an intermolecularly NH---O=C hydrogen bonded amide NH band at 3251 cm^{-1} separately. On the other hand, a free amide CO band and an intermolecularly hydrogen bonded CO were observed at 1679 cm^{-1} and 1644 cm^{-1}, respectively. Lack of these bands for the Co(II) complex indicates the absence of both the free NH and the intermolecularly NH---O=C hydrogen bonded amide NH bands. The band at 3281 cm^{-1} is assignable to the amide NH stretching in the NH---S hydrogen bond.

$(NEt_4)_2[Fe^{II}(2\text{-}t\text{-}BuCONHC_6H_4)_4]$ exhibits a redox potential at -0.29 V vs. SCE in acetonitrile. The positive shift by the NH---S hydrogen bonding is 0.24 V as compared with the value, -0.53 V vs SCE, for $(NEt_4)_2[Fe^{II}(4\text{-}t\text{-}BuCONHC_6H_4)_4]$.

Similarly, an Fe(II) thiolate complex, $[Fe^{II}\{S\text{-}2,6\text{-}(CH_3CONH)_2C_6H_3\}_4]^{2-}$, having intramolecular double NH---S hydrogen bonds was synthesized and the crystal structure was determined. All amide NH groups are directed to the sulfur atoms. The presence of NH---S hydrogen bonds was also confirmed by the IR spectra in the solid state. When compared to that of $[Fe^{II}(SPh)_4]^{2-}$, the Fe-S bond distance of the complex having double NH---S hydrogen bonds is shortened by approximately 0.02 Å. The NH---S hydrogen bonding is considered to decrease the electron density of Fe-S dπ-pπ SOMO as described above.

The Fe(II) complex, $(NEt_4)_2[Fe^{II}\{S\text{-}2,6\text{-}(CH_3CONH)_2C_6H_3\}_4]$, having double NH---S hydrogen bonds exhibits an extremely positive shifted redox potential at +0.03 V vs SCE in acetonitrile. The redox potential shifts according to the number of NH---S hydrogen bonds in a linear fashion. Interestingly, only a slight isotropic shift (0.05 ppm) of amide NH ^1H NMR signal is observed for the doubly NH---S hydrogen bonded complex.

A molybdenum(IV) and a copper(I) complexes, $(PPh_4)_2[Mo^{IV}(S\text{-}2\text{-}CH_3CONHC_6H_4)_4]$ [10] and $(NEt_4)_2[Cu^{I}(S\text{-}2\text{-}t\text{-}BuCONHC_6H_4)_3]$ [12] were synthesized and the crystal structure was determined by the X-ray analysis. The crystal data also indicate the shortening of the M-S (M = Mo(IV), Cu(I)) bond distances by 0.02 Å. The large IR shift of the amide NH stretching in $(NEt_4)_2[Cu^{I}(S\text{-}2\text{-}t\text{-}BuCONHC_6H_4)_3]$ indicates that the NH---S hydrogen bond is stronger than that of a

NH---O=C hydrogen bond. Table 3 lists the shifts of the amide NH bands in various metal complexes from the free amide NH stretching in the corresponding disulfides.

TABLE 3. The amide NH IR band shift by the NH---S hydrogen bond of various intramolecular hydrogen bonded metal complexes (SAr = S-2-*t*-BuCONHC$_6$H$_4$) [25].

	IR/ cm^{-1}			
Complexes	ν(NH)	Δν(NH)[a]	ν(C=O)	Δν(C=O)[a]
[Hg^{II}(SAr)$_2$]	3345	-44	1665	-14
[Mo^{V}O(SAr)$_4$]$^-$	3330	-59	1673	-6
[Co^{II}(SAr)$_4$]$^{2-}$	3281	-108	1667	-12
[Cd^{II}(SAr)$_4$]$^{2-}$	3271	-118	1667	-12
[Cu^{I}(SAr)$_3$]$^{2-}$	3226	-163	1667	-12
cf. NH---O=C	3251	-148	1644	-35

[a] The difference from free ν(NH) (3389 cm^{-1}) and ν(C=O) (1679 cm^{-1}) in ArS-SAr.

6. NH---S Hydrogen Bond in [2Fe-2S] and [4Fe-4S] Ferredoxin Model Complexes

A peptide [2Fe-2S] ferredoxin model complex, (NEt$_4$)$_2$[Fe$_2$S$_2$(20-pep)] (20-pep = Ac-Pro-Tyr-Ser-Cys-Arg-Ala-Gly-Ala-Cys-Ser-Thr-Cys-Ala-Gly-Pro-Leu-Leu-Thr-Cys-Val-NH$_2$) containing an invariant amino acid residues around the active center of plant-type ferredoxins has exhibited a similar redox potential to that of native ferredoxins [8]. The positive shift of the redox potential has been proposed to be ascribed to the presence of single or double NH---S hydrogen bonds as shown by X-ray analysis of *S. platensis* ferredoxin [26].

The ^{1}H NMR spectrum of (NEt$_4$)$_2$[Fe$_2$S$_2$(S-2-*t*-BuCONHC$_6$H$_4$)$_4$] in acetonitrile-d_3 at 30°C shows slightly contact shifted signals of the 2-pivaloylamino-benzenethiolate ligand. The peaks are observed at 10.9 ppm for 3-ArH, 8.9 ppm for 5-ArH, 4.6 ppm (broad) for 6-ArH, 3.1 ppm for 4-ArH and 6.8 ppm (broad) for amide NH. The NH signal was assigned by the disappearance of the signal on deuteration at the amide group. These isotropic shifts are similar to those of the reported model complexes [27, 28]. When unpaired spin density distributes over the conjugated S pπ, aromatic pπ and amide pπ orbitals, the amide NH signal would appear with a plus isotropic shift in low field. The observed minus shift is ascribed to a direct bond formation between amide NH and sulfur atom and the proton sign is governed by the spin sign (α-spin or β-spin) of the sulfur atom.

In the case of double NH---S hydrogen bonded complexes, $(NEt_4)_2[Fe_2S_2\{S\text{-}2,6\text{-}(t\text{-}BuCONH)_2C_6H_4\}_4]$, the isotropic shifts are small and similar in magnitude to those of the singly NH---S hydrogen bonded complexes. It is likely that unpaired electron density on the sulfur atom is distributed into the two amide NH groups to decrease the isotropic shift. Inherently, the spin sign by the NH---S hydrogen bonding competes with the opposite spin sign by the organic π-system comprized of sulfur-arene-amide groups.

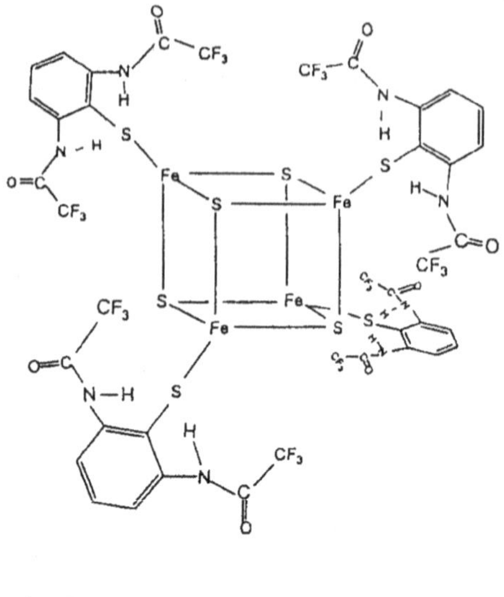

Fig. 9. Crystal Structure of $(NEt_4)_2[Fe_4S_4\{S\text{-}2,6\text{-}(CF_3CONH)_2C_6H_3\}_4]$.

Single and double NH---S hydrogen bonds have been found in bacterial ferredoxins as already reported with the X-ray analysis. Thus, [4Fe-4S] model complexes, $(NEt_4)_2[Fe_4S_4\{S\text{-}2,6\text{-}(RCONH)_2C_6H_3\}_4]$ (R = CH_3, CF_3), are synthesized and the crystal structures were determined as shown in Fig. 9. The positive shifted redox potentials also were found for $(NEt_4)_2[Fe_4S_4\{S\text{-}2,6\text{-}(CH_3CONH)_2C_6H_3\}_4]$ (-0.72 V vs SCE in acetonitrile) and $(NEt_4)_2[Fe_4S_4\{S\text{-}2,6\text{-}(CF_3CONH)_2C_6H_3\}_4]$ (-0.62 V SCE in acetonitrile). The presence of NH---S hydrogen bond was also established by the IR spectral analysis.

A similar [4Fe-4S] model with a single NH---S hydrogen bond, $(NEt_4)_2[Fe_4S_4(S\text{-}2\text{-}CH_3CONHC_6H_4)_4]$, exhibits a broad NH ^1H NMR signal at 8.5 ppm in acetonitrile-d_3. The NH signal shows a small isotropic contact-shift. Also in the case of doubly NH---S hydrogen bonded complex, $(NEt_4)_2[Fe_4S_4\{S\text{-}2,6\text{-}(CH_3CONH)_2C_6H_3\}_4]$, only a slight isotropic-shifted signal is found at 8.23 ppm for the amide NH in acetonitrile-d_3. Other aromatic proton signals are substantially similar to those of *p*-substituted benzenethiolate complexes reported by Holm and his coworkers [27, 29].

Detailed studies for the NH---S hydrogen bond has benn hampared in metalloprotein because of the presence of many kinds of amide groups. Through the previous IR , Raman and NMR spectroscopic analyses, the function has been expected to be significant. Our IR, ^1H and ^2H NMR spectroscopic and electrochemical results for the Cys-containing peptide and simple arenethiolate [1Fe], [2Fe-2S] and [4Fe-4S] model complexes thus revealed a variety of the chemical functions by the NH---S hydrogen bond.

7. References

1. Watenpaugh, K.D., Sieker, L.C. and Jensen, L.H. (1979) The structure of rubredoxin at 1.2 Å resolution, *J. Mol. Biol.* **131**, 509-522.
2. Fukuyama, K., Hase, T., Matsumoto, S., Tsukihara, T., Katsube, Y., Tanaka, N., Kakudo, M., Wada, K. and Matsubara, H. (1980) Structure of *S. platensis* [2Fe-2S] ferredoxin and evolution of chloroplast-type ferredoxins, *Nature,* **286**, 522-524.
3. Adman, E.T., Watenpaugh, K.D. and Jensen, L.H. (1975) NH---S hydrogen bonds in *Peptococcus aerogenes* ferredoxin, *Clostridium pasteurianum* rubredoxin and *Chromtium* high potential iron protein, *Proc. Natl. Acad. Sci. U.S.A.* **72**, 4854.
4. Baker, E.N. (1988) Structure of azurin from *Alcaligenes denificans* refinement at 1.8 Å resolution and comparison of the two crystallograhically independent molecules, *J. Mol. Biol.* **203**, 1071-1095.
5. Ueyama, N., Nakata, M., Fuji, M., Terakawa, T. and Nakamura, A. (1985) Analogues of reduced rubredoxin: positive shifts of redox potentials of cysteine-containing peptide Fe(II) complexes.", *Inorg. Chem.* **24**, 2190-2196.
6. Sun, W., Ueyama, N. and Nakamura, A. (1991) Reduced rubredoxin models containing Z-Cys-Pro-Leu-Cys-Gly-NH-C_6H_4-*p*-X (X = MeO, H, F, CN): Electronic influence by a distant para substitutent through NH---S hydrogen bonds, *Inorg. Chem.* **30**, 4026-4031.
7. Ueyama, N., Sun, W. -Y. and Nakamura, A. (1992) Evidence for intramolecular NH---S hydrogen bonds from ^{2}H-NMR spectroscopy of reduced-rubredoxin model Fe(II) complexes with bidentate peptide ligands, *Inorg. Chem.* **31**, 4053-4057.
8. Ueyama, N., Ueno, S., Nakamura, A., Wada, K., Matsubara, H., Kumagai, S., Sakakibara, S. and Tsukihara, T. (1992) A synthetic analogue for the active site of plant-type ferredoxin: Two different coordination isomers by a four-Cys-containing [20]-peptide, *Biopolymers* **32**, 1535-1544.
9. Ohno, R., Ueyama, N. and Nakamura, A. (1991) Influence of the distal para substituent through NH---S hydrogen bonds on the positive shift of the reduction potentials of $[Fe_4S_4$(Z-cys-Gly-*p*-substituted anilide)$_4]^{2-}$ (X = H, OMe, F, Cl, CN) complexes, *Inorg. Chem.* **30**, 4887-4891.
10. Ueyama, N., Okamura, T. and Nakamura, A. (1992) Structure and properties of molybdenum(IV,V) arenethiolates with a neighboring amide group. Significant contribution of NH---S hydrogen bond to the positive shift of redox potential of Mo(V)/Mo(IV), *J. Am. Chem. Soc.* **114**, 8129-8137.
11. Ueyama, N., Okamura, T. and Nakamura, A. (1992) Intramolecular NH---S hydrogen bond in *o*-Acylamino substituted benzenethiolate iron(II) and cobalt(II) complexes.", *J. Chem. Soc. Chem. Commun.* 1019-1020.
12. Okamura, T., Ueyama, N., Nakamura, A., Ainscough, E.W., Brodie, A.M. and Waters, J.M. (1993) The effect of strong NH---S hydrogen bonds in the copper(I) thiolate complex, $(NEt_4)_2[Cu(o\text{-pabt})_3]$ (*o*-pabt = *o*-pivaloylaminobenzenethiolato), *J. Chem. Soc. Chem. Commun.* 1658-1659.
13. Lovenberg, W. and Sobel, B.E. (1965) Rubredoxin: a new electron transfer protein from *Clostridium pasteurianum*, *Proc. Natl. Acad. Sci.* **54**, 193-199.
14. Eaton, W.A. and Lovenberg, W. (1973) The iron-sulfur complex in rubredoxin in Lovenberg, W (ed), *Iron-Sulfur Proteins II*, Academic Press, New York, pp131-162
15. Sun, W., Ueyama, N. and Nakamura, A. (1993) Spin-lattice relaxation time and temperature depedence of fluorine-19 nuclear magnetic resonance spectra of cysteine-containing peptide iron(II) complexes, *J. Chem. Soc. Dalton Trans.* 1871-1874.

16. Sun, W., Ueyama, N. and Nakamura, A. (1993) An electronic influence of a distant aromatic ring in reduced rubredoxin models. Iron(II) complexes with Z-Cys-Pro-Leu-Cys-Gly-X (X = $NHCH_2C_6H_4$-*p*-F, $NHCH_2CH_2C_6H_4$-*p*-F, and Phe-OMe), *Inorg. Chem.* **32**, 1095-1100.
17. Sun, W., Ueyama, N. and Nakamura, A. (1993) ^1H, ^2H and ^{19}F NMR spectroscopy of a novel iron(II) tetrapeptide cysteine-peptide complex having only one δ isomer. Detection of two isomers in coordination of Cys-X-Y-Cys bidentate peptide ligands to iron(II) ion, *Mag. Res. Chem.* **31**, S34-40.
18. Burley, S.K. and Petsko, G.A. (1988) Weakly Polar Interaction in Proteins, *Adv. Protein Chem.* **39**, 125-189.
19. Yamauchi, O. and Odani, A. (1985) Structure-stability relationship in ternary copper(II) complexes involving aromatic amines and tyrosine or related amino acids. Intramolecular aromatic ring stacking and its regulation through tyrosine phosphorylation, *J. Am. Chem. Soc.* **107**, 5938-5945.
20. Mohammadi, M., Honegger, A., Sorokin, A., Ullrich, A., Schlessinger, J. and Hurwitz, D.R. (1993) Aggregation-induced activation of the epidermal growth factor receptor protein tyrosine kinase, *Biochemistry* **32**, 8742-8748.
21. Camacho, N.P., Smith, D.R., Goldman, A., Schneider, B., Green, D., Young, P.R. and Berman, H.M. (1993) Structure of an interleukin-1β mutant with reduced bioactivity shows multiple subtle changes in conformation that affect protein-protein recognition, *Biochemistry* **32**, 8749-8757.
22. Walters, M.A., Dewan, J.C., Min, C. and Pinto, S. (1991) Models of amide-cysteine hydrogen bonding in rubredoxin: hydrogen bonding between amide and benzenethiolate in $[(CH_3)_3NCH_2CONH_2]_2[Co(SC_6H_5)_4]\cdot 1/2CH_3CN$ and $[(CH_3)_3NHCH_2CONH_2][SC_6H_5]$, *Inorg. Chem.* **30**, 2656-2662.
23. Coucouvanis, D., Swenson, D., Baenziger, N.C., Murphy, C., Holah, D.G., Sfarnas, N., Simopoulos, A. and Kostikas, A. (1981) Tetrahedral complexes containing of the $Fe^{II}S_4$ core. The synthesis, ground-state electronic structure, and crystal and molecular structures of the $[P(C_6H_5)_4]_2Fe(S_2C_4O_2)_2$ complexes. An analogue for the active site in reduced rubredoxins (Rdred), *J. Am. Chem. Soc.* **103**, 3350-3362.
24. Swenson, D., Baenzinger, N.C. and Coucouvanis, D. (1978) Tetrahedral mercaptide complexes. Crystal and molecular structures of $[(C_6H_5)_4P]_2M(SC_6H_5)_4$ complexes (M = Cd(II), Zn(II), Ni(II), Co(II) and Mn(II)), *J. Am. Chem. Soc.* **100**, 1932-1934.
25. Ueyama, N., Taniuchi, K., Okamura, T. A.Nakamura, (1994) Effect of NH---S hydrogen bond on the nature of Hg-S bonding in mono-*o*-acylamino- and di-*o,o'*-diacylaminobenzenethiolato mercury(II) complexes, in preparation.
26. Tsukihara, T., Fukuyama, K., Tahara, H., Katsube, Y., Matsuura, Y., Tanaka, N., Kakudo, M., Wada, K. and Matsubara, H. (1978) X-Ray analysis of ferredoxin from *Spirulina platensis*. II. Chelate structure of active center, *J. Biochem.* **84**, 1645.
27. Holm, R.H., Phillips, W.D., Averill, B.A. and Mayerle, J.J., Herskovitz, T. (1974) Synthetic analogs of the active sites of iron-sulfur proteins. V. Proton resonance properties of the tetranuclear clusters $[Fe_4S_4(SR)_4]^{2-}$. Evidence for dominant contact interactions, *J. Am. Chem. Soc.* **96**, 2109-2117.

28. Hagen, K.S., Reynolds, J.G. and Holm, R.H. (1981) Definition of reaction sequence resulting in self-assembly of $[Fe_4S_4(SR)_4]^{2-}$ clusters from simple reactions, *J. Am. Chem. Soc.* **103**, 4054-4063.
29. Que, Jr. L., Bobrik, M.A., Ibers, J.A. and Holm, R.H. (1974) Synthetic analogs of the active sites of iron-sulfur proteins. V. Ligand substitution reaction of the tetranuclear clusters $[Fe_4S_4(SPh)_4]^{2-}$ and the structure of $[(CH_3)_4N]_2[Fe_4S_4(SC_6H_5)_4]$, *J. Am. Chem. Soc.* **96**, 4168-4178.

3D STRUCTURE OF HiPIPs IN SOLUTION THROUGH NMR AND MOLECULAR DYNAMICS STUDIES

LUCIA BANCI and ROBERTA PIERATTELLI
Department of Chemistry
University of Florence
Via G. Capponi, 7
50121 Florence
Italy

1. Introduction

The detailed structural characterization of a protein is a fundamental step for the understanding of its biological function. Such characterization can be carried out through X-ray studies. However, the latter have to be performed in the solid state and consequently information on the structural and dynamic features of protein in solution is not achieved. Furthermore, in several cases single crystals might not be available due to the difficulties of crystallization or to the instability of the compound, especially when protein-substrate adducts are to be analyzed.

In these cases molecular dynamics (MD) calculations are quite powerful in determining structural models of macromolecules in solution. These calculations are now extensively applied to proteins, nucleic acids and other biological molecules and are used to characterize the reaction mechanisms by modeling the protein-substrate adducts [1-3]. They are also commonly used for the refinement of X-ray structures [4].

MD calculations are based on the determination of the positions and of the trajectories of atoms when they are experiencing the force field produced by all the other atoms in the molecule. Despite the extensive use of MD calculations on biological molecules, their application to metalloproteins is still not common due to the intrinsic difficulties in the treatment of the force-field parameters of the metal ions. Indeed, while extensive tests for the force-field parameters associated with atoms present in amino acids have been performed over more than ten years by several research groups, the problem of the parameters for the metal ions has been tackled only relatively recently [5-10], with the exception of a very few cases [11,12].

In addition, the parameters for describing the metal ion and its ligands strongly depend on the coordination geometries of the metal itself. This means that these parameters cannot be derived only once for one system and then applied without careful checks to different metalloproteins.

The problem becomes even more complex when open-shell metal ions are considered for which accurate calculations are necessary in order to determine the point charges on the metal ion and its ligands. Due to the intrinsic difficulties of the

G.N. La Mar (ed.), Nuclear Magnetic Resonance of Paramagnetic Macromolecules, 281-296.

derivation of the force field parameters, only recently have MD studies been reported on systems containing clusters of metal ions [8,13].

One of the classes of proteins, on which extensive spectroscopic and theoretical studies have been performed in our Laboratory, is the high potential iron sulfur proteins (HiPIPs). They have a MW in the range of 8000-12000 and contain a Fe_4S_4 cluster characterized by a reduction potential ranging between +50 to +450 mV [14].

The electronic structure of such systems has been extensively investigated over the last 20 years by means of Mössbauer [15,16], EPR [17,18] and ENDOR [19-21] spectroscopies, model compounds [21-24] and theoretical studies [25-30]. The oxidized form of HiPIPs formally contains two pairs of iron ions, the mixed valence (Fe^{3+}-Fe^{2+}) pair and the ferric (Fe^{3+}-Fe^{3+}) pair. The total spin of the cluster, S=1/2, is the result of the antiferromagnetic coupling between the larger subspin S_{34}, associated with the mixed-valence pair, and the smaller subspin S_{12}, associated with the ferric pair. In the reduced form the $[Fe_4S_4]^{2+}$ core is formally constituted by $2Fe^{3+}$ and $2Fe^{2+}$ ions. Mössbauer spectroscopy indicates the presence of two mixed-valence pairs, antiferromagnetically coupled to give an S=0 ground state [31,32].

The X-ray structure is available now for four proteins of this class [33-36] while NMR studies have been successfully applied to a much larger number of proteins [13,37-41]. Structural information is needed to rationalize several properties, and in particular the reduction potentials which span over a range relatively large for a series of very similar proteins. With this aim, MD calculations turn out to be quite useful for obtaining structural models.

2. MD calculations on HiPIPs

We have initially applied MD calculations to HiPIP from *Chromatium vinosum* (Cv) [8], for which the X-ray structure was available [33] and for which a detailed NMR characterization was reported for the cysteinic ligands and for other residues around the cluster [37,41,42]. These calculations allowed us to test the force field parameters and the atomic point charges of the cluster and the entire calculation procedure.

As far as the point charges are concerned, we took advantage of *ab initio* Hartree-Fock self-consistent field calculations on model compounds containing a Fe_2S_2 moiety [43]. They were performed on both the oxidized and reduced forms. From these calculations we learned that, upon reduction, the extra electron is essentially delocalized on the iron and on the inorganic sulfur atoms while negligible negative charge is transferred on the cysteine ligands. Among the other parameters, the most relevant is represented by the bending force constants involving the atoms of the cluster which are responsible for maintaining the correct cubane-like structure of the cluster itself. For these force constants, values much larger than those found experimentally from the analysis of Resonance Raman spectra [44] resulted to be necessary. The force constants used (230.1 $kJmol^{-1}rad^{-2}$) are lower than the value (284.7 $kJmol^{-1}rad^{-2}$) for the C-S-S angle of the cystine moiety [45]. No torsional barriers were needed to maintain the cluster structure as the bending forces themselves were sufficient. The non-bonded interactions were correctly described, as shown by the comparison between the MD and the X-ray structures, by parameters already used for iron in other iron-containing proteins [11].

The application of MD calculations to HiPIP from Cv, starting from the X-ray structure, leads to a good refinement of the structure in solution. For this protein extensive NOE studies were available [37,41,42] from which some discrepancies between the X-ray structure and the NOE distances were detected for the region around the cluster [37]. The MD structure is in very good agreement with the experimental structural data obtained through NOEs (Figure 1). This indicates that the MD calculations have been able to reproduce the residue arrangement in solution and that the set of parameters and the overall procedure for the calculations are suitable for the simulation of this class of proteins.

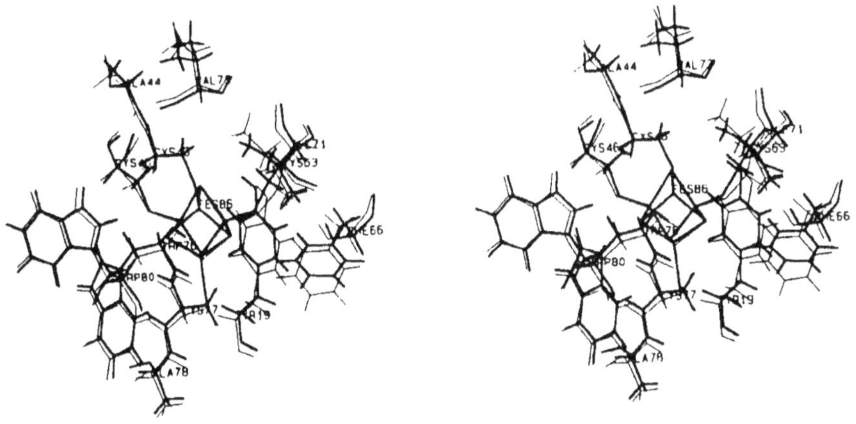

irradiated signal	observed NOE	dist (Å) X-ray	expected NOE (X-ray)	experimental NOE	expected NOE (MD)	dist (Å) MD
Hβ1 Cys 43	Hα Val 73	2.5	medium	very strong	very strong	2.2
Hβ1 Cys 46	Hα Cys 43	2.9	medium	strong	strong	2.4
Hβ1 Cys 63	Hβ2 Phe 66	3.4	weak	very weak	very weak	4.7
Hβ2 Cys 63	Hβ2 Phe 66	1.8	very strong	medium	medium	3.0
Hβ1 Cys 63	Hγ1 Ile 71	1.9	very strong	medium	medium	2.8
Hβ2 Cys 63	Hγ1 Ile 71	3.4	weak	medium	medium	2.9
Hβ1 Cys 77	Hβ1 Tyr 19	2.1	very strong	weak	medium	2.9
Hβ1 Cys 77	Hβ2 Tyr 19	2.3	strong	medium	medium	2.6
Hβ1 Cys 77	HN Ala 70	2.8	medium	strong	strong	2.4

Figure 1 - Upper part. Comparison between some residues in the surrounding of the Fe_4S_4 cluster in the X-ray structure (thin line) and the MD average structure (thick line) of HiPIP from *C.vinosum*. Lower part. Comparison between selected X-ray and MD-derived distances together with experimental and expected NOE intensity. Taken from Ref. [8] with modifications.

Having shown the validity of our approach we then applied MD calculations to several other proteins for which the X-ray structure is not available or it is at low resolution, as in the case of *E. halophila* HiPIP iso-I [13 and L.B. et al., unpublished results]. The resulting structural models from MD calculations were compared with the structural data obtained from NMR in the vicinity of the cluster [13,37-40,46,47]. For all the proteins these models are in complete agreement with all the NMR data. The consistency between the NMR data and the MD structural models is quite fascinating and makes us confident in our models. From the analysis of the MD simulations and of the structures obtained we have learned that these proteins are quite rigid and that the region around the cluster is very similar in all the proteins studied up to now. The cluster moiety seems to "organize" the protein around itself and to determine a well defined structure. In particular, in all the proteins five H-bonds are present between NH protons and sulfur atoms bound to the iron ions. Also the arrangement of some aromatic residues in the vicinity of the cluster results to be similar in all the proteins.

3. Structure determination in solution through NMR

In the last decade NMR spectroscopy has come to be the most powerful experimental technique for obtaining protein structures in solution [48-51]. It is now possible to obtain the solution structure of proteins with MW up to 15000-20000 at high resolution [52,53]. This result can be accomplished through the analysis of multidimensional NMR experiments which lead to the sequence-specific assignment of the spectrum, and the detection and quantification of proton-proton distances and of other structural constraints [54]. Once the structural constraints are obtained, the three-dimensional structure can be determined using distance-geometry (DG) algorithms and then refined using MD methods.

High resolution NMR can be applied also to paramagnetic proteins once the nature and the properties of a paramagnetic molecule (i.e. fast nuclear relaxation rates and large chemical shift spreading) are taken into account.

In paramagnetic metalloproteins the signals from the residues interacting with the paramagnetic center can be sizably broadened and can escape detection in the standard multidimensional NMR experiments [55-57]. To overcome this problem, suitable techniques to detect connectivities involving hyperfine-shifted signals have been developed [58,59]. NMR methodology has now advanced to a point where it is possible to assign residues both in the diamagnetic and paramagnetic parts of the protein, including those residues which experience some paramagnetic effects on relaxation but are not hyperfine shifted outside the diamagnetic region.

The complete sequence specific assignment of the NMR spectrum of a paramagnetic protein can be obtained following these general guidelines, which are sketchily represented in Figure 2:

1) The nuclei which are far from the paramagnetic center are not affected by its presence and their NMR parameters fall in the "usual" range. These nuclei and the connectivities among them can be, therefore, detected by applying the standard multidimensional NMR techniques [60].

2) The nuclei affected by the coupling with the unpaired electrons, which usually are in a sphere of about 8 Å radius around the metal ion, are shifted outside the diamagnetic envelope and can be sizably broadened. Despite these features, multidimensional NMR experiments can be still applied, provided that suitable set-up parameters are used. In particular, short acquisition times in every dimension should be used as magnetization relaxes fast to equilibrium and then only noise is acquired. Furthermore, pulse sequences which do not include too many steps should be preferred in order to limit the extent of magnetization relaxed back to equilibrium.

3) The major problem to be addressed when applying NMR spectroscopy to paramagnetic molecules is constituted by the detection of those signals which are not hyperfine shifted outside the diamagnetic region of the spectrum but which experience some paramagnetic contribution to relaxation and therefore to the linewidths. Due to the properties of these signals, connectivities between them are difficult to be detected and sometimes they can be extracted from the crowd of the diamagnetic cross peaks only playing on the different relaxation properties.

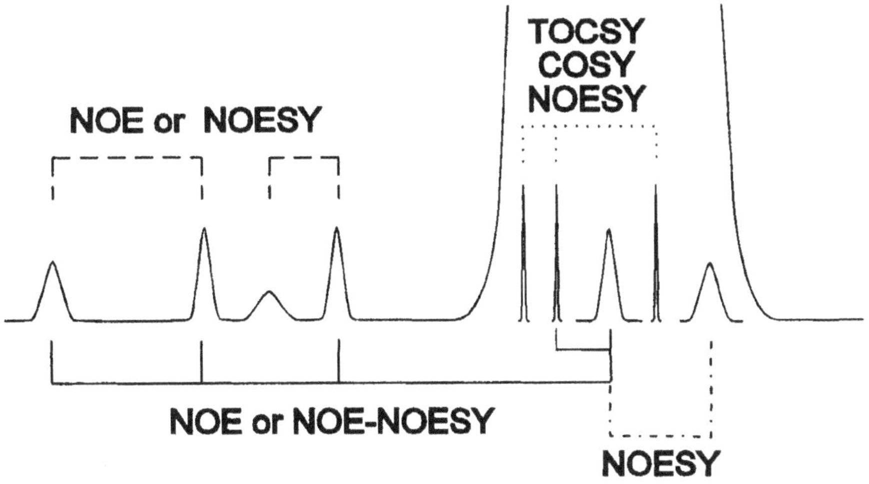

Figure 2 - Sketch of the NMR spectrum of a paramagnetic molecule. The experiments used to [illegible]

4) Finally, connectivities between the fast relaxing and the slow relaxing signals, which are quite relevant for establishing links between the two sets of signals, can be detected with NOESY experiments with the mixing time set close to the reciprocal of the average of the relaxation rates of the two types of signals or with 1D NOE experiments. In the case of very large differences in relaxation rates, 1D NOE experiments, where the fast relaxing signal is saturated and NOE is detected on the slow relaxing signal, are the most useful experiments. It is also useful to edit a NOESY spectrum with a monodimensional NOE difference experiment in a way that, by recording the difference NOESY spectrum, it is possible to obtain information on the signals dipolarly coupled with the fast relaxing ones, despite they are buried in the diamagnetic envelope of the protein [61].

A critical discussion of the experiments for spectral assignment in paramagnetic proteins is reported in the chapter by Luchinat and Piccioli.

Indeed, due to the difficulties in the spectral assignments and the detection of connectivities involving paramagnetic signals, no solution structure for a paramagnetic protein has been solved up to very recently [62].

4. Solution structure of HiPIPs through NMR

We would like to present here an outline of the procedure used in the Nuclear Magnetic Resonance Laboratory of the University of Florence to obtain the first three dimensional structure of a paramagnetic protein through NMR measurements [62]. Our overall procedure can be used by other researchers in the field as a guideline for the determination of the solution structure of other paramagnetic molecules.

We have investigated, through 2D and 3D spectroscopy, the reduced form of the recombinant HiPIP iso-I from *E. halophila* expressed in *E. coli* [63]. The magnitude of the paramagnetism in this protein is such that the T_1 values of the Hβ protons of the iron-bound cysteines are of the order of milliseconds and their linewidths are of the order of hundreds of Hertz [56,64,65]. Despite these features, 71 out of 73 residues were identified, most of which were completely assigned as far as proton resonances are concerned [66].

The conventional sequence specific assignment procedure made use of 2D NMR experiments recorded under the standard conditions employed for diamagnetic macromolecules [48-50,60,67], providing the identification of 62 spin systems in the aliphatic and NH regions. In the aromatic region, cross peak patterns from 10 aromatic rings out of 12 were identified. Most of the spin systems were identified from TOCSY maps obtained with spin-lock times of 45 or 75 ms, performed either in H_2O or D_2O.

Of the 62 identified spin systems, some could be better observed by analyzing TOCSY experiments recorded with shorter spin-lock times (25 ms). Also one of the 10 aromatic patterns was only observed in TOCSY experiments with short spin-lock time (15 ms) and in COSY experiments using appropriate weighting functions. These experiments, therefore, are able to detect connectivities and spin patterns, among residues which are slightly affected by the presence of the paramagnetic center.

The analysis of the dipolar and scalar connectivities observed in the fingerprint region of 2D NOESY and TOCSY, and 3D NOESY-^{15}N HMQC and TOCSY-^{15}N HMQC, provided the identification of six segments of the protein backbone, giving the sequence-specific assignment of 60 residues (Figure 3). The largest segment was constituted by 33 amino acids, and belongs to residues 1-33; the others to residues 40-50, 57-64 and 70-73. The remaining two-residue fragments were then be attributed to 34-35 and to 52-53, since no ambiguities were left once the previous fragments were assigned.

There are four other spin patterns which could be revealed in the aliphatic and in the NH regions and not sequence-specifically assigned with the classical approach. Three of these belong to uniquely remaining amino acids (Ala 37, Thr 51, and Val 68), while the forth is due to a phenolic pattern, which was assigned to the only Tyr not yet

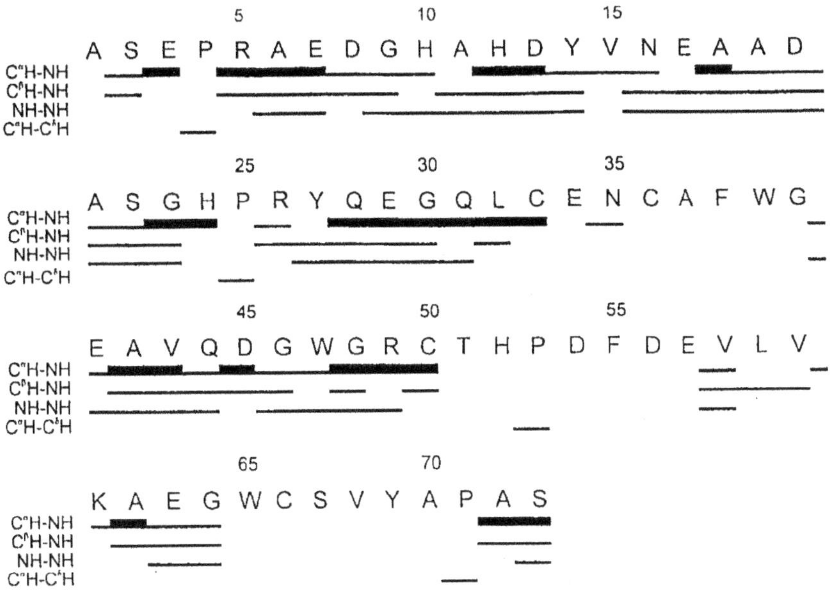

Figure 3 - Schematic representation of the sequential connectivities involving HN, Hα and Hβ protons obtained through standard 2D techniques. NOESY NH-Hα connectivities are divided into strong (thick lines) and weak (thin lines) according to whether the integral of the NOESY cross peak is larger or smaller than 1/4 of the most intense integral value for this class of connectivities. The connectivities involving Hδ protons of Pro residues are also reported. Taken from Ref. [66]

assigned. The aromatic rings of two Trp's (Trp 39 and Trp 65) were identified and distinguished on the basis of their inter-residue NOE's.

A NOESY experiment, optimized to detect connectivities involving broad, fast relaxing, signals, allowed the pairwise identification of the βCH_2 protons of three out of four cysteines.

1D NOE experiments with saturation of the hyperfine shifted signals were quite relevant for the assignment of protons in the proximity of the cluster. Indeed, two additional residues (Cys 66 and Ser 67) were assigned by exploiting the sequential connectivities involving the cysteine residues. These connectivities were observed with 1D NOE experiments, upon selective saturation of the hyperfine shifted, fast relaxing, Hβ protons of the cluster-coordinated cysteines.

Finally, in ^{1}H-^{15}N 3D experiments the signals due to nitrogens close to the cluster, and therefore experiencing a paramagnetic contribution to their linewidths, were detected. The signals of three out of five NH group involved in H-bonds with sulfur atoms bound to the iron ion were detected, despite their large linewidth.

On the basis of the above experiments, which include both standard 2D and 3D NMR experiments and 1D and 2D experiments tailored to detect connectivities involving paramagnetic signals, 68 residues could be unambiguously assigned without reference to any structural model. Three additional residues (Phe 38, Phe 55 and Asp 54) were then assigned by analyzing further 1D NOE data and by referring to the structural models generated from the available NOEs. No signal of the fourth coordinated cysteine (Cys 36) could be detected, because none of its protons are hyperfine shifted out of the diamagnetic region. The other unassigned residue was Asp 56, that is at an average distance of 7Å from the cluster and should not experience a sizable paramagnetic effect.

Using the distance constraints obtained from the NOE and NOESY experiments, the three dimensional structure of this protein in solution was generated using DG calculations.

Most of the NOE constraints (1206) were obtained from the analysis of standard 2D NOESY experiments, which, however, include some connectivities with iron-ligand protons. Forty additional connectivities were obtained from the 1D NOE difference spectra recorded upon selective saturation of the hyperfine-shifted signals, as well as from 2D NOESY experiments optimized to detect connectivities involving fast-relaxing signals [58]. Such connectivities were essential to properly define the cluster environment.

Calibration of NOESY peak volumes and NOE initial slopes for connectivities involving "paramagnetic" signals was performed with the standard procedure used for connectivities involving diamagnetic signals [68]. The NOE volumes and slopes were initially calibrated with respect to connectivities involving diamagnetic signals of protons at known distance. Then, as the structures came out from successive runs of DG calculations, interatomic distances were estimated from these structures and calibration was performed including a larger number of interatomic distances.

A family of structures was generated through the DG program DIANA [68]. In order to describe the cluster during these calculations, we included a non-standard amino acid (Figure 4) to the amino acid coordinates database in the program DIANA. This new amino acid consists of the cysteinyl residue in which the thiol hydrogen was replaced by an iron atom, at the proper distance, on its turn covalently bound to a sulfur atom. Bond lengths and angles used in this amino acid were set equal to the average of these parameters observed in known X-ray structures of HiPIPs and Fe_4S_4 cubane-containing model systems [22,24,33,34]. To maintain the geometry of the cluster, the four iron atoms and the four terminal sulfur atoms were linked to each other through covalent bonds.

The DG calculations provided a family constituted by 40 structures with target function below 2.0 Å^2, which represents a good level of accuracy for the structure. Among these structures, 15 experience a target function lower than 1.0 Å^2. We took these 15 structures as the solution structure of the recombinant HiPIP iso-I from *E. halophila* (Figure 5A).

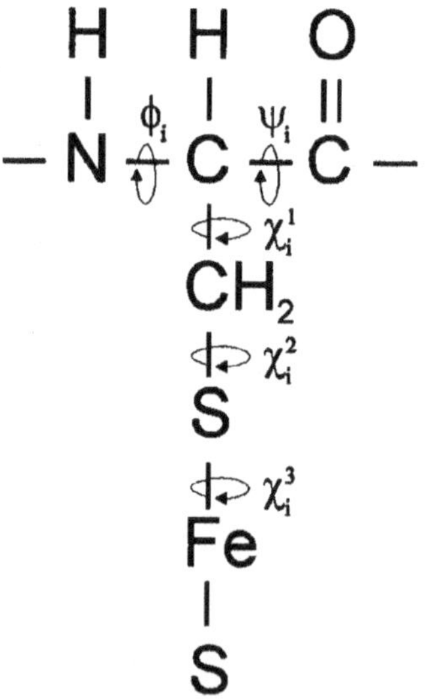

Figure 4 - Schematic representation of the modified amino acid as it has been used in distance geometry calculation. Taken from Ref. [62] with modifications.

A

B

C

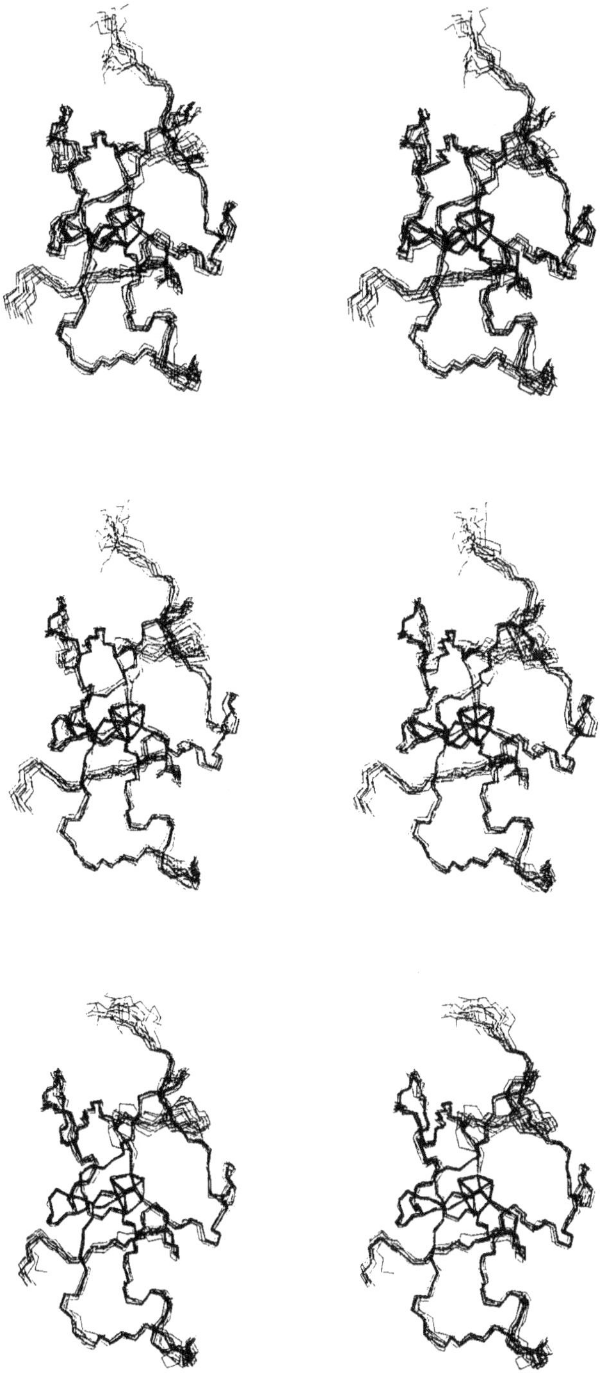

Figure 5 - Stereo drawing of the 15 accepted structures obtained through (A) DG, (B) REM and (C) RMD calculations. Taken from Ref. [62].

The average RMSD values among the structures of the DG family, calculated for residues 4-71, are 0.67 ± 0.11 Å for the backbone and 1.22 ± 0.17 Å for all heavy atoms.

The family of structures generated by DG calculations were then subjected to restrained energy minimization (REM), taking advantage of the force field parameters previously established [8]. The 15 REM structures are reported in Figure 5B. For the residues in the range 4-71, the average RMSD values among the structures of the family are 0.60 ± 0.08 Å for the backbone atoms and 1.23 ± 0.11 Å for all heavy atoms. This REM family was then subjected to restrained MD calculations (RMD), time averaging and subsequent energy minimization to obtain a refined family of structures that is shown in Figure 5C. The average RMSD value for the RMD family is 0.51 ± 0.09 Å for the backbone and 1.20 ± 0.16 Å for the heavy atoms.

The RMD structure with the lowest DG target function was then subjected to a more extensive RMD simulation in a solvent bath to provide the refined structure. From the comparison of the solution with the X-ray structure at 2.5 Å resolution [34], we can see that the overall backbone structure of each of the two independent molecules found in the solid state (molecules I and II) is very similar to that in solution and falls well inside the family of RMD structures. Only three small regions of the protein show some discrepancies which, in the case of two fragments, are essentially due to the limited number of constraints. The third region is constituted by a protein fragment which experiences a large deviation in the two independent molecules found in each cell of the crystal. The NMR structure is well defined with high resolution in this part of the protein, and is in better agreement with one of the two molecules (molecule II). This fact indicates that the structure in solution is reliable and that molecule II of the solid state is more closely related to the solution structure.

As we can see from the above numbers, the quality, in terms of resolution, of the solution structure is quite high, at the level of the best resolved diamagnetic structures. Overall, we have shown that it is therefore possible to overcome the problems involved in the presence of a paramagnetic center. As far as structure determination is concerned, we should consider that, independently of the paramagnetism, the presence of the cluster is expected to decrease the total number of observable NOE's simply because no NOE's are expected across the volume occupied by the cluster itself. In addition, some NOE's are lost because of the paramagnetism. We estimate the number of missing NOE's to be of the order of a few tens, *i.e.* a few percent of the total number of observed NOE's. While the decrease in number of observable NOE's is not dramatic, the missing NOE's tend to concentrate in the restricted area of the protein surrounding the cluster, and therefore may still cause serious problems in the local definition of the structure. However, the presence of well defined constraints due to covalent bonds involving the cluster is determinant for the good definition of the geometry around the cluster, as it is indeed observed.

This overall methodology is quite general and can be successfully applied to other proteins of the same class as well as to other paramagnetic metalloproteins. As an example of this, we would like to briefly comment on the determination of the solution structure of HiPIP from Cv.

In the case of this protein the sequence-specific assignment of most of the residues was already available [42]. By the combined use of standard 2D and 3D NMR experiments with 1D and 2D experiments tailored for the determination of connectivities involving paramagnetic signals three more residues were assigned which lead to the detection of 1461 connectivities [69], again with a good definition of the region around the cluster. By applying the same procedure as discussed above, we obtained a family of 15 DG structures with an average RMSD of 0.73 ± 0.12 Å for the backbone atoms. When REM calculations first, and RMD calculations later, are performed on the DG family, the RMSD values decreased to 0.69 ± 0.11 Å and 0.66 ± 0.13 Å for the backbone atoms. The NMR structure compares very well with the X-ray structure, as already found from the previous unconstrained MD calculations on this HiPIP.

5. Perspectives

The availability of powerful tools for the determination of the structure of metalloproteins in solution and consequently of well refined structural models opens the possibility of tackling several aspects of their biological function.

Among the several still open questions on HiPIPs, one of the most fascinating is about the factors determining the reduction potentials within this class of proteins. Despite the presence of an identical cluster and the very similar structural features of the protein around the cluster, these proteins experience a reduction potential which span from +50 mV up to +450 mV. Therefore, there should be other long range effects, like electrostatic interactions, which should modulate the different reduction properties of the cluster. The availability of structural model for a large number of proteins makes the calculations of the reduction potentials now possible.

Another important aspect to be characterized is the pathway of the electron-transfer from the cluster to other proteins and the factors determining the rate of this process.

These are just two examples of the relevant issues which could be addressed once a detailed structural characterization on these proteins has become available.

In conclusion, we have shown that structure determination in solution can be obtained at high resolution, through NMR and MD studies also on paramagnetic metalloproteins and that these studies can be quite important for the comprehension of the biological functions of these metalloproteins.

6. References

1. McCammon, J.A. and Harvey, S. (1987) *Dynamics of proteins and nucleic acids*, Cambridge University Press, Cambridge.

2. Brooks, C.L,III, Karplus, M., and Petitt, B.M. (1988) *Proteins: a perspective of dynamics, structure and thermodynamics*, J. Wiley and Sons, New York.

3. Kollman, P.A. and Merz, Jr., K.M. (1990) Computer Modeling of the Interactions of Complex Molecules, *Acc. Chem. Res.* **23**, 246-252.

4. Brunger, A.T., Kuriyan, J., and Karplus, M. (1987) Crystallographic R factor refinement by molecular dynamics, *Science* **235**, 458-460.

5. Merz, Jr.,K.M. (1991) CO_2 binding to human carbonic anhydrase II, *J. Am. Chem. Soc.* **113**, 406-411.

6. Bernhardt, P.V. and Comba, P. (1992) Molecular mechanics calculations of transition metal complexes, *Inorg. Chem.* **31**, 2638-2644.

7. Hancock, R.D. (1989) Moleuclar mechanics calculations as a tool in coordination chemistry, in Lippard, S.J. (ed) *Progress in Inorganic Chemistry, Vol. 37*, John Wiley & Sons, Inc., New York, pp.187.

8. Banci, L., Bertini, I., Carloni, P., Luchinat, C., and Orioli, P.L. (1992) Molecular dynamics simulations on HiPIP from *Chomatium vinosum* and comparison with NMR data, *J. Am. Chem. Soc.* **114**, 10683-10689.

9. Banci, L., Carloni, P., La Penna, G., and Orioli, P.L. (1992) Molecular dynamics studies on superoxide dismutase and its mutants : the structural and functional role of Arg 143, *J. Am. Chem. Soc.* **114**, 6994-7001.

10. Banci, L., Bertini, I., and La Penna, G. (1994) The enzymatic mechanism of carboxypeptidase: a molecular dynamics study, *Proteins: Structure,Function,and Genetics* **18**, 186-197.

11. Case, D.A. and Karplus, M. (1979) Dynamics of ligand binding to heme proteins, *J. Mol. Biol.* **132**, 343-368.

12. Shen, J., Wong, C.F., Subrasmaniam, S., Albright, T.A, and McCammon, J.A. (1990) Partial electrostatic charges for the active center of copper, zinc superoxide dismutase, *J. Comp. Chem.* **11**, 346-350.

13. Banci, L., Bertini, I., Capozzi, F., Carloni, P., Ciurli, S., Luchinat, C., and Piccioli, M. (1993) The iron-sulfur cluster in the oxidized high potential iron sulfur protein from *Ectothiorhodospira halophila, J. Am. Chem. Soc.* **115**, 3431-3440.

14. Meyer, T.E., Przysiecki, C.T., Watkins, J.A., Bhattacharyya, A., Simondsen, R.P., Cusanovich, M.A., and Tollin, G. (1983) Correlation between rate constant for reduction and redox potential as a basis for systematic investigation of reaction machanisms of electron transfer proteins, *Proc. Natl. Acad. Sci. USA* **80**, 6740-6744.

15. Middleton, P., Dickson, D.P.E., Johnson, C.E., and Rush, J.D. (1980) Interpretation of the Mossbauer spectra of the high-potential iron protein from Chromatium, *Eur. J. Biochem.* **104**, 289-296.

16. Bertini, I., Campos, A.P., Luchinat, C., and Teixeira, M. (1993) A Mössbauer investigation of oxidized Fe_4S_4 HiPIP II from *E. halophila, J. Inorg. Biochem.* **52**, 227-234.

17. Dunham, W.R., Hagen, W.R., Fee, J.A., Sands, R.H., Dunbar, J.B., and Humblet, C. (1991) An investigation of *Chromatium vinosum* high-potential iron-sulfur protein by EPR and Mössbaur spectroscopy; evidence for a freezing-induced dimerization in sodium cloride solution, *Biochim. Biophys. Acta* **1079**, 253-262.

18. Hagen, W.R. (1992) EPR spectroscopy of iron sulfur proteins, *Adv. Inorg. Chem.* **38**, 165-222.

19. Rius, G.J. and Lamotte, B. (1989) Single-crystal ENDOR study of a ^{57}Fe-enriched iron-sulfur $[Fe_4S_4]^{3+}$ cluster, *J. Am. Chem. Soc.* **111**, 2464-2469.

20. Mouesca, J.M., Lamotte, B., and Rius, G.J. (1991) Comparison between spin population distibutions in two different $[Fe_4S_4]^{3+}$ clusters by proton ENDOR in single crystals of a synthetic model compound, *Inorg. Biochem.* **43**, 251.

21. Mouesca, J.M., Rius, G., and Lamotte, B. (1993) Single-crystal proton ENDOR studies of the $[Fe_4S_4]^{3+}$ cluster: Determination of the spin population distribution and proposal of a model to interpret the ^1H NMR paramagnetic shifts in high potential ferredoxins, *J. Am. Chem. Soc.* **115**, 4714-4731.

22. Holm, R.H., Ciurli, S., and Weigel, J.A. (1990) Subsite-specific structures and reactions in native and synthetic [4Fe-4S] cubane-type clusters, in Lippard, S.J. (ed) *Progress in Inorganic Chemistry: Bioinorganic Chemistry, Vol. 38*, John Wiley & Sons, Inc., New York, pp.1-74.

23. Jordanov, J., Roth, E.K.H., Fries, P.H., and Noodleman, L. (1990) Magnetic studies of the high-potential protein model $[Fe_4S_4(S\text{-}2,4,6\text{-}(i\text{-}Pr)_3C_6H_2)_4]^-$ in the $[Fe_4S_4]^{3+}$ oxidized state, *Inorg. Chem.* **29**, 4288-4292.

24. Carney, M.J., Papaefthymiou, G.C., Spartalian, K., Frankel, R.B., and Holm, R.H. (1988) Ground spin state variability in $[Fe_4S_4(SR)_4]^{3-}$. Synthetic analogues of the reduced clusters in ferredoxins and other iron-sulfur proteins: cases of extreme sensitivity of electronic state and structure to extrinsic factors, *J. Am. Chem. Soc.* **110**, 6084-6095.

25. Belinskii, M.I. (1987) Electronic interaction in trimeric mixed-valence cluster. Heisemberg and double exchange, *Mol. Phys.* **60**, 793-819.

26. Papaefthymiou, V., Girerd, J.-J., Moura, I., Moura, J.J.G., and Munck, E. (1987) Mõssbauer study of *D.gigas* ferredoxin II and spin-coupling model for the Fe_3S_4 cluster with valence delocalization, *J. Am. Chem. Soc.* **109**, 4703-4710.

27. Munck, E., Papaefthymiou, V., Surerus, K.K., and Girerd, J.-J. (1988) , in Que, L., Jr. (ed) *Metal Clusters in Proteins*, American Chemical Society, Washington, D.C., pp.302.

28. Blondin, G. and Girerd, J.-J. (1990) Interplay of electron exchange and electron transfer in metal polynuclear complexes in proteins or chemical models, *Chem. Rev.* **90**, 1359-1376.

29. Noodleman, L. and Case, D.A. (1992) Density-functional theory of spin polarization and spin coupling in iron sulfur clusters, *Adv. Inorg. Chem.* **38**, 424-470.

30. Bertini, I., Ciurli, S., and Luchinat, C. (1994) The electronic and geometric structures of iron-sulfur proteins studied through electron-nuclear hyperfine coupling, *Angew. Chem.* in press

31. Carter, C.W.J., Kraut, J., Freer, S.T., Alden, R.A., Sieker, L.C., Adman, E.T., and Jensen, L.H. (1972) A comparison of Fe4S4 Clusters in High Potential Iron Protein and in Ferredoxin, *Proc. Natl. Acad. Sci. USA* **69**, 3526-3529.

32. Thomson, A.J. (1985) Iron-Sulfur Proteins, in Harrison, P. (ed) *Metalloproteins*, Verlag Chemie, Weinheim, pp.79.

33. Carter, C.W.J., Kraut, J., Freer, S.T., Xuong, N.-H., Alden, R.A., and Bartsch, R.G. (1974) Two-angstrom crystal structure of Chromatium vinosum high-potential iron protein, *J. Biol. Chem.* **249**, 4212-4215.

34. Breiter, D.R., Meyer, T.E., Rayment, I., and Holden, H.M. (1991) The molecular structure of the high potential iron-sulfur protein isolated from *Ectothiorhodospira halophila* determined at 2.5 Å resolution, *J. Biol. Chem.* **266**, 18660-18667.

35. Rayment, I., Wesemberg, G., Meyer, T.E., Cusanovich, M.A., and Holden, H.M. (1992) Three-dimensional structure of the high-potential iron-sulfur protein isolated from the purple phototrophic bacterium *Rhodocyclus tenuis* determined and refined at 1.5 Å resolution, *J. Mol. Biol.* **228**, 672.

36. Benning, M.M., Meyer, T.E., Rayment, I., and Holden, H.M. (1994) Molecular structure of the oxidized High-potential Iron-sulfur protein isolated from *Ectothiorhodospira vacuolata, Biochemistry* **33**, 2476-2483.

37. Bertini, I., Capozzi, F., Ciurli, S., Luchinat, C., Messori, L., and Piccioli, M. (1992) Identification of the iron ions of HiPIP from *Chromatium vinosum* within the protein frame through 2D NMR experiments, *J. Am. Chem. Soc.* **114**, 3332-3340.

38. Bertini, I., Capozzi, F., Luchinat, C., Piccioli, M., and Vicens Oliver, M. (1992) NMR is a unique and necessary step in the investigation of iron-sulfur proteins: the HiPIP from *R. gelatinosus* as an example, *Inorg. Chim. Acta* **198-200**, 483-491.

39. Banci, L., Bertini, I., Ciurli, S., Ferretti, S., Luchinat, C., and Piccioli, M. (1993) The electronic structure of $(Fe_4S_4)^{3+}$ clusters in proteins; an investigation of the oxidized HiPIP II from *Ectothiorhodospira vacuolata, Biochemistry* **32**, 9387-9397.

40. Bertini, I., Capozzi, F., Luchinat, C., and Piccioli, M. (1993) ^{1}H NMR investigation of oxidized and reduced HiPIP from *R. globiformis, Eur. J. Biochem.* **212**, 69-78.

41. Nettesheim, D.G., Harder, S.R., Feinberg, B.A., and Otvos, J.D. (1992) Sequential resonance assignments of oxidized high-potential iron-sulfur protein from *Chromatium vinosum, Biochemistry* **31**, 1234-1244.

42. Gaillard, J., Albrand, J.-P., Moulis, J.-M., and Wemmer, D.E. (1992) Sequence-specific assignment of the ^{1}H nuclear magnetic resonance spectra of reduced high-potential ferredoxin (HiPIP) from *Chromatium vinosum, Biochemistry* **31**, 5632-5639.

43. Carloni, P. and Corongiu, G. (1994) *Model Calculations on 2Fe 2S ferredoxin*, submitted.

44. Czernuszawicz, R.S., Macor, K.A., Johnson, M.K., Gevirth, A., and Spiro, T.S. (1987) Vibrational mode structure and symmetry in proteins and analogues containing Fe_4S_4 clusters: resonance raman evidence for different degrees of distortion in HiPIP and ferredoxin, *J. Am. Chem. Soc.* **109**, 7178-7187.

45. Pearlman, D.A., Case, D.A., Caldwell, G.C., Siebel, G.L., Singh, U.C., Weiner, P., and Kollman, P.A. (1991) *AMBER 4.0, University of California*, S. Francisco.

46. Bertini, I., Gaudemer, A., Luchinat, C., and Piccioli, M. (1993) Electron self-exchange in HiPIPs. A characterization of HiPIP I from *Ectothiorhodospira vacuolata*, *Biochemistry* **32**, 12887-12893.

47. Banci, L., Bertini, I., Briganti, F., Luchinat, C., Scozzafava, A., and Vicens Oliver, M. (1991) ^1H NMR spectra of oxidized high-potential iron-sulfur protein (HiPIP) from *Rhodocyclus gelatinosus*. A model for oxidized HiPIPs, *Inorg. Chem.* **30**, 4517-4524.

48. Wüthrich, K.J. (1986) *NMR of protein and nucleic acids*, Wiley, New York.

49. Wagner, G. (1990) NMR investigations of protein structure, *Progr. Nucl. Magn. Reson. Spectrosc.* **22**, 101-139.

50. Clore, G.M. and Gronenborn, A.M. (1991) Aplications of three- and four-dimensional heteronuclear NMR spectroscopy to protein structure determination, *Progr. Nucl. Magn. Reson. Spectrosc.* **23**, 43-92.

51. Kaptein, R., Boelens, R., and Koning, T.M.G. (1991) NMR Studies of Proteins, Nucleic Acids and their Interactions, in Bertini, I., Molinari, H., and Niccolai, N. (eds) *NMR and Biomolecular Structure*, VCH, Weinheim, pp.113-139.

52. *Macromolecular Structures*, Hendrickson, W.A. and Wüthrich, K. (eds), Current Biology Ltd., London, (1993).

53. Clore, G.M. and Gronenborn, A.M. (1994) Structures of larger proteins, protein-ligand and protein-DNA complexes by multidimensional heteronuclear NMR, *Protein Sci.* **3**, 372-390.

54. Wüthrich, K. (1989) The development of nuclear magnetic resonance spectroscopy as a technique for protein structure determination, *Acc. Chem. Res.* **22**, 36-44.

55. Bertini, I. and Luchinat, C. (1986) *NMR of paramagnetic molecules in biological systems*, Benjamin/Cummings, Menlo Park, CA.

56. Banci, L., Bertini, I., and Luchinat, C. (1991) *Nuclear and electron relaxation. The magnetic nucleus-unpaired electron coupling in solution*, VCH, Weinheim.

57. Bertini, I., Turano, P., and Vila, A.J. (1993) NMR of paramagnetic metalloproteins, *Chem. Rev.* **93**, 2833-2932.

58. Banci, L., Bertini, I., and Luchinat, C. (1994) 2D NMR spectra of paramagnetic systems, in James, T.L. and Oppenheimer, N.J. (eds) *Methods in enzymology*, Academy press, Inc., Florida.

59. *Biological Magnetic Resonance, Vol. 12: NMR of Paramagnetic Molecules*, Berliner, L. J., Reuben, J. (eds), Plenum Press, New York, (1993).

60. Ernst, R.R., Bodenhausen, G., and Wokaun, A. (1987) *Principles of Nuclear Magnetic Resonance in one and two dimensions*, Oxford University Press, London.

61. Bertini, I., Dikiy, A., Luchinat, C., Piccioli, M., and Tarchi, D. (1994) NOE-NOESY: a further tool in NMR of paramagnetic metalloproteins, *J. Magn. Reson. Ser. B* **103**, 278-283.

62. Banci, L., Bertini, I., Eltis, L.D., Felli, I., Kastrau, D.H.W., Luchinat, C., Piccioli, M., Pierattelli, R., and Smith, M. (1994) The three dimensional structure in solution of the paramagnetic protein HiPIP I from *E. halophila* through nuclear magnetic resonance, *Eur. J. Biochem.*, in press.

63. Eltis, L.D., Iwagami, S.G., and Smith, M. (1994) Hyperexpression of a synthetic gene encoding high potential iron sulfur Iso I from E. Halophila, *Protein Eng.*, in press

64. Luchinat, C. and Ciurli, S. (1993) NMR of polymetallic systems in proteins, *Biological Magnetic Resonance* **12**, 357-420.

65. Banci, L. (1993) Nuclear relaxation in paramagnetic metalloproteins, *Biological Magnetic Resonance* **12**, 79-111.

66. Bertini, I., Felli, I., Kastrau, D.H.W., Luchinat, C., Piccioli, M., and Viezzoli, M.S. (1994) Sequence-specific assignment of the ^{1}H and ^{15}N nuclear magnetic resonance spectra of the reduced recombinant high potential iron sulfur protein (HiPIP) I from *Ectothiorhodospira halophila*, *Eur. J. Biochem.*, in press.

67. Scheek, R.M., van Gunsteren, W.F., and Kaptein, R. (1989) Molecular dynamics techniques for determination of molecular structures from nuclear magnetic resonance data, *Methods Enzymol.* **177**, 204-218.

68. Güntert, P., Braun, W., and Wüthrich, K. (1991) Efficient computation of three-dimensional protein structures in solution from Nuclear Magnetic Resonance data using the program DIANA and the supporting programs CALIBA, HABAS and GLOMSA, *J. Mol. Biol.* **217**, 517-530.

69. Banci, L., Bertini, I., Dikiy, A., Kastrau, D., Luchinat, C., Sompornpisut, P., The three-dimensional solution structure of the reduced high potential iron-sulfur protein from *Chromatium vinosum* through NMR, submitted.

MULTINUCLEAR MAGNETIC RESONANCE AND MUTAGENESIS STUDIES OF STRUCTURE-FUNCTION RELATIONSHIPS IN [2FE-2S] FERREDOXINS

YOUNG KEE CHAE, BIN XIA, HONG CHENG, BYUNG-HA OH, LARS SKJELDAL, WILLIAM M. WESTLER, and JOHN L. MARKLEY

Department of Biochemistry, University of Wisconsin - Madison, 420 Henry Mall, Madison, WI 53706

1. Introduction

Iron-sulfur proteins are present in virtually all living organisms. Their function is to transfer electrons to various partners. They have iron and sulfur in their chromophore and rather low molecular weights (5 to 25 kDa). They usually have the same ratio of iron to sulfur except for rubredoxin which has only one iron ligated by four cysteine residues. Several cluster classes have been identified in iron-sulfur proteins, including [1Fe], [2Fe-2S], [4Fe-4S], and [3Fe-4S] types (Figure 1). Some iron-sulfur proteins contain more than one cluster of a given type or of different types. [2Fe-2S] ferredoxins contain two high-spin ferric ions that are antiferromagnetically coupled resulting a total spin number of zero in the oxidized state; thus they are EPR[1] silent. Population of higher electronic states, however, makes the cluster paramagnetic at room temperature (and at all temperature studied by NMR). Upon reduction, one of the two ferric ions (Fe^{3+}) is changed to a ferrous ion (Fe^{2+}), and the other remains ferric. Reduced [2Fe-2S] ferredoxins have a total spin number of one-half and produce an EPR signal. [4Fe-4S] cluster types have three possible oxidation states: $3Fe^{3+}$-Fe^{2+}, $2Fe^{3+}$-$2Fe^{2+}$, and Fe^{3+}-$3Fe^{2+}$. Typically in iron-sulfur proteins, only two of the three states are easily accessible: the first two as in high-potential iron-sulfur proteins (HiPIPs) or the last two as in [4Fe-4S] ferredoxins. We have recently investigated three different [2Fe-2S] ferredoxins by NMR spectroscopy: *Anabaena 7120* vegetative

[1] Abbreviations used: 1D, one-dimensional; 2D, two-dimensional; CHOPPER, computer program for excising a portion of a two-dimensional spectrum; DQC, double-quantum coherence; DQF-COSY, double-quantum filtered correlated spectroscopy; DTT, dithiothreitol; E.COSY, exclusive correlated spectroscopy; EPR, electron paramagnetic resonance; HOHAHA, homonuclear Hartmann-Hahn; IPTG, isopropyl thiogalactoside; MBC, multiple-bond correlation; NMR, nuclear magnetic resonance; NOE, nuclear Overhauser effect; NOESY, nuclear Overhauser effect spectroscopy; OD, optical density; SBC, single-bond correlation; UL, uniformly labeled; ZOOM, computer program for inverse Fourier transform, zerofilling, and Fourier transform of NMR data.

G.N. La Mar (ed.), Nuclear Magnetic Resonance of Paramagnetic Macromolecules, 297-317.

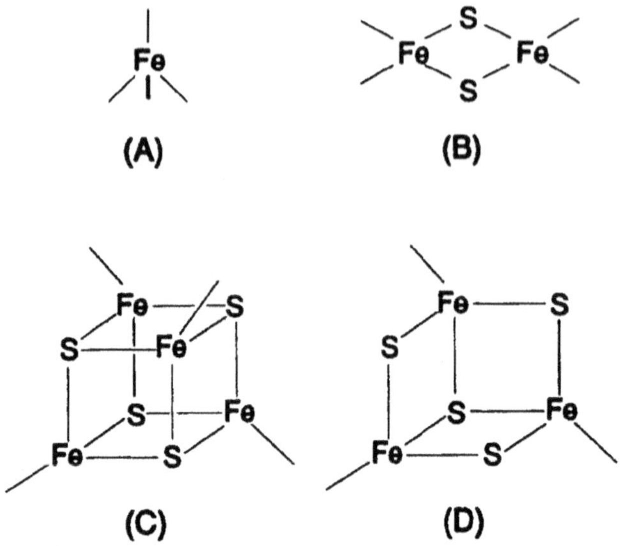

Figure 1. Four different cluster types appearing in iron-sulfur proteins. The first three involve ligation of the cluster to four cysteine residues of the protein chain. One cysteine is absent in the fourth cluster type. (A) One Fe ligated by four cysteines; (B) [2Fe-2S] cluster; (C) [4Fe-4S] cluster; (D) [3Fe-4S] cluster.

ferredoxin, *Anabaena 7120* heterocyst ferredoxin, and human placental ferredoxin. The first two are plant-type ferredoxins from a cyanobacterium, and the last is vertebrate-type ferredoxin. *Anabaena 7120* vegetative ferredoxin is involved in photosynthesis; it donates one electron to ferredoxin-$NADP^+$ oxidoreductase (FNR) which catalyzes the reduction of $NADP^+$ to NADPH [1]. Extensive studies of this ferredoxin in our laboratory have led to assignments of the diamagnetic resonances [2, 3] and assignments of the paramagnetically shifted resonances [4, 5, 6]. *Anabaena 7120* heterocyst ferredoxin donates one electron to nitrogenase component II (Fe protein), which reduces nitrogenase component I (Mo-Fe protein), which finally reduces N_2 to NH_3 [7, 8]. The vegetative and heterocyst ferredoxins have 51% sequence homology, but subtle differences in the structure allow the heterocyst ferredoxin to interact with both FNR and nitrogenase component II, whereas the vegetative ferredoxin can interact only with FNR [9]. We have made extensive assignments of the diamagnetic ^1H, ^{13}C, and ^{15}N resonances of the heterocyst ferredoxin, and determined its secondary structure [10], and have analyzed its hyperfine ^{15}N signals [11]. In addition, the x-ray structure of the oxidized form of the protein has been solved [12] and refined to 1.7 Å [13]. Human placental ferredoxin is involved in electron transfer to mitochondrial cytochrome P-450 [14, 15], and it is slightly larger (M_r = 13,500) than the vegetative and heterocyst ferredoxins (M_r =

11,000). NMR studies of this protein show that its hyperfine ^1H NMR signals resemble those of adrenodoxins [16]. We have recently investigated the properties of the histidines of human ferredoxin [17], and a more complete NMR analysis of this protein is underway. This review will concern only the NMR studies of *Anabaena 7120* vegetative ferredoxin and comparisons with x-ray structures of the same protein.

The x-ray structure of the wild-type vegetative ferredoxin was first determined at 2.5 Å [18] and was refined subsequently to 1.7 Å [19]. Several single-site mutant ferredoxins have been crystallized and their structures determined [20]. Among these is the C49S mutant whose structure is identical to that of wild-type except for the Fe-O^{γ} distance which in the mutant is considerably shorter than the Fe-S^{γ} distance in the wild-type ferredoxin (1.8 Å vs 2.3 Å).

Functional properties of the wild-type and mutant ferredoxins have been characterized [21, 22]. Several mutants were prepared to investigate the interaction of the vegetative ferredoxin with FNR. Most mutants did not show appreciable changes in the electron transfer activity. However, mutations at Glu-94 and Phe-65 (surface residues) showed dramatic decreases. Interestingly, Phe-65 can be replaced by other aromatic amino acids (Tyr or Trp) without loss of activity, but not by non-aromatic hydrophobic residues [22].

2. Methodology

2.1. PROTEIN PRODUCTION AND LABELING

Earlier studies from our laboratory made use of protein obtained directly from the cyanobacterium *Anabaena 7120*. Seventy-liter batches of *Anabaena 7120* were grown on a slightly modified medium C. Each batch was grown for six days under optimal illumination and agitation [23]. To obtain ferredoxin uniformly enriched with ^{15}N, [^{15}N, 98 %] $K^{15}NO_3$ was used as the sole nitrogen source. Double-labeled ferredoxin with ^{13}C and ^{15}N was produced by growing *Anabaena 7120* on [^{13}C, 26 %] $^{13}CO_2$ and [^{15}N, 98 %] $K^{15}NO_3$ as the sole carbon and nitrogen sources [3].

More recently, after the gene of the *Anabaena 7120* vegetative ferredoxin became available, the protein could be heterologously expressed in *Escherichia coli*. The original plasmid containing vegetative ferredoxin gene, *petF*, was kindly provided by Prof. S. Curtis at North Carolina State University. The upstream and downstream sequences of *petF* gene were modified by using PCR to create suitable restriction enzyme sites (NdeI and BamHI) [6]. The resulting PCR fragment was inserted into pET3a and pET9a vectors (Novagen, WI); the resulting plasmids were named pET3a/F and pET9a/F, respectively. Three host systems were used: BL21(DE3) for uniform labeling, JM15 for cysteine selective labeling, and PA200 for arginine selective labeling.

BL21(DE3) is used widely as an overexpression host because the T7 RNA polymerase gene in its chromosome is under the control of the lac operator, which can be induced conveniently by adding IPTG. The plasmid pET3a/F or pET9a/F was transformed into BL21(DE3) along with another plasmid, pLysS, which carried the T4 lysozyme gene. T4 lysozyme, which is known as a natural inhibitor of T7 RNA polymerase, reduces the basal level expression of the desired protein before induction; this can prevent toxic effects of the protein on the host cell.

To one liter of LB medium, a 10 ml inoculum was added of an overnight culture of BL21(DE3) containing pET3a/F or pET9a/F along with pLysS. Cells grew at 37 °C until the OD_{600} = 1.0 - 1.2. Then the culture was induced by adding 100 mg of IPTG, which caused the induction of the T7 RNA polymerase gene and production of T7 RNA polymerase which, in turn, bound to the T7 promoter in pET3a/F or pET9a/F and induced the production of large amounts of the mRNA coding for the vegetative ferredoxin. The culture was harvested by centrifugation 2 - 3 h after the induction. The cell pellets were resuspended in 20 ml of 50 mM phosphate buffer pH 8.0, and frozen at -20 °C. For ^{15}N uniform labeling, M9 minimal medium [24] with $^{15}NH_4Cl$ was used.

The production of selectively labeled vegetative ferredoxin from JM15 or PA200 is similar to that above, but a different induction method was used. Since neither of these strains has the T7 RNA polymerase gene in its chromosome, a separate plasmid, pGp1-2, carrying that gene had to be used. For selective labeling, the recipe described in Muchmore *et al.* [25] was used. 60 mg and 125 mg of labeled (^{2}H, ^{15}N, or ^{13}C) cysteine and arginine, respectively, were added into the medium [6]. The auxotrophic cells (JM15 or PA200) containing pET3a/F, and pGp1-2 were grown at 30 °C until OD_{600} = 1.0 - 1.2. Then the temperature was shifted to 42 °C so that T7 RNA polymerase could be produced. The subsequent procedures were identical to those described above.

2.2. CLUSTER RECONSTITUTION

The previously frozen, resuspended cell pellets were thawed. The T4 lysozyme (produced by pLysS) caused the thawed cells to lyse, and the lysate became very viscous. The viscosity was lowered by adding small amounts of DNase I and RNase A. In the case of JM15 or PA200, which do not contain pLysS, sonication was used to disrupt the cells. At this stage, about half of the total vegetative ferredoxin was soluble (either as holo- or apoferredoxin); the other half was insoluble, perhaps in "inclusion bodies." Therefore, it proved convenient to denature the whole lysate and reconstitute the cluster [12]. Urea was added to the lysate (final concentration of 8 M), and the resulting mixture was degassed to remove oxygen. Then, DTT was added (final concentration of 100 mM) to reduce all cysteine residues and prevent the formation of intra- or inter-molecular disulfide bonds. After that, with argon bubbling, an excess amount of Fe^{3+} (the added Fe^{3+} is reduced immediately to Fe^{2+} by DTT in the solution) and S^{2-} were added in the form of $FeCl_3$ and Na_2S, respectively. After about 15 to 20 min of stirring and argon flushing, the reaction mixture was diluted 8 fold with degassed 50 mM phosphate buffer of pH 7.4. The mixture was quickly loaded onto an anion exchange column (DE53, 3 cm x 5 cm), washed, and eluted with 50 mM phosphate buffer of pH 7.4 with 1.0 M NaCl. The vegetative ferredoxin was purified by anion exchange and gel filtration column chromatography as described by Cheng *et al.* [6]. Typical yields of holoprotein were 30 mg/liter of the original culture.

2.3. SITE-DIRECTED MUTAGENESIS

Mutagenesis was performed according to the Kunkel procedure [26]. Mutagenic oligonucleotides were prepared by chemical synthesis. Standard mutagenesis tools were used: M13 mp18 phage for synthesis of the template single-stranded DNA strand, RZ1032 for preparation of uracil-incorporated single-stranded DNA, and JM103 for selecting mutant DNA by disrupting the uracil-incorporated DNA strand. After mutagenesis, the proper portion was cut, and inserted into either the pET3a or pET9a vector [27].

2.4. REDUCTION OF PROTEIN SAMPLES

The protein sample was dissolved and placed in a small vial which was degassed and flushed repeatedly with argon. Solid sodium dithionite (2 mg) was added to a 5 mm NMR tube, and the NMR tube was evacuated. The degassed and argon-flushed protein solution was transferred by a syringe through a rubber septum to the NMR tube. The degassing and argon-flushing steps were then repeated several times with the solution in the NMR tube. Finally the NMR tube was sealed under vacuum with a torch.

3. ASSIGNMENT STUDIES

3.1. ASSIGNMENT OF DIAMAGNETIC RESONANCES

The results of several NMR experiments were used in analyzing the diamagnetic region of oxidized *Anabaena 7120* vegetative ferredoxin [3, 4]. These included homonuclear 2D experiments (DQF-COSY [28], NOESY [29], HOHAHA [30], etc.) and heteronuclear 2D experiments ($^1H\{^{15}N\}$SBC [31], $^1H\{^{15}N\}$MBC [32, 33], $^1H\{^{13}C\}$SBC [31], $^1H\{^{15}N\}$SBC-NOE [34, 35], $^{13}C\{^{15}N\}$SBC [36, 37], $^{13}C\{^{13}C\}$DQC [38], etc.). Sequential assignments relied on $H^{\alpha}_{i}/H^{N}_{i+1}$ NOE connectivities (Figure 2) [3]. Side-chain assignments were deduced from the results of several experiments including COSY, $^1H\{^{13}C\}$SBC, and $^{13}C\{^{13}C\}$DQC. The most distinctive experiment was $^{13}C\{^{13}C\}$DQC (Figure 3), which exploits direct one-bond couplings of neighboring carbons [38]. The ^{13}C experiments utilized 26 % UL ^{13}C labeling, which reduced the effects of long-range ^{13}C - ^{13}C splitting by scalar coupling. The approach led to assignments of nearly all the diamagnetic ^1H, ^{13}C, and ^{15}N resonances. However, resonances from 18 residues near the paramagnetic [2Fe-2S] cluster were not detected in these data sets.

3.2. ASSIGNMENT OF RESONANCES BROADENED OR SHIFTED BY HYPERFINE INTERACTIONS

3.2.1. *Proton Assignments*

In its oxidized state, the vegetative ferredoxin shows a very broad ^1H NMR signal at 37 ppm which arises from several overlapping resonances [39]. Upon reduction,

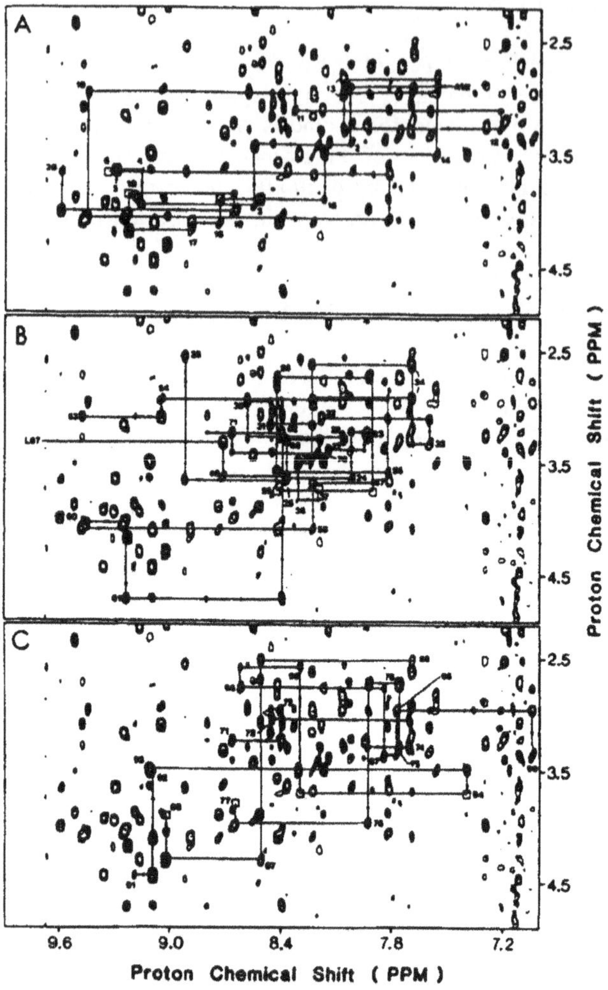

Figure 2. Fingerprint region of the NOESY spectrum of *Anabaena 7120* vegetative ferredoxin is shown three times with different sequential connectivities indicated: (A) 1-5 and 8-20; (B) 22-26, 29-37, 53-63, and 67-71; (C) 71-77, 86-88, and 91-99 (reproduced with permission from Oh and Markley [3]).

eleven well-resolved paramagnetically shifted proton resonances can be detected (Figure 4) [39]. Initial assignments of the ^1H NMR signals from H^{α} and H^{β} of the cysteines that ligate the reduced [2Fe-2S] cluster were made on the basis of theoretical predictions [40]. Hyperfine shifts are predicted to be larger for cysteinyl β-protons than for α-protons. The β-proton of the two cysteines bound to Fe^{3+} are predicted to

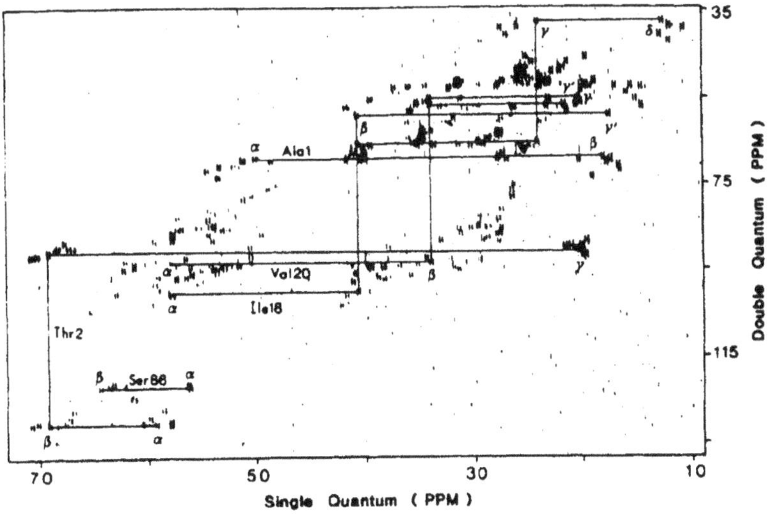

Figure 3. Representative region of the $^{13}C\{^{13}C\}$ DQC spectrum of oxidized [UL ^{13}C, 26 %] *Anabaena 7120* vegetative ferredoxin. Solid lines show the assignments tracing from the α-carbons to the terminal side-chain carbons through ^{13}C-^{13}C coupling (reproduced with permission from Oh *et al.* [4]).

have Curie-type temperature dependence; those from the two cysteines bound to $Fe^{2+}$are predicted to have anti-Curie-type dependence. By combining information from the temperature dependence (Figure 5), 1D NOE data [41], 2D NOESY data (Figure 6), and the x-ray structure of the same protein [18], tentative assignments of the hyperfine-shifted resonances could be determined (Table 1) [5].

Certain aspects of these assignments have been tested recently by selective deuteration. $[^2H^{\alpha}]$Cys and $[^2H^{\beta 2,\beta 3}]$Cys were incorporated into separate samples of *Anabaena 7120* vegetative ferredoxin [27]. Figure 7 shows ^1H NMR spectra of reduced vegetative ferredoxin: unlabeled (bottom), selectively labeled with $[^2H^{\alpha}]$Cys (top), and selectively labeled with $[^2H^{\beta 2,\beta 3}]$Cys (middle). Since the H^{α} position of the incorporated cysteine was deuterated only partially, the $^1H^{\alpha}$ signals (e and j) of the $[^2H^{\alpha}]$Cys sample (top trace) are attenuated but still present. From these spectra of deuterated samples, eight peaks (a-d, f-i) could be classified as cysteinyl β-protons, while peaks e and j were classified as cysteinyl α-protons. These results are consistent with the previous α- and β-proton assignments [5].

The recent selective labeling studies, however, show that the assignment of peak k to R42H^{α} [6] was incorrect. Selective deuteration confirms that peak k does not arise from cysteine. The proton spin system of R42 was unambiguously assigned by incorporation of [26 % UL ^{13}C]Arg and heteronuclear correlation [27]. The spin system so assigned did not include peak k. Moreover, no loss of intensity was found for peak k in the ^1H NMR spectrum of $[^2H^{\alpha,\beta 2,\beta 3}]$Arg ferredoxin [6]. The origin of peak k remains a mystery.

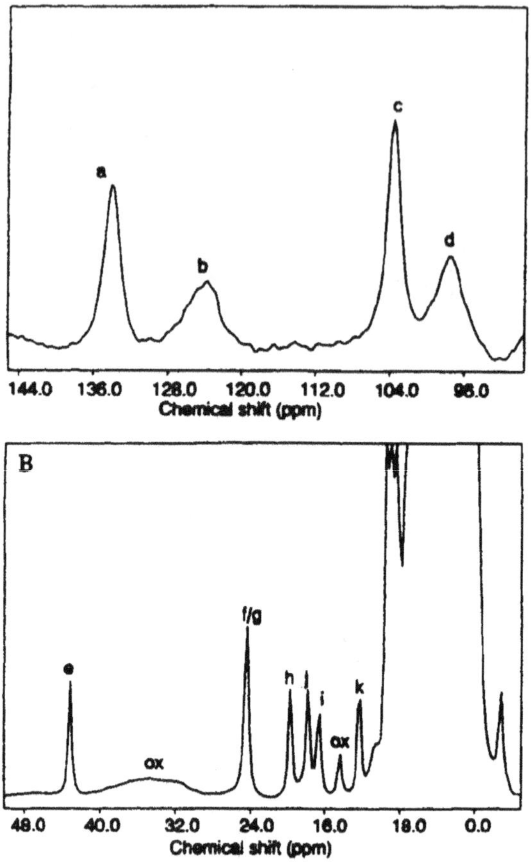

Figure 4. Downfield region of the one-dimensional ^{1}H NMR spectra of reduced *Anabaena 7120* vegetative ferredoxin. The labeling scheme is consistent with that of Dugad *et al.* [41] (reproduced with permission from Skjeldal *et al.* [5]). (A) Four peaks shifted farthest downfield; (B) Remainder of the spectrum.

3.2.2. *Carbon Assignments*

All four cysteine residues were labeled selectively with [$^{13}C^{\beta}$]Cys [6]. In the oxidized state, the ^{13}C spectrum showed 3 signals which separated into four resolved peaks upon reduction (Figure 8). By utilizing very efficient ^{1}H-^{13}C one bond coupling, selective ^{13}C-decouped one-dimensional ^{1}H spectra revealed clear correlations between H^{β} signals assigned to C49 and C79 and their attached carbons (Figure 8, inset). Furthermore, with three peaks, two with one-carbon intensity and one with two-carbon intensity, the HMQC spectrum showed two cross peaks which correlated H^{β} and C^{β} signals from C41 and C46. In this fashion, all four β-carbons of four cysteines were assigned (Table 1).

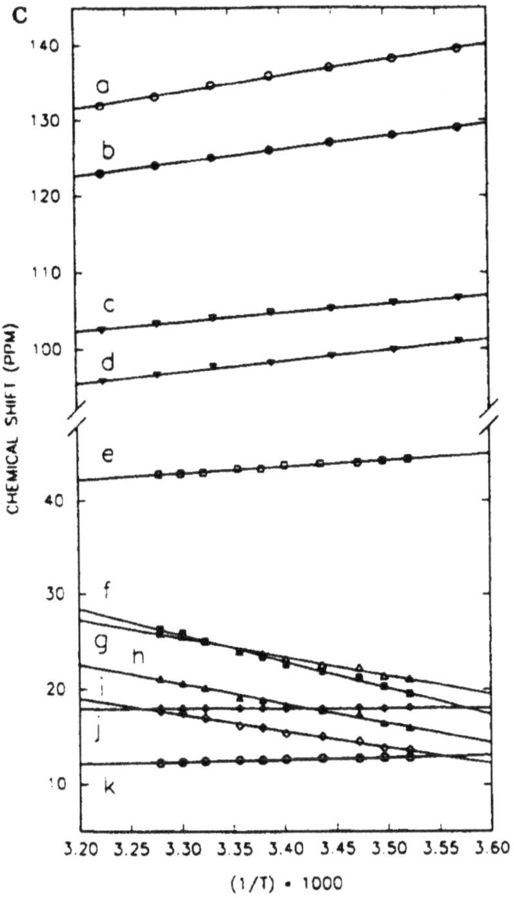

Figure 5. Temperature dependence of hyperfine proton signals (a-k) of *Anabaena 7120* vegetative ferredoxin (reproduced with permission from Skjeldal *et al.* [5]).

3.2.3. *Nitrogen Assignments*

Cysteine residues were also labeled selectively with [^{15}N]Cys. Four separate signals were resolved in ^{15}N NMR spectra of both oxidized and reduced vegetative ferredoxin (Figure 9). The hyperfine shifts were larger and more diverse in ^{15}N spectra of the reduced ferredoxin than of the oxidized ferredoxin. The ^{15}N signals from C46 was assigned by a double labeling ($^{13}C_i$ - $^{15}N_{i+1}$) strategy [6].

Table 1. Tentative assignments of hyperfine-shifted cysteinyl ^{1}H, ^{13}C, and ^{15}N resonances of oxidized and reduced *Anabaena 7120* vegetative ferredoxin.[a]

Cysteine Nucleus	oxidized state		reduced state	
	chemical shift (ppm)[b]	assignments	chemical shift (ppm)[b]	assignments
α-protons[c]	15.0	$C49H^{\alpha}$	43.5 (e)	$C49H^{\alpha}$
	9.0	$C79H^{\alpha}$	16.7 (j)	$C79H^{\alpha}$
			5.05	$C46H^{\alpha}$
			3.20	$C41H^{\alpha}$
β-protons[c]	48.2		134.6 (a)	$C79H^{\beta'}$
	36.0		125.0 (b)	$C79/C49H^{\beta}$
	28.0		104.2 (c)	$C49H^{\beta'}$
			98.0 (d)	$C49/C79H^{\beta'}$
			24.4 (f)	$C41H^{\beta'2}$
			24.4 (g)	$C41H^{\beta'3}$
			19.7 (h)	$C46H^{\beta'2}$
			16.7 (i)	$C46H^{\beta'3}$
β-carbons[d]	89.3 (1,2)		150.2 (1')	$C79C^{\beta}$
	89.3 (1,2)		123.4 (2')	$C49C^{\beta}$
	62.7 (3)	$C46\ C^{\beta}$	63.1 (3')	$C46C^{\beta}$
	56.9 (4)		-32.7 (4')	$C41C^{\beta}$
amide nitrogens[d]	155.0 (1)		362.1 (1')	
	152.9 (2)		324.5 (2')	
	131.3 (3)	C46N	155.6 (3')	C46N
	113.7 (4)	C41N	61.9 (4')	C41N

[a] Peak k was assigned originally to $R42H^{\alpha}$ (Skjeldal *et al.*, 1991b), but recent results rule this out [6].

[b] Chemical shifts are reported from spectra recorded at 298 K; peak lables are as in Figures 4 and 5.

[c] From Skjeldal *et al.* [5] and Cheng *et al.* [6].

[d] From Cheng *et al.* [6].

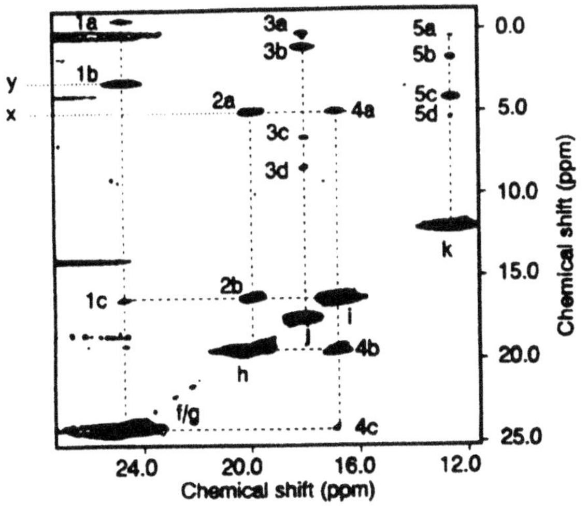

Figure 6. Two-dimensional NOESY spectrum of reduced *Anabaena 7120* vegetative ferredoxin. Only the hyperfine-shifted region is shown (reproduced with permission from Skjeldal *et al.* [5]).

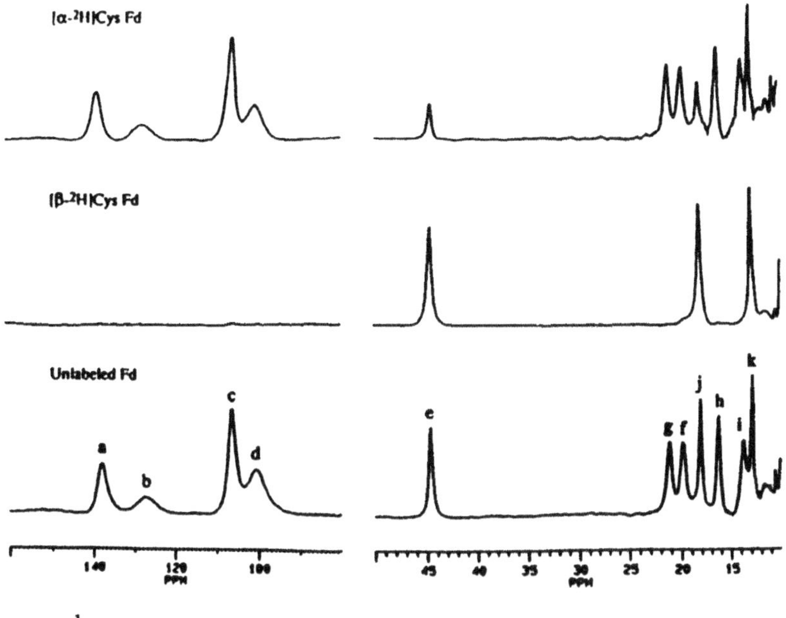

Figure 7. ^{1}H NMR spectra (499.84 MHz) of reduced *Anabaena 7120* vegetative ferredoxin: (top) selectively labeled with [^{2}H$^\alpha$]Cys ; (middle) selectively labeled with [^{2}H$^{\beta2,\beta3}$]Cys; (bottom) unlabeled (adapted from Cheng *et al.* [6]).

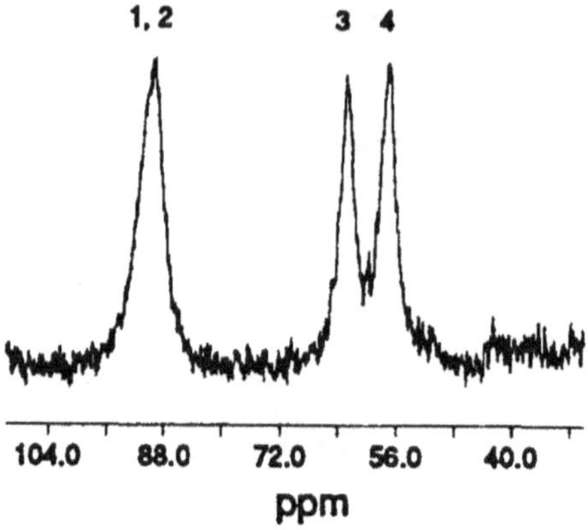

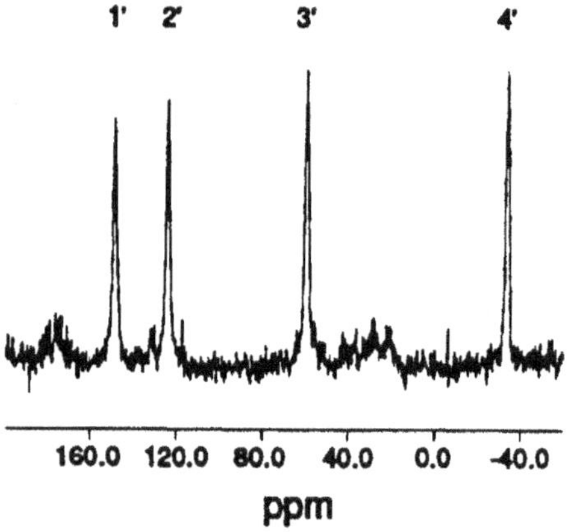

Figure 8. Hyperfine shifted ^{13}C signals of oxidized and reduced [$^{13}C^{\beta}$]Cys *Anabaena 7120* vegetative ferredoxin. Inset showing ^{13}C-decoupled hyperfine-shifted ^{1}H difference spectra of reduced [$^{13}C^{\beta}$] vegetative ferredoxin (adapted from Cheng *et al.* [6]).

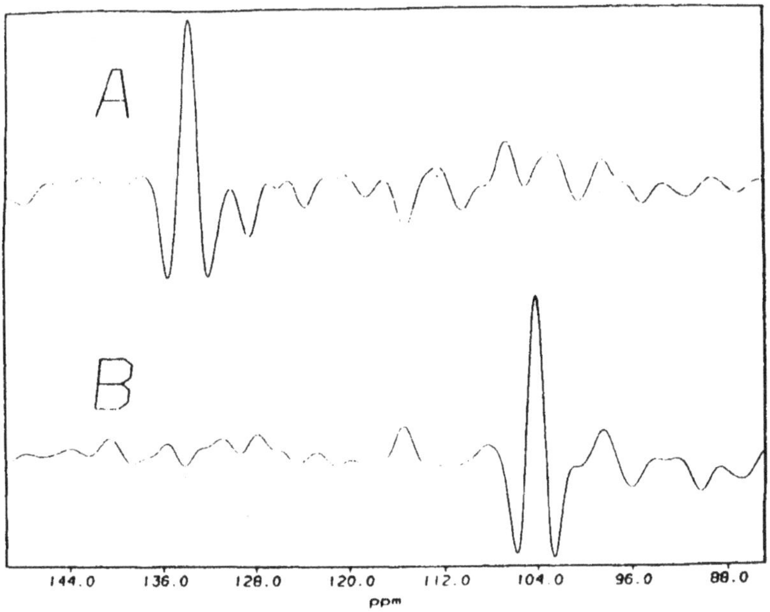

Figure 9. ^{13}C-decoupled hyperfine-shifted ^{1}H difference spectra of reduced [^{13}C$^{\beta}$] vegetative ferredoxin (adapted from Cheng *et al.* [6]).

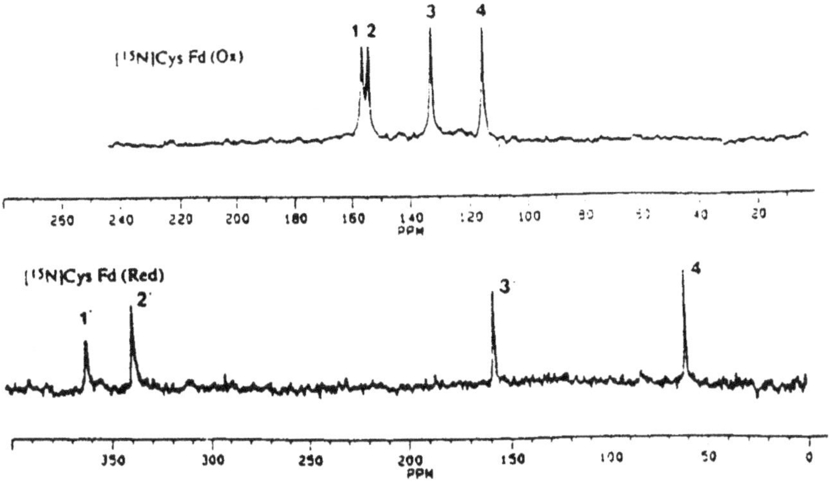

Figure 10. Hyperfine-shifted ^{15}N signals of oxidized and reduced [^{15}N]Cys *Anabaena 7120* vegetative ferredoxin (adapted from Cheng *et al.* [6]).

4. Solution Structure of Oxidized Wild-Type Ferredoxin

4.1. SECONDARY STRUCTURE ANALYSIS

The secondary structure of the *Anabaena 7120* vegetative ferredoxin was determined from NMR data by Oh and Markley [3]. It contains two α-helices (residues 25-32 and 68-76) and four β-strands (residues 2-8, 16-21, 51-54, and 85-91). The secondary structure was deduced from NOE connectivities and amide proton exchange rates. One α-helix showed the expected slow amide proton exchange rates and NOE connectivities, which implied a quite rigid α-helix. However, the other helix showed non-uniform amide proton exchange rates, which suggested that this helix is less rigid. An α-helical turn was detected at the C-terminus on the basis of strong H^N/H^N-type NOEs. The β-strands form a mixed-type β-sheet structure. Residues 2-8 serve as a central strand which is antiparallel to residues 16-21, and parallel to 85-91. A short β-strand (residues 51-54) is antiparallel to residues 85-91. The predicted secondary structure was confirmed by an analysis of H^α chemical shifts [10]. The NMR results on secondary structure are fully consistent with the x-ray structure [19].

4.2. ASSIGNMENT OF PROCHIRAL METHYLENES

The raw E. COSY [42] data were processed to final 4K x 4K frequency domain data points. Individual (H^α, H^β) and (H^α, $H^{\beta'}$) cross peaks were excised from the spectrum with the CHOPPER program [43], inverse Fourier transformed, zerofilled and Fourier transformed with the ZOOM program [44]. The final digital resolution was 0.34 Hz/point (ω_1) and 0.7 Hz/point (ω_2). The coupling constants were measured by the method of Kim and Prestegard [45]. Stereospecific assignments were determined by combining the coupling constant measurements and relative intensities of the (H^α, H^β) and (H^α, $H^{\beta'}$) cross peaks in NOESY spectrum [46].

4.3. INPUT DATA FOR STRUCTURE DETERMINATION

Distance constraints were derived from the NOESY spectrum. 415 interresidue and 101 intraresidue NOEs were identified from NOESY spectra. Cross peak intensities were scaled into three distinct distance ranges. These were scaled to distances by comparison with the number of contour levels of peaks from amino acids already known to adopt a regular secondary structure (α-helix or β-strand) as determined from NMR data as discussed above; in these, standard interproton distances [47] were assumed. Dihedral angle constraints were determined from H^α/H^N coupling constants. The measured constants from 55 residues were converted into intraresidue distances. Coupling constants between 6.5 to 8.0 Hz were not interpreted, however, because four different dihedral angles are compatible with such values. Forty dihedral angle constraints were used in the structure determinations. Hydrogen bond constraints were

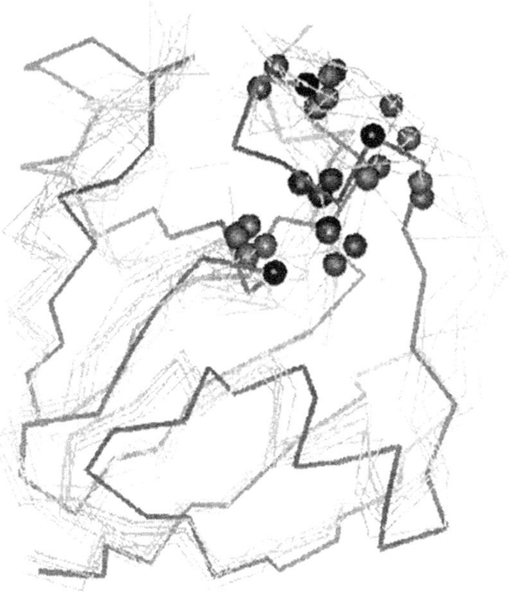

Figure 11. NMR structures of *Anabaena 7120* vegetative ferredoxin (adapted from Chae [46]).

used for all backbone amides whose signals were observed in the HMQC spectrum of [98 % UL ^{15}N] ferredoxin in $^{2}H_2O$. For those amide protons, the NH-O distance was given to an upper limit of 2.1 Å. Four covalent bonds to the cluster were also included in the structure calculation. Furthermore, 32 "paramagnetic constraints" were added to the constraint set: 16 nitrogen backbone signals were not detected in $^{1}H\{^{15}N\}$SBC, and close to [2Fe-2S] cluster as deduced from the primary sequence, which might be interpreted that they were close to the [2Fe-2S] cluster in the three dimensional space. The six best NMR structures are superimposed and compared with the x-ray structure in Figure 11.

5. Mutagenesis Studies

Each of the four cysteine residues of *Anabaena 7120* vegetative ferredoxin was individually mutated to serine: C41S, C46S, C49S, and C79S. Although all four mutants could be reduced by sodium dithionite, C41S and C79S showed very low stability at room temperature, particularly when reduced, which prevented the investigation of these two mutants by NMR. The other two mutants were studied by NMR in their reduced state [27].

The ^1H spectrum of the C46S mutant showed a hyperfine chemical shift pattern similar to that of the wild-type, but the magnitudes of the hyperfine shifts were much larger (Figure 12). The hyperfine peaks were assigned on the basis of chemical shift and T_1 relaxation comparisons with the wild-type ferredoxin. The most interesting result concerned peaks F and G (Figure 12) which were assigned to the β-protons of S46. These signals relaxed much faster than the β-protons of C46, which suggests that the distance between the β-protons and iron is shorter. This finding is consistent with x-ray results for the C79S mutant which show that the Fe-O distance (1.8Å) in the mutant is shorter than the Fe-S distance (2.3Å) in the wild-type ferredoxin [20].

When reduced, the C49S mutant showed a remarkably different hyperfine-shifted ^1H NMR spectrum from the others (Figure 13). This mutant showed a vertebrate-type pattern of the kind reported by Skjeldal *et al.* [16] for adrenodoxin and human ferredoxin. A broad signal was detected at the high field edge of the diamagnetic region. Its origin is unknown [27].

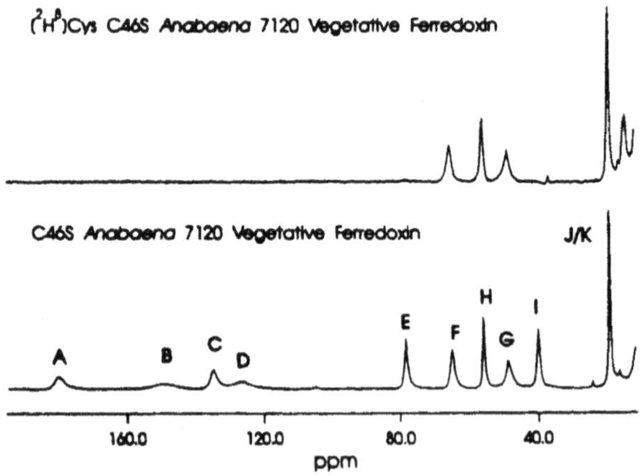

Figure 12. Hyperfine-shifted ^1H resonances of reduced C46S mutant: (top) selectively labeled with $[^1H^\beta]$Cys; (bottom) unlabeled (reproduced with permission from Cheng *et al.* [27]).

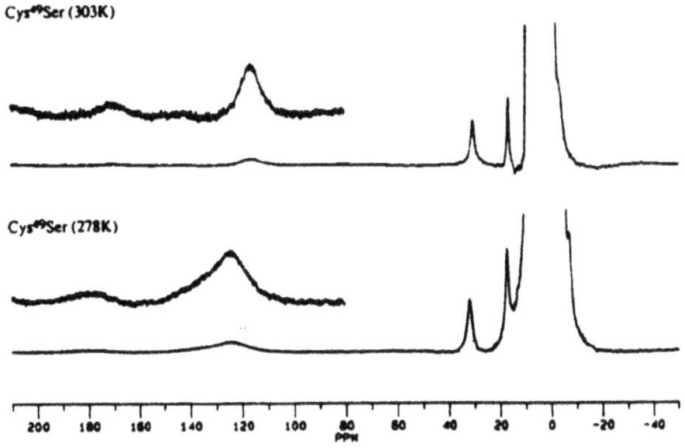

Figure 13. Hyperfine-shifted ^1H resonances of reduced C49S mutant at two different temperatures: 278K (bottom) and 303K (top) (reproduced with permission from Cheng *et al.* [27]).

6. Discussion

NMR spectroscopy is the most versatile tool available at present for studying macromolecules in solution since it reports on structure and dynamics. An x-ray structure represents a static structure, rather than a dynamic model. In addition, since crystal packing can alter structures in the solid state, it is useful to obtain structures in solution for reference. High-resolution x-ray structures have been solved for two of the [2Fe-2S] wild type ferredoxins we have studied by NMR: *Anabaena 7120* vegetative ferredoxin [18, 19] and heterocyst ferredoxin [13]. Difference between these two structures are minor. The RMSD between the α-carbon traces of the two x-ray structures is 1.0 Å, and the only significant differences in the polypeptide backbone between these ferredoxins occur in the loop from residue 9 to 24 and in the three C-terminal residues [13]. Comparison of NMR data for the two *Anabaena 7120* ferredoxins showed that the heterocyst ferredoxin is much less rigid than the vegetative in solution [10].

Functional studies showed that mutations at positions E94 and F65 reduce electron transfer to FNR by factors of 10^4 without disrupting the binding of ferredoxin to FNR that precedes electron transfer [21]. Thus, local structure and particular functional groups on the protein surface are required for specific electron transfer.

The electronic distribution around the cluster also must be considered in order to understand functional properties of ferredoxins. NMR provides a way of exploring the region around the paramagnetic cluster since the hyperfine-shifted NMR signals yield a "fingerprint" of the electronic distribution around the cluster. These signals contain information about differences in biophysical and biochemical properties (redox potential, EPR, stability, etc.) of various [2Fe-2S] ferredoxins.

Mutagenesis studies of *Anabaena 7120* vegetative ferredoxin [27] have shown that serine can replace cysteine as a cluster ligand and can support self assembly of the cluster.[2] However, the stabilities of the Cys → Ser mutants are lower than that of the wild-type ferredoxin. Thus, the main reason for the presence of cysteinyl ligands is to afford the stability of the protein both in the reduced and oxidized states. Metal substitution experiments have been attempted to see if the polypeptide backbone can accommodate metals other than Fe (Y. K. Chae, unpublished). Several metals were tried: Zn, Co, Ru, Al, Ga, and Cd. None of these gave consistent results in supporting cluster formation. Therefore, cysteinyl ligands and Fe ion can be thought of as an "optimal match" in the sense of overall protein folding and cluster formation.

Protein structure determination by NMR has been a great quest to many NMR spectroscopists. Methods developed with diamagnetic proteins for providing structural information are not applicable to regions surrounding paramagnetic centers like those of oxidized [2Fe-2S] ferredoxins where no NOE and coupling parameters can be detected.[3] Although the paramagnetic cluster broadens and shifts signals from nearby atoms, the hyperfine shifts and relaxation parameters contain information about local geometry. Improvements need to be made in the structural interpretation of the chemical shifts and relaxation parameters of hyperfine ^{13}C and ^{15}N signals. Recent studies of ^{15}N relaxation in [2Fe-2S] ferredoxins [11] indicate that the relaxation mechanism is predominantly dipolar and that such studies may provide the added information needed to refine the local structure around the iron-sulfur cluster.

7. References

1. Masaki, R., Yoshikawa, S., and Matsubara, H. (1982) Steady-state kinetics of oxidation of reduced ferredoxin with ferredoxin NADP reductase EC 1.18.1.2. *Biochim. Biophys. Acta* **700**, 101-109.
2. Oh, B.-H., Mooberry, E. S., and Markley, J. L. (1990) Multinuclear magnetic resonance studies of the 2Fe-2S* ferredoxin from *Anabaena* sp. strain PCC 7120: 2. Sequence-specific carbon-13 and nitrogen-15 assignments of the oxidized form, *Biochemistry* **29**, 4004-4011.
3. Oh, B.-H. and Markley, J. L. (1990) Multinuclear magnetic resonance studies of the 2Fe-2S* ferredoxin from *Anabaena* sp. strain PCC 7120: 1. Hydrogen-1 resonance assignments and secondary structure in solution of the oxidized form, *Biochemistry* **29**, 3993-4004.
4. Oh, B.-H. and Markley, J. L. (1990) Multinuclear magnetic resonance studies of the 2Fe-2S* ferredoxin from *Anabaena* sp. strain PCC 7120: 3. Detection and characterization of hyperfine-shifted nitrogen-15 and hydrogen-1 resonances of the oxidized form, *Biochemistry* **29**, 4012-4017.
5. Skjeldal, L., Westler, W. M., Oh, B.-H., Krezel, A. M., Holden, H. M., Jacobson, B. L., Rayment, I., and Markley, J. L. (1991) Two-dimensional magnetization exchange spectroscopy of *Anabaena 7120*

---

[2] Similar Cys to Ser mutations have been reported for the [2Fe-2S] ferredoxin from *Clostridium pasteurianum*. The mutant proteins have been studied by optical and EPR spectroscopy [48].

[3] This situation is quite different for heme proteins and high-potential iron-sulfur proteins (see paper in this volume by Banci *et al.*) whose signals are considerably sharper.

ferredoxin. Nuclear Overhauser effect and electron self-exchange cross peaks from amino acid residues surrounding the 2Fe-2S* cluster, *Biochemistry* **30**, 7363-7368.

6. Cheng, H., Westler, W. M., Xia, B., Oh, B.-H., and Markley, J. L. (1994) Protein expression, selective labeling, and analysis of hyperfine-shifted NMR signals of *Anabaena 7120* [2Fe-2S] ferredoxin, submitted.
7. Schrautemeier, B. and Böhme, H. (1985) A disttinct ferredoxin for nitrogenfixation isolated from heterocysts of the cyanobacterium *Anabaena variabilis*, *FEBS Lett.* **184**, 304-308.
8. Böhme, H. and Schrautemeier, B. (1987) Electron donation to nitrogenase in a cell-free system from heterocysts of *Anabaena variabilis*, *Biochim. Biophys. Acta* **891**, 115-120.
9. Böhme, H. and Schrautemeier, B. (1987) Comparative chracterization of ferredoxins from heterocysts and vegetative cells of *Anabaena variabilis*, *Biochim. Biophys. Acta* **891**, 1-7.
10. Chae, Y. K., Abildgaard, F., Mooberry, E. S., and Markley, J. L. (1994) Multinuclear, multidimensional NMR studies of *Anabaena 7120* heterocyst ferredoxin. Sequence-specific resonance assignments and secondary structure of the oxidized form in solution, Biochemistry **33**, 3287-3295.
11. Chae, Y. K. and Markley, J. L. (1994) Analysis of hyperfine-shifted nitrogen-15 resonances of the oxidized form of *Anabaena 7120* heterocyst ferredoxin, *Biochemistry*, submitted.
12. Jacobson, B. L., Chae, Y. K., Böhme, H., Markley, J. L., and Holden, H. M. (1992) Crystallization and preliminary analysis of oxidized recombinant heterocyst [2Fe-2S] ferredoxin from *Anabaena 7120*, *Arch. Biochem. Biophys.* **294**, 279-281.
13. Jacobson, B. L., Chae, Y. K., Markley, J. L., Rayment, I., and Holden, H. M. (1993). Molecular structure of the oxidized, recombinant heterocyst [2Fe-2S] ferredoxin from Anabaena 7120 determined to 1.7 Å resolution, *Biochemistry* **32**, 6788-6793.
14. Mason, J. I. and Boyd, G. S. (1971) The cholesterol side-chain cleavage enzyme system in mitochondria of human term placenta, *Eur. J. Biochem.* **21**, 308-321.
15. Simpson, E. R. and Miller, D. A. (1978) Cholesterol side chain cleavage cytochrome P-450 and iron sulfur protein in human placental mitochondria, *Arch. Biochem. Biophys.* **190**, 800-808.
16. Skjeldal, L., Markley, J. L., Coghlan, V. M., and Vickery, L. E. (1991) Hydrogen-1 NMR spectra of vertebrate [2Fe-2S] ferredoxins. Hyperfine resonances suggest different electron delocalization patterns from plant ferredoxins, *Biochemistry* **30**, 9078-9083.
17. Xia, B., Cheng, H., Skjeldal, L., Coghlan, V. M., Vickery, L. E., and Markley, J. L. (1994) Multinuclear magnetic resonance and mutagenesis studies of the histidine residues of human placental ferredoxin, submitted.
18. Rypniewski, W. R., Breiter, D. R., Benning, M. M., Wesenberg, G., Oh, B.-H., Markley, J. L., Rayment, I., and Holden, H. M. (1991) Crystallization and structure determination to 2.5 Å resolution of the oxidized [2Fe-2S] ferredoxin isolated from *Anabaena 7120*, *Biochemistry* **30**, 4126-4131.
19. Holden, H. M., Jacobson, B. L., Hurley, J. K., Tollin, G., Oh, B.-H., Skjeldal, L, Chae, Y. K., Cheng, H., Xia, B., and Markley, J. L. (1994) Structure-function studies of [2Fe-2S] ferredoxins, *J. Bioenerg. Biomemb.* **26**, 67-88.
20. Jacobson, B. L., Holden, H. M., Cheng, H., Xia, B., and Markley, J. L. (1993) Crystallization and structure determination of cysteine to serine mutants of *Anabaena 7120* vegetative ferredoxin, in preparation.
21. Hurley, J. K., Salamon, Z., Meyer, T. E., Fitch, J. C., Cusanovich, M. A., Markley, J. L., Cheng, H., Xia, B., Chae, Y. K., Medina, M., Gomez-Moreno, C., and Tollin, G. (1993) Amino acid residues in *Anabaena* ferredoxin crucial to interaction with ferredoxin-NADP reductase: site-directed mutagenesis and laser flash photolysis, *Biochemistry* **32**, 9346-9354.

22. Hurley, J. K., Cheng, H., Xia, B., Markley, J. L., Medina, M., Gomez-Moreno, C., and Tollin, G. (1993) An aromatic amino acid residue is required at position 65 in *Anabaena* ferredoxin for rapid electron transfer to ferredoxin:NADP$^+$ reductase, *J. Am. Chem. Soc.* **115**, 11698-11701.

23. Stockman, B. J., Westler, W. M., Mooberry, E. S., and Markley, J. L. (1988) Flavodoxin from Anabaena 7120: Uniform nitrogen-15 enrichment and hydrogen-1, nitrogen-15, and phosphorus-31 NMR investigations of the flavin mononucleotides binding site in the reduced and oxidized states, *Biochemistry* **27**, 136-142.

24. Sambrook, J., Fritsch, I, and Maniatis, E. F. (1989) *Molecular Cloning, A Laboratory Manual*, Cold Spring Harbor Laboratory Press, Cold Spring Harbor, NY, pp. A.3.

25. Muchmore, D. C., McIntosh, L. P., Russell, C. B., Anderson, D. E., and Dahlquist, F. W. (1989) Expression and nitrogen-15 labeling of proteins for proton and nitrogen-15 nuclear magnetic resonance, *Methods Enzymol.* **177**, 44-73.

26. Kunkel, T. A., Roberts, J. D., and Zakour, R. A. (1987) Rapid and efficient site-specific mutagenesis without phenotypic selection, *Methods Enzymol.* **154**, 267-381.

27. Cheng, H., Xia, B., Reed, G. H., and Markley, J. L. (1994) Optical, EPR, and ^{1}H NMR spectroscopy of serine-ligated [2Fe-2S] ferredoxin produced by site-directed mutagenesis of cysteine residues in recombinant *Anabaena 7120* vegetative ferredoxin, *Biochemistry* **33**, 3155-3164.

28. Piatini, U., Sørenson, O. W., and Ernst, R. R. (1983) Multiple quantum filters for elucidating NMR coupling network, *J. Am. Chem. Soc.* **104**, 6800-6801.

29. Jeener, J., Meier, B. H., Bachmann, P., and Ernst, R. R. (1979) Investigation of exchange process by two-dimensional NMR spectroscopy, *J. Chem. Phys.* **71**, 4546-4553.

30. Davis, D. G. and Bax, A. (1985) Assignments of complex ^{1}H NMR spectra via two-dimensianl homonuclear Hartman-Hahn specroscopy, *J. Am. Chem. Soc.* **107**, 2820-2821.

31. Griffey, R. H. and Redfield, A. G. (1987) Proton-detected heteronuclear edited and correlated NMR and nuclear Overhauser effect in solution, *Q. Rev. Biophys.* **19**, 51-82.

32. Bax, A. and Summers, M. F. (1986) Proton and carbon-13 assignments from sensitivity-enhanced detection of heteronuclear multiple-bond connectivity by 2-dimensional multiple quantum NMR, *J. Am. Chem. Soc.* **108**, 2093-2094.

33. Bax, A. and Marion, D. (1988) Improved resolution and sensitivity in ^{1}H-detected heteronuclear multiple-bond correlation spectroscopy, *J. Magn. Reson.* **78**, 186-191.

34. Shon, K. and Opella, S. J. (1989) Detection of ^{1}H homonuclear NOE between amide sites in proteins with ^{1}H/^{15}N heteronuclear correlationspectroscopy, *J. Magn. Reson.* **82**, 193-197.

35. Gronenborn, A, M., Bax, A., Wingfield, P. T., and Clore, G. M. (1989) A powerful method of sequential proton resonance assignment in protein using relayed ^{15}N-^{1}H multiple quantum coherence spectroscopy, *FEBS Lett.* **243**, 93-98.

36. Westler, W. M., Stockman, B. J., Markley, J. L., Hosoya, N., Miyake, Y., and Kainosho, M. (1988) Correlation of carbon-13 and nitrogen-15 chemical shifts in selectively and uniformly labeled proteins by heteronuclear two-dimensional NMR spectroscopy, *J. Am. Chem. Soc.* **110**, 6256-6258.

37. Mooberry, E. S., Oh, B.-H., and Markley, J. L. (1989) Improvement of ^{13}C-^{15}N chemical shift correlation spectroscopy by implementing time proportional phase incrementation, *J. Magn. Reson.* **85**, 147-149.

38. Oh, B.-H., Westler, W. M., Darba, P., and Markley, J. L. (1988) Protein carbon-13 spin systems by a single two-dimensional nuclear magnetic resonance experiment, *Science* **240**, 908-911.

39. Skjeldal, L., Westler, W. M., and Markley, J. L. (1990) Detection and characterization of hyperfine shifted resonances in the proton NMR spectrum of *Anabaena 7120* ferredoxin at high magnetic fields, *Arch. Biochem. Biophys.* **278**, 482-485.

40. Dunham, W. R., Palmer, G., Sands, R.H., and Bearden, A. J. (1971) On the structure of the iron-sulfur complex in the two-iron ferredoxins, *Biochim. Biophys. Acta* **253**, 373-384.

41. Dugad, L. B., La Mar, G. N., Banci, L., and Bertini, I. (1990) Identification of localized redox states in plant-type two-iron ferredoxins using the nuclear Overhauser effect, *Biochemistry* **29**, 2263-2271.

42. Griesinger, C., Sørenson, O. W., and Ernst, R. R. (1986) Correlation of connected transitions by two-dimensional NMR spectroscopy, *J. Chem. Phys.* **85**, 6837-6852.

43. Zolnai, Z., Macura, S., and Markley, J. L. (1989) Spline method for correcting baseplane distortions in the two-dimensional NMR spectra, *J. Magn. Reson.* **82**, 496-504.

44. Zolnai, Z., Macura, S., and Markley, J. L. (1989) ZOOM: a computer program for inverse Fourier transform, zerofilling, and Fourier transform of NMR data, *unpublished*.

45. Kim, Y.-M. and Prestegard, J. H. (1989) Measurements of vicinal couplings from cross peaks in COSY spectra, *J. Magn. Reson.* **84**, 9-13.

46. Chae, Y. K. (1994) Ph.D. thesis, University of Wisconsin - Madison.

47. Wüthrich, K. (1986) *NMR of Proteins and Nucleic Acids*, Wiley, NY, pp. 125-127.

48. Fujinaga, J., Gaillard, J., and Meyer, J. (1993) Mutated forms of [2Fe-2S] ferredoxin with serine ligands to the iron-sulfur cluster, *Biochem. Biophys. Res. Commun.* **194**, 104-111.

1D AND 2D PROTON NMR STUDIES ON [3Fe-4S] AND [4Fe-4S] FERREDOXINS ISOLATED FROM *Desulfovibrio gigas*

ANJOS L. MACEDO and JOSÉ J. G. MOURA*
Centro de Química Fina e Biotecnologia, Departamento de Química, Faculdade de Ciências e Tecnologia, Universidade Nova de Lisboa, 2825 Monte de Caparica-Portugal

ABSTRACT

Two-dimensional Proton NMR was used to identify the four cysteinyl cluster ligands (Cys 8, Cys 11, Cys 14 and Cys 50) in [4Fe-4S] *Desuflovibrio gigas (Dg)* FdI and to structuraly assign the three cysteinyl ligands (Cys 8, Cys 14 and Cys 50) of the [3Fe-4S] core in *Dg* FdII. The other two cysteinyl residues present (Cys 18 and Cys 42) are connected forming an intramolecular disulfide bridge in *Dg* FdII. The use of NMR to follow cluster interconversion is also demonstrated and Cys 11 is indicated to have a crucial role on this process. 2D NOE measurements in native *Dg* FdII, in conjunction with the available X-ray crystallographic coordinates and the temperature dependence studies of a pair of β-CH_2 protons of cluster bound cysteines were used to calculate the coupling constant between the iron sites. A combined NMR and Mössbauer study on native and redox intermediate states of *Dg* FdII suggest that a total of three electrons are transfered to the protein indicating redox activity associated with a conformational change occuring when forming or breaking the S-S bridge.

Key words: Sulfate Reducing Bacteria, *D. gigas* ferredoxins, [3Fe-4S] and [4Fe-4S] clusters, redox active disulfide bridge, conformational changes, cluster interconvertions, 1D and 2D NMR, assignment of iron-sulfur cysteinyl ligands.

---

*Correspondent author
ABBREVIATIONS: NMR- nuclear magnetic resonance; EPR- electron paramagnetic resonance; COSY- two dimensional correlated spectroscopy; M-COSY- magnitude COSY; NOE- nuclear Overhauser effect; NOESY- two dimensional nuclear Overhauser spectroscopy; Fd- ferredoxin; *D.- Desulfovibrio*

G.N. La Mar (ed.), Nuclear Magnetic Resonance of Paramagnetic Macromolecules, 319-338.

1. Introduction

Iron-sulfur proteins are defined as mono and multi-iron complexes generally tetrahedral coordinated to inorganic and cysteinyl thiol sulfur. Established basic structures are [Fe(4Cys), [2Fe-2S], [3Fe-4S] and [4Fe-4S] cores [1-3]. A recent addition is the 6Fe structure [3]. During the last years, the large effort devoted to the study of iron-sulfur proteins indicate that the simple chemical constitution of the clusters contrasts with an high versatility in terms of structural features and reactivity. The research in the field faces different challenges: new structural arrangements some implying higher coordination numbers than four at the iron site, different coordinating atoms at the cluster than sulfur, direct substrate binding to the cluster, cluster interconversions, unexpected catalytic capabilities and other roles than redox such as participation of the iron-sulfur cores in regulation of metabolic pathways [4-9].

In this article we review the application of 1D and 2D NMR as a valuable tool for the study of *Dg* Fds, where most of these points have been addressed. *Dg* Fds represent a particular situation where the same polypeptide chain accomodates two types of clusters forming two ferredoxins: a single [3Fe-4S] cluster similar to a [4Fe-4S] core but lacking one iron atom at one corner of the cube in *Dg* FdII or a [4Fe-4S] cluster in *Dg* FdI. The two ferredoxins were reported to have different molecular mass due to oligomerization of the same monomeric unit and indicated to have different redox and magnetic properties and distinct physiological roles [10-13]. The primary structure contains 58 amino acids and six cysteinyl residues arranged in the following sequence:

...-CYS8-X-X-CYS11-X-X-CYS14-X-X-CYS18-PRO-.l-CYS42-..-CYS50-PRO-...

The X-ray structural analysis indicated that in *Dg* FdII the cluster is bound to the polipeptide chain by three cysteinyl residues (Cys 18, 14 and 50) [14-16]. Cys 18 and Cys 42 form a disulfide bridge (Figure 1). Cys 11 was found to be tilted towards the solvent away from the cluster. Experiments have shown that the two forms can be interconverted by adding or removing one iron atom. Then, Cys 11 has been indicated to play a role on the interconversion of the [3Fe-4S] core into a [4Fe-4S] core becoming the fourth ligand. Other ligands may be available for coordination and other ligands could be proposed such as Glu 12 or an exogenous ligand such as water (or OH-). This is not an unexpected situation since Cys 11 is not a conserved residues when bacterial ferredoxins amino-acid sequences are compared and can be replaced by an aspartate. Carboxylic cluster ligation has been postulated in addition to the three cysteinyl residues in the *D. africanus* FdIII [17] and a OH^- as a fourth ligand in *Pyrococcus furiosus* Fd [18].

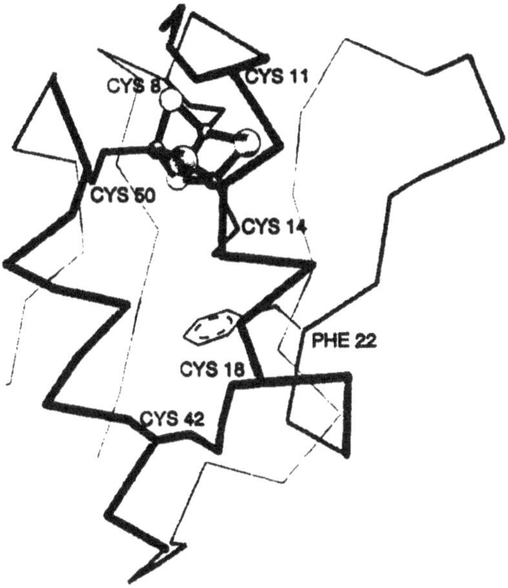

Figure 1. Structural features of [3Fe-4S] cluster *D. gigas* FdII, based on the X-ray coordinates [14-16]. The figure gives relevance to the coordinated cysteines as well as the relative position of the aromatic ring of the single aromatic residue (Phe 22) and the localization of the disulfide bridge.

Dg FdI contains mainly a [4Fe-4S] cluster and in the absence of a X-Ray structure the putative ligands are Cys 8, Cys 11, Cys 14 and Cys 50, and Cys 18 and Cys 42 are assumed to be connected forming a disulfide bridge.

The indicated features made the [3Fe-4S] core in FdII a prototype for probing parameters controlling the cluster interconversion process and the formation of heterometal containing clusters of the type [M,3Fe-4S]. The interconversion step enabled the specific isotopic labelling of the core, which is a relevant tool when site differentiation within the core is addressed [11, 19-21].

2. [3Fe-4S] and [4Fe-4S] Cluster Oxidation States

The [3Fe-4S] cluster can be stabilized in solution in two oxidation states. The oxidized state, $[3Fe\text{-}4S]^{1+}$, contains three high-spin ferric ions spin-coupled to form a S=1/2 ground state [22]. Upon one electron reduction of the cluster, $[3Fe\text{-}4S]^{0}$, a S=2 state is attained resulting from the antiferromagnetic coupling between an high spin ferric site (S=5/2) and a delocalized iron pair (S=9/2), that share the incoming electron [21-25]. In addition, it has been postulated based on electrochemical studies, that this cluster can be further reduced (2e step) into a all ferrous state, [3Fe-4S]2- [26,27]. These observations indicate that the low molecular mass [3Fe-4S] *Dg* FdII would be able to transfer more than one electron. In addition, the active role of the disulfide bridge has been implied (see section 8.) [28].

Three stable oxidations states (+3, +2 and +1) have been observed for the [4Fe=4S] clusters. The +3 and +2 states are stable in high-potential iron-sulfur proteins (HiPIP), while the +2 and +1 states are detected for bacterial ferredoxins and other tetranuclear iron cluster containing proteins [29]. Both the +3 and +1 states are paramagnetic having a cluster spin of 1/2. The +2 state is diamagnetic. In all three oxidation states, valence delocalized pairs are observed. Theoretical calculations indicate that the concept of valence localization (as well as delocalization) is important for the understanding of iron-sulfur core electronic states [30,31].

3. NMR and Iron-Sulfur Clusters

Earlier NMR work has shown the potentialities of the use of one- and two-dimensional techniques to study contact shifted resonances from the coordinated cysteinyl residues to elucidate the cluster electronic and magnetic properties and to obtain protein structural information [32-38]. Even the clusters being diamagnetic, S=0, ground states have sufficient fraccional population of paramagnetic excited states at 300 K and hyperfine shifted resonances are observed. However, the observation of NMR spectra of paramagnetic systems is not an easy task and most of the studies have been using 1D NOE routines. Methods based on bond and spatial correlations (spin-spin connectivity), M-COSY and NOESY techniques were only recently used to assign β-CH_2 cysteinyl proton pairs of [2Fe-2S] cluster cysteinyl ligands in trapped valence situation [39-40], *Azotobacter vinelandii* and *Pseudomonas putida* [3Fe-4S]+[4Fe-4S] Fd [41], *Clostridium pasteurianum* 2x[4Fe-4S] Fd [42-45], several HIPIPs [46-52] and [4Fe-4S] Fd interconverted into a [3Fe-4S] containing protein from *Pyrococcus furiosus* and *Thermococcus litoralis* [53]. Also, the study of the temperature dependence of the contact shifted cysteinyl resonances (cluster ligands) have been extensively used to probe the electronic structure of the cores [39,42,45,47].

Combined use of molecular dynamics simulations and NMR were recently introduced as an approach to the study of iron-sulfur containing proteins. The solution structure of Chromatium HiPIP was modeled and compared with the X-Ray data [54].

We will summarize, next, the main features of the spectra of native (oxidized) and dithionite reduced *Dg* FdI and *Dg* FdII relevant for cluster ligand assignments and for probing the cluster electronic structure, as well as monitoring cluster conversion and conformational changes that take place during redox cycling of the proteins [30,55].

4. Two-Dimensional NMR Studies of the [4Fe-4S] Cluster in *Dg* FdI

Details of the 500 MHz Proton NMR spectra of native (oxidized, S=0) and dithionite reduced (S=1/2) Dg FdI are shown in Figure 2-A,B. Native Dg FdI present six well resolved resonances (labeled from a-f) in the low field region from 18 to 10 ppm. Several resolved resonances (g-o, w and y) can also be detected in the aromatic region. These resonances have characteristic sh ort T1 values and large line widths.

Eight well resolved peaks (a'-h') can be detected in the region between 11 and 50 ppm in the reduced state. They have also a characteristic relaxation behavior (Table I). The aromatic region is again very well resolved (Figure 2-B). The indicated resonances correspond to one proton intensity and their chemical shifts are temperature dependent. The oxidized state shows an uniform anti-Curie dependence and the reduced form a mixed behavior, Curie and anti-Curie.

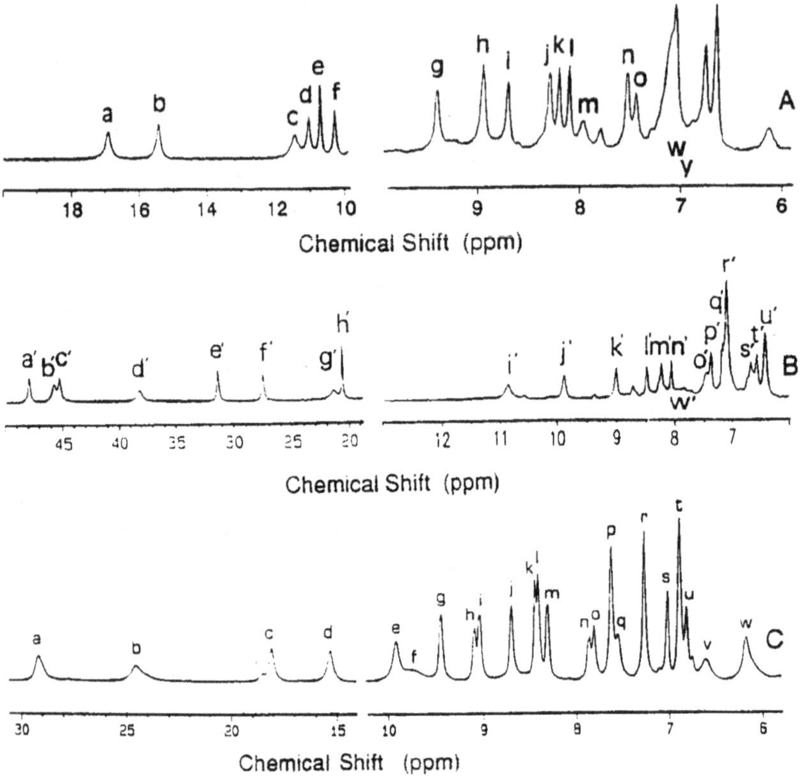

Figure 2 **500 MHz ^{1}H NMR spectra of *D. gigas* Fds in D_2O, pH 7.6 at 293 K.**
Resonances are labeled from a,b,... starting from the low-field region.
A - *Dg* FdI (oxid); B - *Dg* FdI (reduced); C - *Dg* FdII (oxidized)
The figure show an expanded low-field region and details on the aromatic spectral region.
Resonances labelled with a prim refer to reduced state.

1D nuclear Overhauser, M-COSY and NOESY experiments were used in order to identify the geminal pairs of β-CH protons and its vicinal α-CH partner from the cysteine cluster ligands. 2D NOESY experiments (Figure 3-A), with short mixing time (10 ms) detect connectivities between four pairs of β-CH protons: resonances a and f; b and o; c and m; and d and y.

1D NOE experiments confirmed and extended these results: saturating peaks a and b also shown correlation with resonances e and w, respectively, in addition to the detected NOESY cross peaks described for the geminal partners, leading to the attribution of these resonances to α-CH protons. Assignment to β-CH protons (i.e., resonances f and o) were based on the relative intensity (more intense) NOE observed.

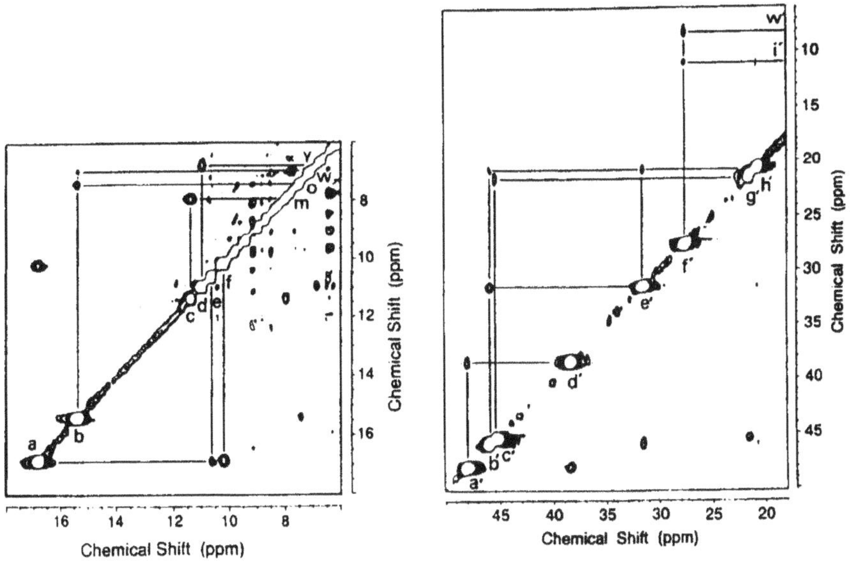

Figure 3 Low field region of the 500 MHz ^1H NMR NOESY spectrum of *Dg* FDI.
PANEL A - native (oxidized) *Dg* FdI in D_2O, pH 7.6 at 293 K, 10 ms mixing time. PANEL B - dithioniate reduced *Dg* FdI in D_2O, pH 8.2 at 298 K, 5 ms mixing tim.
The map illustrates the geminal connectivities of the Cys β-CH_2 signals and some of the vicinal α-CH. The peaks are labeled as in Figure 2.

The 5 ms mixing time NOESY experiment of a dithionite reduced sample of FdI is presented in Figure 3-B. Several connectivities can be detected in the region 50-10 ppm, where three pairs of resonances are assigned: a' and d'; b' and e'; and c' and g'. Resonances b' and e' also present connectivities with resonance h'. Resonance f' shows cross-peaks with two resonances out of that envelop (i' and w') around 11 and 8 ppm. Table I gives a summary of these assignments.

TABLE I - NMR spectral parameters and assignment of the cluster cysteines protons of *Dg*FdI

Oxidized State (native)				Reduced State			
Cys	Signal	T1(ms)	Assignment	Cys	Signal	T1(ms)	Assignment
A	a	7	β-CH	A′	a′	18	β-CH
	f	11	β-CH		d′	4	β-CH
	e	31	α-CH				
B	b	11	β-CH	B′	b′	17	β-CH
	o	35	β-CH		e′	15	β-CH
	w	36	α-CH		h′	29	α-CH
C	c	6	β-CH	C′	c′	17	β-CH
	m	12	β-CH		g′	4	β-CH
D	d	13	β-CH	D′	f′	11	β-CH
	y	35	β-CH		w′	16	β-CH
					i′	29	α-CH

The protons peaks are labeled as indicated in Figure 2.

A remarkable difference was observed when comparing oxidized and reduced state plots of the chemical shifts in function of the reciprocal temperature [55]. The oxidized state shows a uniforme anti-Curie law dependence. The same type of behavior was observed for the equivalent magnetic states of reduced HIPIP proteins and oxidized bacterial [4Fe-4S] ferredoxins in a diamagnetic state (S=0) [42,47].

Such a behavior as been taken as the demonstration that electronic delocalization takes an important role and although the irons of the core contain formal oxidations states 2 Fe^{2+} and 2 Fe^{3+} it is not possible to distinguish the ferric from the ferrous atoms, as it is the case of trapped valence systems such as [2Fe-2S] ferredoxins [39,40]. Both Curie and anti-Curie behavior is observed upon one electron reduction of the [4Fe-4S] cluster of *Dg* FdI (formal valences 1 Fe^{3+} and 3 Fe^{2+}). Again, prototypes of this magnetic state has been explored such as reduce *B. polymyxa*, *B. stearothermophylus*, *B. thermoproteolyticus*, *C. pasteurianum* Fds [45, 56, 58]. Oxidized HIPIPs have an equivalent magnetic state (S=1/2) but correspond to a two electron more oxidized state (formal valences 3 Fe^{3+} and 1 Fe^{2+}) [47, 48, 51, 58] and also show a mixed Curie and anti-Curie law dependence. Due to the difficulties of considering the simple frame of the Heisenberg model that could not reproduce neither the temperature dependence nor the correct electronic ground state, Bertini and co-workers proposed a model (based on Noodlemann approach [30]) for the interpretation of the reduced state of the [4Fe-4S] core assuming a $S_{12}S_{34}S$ coupling system (site 4 correspondes to the ferric ion) and impose a small decrease on J34 in order to get a non degenerated ground state with S=1/2 (resulgint from the antiferromagnetic coupling between S12=4 and S34=9/2). The analysis of the temperature dependence of the isotropic shifted resonances indicate that Cys A' and Cys D' (Curie) and Cys B' and Cys C' (anti-curie) show a similar behavior as observed for other reduced [4Fe-4S] cores and is tempting to assign the special pair S=9/2 to the iron sites coordinated by Cys A' and Cys D' [55].

It is clear from the data of Table I that bond correlation provides unambiguous identification of all the methylene protons for each of the four cysteines in *Dg* FdI. This is a direct evidence that four cysteinyl residues bind the [4Fe-4S] cluster. Cys 11 participates as a cluster ligand when the fourth site of the core is occupied.

5. 2D NMR Studies of the [3Fe-4S] Cluster in Dg FdII

A 500 MHz Proton NMR spectrum of oxidized *Dg* Fd II from shown in Figure 2-C. Isotropically shifted resonances (labeled a-z) are detected in the 6-30 ppm region. Their lage down-field shifts indicate again that they are most likely originating from protons of the cysteinyl residues coordinated to the [3Fe-4S] cluster. The low-field spectral region of native *Dg* FdII, resolves four broad resonances (a-d) between 30-10 ppm [55,59]. In the aromatic region the resonances are sharper and labeled from e-w. Table II indicates T1 values for relevant resonances. Cross-peaks are detected on the NOESY spectrum between peaks a and b; peak d and peaks e and y (~8.5 ppm); and peak c and peaks at ~3.5 and 1.5 ppm. The smaller linewidth and longer T1 of peak e indicates its origin from an α-CH proton counterpart, supported by the relative intensity of the observed NOESY map. Irradiation of peak d has effect on the resonances of the single aromatic residue (Phe 22), assigned in TOCSY experiments [55].

The temperature dependence study of oxidized *Dg* FdII indicate that except for the peaks a and b, which show temperature dependence of the Curie type, the others exhibit anti-Curie type dependence. The observed different types of temperature dependences for these resonances are indicative of the presence of a nearby paramagnetic spin-coupled iron cluster, and reflect the intrinsic spin orientations of the iron atoms. Consequently, peaks a and b are attributed to the β-CH_2 protons of a same cysteine residue. Further evidence in support of this assignment was obtained from 1D and 2D proton nuclear Overhauser effect (NOE) [55].

Two resonances at 18 and -12 ppm can be resolved from the diamagnetic envelope, that characterized the reduced state of *Dg* FdII (S=2). Other broad resonances can be detected in the very low-field region, at around 200 ppm, as well as in the high field region around -80 ppm [55 and our unpublished results]. These resonances are difficult to detected and hvae line widths around 10 ppm.

5.1. SPIN COUPLING MODEL FOR OXIDIZED [3Fe-4S] CLUSTER

In order to gain information concerning the electronic structure of and the spin-spin interactions within the [3Fe-4S] cluster in oxidized *Dg* FdII, a spin-coupling model was proposed, following a model introduced by Kent *et al.* [22] to calculate the energy level scheme of the cluster, and employing the theory developed by Bertini *et al.* [47] to estimate the contact shifts of the coordinated cysteine protons [59]. Based on Mössbauer and electron paramagnetic resonance data, Kent *et al.* had shown that the electronic structure of the oxidized [3Fe-4S] cluster can be characterized as a system containing three exchange-coupled high-spin ferric ions (S = 5/2) with the exchange interaction described by the Heisenberg-Dirac-Van Vleck spin Hamiltonian:

$$H = J_{12}S_1 \bullet S_2 + J_{13}S_1 \bullet S_3 + J_{23}S_2 \bullet S_3 \quad (1)$$

where $S_1 = S_2 = S_3 = 5/2$ are the intrinsic spin numbers of the ferric ions, and J_{ij} is

cm^{-1} for J has been determined for the oxidized [3Fe-4S] cluster [63]. In consistent with the magnetization measurements, our analysis of the proton NMR isotropic shifts indicates that $J \approx 300$ cm^{-1}. This is still a non solved question.

5.2. STRUCTURAL ASSIGNMENT OF CYSTEINYL LIGANDS IN *Dg* FdII

5.2.1. *Cysteine 50*

The pattern of the contact shifts of the β-CH_2 protons are determined by the orientation of the cysteine ligand with respect to its coordinated Fe, and the ratio of their contact shifts can be estimated based on the calculation of the dihedral angle defined by the planes Fe-S-C_β and S-C_β-C_α [64, and references there in]. The unique temperature dependence of the resonances a and b and has provided the means for their unambiguous assignment. Since the crystallographic staructure of the oxidized *Dg* FdII has been determined to high resolution [16], attempts can be made to identified the cysteine residue responsible for the resonances a and b. The cluster is bound to the polypeptide chain by three cysteine residues: Cys 8, Cys 14, and Cys 50 and the dihedral angles could be calculated. The experimental determined ratio of the observed contact shifts for the peaks a and b compared with the theoretical data, suggested that the resonances a and b may be attributed to Cys 50 [59].

Although there is no apparent reason for the iron sites to retain their symmetry upon reduction, it is intriguing to note that the reduced [3Fe-4S] cluster does contain a distinguishable ferric site and the additional electron is shared by the two other iron atoms [65]. Since the temperature dependence of proton NMR shifts contain detailed structural and electronic information about the cluster, as demonstarted in this report, it would be interesting to perform a similar measurement on the reduced *Dg* FdII and to attempt to identify the ferric and the paired ferric-ferrous sites.

5.2.2. *Cysteines 8 and 14 - Phenylalanine 22 as a structural marker*

The assembly of results obtained by the described 2D NMR experiments conducts to the fully recognition of the β-CH_2 pairs of protons from the cysteinyl ligands (see Table II). In addition, the data collected reveal part of a set of NOE connectivities occuring on the whole protein, in particular in the vicinity of the cluster. Following the assignment of the α-CH and β-CH_2 cysteinyl protons it is clear that additional effects are observed on other resonances. Irradiation of resonance d shows effects on e and y as indicated and additional effects on Phe 22 aromatic proton resonances. The X-ray structure available for Dg FdII was used as a starting point and the distances of all the residues close to the cysteine ligands of the cluster were compared (x). Phe 22 is close to Cys 14 (see Figure 1). The experimental data support the assignment of resonances d (b), e (a) and y (b) to Cys 14. Cys 50 was previously assigned to resonances a (b) and b (b). This leaves the final structural assignment of Cys 8 to resonances c (b), ~3.5 ppm (b) and ~1.5 ppm (a) (see Table II) [55].

Table II - NMR spectral parameters and assignment of coordinated cysteinyl protons of native *Dg*FdII

Cys	Signal	T1 (ms)	Assigments
A (Cys 50)	a b	4.3 3.1	β-CH β-CH
B (Cys 14)	c ~3.5 ppm ~1.5 ppm	4.0 n.d. n.d.	β-CH β-CH α-CH
C (Cys 8)	d y e	7.0 n.d. n.d.	β-CH β-CH α-CH

n.d. not determined
The protons peaks are labeled as indicated in Figure 2.

6. NMR Detection of Cluster Alterations Upon Redox Cycling - Cluster Interconversions

The NMR spectrum of oxidized *Dg* FdI shows variability depending if the protein is isolated anae- or aerobically. Figure 4-A indicates the 300 MHz NMR spectrum of the native oxidized aerobocally purified *Dg* FdI. A few resonances are also observed in the low-field region but with a distinct pattern of the previously described one (Figure 2). The resonances appear at 16, 13 and 12 ppm. The sample can be reduced by dithionite yield the spectrum indicated. It is clear that this sample is only 90% reduced. The reduced and oxidized forms are in slow exchange regime in the NMR time scale as observed upon reoxidizing the sample (Figure 4-B,C and D). The fully reoxidized state (Figure 4-E) shows a distinct pattern when compared with the starting material but consistent with the spectral features always observed in different aerobic preparations of native *Dg* FdI and interconverted *Dg* FdII. These results suggests that the [4Fe-4S] can adopt different conformational states. In addition we observe that the EPR spectrum of reduced 4Fe core in *Dg* FdI can be interpreted as being constituted by two rhombic "g=1.94" species at 2.05 (80%) adn 2.03 (20%) [our unpublished results]

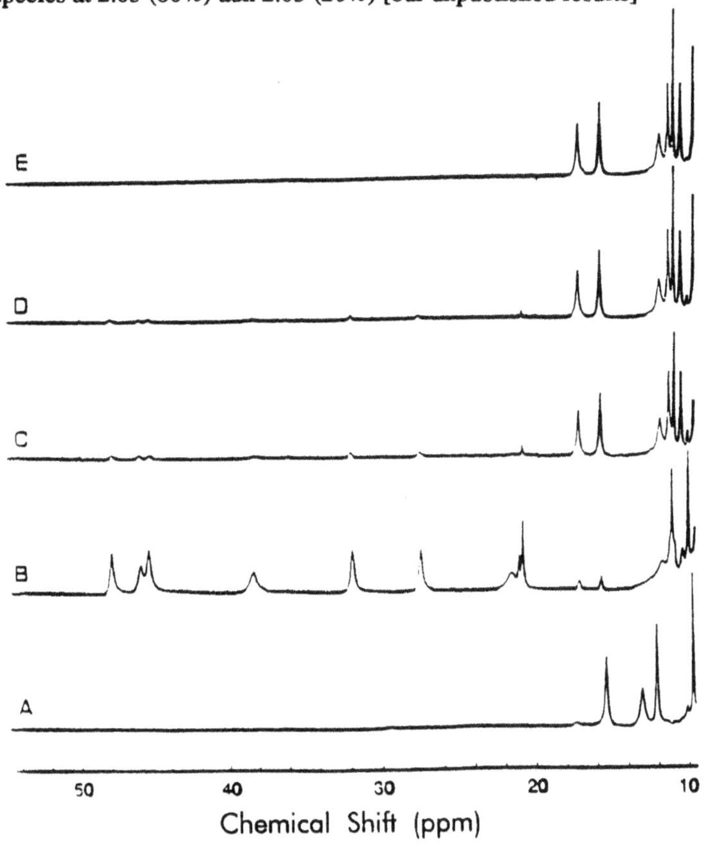

Figure 4 Low field region of the 300 MHz ^{1}H NMR spectra of *Dg* FdI at 303 K.
A- aerobic preparation, native oxidized state; B to D- stepwise reoxidation; E- fully reoxidized sample. Adapted ref. 28.

As indicated before, the interconversion between these clusters is facile. Under reducing conditions, in the presence of dithionite excess, the redox cycling of *Dg* FdII led to the interconversion of the [3Fe-4S] center into a [4Fe-4S] one (Figure 5).

The "Finger print" spectra previously described can be used to visualize the cluster transformationl. In particular the resonances at 18 and 16 ppm are a good reference point for the detection of [4Fe-4S] centers in the oxidized state. The EPR analysis of this samples after three oxidation/reduction cycles supports these conclusions.

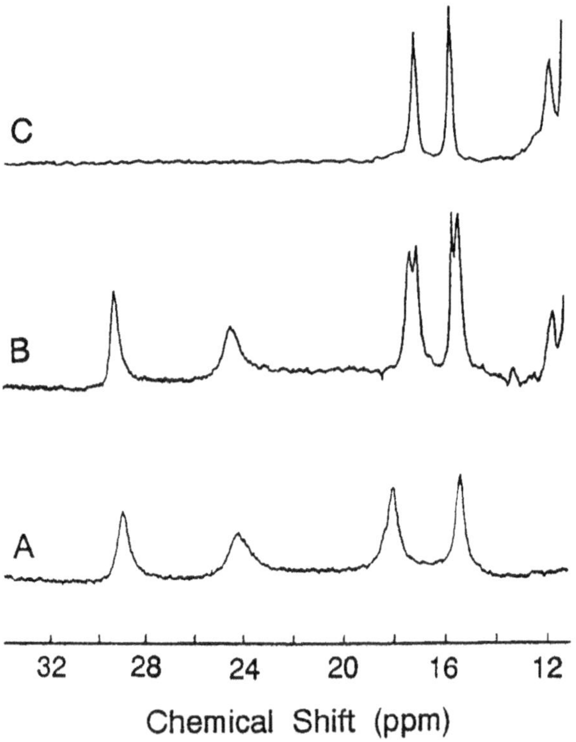

Figure 5 - Low field region of the 300 MHz ^{1}H NMR spectra of *Dg* FdII at 303 K. A- anative oxidized state; B- air reoxidized sample after dithionite reduction (one redox cycle); C- conversion of 3Fe into 4Fe core after three redox cycles. Adapted ref. 28.

7. Thiol/Disulfide Formation Associated with Redox Activity of *Dg* FdII [3Fe-4S] Cluster - Mössbauer and NMR Correlations

As indicated before, the ^{1}H NMR spectra of *Dg* FdII in the oxidized state exhibited four low-field contact shifted resonances at 29, 24, 18 and 15.5 ppm, whereas the reduced state yields two features at 18.5 and -11 ppm. In the course of studying the redox behaviour of *Dg* FdII, a stable intermediate was detected that yields a distinct NMR spectra with features at 24, 21.5, 21 and 14 ppm [28]. This intermediate is generated in the potential range where the cluster (E_0=-130mV) is reduced from 1+ to 0 state. The intermediate could be generated independently of the reducing agent used (dithionite, H2/Hydrogenase system) or after partial reoxidation (2,6 dichloro-indophenol or air). Mössbauer and EPR demonstrate unambiguously that the clusters in the native and in the intermediate redox species are in the same oxidation state. However, Mössbauer spectroscopy can detect modifications of the exchange couplings among the three ferric atoms. Quantitative titrations indicated that the protein can accept three electrons when the solution potential is lowered from -50 to -200 mV. One electron is taken by the metal core. The two other electrons are proposed to reduced the disulfide bridge (Cys 18-Cys 42), a result supported by titrations using disulfide forming reagents [28]. Square wave voltametric measurements also indicated a complex process associated with the cluster redox transition [x]. The differences in the NMR and Mössbauer spectra of these two species where interpreted as resulting of a conformational change following the break and formation of the disulfide bridge that is located 1.3 nm away from the cluster. A pathway for transmitting the change in the conformation of Cys 18 to the cluster is clearly indicated by inspection of the X-Ray structure which show that the carbonyl group of Cys 14 (a cluster ligand) is hydrogen bonded to the NH group of Cys 18 [67]. The reduction of the disulfide seems to be mediated by the [3Fe-4S] cluster.

8. NMR Data on Heterometal Containing Clusters

Moura *et al.* demonstrated in 1982 the facile conversion process ([3Fe-4S] → [4Fe-4S]) in *Dg* FdII [21]. Using this concept, the same authors could then synthesize heterometal cluster, within *Dg* FdII core, of the type [M,3Fe-4S], M=Co, Zn, Cd, Ga and Ni [65-70]. Mössbauer and EPR studies define electronic distributions and spin states of this new compounds. The extension of these studies using NMR is anticipated with interest. A preliminary 1D study was already initiated for the heterometal clusters $[Co,3Fe\text{-}4S]^{2+}$ (S=1/2) and $[Zn,3Fe\text{-}4S]^{+1}$ (S=5/2) [our unpublished results).

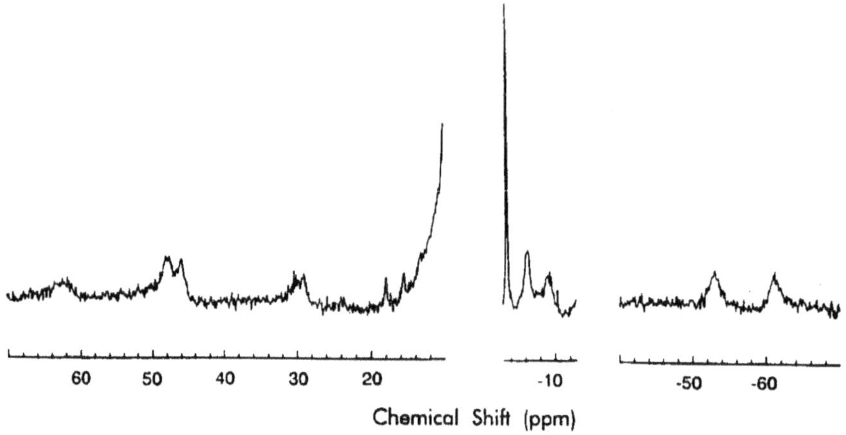

Figure 6 Low field region of the 300 MHz ^1H NMR spectra of $[Co,3Fe\text{-}4S]^{2+}$ formed in *Dg* FdI at 303 K.

9. Conclusions - Perspectives

NMR methods were used to probe the chemical environment of the iron-sulfur core in *Dg* Fds and derived herometal clusters. It was shown that NMR spectra can now be used as a finger print of each cluster in both oxidation states and at intermediate redox levels. Also, the interconversion process of [3Fe-4S] into [4Fe-4S] cores, during the redox cycles performed, can be followed. The possibility for accomodation either a 3Fe or a 4Fe cluster by the same polypeptide chain was anticipated by comparison of FdI and FdII sequences and spectroscopic data. The disulfide group, in this case, replaces a second cluster, when amino acid sequences of two 4Fe clusters ferredoxins are compared, and indicated to be redox active. Molecular modeling of homologous structures build up from FdII crystal structure with and without structural constraints imposed by the disulfide bridge are under consideration in order to infer on the structural role of this internal chemical bond. Well resolved data has now been collected by 2D-NMR on FdI and FdII. NOESY, COESY and TOCSY are currently being analysed in D_2O and H_2O, using complementary sequence data and X-Ray coordinates.

10. Acknowledgments

This work was funded by Grants from STRIDE-Junta Nacional de Investigação Científica e Tecnológica, National Science Foundation Grant DMB-9005734 and EC NETWORK. We want to thank Drs. I.Moura, J.LeGall, E.Münck, B.H.Huynh, K.K.Surerus and P.N.Palma for many contributions.

REFERENCES

1. Holm, R.H., Ciurli, S., Weigel, J.A. in "Progress in Inorganic Chemistry: Bioinorganic Chemistry", Ed. Lippard, S.J., John Wiley and Sons, New York, 1990, vol.38, pp.1-.

2. Moura, I., Macedo, A.L. and Moura, J.J.G. (1989) in "Advance EPR and Applications in Biology and Biochemistry", Ed. Hoff, A.J., Elsevier.

3. Moura, J.J.G. (1986) in "Iron-Sulfur Proteins Research", Matsubara eds, Japan Science Society Press, Tokyo.

4. Lovenberg, W., in "Iron-Sulfur Proteins", Acad. Press, New York, 1973-1977, vols.I-III.

5. Spiro, T.G. in "Iron-Sulfur Proteins", John Wiley and Sons, New York, 1982, vol.IV.

6. Switzer, R.L. (1989) *Biofactors* **2**, 77.

7. Beinert, H. (1990) *FASEB J.* **4**, 2483.

8. Beinert, H. and Kennedy, M.C. (1989) *Eur.J.Biochem.* **186**, 5.

9. Haile, D.J., Rouault, T.A., Harford, J.B., Kennedy, M.C. , Blodin, G.A., Beinert, H. and Klausner, R.D. (1992) *Proc.Nat.Acad.Sci., USA* **89**, 11735.

10. Bruschi, M., Hatchikian, E.C., LeGall, J., Moura, J.J.G.and Xavier, A.V. (1976) *Biochim.Biophys.Acta* **449**, 275.

11. Cammack, R., Rao, K.K., Hall, D.O., Moura, J.J.G., Xavier, A.V., Bruschi, M. and LeGall, J. *Biochim.Biophys.Acta* **490**, 311.

12. Moura, J.J.G., Xavier, A.V. Bruschi, M. and LeGall, J. (1977) *Biochim.Biophys.Acta* **459**, 278.

13. Moura, J.J.G., Xavier, A.V., Hatchikian, E.C. and LeGall, J. (1977) *FEBS Letts* **89**, 177.

14. Kissinger, C.R., Adman, E.T., Sieker, L.C. and Jensen, L.H. (1988) *J.Am.Chem.Soc.* **110**, 8721.

15. Kissinger, C.R., Adman, E.T., Sieker, L.C., Jensen, L.H. and J. LeGall (1989) *FEBS Lett.*, **244**, 447.

16. Kissinger, C.R., Sieker, L.C., Adman, E.T. and Jensen, L.H. (1991) *J.Mol.Biol.*, **219**, 693.

17. Armstrong, F.A., George, S.J., Cammack, R., Hatchikian, E.C. and Thomson, A.J. (1987) *Biochem.J.* **264**, 265.

18. Conover, R.C., Kowal, A.T., Fu, W., Park, J.-B., Aono, S., Adams, M.W.W. and Johnson, M.K. (1990) *J.Biol.Chem.* **265**, 8533.

19. Stack T.D.P., Weigel, J.A. and Holm, R.H., *Inorg.Chemistry* (1990) **29**, 3745.

20. Kent, T.A., Moura, J.J.G., Moura, I., Lipscomb, J.D., Huynh, B.H., LeGall, J., Xavier, A.V. and Münck, E. (1982) *FEBS Letts* **138**, 55.

21. Moura, J.J.G., Moura, I., Kent, T.A., Lipscomb, J.D., Huynh, B.H., LeGall, J., Xavier, A.V. and Münck, E. (1982) *J.Biol.Chem.* **257**, 6259.

22. Kent, T.A., Huynh, B.H. and Münck, E. (1980) *Proc.Nat.Acad.Sci.USA* **77**, 6574.

23. Huynh, B.H., Moura, J.J.G., Moura, I., Kent, T.A., LeGall, J., Xavier, A.V. and Münck, E. (1980) *J.Biol.Chem.* **255**, 3242.

24. Xavier, A.V., Moura, J.J.G. and Moura, I. (1981) *Struct. and Bonding* **43**, 187.

25. Thomson, A.J., Robinson, A.E., Jonhson, M.K., Moura, J.J.G., Moura, I., Xavier, A.V. and LeGall, J. (1981) *Biochim.Biophys.Acta* **670**, 93.

26. Armstrong, F.A., Butt, J.N., George, S.J., Hatchikian, E.C. and Thomson, A.J. (1989) FEBS Lett. **259**, 15.

27. Moreno, C., Macedo, A.L., Moura, I., LeGall and Moura, J.J.G. (1994) *J. Inorg. Biochemistry* **53**, 219.

28. Macedo, A.L., Moura, I., Surerus, K.K., Papaefthymiou, Liu, M.Y., LeGall, J., Münck, E. and Moura, J.J.G. (1994) *J.Biol.Chem.* **269**, 8052.

29. Carter, C.W. Jr. in "Iron-Sulfur Proteins", Ed. W. Lovenberg, vol.III, p.157, Acad. Press, New York (1977).

30. Noodlemann, L. (1988) *Inorg.Chem.* **27**, 3677.

31. Jordanov, J., Roth, E.K.H., Fries, P.H. and Noodlemann, L. (1990) *Inorg.Chem.* **29**, 4288.

32. Poe, M., Phillips, W.D., McDonald, C.C. and Lovenberg, W. (1970) *Proc. Nat. Acad. Sci. USA* 65, 797.

33. Dunham, W.R., Palmer, G., Sands, R.H. and Bearden, A.J. (1971) *Biochim.Biophys. Acta* **253**, 373.

34. Phillips, W.D. (1973) in "NMR of Paramagnetic Molecules", Eds. La Mar, G.N., Horrocks,Jr., W.D. and Holm, R.H., Acad. Press, N.Y., Chapter 11.

35. Phillips, W.D. and Poe, M. (1977) in "Iron-Sulfur Proteins", Ed. Lovenberg, W., Acad. Press, N.Y., vol. II, Chapter 7.

36. Packer, E.L., Sweeney, W.V., Rabinowitz, J.C., Sternlicht, H. and Shaw, E.N. (1977) *J.Biol.Chem.* **252**, 2245.

37. LeGall, J., Moura, J.J.G., Peck, H.D., Jr. and Xavier, A.V., in "Metal Ions in Biology, Iron-Sulfur Proteins", Ed. Spiro, T.G., vol.4, p.177, Wiley and Sons, New York, 1982.

38. Markley, J.L., Chan, T.-M.,Krishnamoorthi, R. and Ulrich, E.L. (1986) in "Iron-Sulfur Proteins Research", Matsubara eds, Japan Science Society Press, Tokyo.

39. Dugad, L.B., La Mar, G.N., Banci, L. and Bertini, I. (1990) *Biochemistry* **29**, 2263.

40. L. Skjeldal, L., Westler, W.M., Oh, B.-H., Krezel,, A.M., Holden, H.M., B.L. Jacobson B.L., Rayment, I. and Markley, J.L. (1991)*Biochemistry* **30**, 7363.

41. Cheng, H., Grohmann, K. and Sweeney, W.V. (1990) *J.Biol.Chem.*, **265**, 12388.

42. Bertini, I., Briganti, F., Luchinat, C., and Scozzafava, A. (1990) *Inorg.Chem.* **29**, 1874.

43. Bertini, I., Briganti, F., Luchinat, C., Messori, L., Monnanni, R., Scozzafava, A. and G. Vallini, G. (1991) *FEBS Lett.*, **289**, 253.

44. Busse, S.C., La Mar, G.N. and J.B. Howard (1991) *J.Biol.Chem.*, **266**, 23714.

45. Bertini, I., Briganti, F., Luchinat, C., Messori, L., Monnanni, R., Scozzafava, A. and Vallini, G. (1992) *Eur.J.Biochem.* **204**, 831.

46. Cowan, J.A. and Sola, M. (1990) *Biochemistry*, **29**, 5633.

47. Bertini, I., Briganti, F., Luchinat, C., Scozzafava, A. and Sola, M. (1991) *J. Am. Chem. Soc.*, **113**, 1237.

48. Banci, L., Bertini, I., Briganti, F., Scozzafava, A., Oliver, M.V. and Luchinat, C. (1991)*Inorg.Chimica Acta* **180**, 171.

49. Banci, L., Bertini, I., Briganti, F., Luchinat, C. Scozzafava, A. and Oliver, M.V. (1991) *Inorg.Chem.* **30**, 4517.

50. Nettesheim, D.V., Harder, S.R., Feinberg, B.A. and Otvos, J.D. (1992)*Biochemistry* **31**, 1234.

51. Bertini, I., Capozzi, F., Ciurli, S., Luchinat, C., Messori, L. and Piccioli, M. (1992) *J.Am.Chem.Soc.*, **114**, 3332.

52. Bertini, I., Capozzi, F., Luchinat, C. and Piccioli, M. (1993)*Eur.J.Biochem.*, **212**, 69.

53. Busse, S.C., La Mar, G.N., Yu, L.P., Howard, J.B., Smith, E.T., Zhou, Z.H. and Adams, M.W.W. (1992) *Biochemistry*, **31**, 11952.

54. Banci, L., Bertini, I., Carloni, P., Luchinat, C. and Oriolo, P.L. (1992) *J.Am. Chem. Soc.* **114**, 10683.

55. Macedo, A.L., Palma, P.N., Moura, I., LeGall, J., Wray, V. and Moura, J.J.G. (1993) *Magn.Res.Chem.* **31**, S59.

56. Phillips, W.D., McDonald, C.C., Stombaugh, N.A. and Orme-Johnson, W.H. (1974) *Proc.Nat.Acad.Sci., USA* **71**, 140.

57. Nagayama, K., Ozaki, Y., Kyogoku, Y., Hase, T. and Matsubara, H. (1983) *J.Biochem.* **94**, 893.

58. Krishnamoorthi, R., Markley, J.L., Cusanovich, M.A., Przysiecki, C.T. and Meyer, T. (1986) *Biochemistry* **25**, 60.

59. Macedo, A.L., Moura, I., Moura, J.J.G., LeGall, J. and Huynh, B.H. (1993) *Inorg. Chem.* **32**, 1101.

60. Papaephtymiou, V., Girerd, J.J., Moura, I., Moura, J.J.G. and Münck, E., (1987) *J.Biol.Chem.* **255**, 3242.

61. Gayda, J.P., Bertrand, P., Theodule, F.X. and Moura, J.J.G. (1982) *J.Chem.Phys.* **77**, 3387.

62. Day, E.P., Peterson, J., Bonvoisin, J.J., Moura, I. and Moura, J.J.G. (1988) *J.Biol.Chem.* **263**, 3684.

63. Moura, I., Moura, J.J.G., Münck, E., Papaephthymiou, V. and LeGall, J. (1986) *J.Am.Chem.Soc.* **108**, 349.

64. Surerus, K., Münck, E., Moura, I., Moura J.J.G. and LeGall, J. (1987) *J.Am.Chem.Soc.* **109**, 3805.

65. Surerus, K.K., *PhD Dissertation*, University of Minnesota (1989).

66. Moreno, C., Macedo, A.L., Surerus, K.K., Münck, E., LeGall, J. and Moura, J.J.G., *Abst. 3th Int. Conference on Molecular Biology of Hydrogenases*, Tróia, Portugal, 156 (1991).

67. Münck, E., Papaephtymiou, V., Surerus, K.K. and Girerd, J.J. (1988) in"Metal Clusters in Proteins" (Ed. Que, L. Jr.) p.302, American Chemical Society, Washington, D.C.

ISOTROPIC PROTON HYPERFINE COUPLING IN HIGH POTENTIAL $[Fe_4S_4]^{3+}$ MODELS

Comparison of Calculations with Experimental ENDOR and Paramagnetic NMR Results

L. NOODLEMAN, J.-L. CHEN and D.A. CASE
Department of Molecular Biology
The Scripps Research Institute
La Jolla, California 92037 USA

C. GIORI
Instituto di Scienze Fiziche
Universitá degli Studi di Parma (Italy)

G. RIUS, J.-M. MOUESCA and B. LAMOTTE
Dept. de Recherche Fondamentale sur la Matière Condensée
Centre d'Etudes Nucléaires de Grenoble, CEA
85 X 38041 Grenoble, France

1. Introduction

A wide variety of iron-sulfur clusters and related mixed-metal species are found as active sites in metalloenzymes. In some cases these proteins form part of electron transport chains, and in other cases they serve as catalytic centers for quite unusual chemistry. Functionally, these play a vital role in coupling electron transfer to energy conserving processes in cells, and in performing various biosynthetic tasks. From the viewpoint of electronic structure, all possess active sites where high-spin transition metal atoms are spin coupled via bridging and terminal ligands. The catalytic and electron transfer events that occur when the protein interacts with its reaction partner are quite different from the more familiar processes of Lewis acid-base chemistry. Both oxidation-reduction and ligand binding events in iron sulfur proteins are associated with

G.N. La Mar (ed.), Nuclear Magnetic Resonance of Paramagnetic Macromolecules, 339-367.

major changes in charge distribution, spin distribution, and spin states. These changes can profoundly influence the appearance of spectra, and understanding these can be a key to piecing together accounts of catalytic or redox cycles. In particular, paramagnetic NMR, along with Mossbauer spectroscopy and ENDOR measurements, can often provide detailed probes of spin coupling among Fe sites, and the asymmetry of the electronic spin distribution about the metal sites.

An essential feature of the proton NMR spectra of iron-sulfur proteins is the presence of relatively broad lines with large positive (and sometimes negative) shifts, outside of the ordinary diamagnetic envelope [1]. These have been attributed to the protons of the cysteine amino acids ligating the iron atoms of the iron-sulfur cluster(s) that constitute the active sites of these proteins. In many cases it has been possible to make sequence-specific assignments to particular β and α protons of the cysteine side chains [2-7]. The observed shifts are especially large for the $\beta-CH_2$ protons, being essentially due to the unpaired electron spin population that is partially delocalized from the iron atoms to the cluster ligands. The hyperfine interactions that give rise to paramagnetic shifts can also be explored through EPR and ENDOR measurements. In this paper we compare experimental and calculated ligand hyperfine interactions for clusters that are models for the active site in high-potential iron proteins (HiPIP), and discuss some of the implications of these results for the interpretation of NMR spectra.

A useful additional source of information comes from clusters that have been chemically synthesized using organic thiolates in place of cysteine from the protein [8]. Compared to the native FeS clusters in proteins, the synthetic structures can be accurately determined by X-ray diffraction, the magnetic measurements can be more precise [9], and, in some cases, the orientation of the g–tensor and the Fe hyperfine (A) tensors determined by EPR and ENDOR can be related to the geometric orientation of the cluster [10,11]. The properties of the synthetic clusters can be usefully compared with those of FeS proteins, and considerable insight is then gained into which properties are intrinsic to the FeS clusters themselves, and which depend on the protein polypeptide environment. The proton isotropic hyperfine coupling provides the effective electron spin densities at the hydrogen nuclei. The anisotropic hyperfine coupling largely monitors the through-space dipolar hyperfine interactions of the cluster spin density with the hydrogen nuclei, and is sensitive to the spin densities at the iron and sulfur sites. The advantage of ENDOR spectroscopy is this access to the full hyperfine tensors at the protons, as well as the ability to examine low temperatures, so that only the system spin ground state is occupied. However, maximum information is obtained only when single crystals are present, making the study of proteins themselves more difficult. Furthermore, the temperature dependence of the hyperfine coupling is also of great interest, as this reflects the occupation of excited spin states, and other effects such as electron detrapping. For these reasons, paramagnetic NMR and proton ENDOR provide important complementary techniques.

2. Experimental methods

Our studies on paramagnetic states of Fe_4S_4 cubanes have focussed on single crystals of the $[Et_4N]_2$ $[Fe_4S_4(SCH_2C_6H_5)_4]$ model compound (**1**), whose crystallographic structure is shown in Fig. 1 and is described by Averill *et al.* [12]. An attractive feature of this synthetic model is that the thiolate ligands have CH_2 groups in the same position as the β–CH_2 protons of cysteines in proteins.

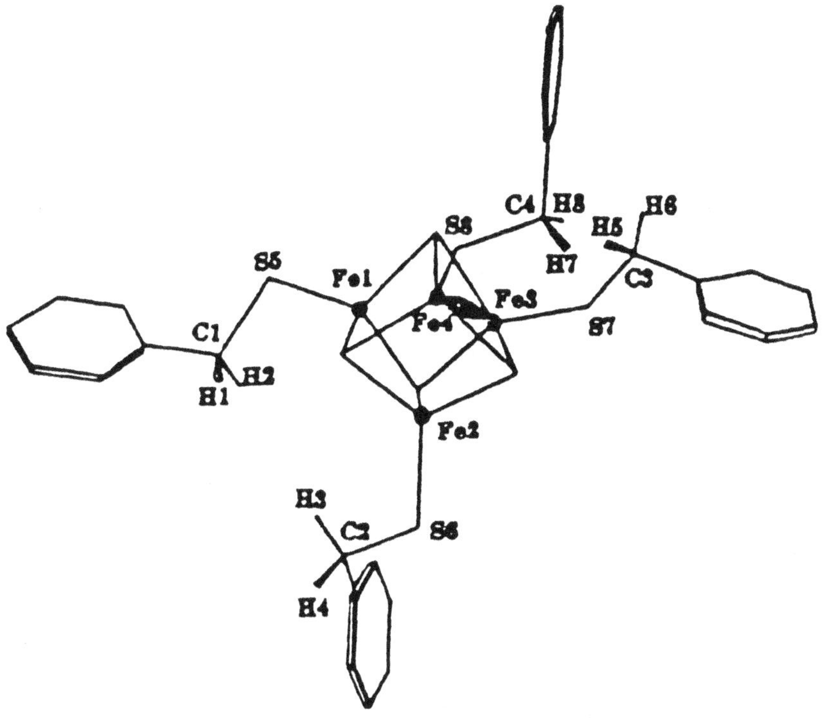

Figure 1. Structure of (**1**)

Since the crystals correspond to the $[Fe_4S_4]^{2+}$ state have an S=0 ground state, they do not give EPR signals. Paramagnetic $[Fe_4S_4]^{3+}$ and $[Fe_4S_4]^{1+}$ states are then created by irradiation of the crystals with gamma rays: the former can be considered as "holes" corresponding to the active sites of oxidized HiPIP's, and the latter as trapped electrons in states analogous to the reduced states of ferredoxins. This irradiation method thus permits the study of both states in high-resolution conditions by both EPR and ENDOR, since each paramagnetic center is both oriented in the crystal lattice and is "diluted" at low concentration in the diamagnetic crystalline matrix. The single-crystal environment permits measurements of the complete g and hyperfine interaction tensors, providing the basis for a detailed interpretation of the spin distribution in these clusters.

2.1 EPR STUDIES

The different paramagnetic centers created by irradiation of single crystals of (1) have been studied by EPR as a function of orientation of the sample with respect to the magnetic field. These centers are more numerous than expected, and the curves of angular variation of the most intense (*i.e.*, the most representative) features are shown in Fig. 2. They have been characterized by their g–tensors, given in Table 1. They can effectively be classified into two categories, the first with $g_{av} > 2$ (with principal values in agreement with those found in oxidized HiPIP proteins,) and the second class with $g_{av} < 2$ (as found in reduced ferredoxins). This lead to the identification of the the sites with $[Fe_4S_4]^{3+}$ and $[Fe_4S_4]^{1+}$ states, respectively.

The most remarkable and unexpected characteristic of these results is that several different centers are simultaneously observed for a given paramagnetic redox state. This is most striking for the $[Fe_4S_4]^{3+}$ state, for which we have been able to identify six sites, numbered I to VI. Centers IV and I have large intensities, whereas II and III are, on average, three to four times less intense. Centers V and VI have very small intensity. Finally, we have observed two "reduced" $[Fe_4S_4]^{1+}$ states, labelled I_R and II_R, whose angular variations are also reported in Fig. 2.

The analysis of the directions of the principal axes of the tensors of these different centers with respect to the Fe–Fe bond directions of the cubane cluster has given us the key to explain the variety of paramagnetic centers which can be observed for a given redox state. As demonstrated originally by Mössbauer spectroscopy [13,14], partial delocalization of the charge takes place in the mixed-valence sites of cubane-like clusters, so that dimers that formally have one Fe^{2+} and one Fe^{3+} site are best described as a delocalized mixed-valence

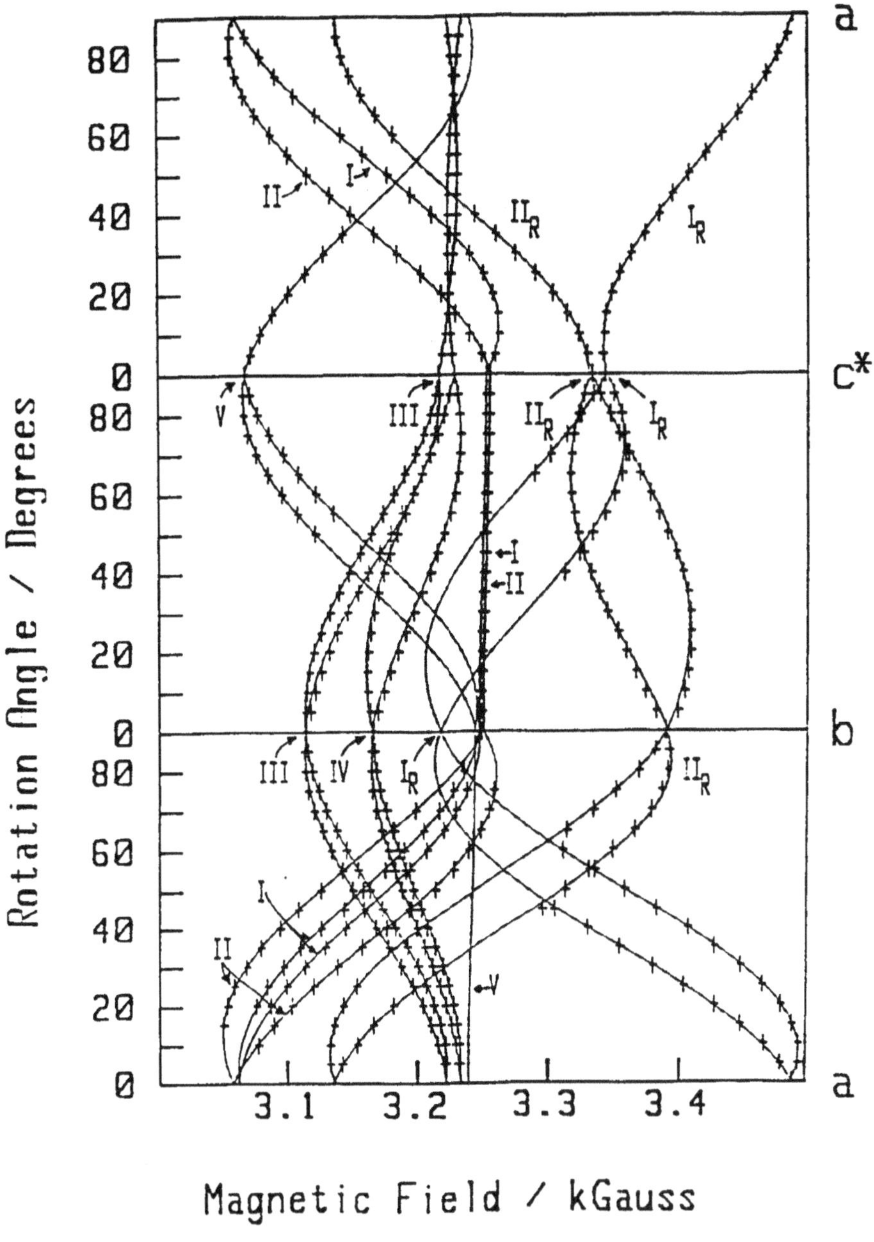

Figure 2. Angular variation of intense ENDOR features.

Table 1. EPR data for various centers.

Paramagnetic centers	Isotropic g - values ($g_{av.}$)	Principal g - values	Principal direction cosines with respect to :		
			a	b	c*
		g_1 = 2.142	+0.979	−0.048	−0.201
I	2.053	g_2 = 2.013	+0.020	+0.990	−0.140
		g_3 = 2.004	+0.205	+0.133	+0.970
		g_1 = 2.146	+0.974	−0.204	+0.095
II	2.053	g_2 = 2.009	+0.048	+0.600	+0.798
		g_3 = 2.003	+0.220	+0.774	−0.595
		g_1 = 2.101	−0.059	+0.996	−0.071
III	2.054	g_2 = 2.039	−0.629	+0.018	+0.777
		g_3 = 2.023	+0.775	+0.090	+0.625
		g_1 = 2.070	−0.089	+0.962	−0.259
IV (a)	2.038	g_2 = 2.026	+0.680	+0.248	+0.690
		g_3 = 2.018	+0.727	−0.115	−0.676
		g_1 = 2.135	−0.137	+0.087	+0.987
V	2.055	g_2 = 2.017	+0.862	+0.501	+0.075
		g_3 = 2.014	−0.488	+0.861	−0.144
		g_1 = 2.043	−0.142	+0.947	−0.288
I_a	1.954	g_2 = 1.948	+0.048	+0.297	+0.954
		g_3 = 1.871	+0.989	+0.121	−0.088
		g_1 = 2.087	+0.990	−0.133	+0.039
II_a	1.992	g_2 = 1.971	−0.090	−0.401	+0.911
		g_3 = 1.917	+0.105	+0.906	+0.409

pair, with effective oxidation state $Fe^{2.5+}$. The theoretical origins of this effect are discussed below. Hence, the $[Fe_4S_4]^{3+}$ state can be described as consisting of a ferric pair and a mixed valence pair, whereas the $[Fe_4S_4]^{1+}$ state is comprised of a ferrous pair and a mixed-valence pair. In principle, there are then six different possible ways in which the mixed-valence pair might be situated on two of four irons in an approximate tetrahedral symmetry. Our analysis of the principal directions of the g–tensor associated with the different species in the $[Fe_4S_4]^{3+}$ state has shows that, indeed, the centers I to VI correspond to all of the possible localizations of the mixed-valence pair, such that the principal direction associated with g_1 ($g_1 > g_2 > g_3$) is approximately colinear with the direction perpendicular to both the $Fe^{2.5+} - Fe^{2.5+}$ and $Fe^{3+} - Fe^{3+}$ directions (Table 2). We have also found that the two $[Fe_4S_4]^{1+}$ centers I_R and II_R can also be explained in the same way, the second direction now being replaced by the $Fe^{2+} - Fe^{2+}$ direction (Table 2). Consequently, our general explanation of the fact that we observe different sites with the same overall charge is that they correspond to different ways of localizing the mixed-valence pair. The fact that all possibilities are observed for $[Fe_4S_4]^{3+}$ suggests that they must lie rather close in energy.

2.2 PROTON ENDOR STUDIES ON CENTER IV

The $[Fe_4S_4]^{3+}$ state at center IV is the most intense of all the species created by irradiation of our crystals, and was the first to be the subject of a detailed proton ENDOR investigation. We measured the complete hyperfine tensors of the eight protons of the CH_2 groups that are part of benzylthiolate ligands surrounding the cubane core. In these experiments, we had two principal objectives:

(a) to determine a map of the distribution of the unpaired electron spin population on various atoms in the cluster, through an analysis of the anisotropic part of the proton hyperfine tensors;

(b) to analyze the isotropic portion of these tensors in order to make comparisons to theoretical calculations and to paramagnetic shifts observed by NMR for the $\beta-CH_2$ protons of cysteines in proteins.

Table 2. Orientations of largest g-values.

Paramagnetic centers	Greatest principal g-value	(Fe N°	Fe) N°	(Fe N°	Fe) N°	Θ angles
I	g_1 - 2.142	1	4	2	3	10°
II	g_1 - 2.146	1	4	2	3	10°
III	g_1 - 2.101	1	2	3	4	11°
IV	g_1 - 2.070	1	2	3	4	16°
V	g_1 - 2.135	1	3	2	4	11°
I_R	g_1 - 2.043	1	2	3	4	20°
II_R	g_1 - 2.087	1	4	2	3	6°

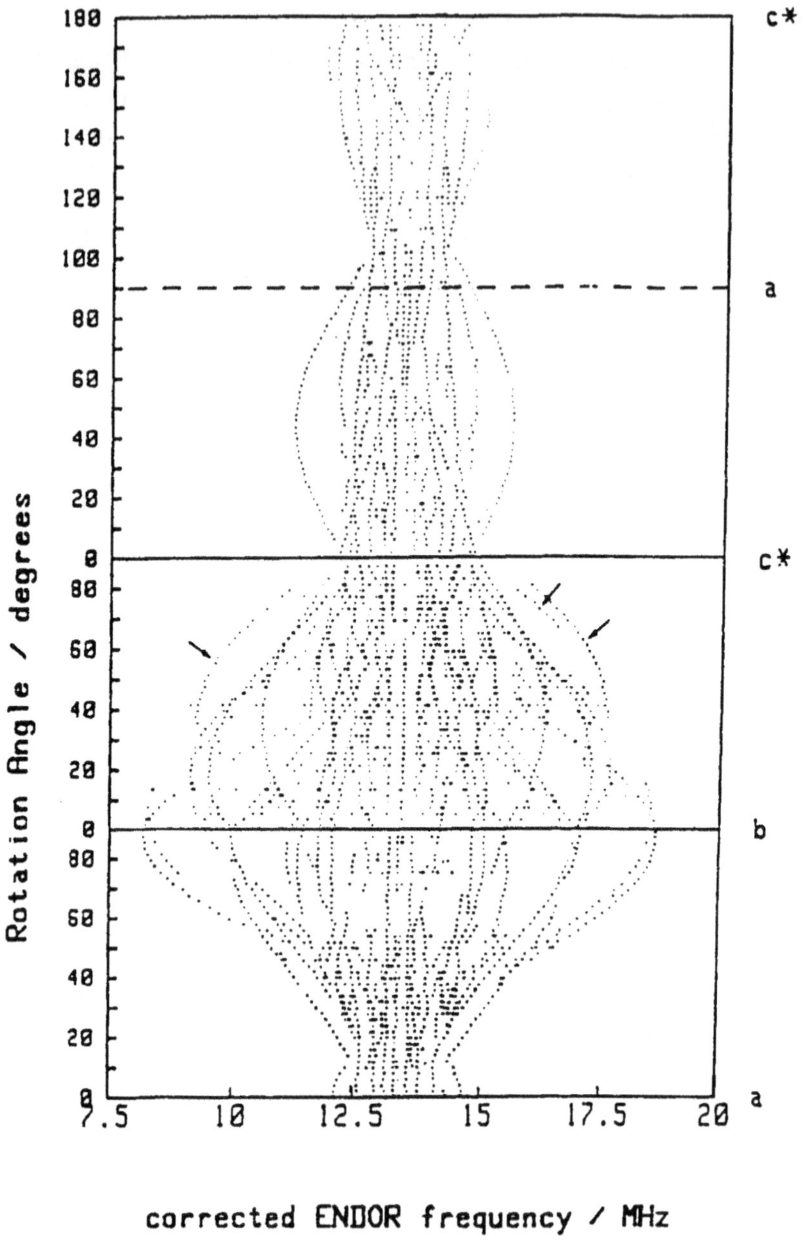

Figure 3a. ENDOR transitions for Center IV

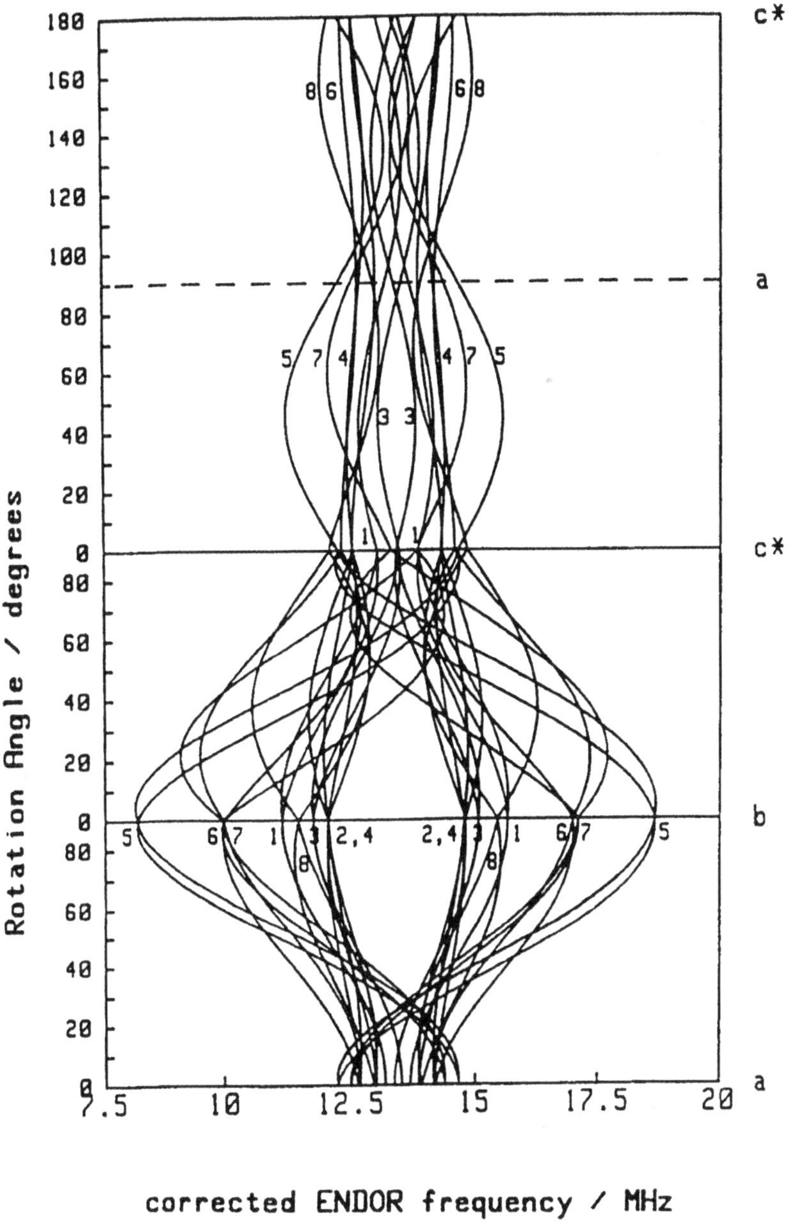

Figure 3b. ENDOR transitions for Center IV, with fits.

By experience, we learned how difficult a task it was to follow all of the thirty-two ENDOR lines associated with these eight protons (since we have two proton ENDOR transitions and, here, two inequivalent molecules in the unit cell.) This is why it has been absolutely necessary to eliminate the signals of all other protons by carrying out experiments in crystals for which the phenyl groups of the benylthiolate ligands and the Et_4N^+ counterions were fully deuterated.

Figure 3 shows the angular dependencies of the proton ENDOR transitions and their corresponding fits. The tensors were calculated from these fits. Then, each tensor was split into its anisotropic and isotropic parts. The anisotropic parts are essentially due to through-space electron–nuclear dipole–dipole interactions, and were first used in order to assign each of the tensors to a definite proton in the crystallographic structure. We have shown that the point-dipole model is a valid approximation to make these assignments, and to calculate "theoretical" tensors, both necessary steps to obtain information on the distribution of unpaired spin population. This semi-empirical distribution of spin populations on the various atoms is obtained by a least-squares fitting procedure that minimizes the differences between calculated and observed anisotropic tensors. The populations obtained on Fe_1 through Fe_4 are −0.72, −0.62, +1.29 and +1.36, respectively, with smaller populations on the sulfur atoms. These results clearly show negative spin excess on the ferric pair (sites Fe_1 and Fe_2) and larger, positive spin excess at the mixed-valence pair (sites Fe_3 and Fe_4). A further discussion of these values will be presented below, in conjunction with comparisons to theoretical calculations and spin coupling models.

2.3 PROTON ENDOR STUDIES ON OTHER CENTERS

Following this study of the center IV, similar proton ENDOR work has been continued on other sufficiently intense centers described in Section 2.1, i.e. the $[Fe_4S_4]^{3+}$ centers I and III and the $[Fe_4S_4]^{1+}$ center I_R. These results are at the moment unpublished because they are incomplete, since large overlaps of ENDOR lines prevent us from measuring without ambiguity the tensors of all eight protons of the four CH_2 groups, and especially the weakest. In practice, typically six out of eight tensors are obtained in these cases. In spite of the incompleteness of the experimental results, it is, however, possible to evaluate the spin population distribution on the iron atoms for these different centers relatively well from the anisotropic components of the tensors, using the point-dipole approximation. In each case, the distribution obtained indicates without ambiguity the positions of the two iron atoms with large positive spin

Table 3. ENDOR data for CH_2 groups adjacent to mixed-valence pairs

Center #	PROTON #							
	H_1 θ_1=-43	H_2 θ_2=+80	H_3 θ_3=+11	H_4 θ_4=+126	H_5 θ_5=-1	H_6 θ_6=+120	H_7 θ_7=+22	H_8 θ_8=+146
IV					+1.86[a]	+3.63[a]	+1.60[a]	+2.60[a]
"oxidized"					d_3=+1.29[b]		d_4=+1.36[b]	
I			+1.59[a]	+2.70[a]	+1.91[a]	+1.78[a]		
"oxidized"			$d_2 \approx$ +1.3[b]		$d_3 \approx$ +1.7[b]			
III	+2.24[a]	+2.92[a]	+1.32[a]	+2.55[a]				
"oxidized"	$d_1 \approx$ +1.1[b]		$d_2 \approx$ +1.4[b]					
I_R	+1.65[a]	+2.33[a]	+2.45[a]	+2.82[a]				
"reduced"	$d_1 \approx$ +1.1[b]		$d_2 \approx$ +1.3[b]					

a: Experimental isotropic hyperfine coupling constant for the proton H_i (in MHz);

b: Spin densities d_i on the iron atom i, as deduced from the anisotropic hyperfine tensors of the protons (within the point-dipole approximation).

populations constituting the mixed-valence pair: Fe_2 and Fe_3 for center I, Fe_1 and Fe_2 for center III, and Fe_1 and Fe_2 for center I_R. In fact, we consider that the spin populations on these atoms are quite well determined because they depend primarily on the tensors of the CH_2 protons placed on the side of the mixed-valence pair, which, being larger than the others, are the easiest to measure. By contrast, we consider that the spin populations of the alternate iron atoms, on Fe_1 and Fe_4 for center I, and on Fe_3 and Fe_4 for center III and center I_R are presently more uncertain, being primarily determined by a smaller number of proton tensors in good position to probe them, *i.e.* CH_2 protons of the ligands placed on the side of the pair of iron atoms opposite to the mixed-valence pair. Consequently, we limit our discussion of the iron spin populations and their related isotropic proton hyperfine couplings in this section to the mixed-valence pairs on the different centers, which are reported in Table 3. It is apparent from this table that, among the different $[Fe_4S_4]^{3+}$ centers studied, center I has larger spin populations on the mixed valence pair than centers III and IV, which have comparable values. This suggests that center I has a ground state which corresponds to $|S_{mv}\ S_{ferric}\ S>$ of $|\ 9/2\ 4\ 1/2>$ while the ground state of the centers IV and III correspond to $|\ 7/2\ 3\ 1/2>$ [10], as discussed below.

2.4 Isotropic Hyperfine Couplings for the Different Centers

For each of the centers studied, we also possess the isotropic couplings of the protons of the CH_2 groups of the ligands bonded to the iron atoms of the mixed-valence pairs, which are also reported in Table 3. This relatively large set of data gives us the opportunity (already explored before for center IV alone [10]) to test on a wider and more homogeneous set of data whether we can derive a unique law relating the isotropic hyperfine values to two parameters: the spin population on the adjacent iron atom and the Fe-S-C-H dihedral angles. The rescaled data, obtained by dividing each isotropic hyperfine coupling parameter by the effective spin population (d_i) of the nearest Fe site (i) is presented in Figure 4. This Figure shows the rescaled experimental data for the mixed-valence

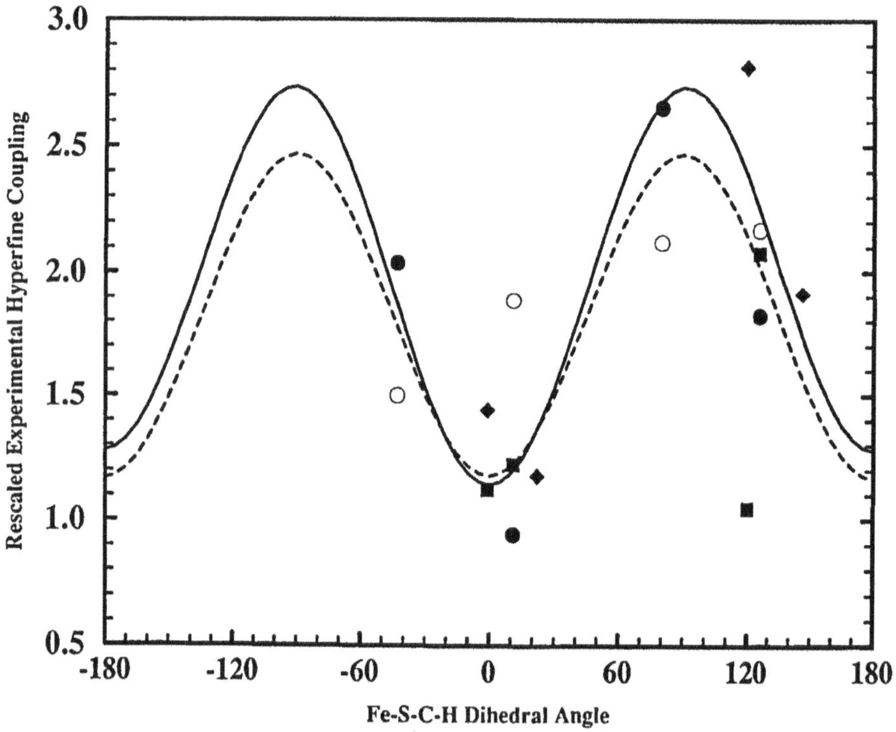

Figure 4. Rescaled isotropic hyperfine couplings.

pair of Center IV (filled diamond), III (filled circle), I (filled square), and reduced center I_R (open circles). Two fits to the data for the oxidized high potential centers are presented. The filled symbols for the three oxidized centers were fit to the empirical function proposed by Bertini *et al.* [7]:

$$f(\theta) = A\sin^2(\theta) + B\cos(\theta) + C \tag{1}$$

giving the dashed curve. One point from Center I is well off the "best fit" curve (solid square at $\theta = 120^0$); removing this point gives the solid curve. The two fits give parameters: A=+1.29, B=0.0, C=+1.17 (erf=0.12) for the dashed line; A=+1.52, B=-0.07, C=+1.21 (erf=0.07, solid line). Clearly, the main contribution to the error function comes from the single point from Center I at $\theta = 120^0$. The observation that two protons from Center I, having very similar θ values near 120^0 have very different isotropic couplings (Table 3) requires further examination. We can say that Center I has a location for the mixed valence pair which is less symmetric than those of Centers III and IV (see Section 3), so that the Fe-S-C-H dihedral angle may not be the only relevant geometric parameter for Center I.

3. Computational studies

We have been working for a number of years to develop a theoretical framework for analysis of spin-dependent properties like those presented above [15]. Our approach combines density functional quantum mechanical calculations with more phenomenological spin Hamiltonian models, in order to develop a picture of the ground and low-lying electronic excited states for a variety of cluster oxidation states and geometries. A more complete description of these phenomena would include consideration of vibronic coupling effects [16], which are discussed elsewhere in this volume. Here we give some details of our calculations of ligand hyperfine interactions in iron-sulfur cubanes.

3.1 DENSITY FUNCTIONAL CALCULATIONS

Density functional theory is now becoming widely recognized as a high-level method for carrying out quantum chemistry calculations, particularly for transition-metal clusters, which are difficult to handle by more conventional *ab initio* techniques [17,18]. As with any quantum chemistry method, though, the results depend in an important fashion on the quality and flexibility of the basis set; the development over the past few years of non-local corrections to the exchange-correlation functional has led to studies of the appropriateness of the Hamiltonian as well. When non-local corrections are included, density functional calculations are systematic improvements on Hartree-Fock theory, especially for bond strength calculations; for systems of the size we are dealing with here, these methods are the most accurate that are currently feasible [19-21].

We have carried out our calculations with the Amsterdam Density Functional (ADF) codes developed in the laboratory of E.J. Baerends [22]. The ADF codes expand orbitals in a basis of Slater-type orbitals (STO's), which generally allows a more rapid convergence than Gaussian basis expansions, at the expense of some computation time. We have successfully converged ADF calculations with up to 500 basis functions, enabling, for example, calculations on models of iron porphyrins or 4Fe iron-sulfur clusters to be carried out at the triple-zeta plus polarization level of basis set (with exponents optimized for density functional calculations), which allows for considerable flexibility in the description of details of the electron density.

Some special techniques are required to apply density functional calculations to spin-coupled systems where there are (formally) a large number of unpaired electrons, and we have used density functional methods to make electronic structure calculations on $[Fe_2S_2]$, $[Fe_3S_4]$, $[ZnFe_3S_4]$, and $[Fe_4S_4]$ clusters. We have constructed a novel broken-symmetry method which is well adapted for treating high spin transition metal sites that are spin coupled via bridging ligands, and where metal-ligand covalency must be realistically represented [15]. These may also be referred to as density functional valence bond calculations since the energies and properties of pure spin states can be extracted from the theory by the use of spin coupling algebra.

The model system used for the $[Fe_4S_4]^{3+}$ calculations reported here is $[Fe_4S_4(SR)_4]^{1-}$ where $R = CH_3$. The model geometry (with D_{2d} symmetry) is based on an idealization of the experimental structure of the $[Et_4N]_2$ $[Fe_4S_4(SCH_2C_6H_5)_4]$ synthetic model compound [12]. It has an axis of compression with four short (2.24 Å) and eight long (2.31 Å) Fe-S* bonds; the Fe-Fe distances fall into two groups, 2.775 Å (x2), and 2.733 Å (x4). To keep the closest correspondence with the positions of the mixed valence and ferric pairs from Center IV and Center III, the mixed valence and ferric pairs (and the Fe_2S_2 planes containing these) were located perpendicular to the axis of compression. For the ferric pair and the bonded thiolate groups, the C-S-Fe-Fe-S-C unit was constrained to be planar, and similarly, for the mixed valence pair and its bonded thiolates, again in correspondence with Centers III and IV. These two centers are distinguished by switching the opposite $Fe_2S^*_2$ planes (rhombs) containing the mixed valence and ferric pairs respectively. (Center I has the mixed valence and ferric pairs on different faces of the cubane, parallel to the axis of compression, so that neither pair has a coplanar C-S-Fe-Fe-S-C unit.) The electronic symmetry of the broken symmetry state is C_{2v}, and the planar C-S-Fe-Fe-S-C units each bisect the cone of the respective methyl group protons. The simple methyl ligand provides 3 sites for evaluating the proton coupling; we examined two different methyl group orientations related by a 180 degree rotation, so the Fe-S-C-H dihedral angles examined are $0, \pm 60, \pm 120$, and 180°. Further details of the results will be presented elsewhere [23].

3.2 SPIN HAMILTONIAN MODELS

The 2Fe2S ferredoxins exhibit trapped $Fe^{2+} - Fe^{3+}$ valence sites in the mixed valence state [24], whereas 4Fe4S clusters exhibit delocalized valence electron distributions over specific pairs of sites $Fe^{2.5+} - Fe^{2.5+}$ [25,26]. 3Fe4S clusters also have a valence delocalized pair in the reduced form [27]. Our work [15] and that of others [16,28,29] shows that the spin state of a system is often the result of competition between Heisenberg exchange and resonance delocalization (double exchange).

Iron-sulfur proteins exhibit a large variety of spin states, and spin coupling patterns among sites. This variety of spin states has a profound effect on the observed properties. For example, in 4Fe4S clusters with oxidation states 3+,2+, the ground spin states having $S=\frac{1}{2}$ and S=0 respectively are well separated in energy from states of higher spin. In contrast, the 1+ oxidation state may have ground states with S=1/2, 3/2, 7/2 or statistical mixtures of these in different proteins or synthetic analogs. Excited spin states appear much closer in energy than in the 3+ or 2+ oxidation states. Both experimental studies of magnetic susceptibility and our recent theoretical analysis show that there is a significant decrease in the size of all Heisenberg antiferromagnetic coupling constants J when the cluster oxidation state is reduced from 3+ to 2+ to 1+. Valence delocalization also strongly affects spin coupling. The valence delocalization energy depends on the effective spin quantum number (S_{34}) of the mixed valence pair as $B(S_{34} + \frac{1}{2})$. Because of this, the Heisenberg Hamiltonian typically used for exchange coupled systems is inadequate, and must be replaced by a Hamiltonian containing both Heisenberg coupling terms and resonance delocalization terms (also called double exchange).

We will describe the spin coupled states of the $[Fe_4S_4]^{3+}$ system within a pairwise scheme. We distinguish a ferric pair $Fe^{3+} - Fe^{3+}$ and a mixed valence pair $Fe^{2+} - Fe^{3+}$ (since the latter is delocalized, this is equivalent to $Fe^{2.5+} - Fe^{2.5+}$). We define the corresponding pair spin quantum numbers as $S_{12} = S_{ferric}$ and $S_{34} = S_{mv}$ (mixed valence) which couple to generate a total spin S. The site spins are S_1, S_2 for the ferric pair sites, and S_3, S_4 for the mixed valence pair. This description is consistent with Mossbauer spectroscopy on HiPIP complexes, which shows that the four Fe sites occur in two internally equivalent pairs [13,14]. Further, this pairwise description is followed also by the lowest energy broken symmetry state within density functional (DF) theory, from which quantitative calculations of the spin dependent properties of the system can be obtained. The strong delocalization of the mixed valence pair is directly predicted by DF calculations, and the magnitude of the resonance delocalization ("double exchange") parameter is large.

Within a spin Hamiltonian of the form

$$H = J(S_1 \cdot S_2 + S_1 \cdot S_3 + S_1 \cdot S_4 + S_2 \cdot S_3 + S_2 \cdot S_4 + S_3 \cdot S_4)$$

$$\pm B(S_{34} + 1/2) + \Delta J_{12}(S_1 \cdot S_2) + \Delta J_{34}(S_3 \cdot S_4) \qquad (2)$$

our most recent calculations on the model compound $[Fe_4S_4(SR)_4]^{1-}$, where R = CH_3 (using DF methods with a Vosko-Wilk-Nusair (VWN) parameterization plus Becke exchange and Stoll energy corrections added) gives J = 670 cm^{-1}, B/J = 1.3, $\Delta J_{12}/J$ = +0.24, where we assume ΔJ_{34} = 0. These results are in good agreement with fits to experimental magnetic susceptibility data using the same spin Hamiltonian [30].

While in principle, both B and ΔJ_{34} can be different from zero, it is difficult to determine both of these simultaneously either by theory or with experimental fits to temperature dependent NMR data. Since from our quantum chemistry calculations, B is quite large (see above), and this is unambiguous, we have chosen to set $\Delta J_{34} = 0$ when we determine spin Hamiltonian parameters from broken symmetry and high spin states. Some researchers, in fitting the temperature dependence of proton paramagnetic shifts by NMR, have chosen to set $B = 0$, and $\Delta J_{34} \neq 0$, since then a completely Heisenberg Hamiltonian can be used. We consider this to be ill-advised, since B is large theoretically, and a proper fit to T dependent NMR or magnetic susceptibility should include this parameter; a ΔJ_{34} term can then be introduced as a further refinement if necessary. It has also been argued that by adjusting ΔJ_{34} with $B = 0$, the same ground state can be achieved as is found with $B \neq 0$. While this is usually true, one of the main practical uses of the spin Hamiltonian above resides in temperature dependent properties. The ΔJ_{34} term takes the form $\Delta J_{34} S_{34}(S_{34} + 1)/2$, and has a quadratic dependence on $S_{34} = S_{mv}$, while $\pm B(S_{34} + 1/2)$ has linear dependence on S_{34} so the the energies of the spin ladder will differ significantly.

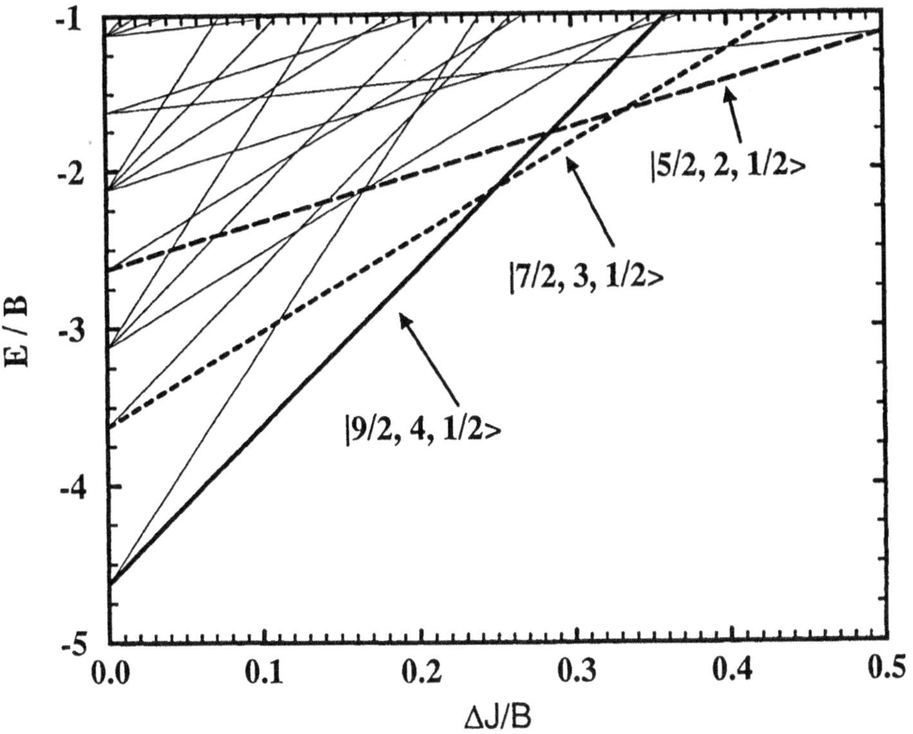

Figure 5. Low lying states of $[Fe_4S_4]^{3+}$ *clusters*

The consequences of this range of parameters are, first, that the Heisenberg AF coupling constant is largest for the ferric pair ($J + \Delta J_{12}$) while all other Fe-Fe pairwise interactions are smaller (J); this leads to spin frustration within the ferric pair, and the ferric pair spin is less than maximal $S_{12} = S_{ferric} < 5$; second, the large B term enforces site equivalence within one specific mixed valence pair (sites 3,4), and drastically lowers the spin degeneracy of the problem, making the lowest states have large values of $S_{mv} = 7/2, 9/2$. The two lowest energy spin states derived from Eq.(1) with the parameters above are: $|S_{mv} S_{ferric} S> = |9/2\ 4\ 1/2>$ and $|7/2\ 3\ 1/2>$. A representative spin state energy diagram is given in Figure 5. We plot E/B versus $\Delta J/B$, at the fixed ratio B/J=1. We have chosen $J = B = 600$ cm^{-1}. The next two states are: $|\ 5/2\ 2\ 1/2>$ and $|\ 9/2\ 3\ 3/2>$. From the calculations above, the predicted $\Delta J_{12}/B$ ratio is 0.18, while the experimental magnetic susceptibility fit gives about 0.24 [30].

The results that we will discuss here correspond to an electronic state where the sites of the ferric pair are each high spin ($S_1 = S_2 = 5/2$), as originally described in our 1988 paper [31]. We have recently found that in our model calculations where R = CH_3, there are two other low lying states, with different electronic character; for these, the ferric pair site spins can be considered as a quantum mixture of high spin and intermediate spin ($S_1 = S_2 = 5/2, 3/2$), mixed with spin forbidden ligand $\rightarrow$ metal charge transfer excitations These three electronic states lie within 0.3 to 0.6 eV of one another depending on detailed geometry when R = CH_3; the J, B parameters are similar for all of these, as are net atom spin populations. Our preliminary results suggest that two of these three electronic states display similar angular dependence for proton paramagnetic shifts, (specifically the one with high spin ferric sites and one of the two having a high spin/intermediate spin mixture), while the remaining electronic state shows different angular dependence. Further work is ongoing on this problem.

To calculate the hyperfine coupling at the protons, we note that for a spin coupled system in the strong coupling limit ($J \gg D$, where D is the zero field splitting parameter), the observed isotropic hyperfine coupling (A_p) is related to the intrinsic site hyperfine (a_p) by

$$A_p = (< S_{iz} > / < S_z >)a_p = K_i a_p \quad (3)$$

The indices i label the Fe sites, while p labels those proton sites closest to the associated Fe_i site. The K_i are spin projection coefficients, determined by the site spins, pair spins, and total spin of the system. These have a simple closed form when the pair spins are good quantum numbers [31]. The intrinsic site hyperfine values at the protons (site p) were determined by direct evaluation of the spin density at the appropriate nucleus (N_p) from the broken symmetry state wave function.

$$a_p = (8\pi/3) g_N \beta_N g_e \beta_e \rho_s(N_p)/(\pm 2S_i)$$

$$= (7.998 MHz)(g_N)\rho_s(N_p)/(\pm 2S_i) \quad (4)$$

where g_e is the free electron g factor, and $g_N = 5.586$ is the nuclear g factor for proton. $\rho_s(N_p)$ is the calculated spin density at the nucleus, S_i is the formal site spin ($S_i = 5/2$ for ferric sites, and $S_i = 9/4$ for mixed valence sites). The $\pm$ sign arises because of the opposite alignment of spins in the broken symmetry wave function. We have used the nonlocal fully self-consistent Vosko-Becke-Perdew (VBP) potential as well as the $X\alpha$ density functional potential, and noted only small differences between these.

4. Comparison of Calculations with Proton ENDOR Spectroscopy

4.1 ANGULAR DEPENDENCE OF PROTON HYPERFINE COUPLINGS

Figure 6 shows a direct comparison of the calculated proton isotropic hyperfine coupling parameters A^p_{iso} using a VBP potential with those experimentally observed in Center IV for all eight methylene protons [10]. The four protons associated with the mixed valence pair have positive couplings, and the four of the ferric pair have negative couplings. We have calculated these proton hyperfine couplings from density functional theory, using Eqs. 2 and 3. The solid diamonds give the experimental results for the mixed-valence (positive) and ferric (negative) pairs. The corresponding calculated results for the two lowest

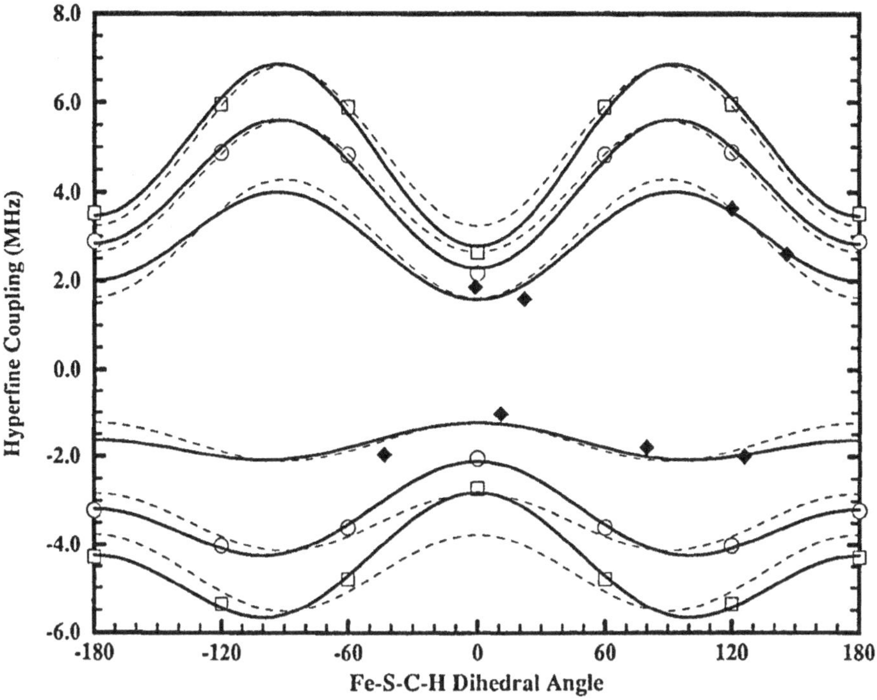

Figure 6. Calculated proton isotropic hyperfine values

states | 9/2 4 1/2 > and | 7/2 3 1/2 > are given by the open squares and open circles respectively. The best fit solid lines were again obtained with the functions $f(\theta)$ fitted for each dataset separately, and the dashed lines are fits to the simpler function $A\sin^2(\theta) + C$.

Similar fits were performed on the theoretical rescaled isotropic hyperfine data, which are compared with the rescaled experimental data in Figure 7. Here the theoretical ferric data is given by the open circles, the theoretical mixed valence data by the open squares, and the corresponding experimental ferric and mixed valence data from Center IV by the solid circles and solid squares respectively.

The rescaling of the theoretical results requires effective spin populations at the irons, analogous to the experimental d_i. We have called these $d_S(Fe_i)$, to emphasize that they are spin state dependent. Then the appropriate theoretical equation for the rescaled results is given by

$$A^{*p}_{iso} = A^{p}_{iso}/d_S(Fe_i) \tag{5}$$

where the $d_S(Fe_i)$ were generated by the simple method described in the next section based on Mulliken spin populations and spin projection coefficients. The theoretical rescaled data in Figure 7 is the same for both | 9/2 4 1/2 > and | 7/2 3 1/2 > because K_i appears as a factor in both A^{p}_{iso} and $d_S(Fe_i)$.

For the experimental fit in Figure 7, in contrast to the fitting functions used in Figure 4, we allow for more asymmetry in the fit to the experimental rescaled hyperfine coupling data by using the empirical equation from Mouesca *et al.* [10]

$$A^{*p}_{iso}(exp) = A + B\cos(\theta_i + \theta_0) + C\cos^2(\theta_i + \theta_0) \tag{6}$$

where θ_i is the dihedral angle, and θ_0 is a constant offset angle, allowing the fit to be asymmetric with respect to $\theta_i \rightarrow -\theta_i$ transformation. The best fit parameters are: A=3.03, B=0.52, C=−2.28, and $\theta_0 = -21^0$.

Calculated proton hyperfine couplings were generated by DF calculations, and the spin projection coefficients for the possible states |7/2 3 1/2 >: $K_1 = K_2 = -1$; $K_3 = K_4 = 1.5$; and |9/2 4 1/2 >: $K_1 = K_2 = -1.333$; $K_3 = K_4 = 1.833$, using Eq. 2. (For explicit equations, see [31] and note that the K_i must be averaged over the mixed-valence pair.) From Figs. 6 and 7, the angular dependence of the experimental scaled or unscaled hyperfine couplings is well represented by the DF calculations. The magnitudes of the |7/2 3 1/2 > couplings are smaller than those for |9/2 4 1/2 > because of the smaller K_i, and the former are in better agreement with experiment. Previously, we have presented an argument [10] that |7/2 3 1/2 > is the best representation of Center IV

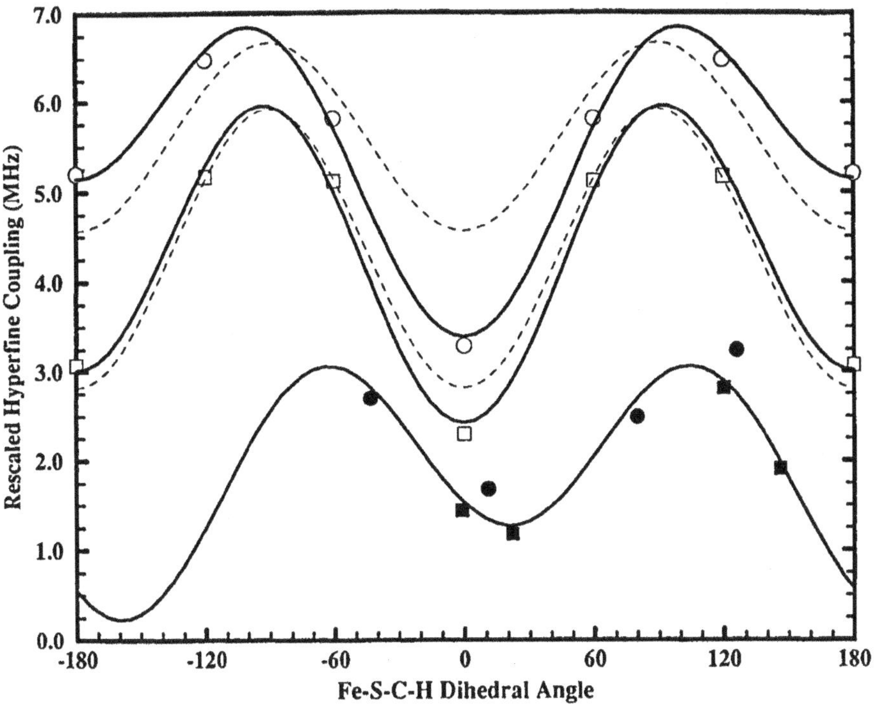

Figure 7. Rescaled isotropic hyperfine couplings.

based on performing partial sums of the effective spin populations of each Fe with four sulfur neighbors, and comparing this with K_i. In any event, while the calculated hyperfine couplings are larger than experiment, the angular dependence largely reproduces the experimental behavior. However, while the theoretical data is symmetric about $\theta = 0^0$, the experimental data has some asymmetry as shown by the angular offset $\theta_0 = -21^0$. The electronic symmetry in the Broken Symmetry state is C_{2v} by construction. By contrast, the experimental phase angle must reflect some asymmetry of the methylene-S-Fe-Fe-S-methylene units, for example deviation of C-S-Fe-Fe-S-C from planarity, or perhaps this plane does not precisely bisect the cone of the proton positions. We notice also that before rescaling, the mixed-valence pair hyperfine couplings are larger both experimentally and from theory, while after rescaling, the ferric site values are larger, again both from experiment and theory. Finally, we note the calculated proton hyperfine is much closer to the experimental proton hyperfine without rescaling. This is because the calculated effective spin densities at the Fe's are considerably different from those obtained experimentally from the point-dipole fit of the proton anisotropic hyperfine coupling.

4.2 EFFECTIVE SPIN DENSITIES ON IRON AND SULFUR SITES

In the discussion above, we have made reference to the effective spin populations at the Fe sites. From the Wigner-Eckard theorem for spin coupled systems, and the theory of reduced density matrices, one can obtain the integrated effective spin density $d_S(Fe_i)$ (or on sulfurs) from the corresponding spin populations (Mulliken analysis) ΔP_i of the sites in the broken symmetry state by:

$$d_S(Fe_i) = K_i|\Delta P_i|/2S_i \tag{7}$$

These values, calculated from DF theory can be compared directly with those obtained from the point dipole fit to the proton anisotropic hyperfine tensors obtained by ENDOR. This is done in Table IV. While the sizes of $d_S(i)$ at Fe and thiolate sulfur are broadly comparable from theory vs. experiment, overall the experimental values are larger, and these sum to 1.28. By contrast, the calculated values sum only to about 0.6. From basic theory, the $\sum_i K_i = 1$. The transferred spin population from Fe to thiolate and bridging sulfur results in some decrease of the overall spin population, as does antiferromagnetic coupling among the Fe sites. One then expects that

$$\sum_{i=all\ sites} d_S(i) \leq 1 \tag{8}$$

Table IV. Effective Spin Densities at Fe and S Sites
Density Functional VBP *vs.* ENDOR Point Dipole Fit

Atom type	| 7/2 3 1/2>	| 9/2 4 1/2 >	*Experimental fit*
Fe_a Ferric	-0.619	-0.825	−0.722,−0.618 (2 sites)
Fe_b MV	+0.942	+1.151	+1.293,+1.359 (2 sites)
S_a Ferric	-0.060	-0.080	−0.056,−0.048 (2 sites)
S_b MV	+0.077	+0.094	+0.037,+0.039 (2 sites)
S^*_a Ferric	-0.022	-0.029	0.0 , 0.0 (2 sites)
S^*_b MV	-0.015	-0.018	0.0 , 0.0 (2 sites)
Total	0.606	0.586	+1.284

while the experimental ENDOR fit gives a sum greater than 1. This is currently under study from both experimental and theoretical sides. From the experimental side, the point dipole approximation may fail for the bonded thiolate sulfurs which are quite close to the protons. From the theoretical side, we seek a better analysis than that provided by Mulliken spin populations, and we want to examine environmental effects, such as solvent polarization, on the calculated spin densities.

4.3 SPIN DISTRIBUTION IN THIOLATE SULFUR P ORBITALS

The Mulliken S (3p) population analysis from the density functional calculation confirms that the thiolate S (3p) spin density is π to the respective Fe-S bond for both the mixed-valence and ferric pairs. This was proposed based on the angular dependence of the methylene proton hyperfine couplings as found by ENDOR spectroscopy in the synthetic HiPIP *ox* complex [10] and from paramagnetic NMR on oxidized 4Fe ferredoxin proteins [7]. Specifically, we define the Fe-Fe bond of the diferric pair as the global x axis, and the Fe-Fe bond of the mixed-valence pair as y, with z perpendicular to these (corresponding to the axis of compression). For the theoretical model geometry, the thiolate sulfurs bonded to the diferric sites are in the S-Fe-Fe-S xz plane, while those sulfurs bonded to the mixed-valence iron pair are in the yz plane. Then, the broken symmetry state with the VBP potential gives the following thiolate S(3p) spin populations: on the $Fe^{3+} - Fe^{3+}$ pair, $S(3p_x)$=0.05, $S(3p_y)$=0.20, $S(3p_z)$=0.02, while for the mixed-valence $Fe^{2.5+} - Fe^{2.5+}$ $S(3p_x)$=-0.17, $S(3p_y)$=-0.03, $S(3p_z)$=-0.02. Clearly, the thiolate $S(3p)$ spin populations are greatest in magnitude perpendicular to the Fe-S bonds, as well as perpendicular to the diferric, and mixed-valence Fe-Fe pair vectors, respectively.

5. Comparisons to Paramagnetic NMR Measurements

The result for the paramagnetic shift of protons near the spin coupled Fe sites in 4Fe4S systems is

$$(\Delta\nu_p/\nu_0) = (g_e\beta_e/g_N\beta_N 3kT)a_p$$

$$\times \ [\sum_n K_{in}S_n(S_n+1)(2S_n+1)e^{-E_n/kT}] \, / \, [\sum_n (2S_n+1)e^{-E_n/kT}] \qquad (9)$$

where $E_n = E_n(S_{12}, S_{34}, S)$ are the distinct spin state energies of the spin Hamiltonian, (taking the ground spin state $E_0 = E_0(S_{12}, S_{34}, S)$ as the zero of energy) in the absence of a magnetic field [32]. Eq. (9) already contains the sum over magnetic sublevels of each spin state [1]. If only the lowest energy spin state is occupied, or equivalently, if for a specified temperature T, all excited states lie well above the ground state, $(E_n - E_0)/kT \gg 1$ (with E_0 as the ground spin state energy), then this equation simplifies to

$$(\Delta\nu_p/\nu_0) = (g_e\beta_e/g_N\beta_N 3kT)a_p[K_{i0}S_0(S_0+1)]$$

$$= (g_e\beta_e/4g_N\beta_N kT)a_p K_{i0} \tag{10}$$

In the last expression, we assume an S=1/2 ground state, and the temperature dependence is that for a Curie Law. The values approach zero linearly as (1/T) for increasing T. Positive K_{i0} values give positive (downfield) shifts, negative values negative (upfield) shifts, for positive a_p (as we have found). At 300^0 K, this frequency shift amounts to (in ppm) [10]

$$(\Delta\nu_p/\nu_0) = (\Delta\nu_p/\nu_0)_{dia} + (\Delta\nu_p/\nu_0)_{para} = 3 + 26.3K_{io}a_p \tag{11}$$

Occupation of excited spin states leads to a deviation from Curie Law dependence. There can be downfield peaks which shift further downfield with increasing temperature (anti-Curie type). These originate from upfield shifted peaks at lower temperature which cross over to downfield as the temperature is increased. This can, be a consequence of two effects: *(1)* occupation of excited spin states as mentioned above; *(2)* detrapping of mixed valence pair, and ferric pair sites, so that at relevant NMR temperatures, a given proton may experience a statistical mixture of Fe site valences on the nearest neighbor Fe. The first effect is represented by Eq.(9) , and from this, it is clear that each proton shift arises from a weighted average of the spin projection coefficients K_i for that type of site. We have found that K_i does not change sign for the low lying spin states of the spin Hamiltonian Eq. (2), and that the same set of low lying spin states is found whether from calculation, or from the experimental fit [30]. Specifically, the low lying states, in addition to | 7/2 3 1/2 >, and | 9/2 4 1/2 > are | 9/2 3 3/2 >, with $K_1 = K_2 = -0.600$, $K_3 = K_4 = 1.10$, | 5/2 2 1/2 >, with $K_1 = K_2 = -0.667$, $K_3 = K_4 = 1.167$, and | 7/2 2 3/2 > with $K_1 = K_2 = -0.40$, $K_3 = K_4 = 0.90$.

A more detailed picture of the K_i weighted average and its temperature dependence from Eq.(9) is shown in Figure 8. The curves labelled "10K all" (thick,dashed lines) show the weighted average over all spin states for low temperature (T=10K), but since the ground state always has $S = 1/2$ over the entire parameter range, it is dominated by whichever $S = 1/2$ spin state is lowest for the particular value of $\Delta J/B$ (see also Figure 5), and shows a stepwise behavior as the ground state changes from | 9/2 4 1/2 > to | 7/2 3 1/2 > to | 5/2 2 1/2 > with increasing $\Delta J/B$. The high temperature behavior "300K all" (thick, solid

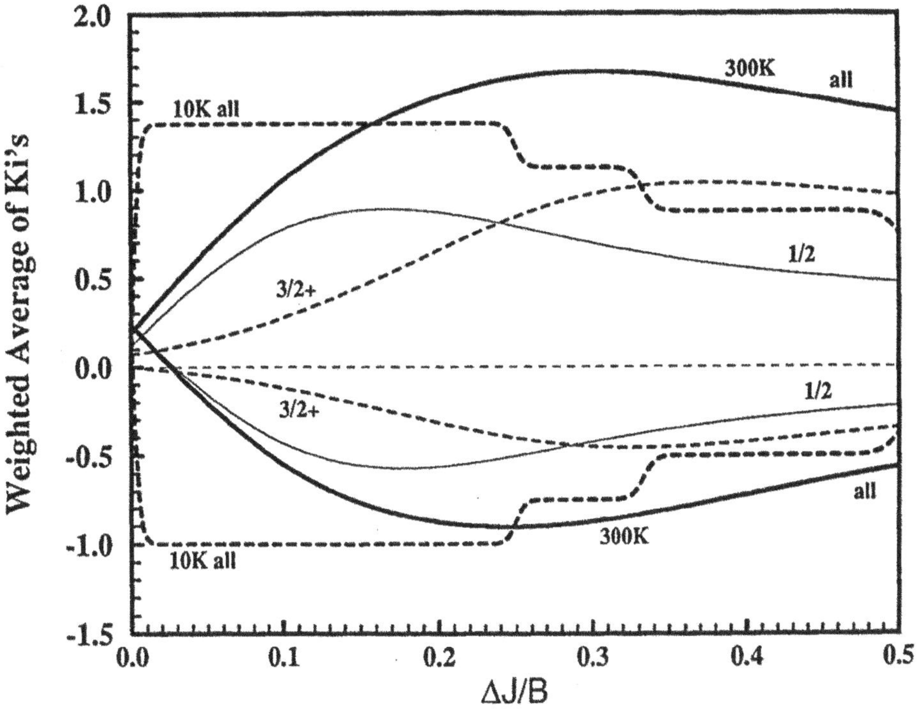

Figure 8. Thermally-averaged spin projection coefficients.

lines) again shows the weighted average over all spin states, but now at T=300K, thermal averaging smoothes the curves where the spin ground state changes as a function of $\Delta J/B$. The separate contributions to the 300K curves from the S=1/2 states (thin, solid lines) and from S=3/2 and higher spins (labelled 3/2+, dashed lines) is shown. Both S=1/2 and S=3/2 make significant contributions to the total K_i weighted average, while the contribution of higher spin states is very small. The contribution of the S=3/2 states, despite their high energies [30], is due to the degeneracy factor $S_n(S_n + 1)(2S_n + 1)$ which is 15 for S=3/2, but only 1.5 for S=1/2. We note also that the energies of the lowest S=3/2 states decrease relative to those of S=1/2 states as $\Delta J/B$ increases.

Figure 8 shows that thermal occupation of different spin states can certainly lead to deviations from Curie law temperature dependence for the proton paramagnetic shifts, particularly when a wide range of temperatures is considered (from ENDOR to NMR temperature ranges), but the anti-Curie behavior (cross-over from upfield to downfield with increasing temperature) is not found. Because all of the weighted K_i have K_1, K_2 negative, and K_3, K_4 positive, crossover cannot occur via mechanism 1 for attainable temperatures; therefore, detrapping is the most likely source for the anti-Curie temperature dependence of some peaks. The experimental NMR temperature dependence on various HiPIP proteins shows a particular type of detrapping is most common [6,33]. Sites 3 and 4 of the mixed valence pair do not simply interchange with sites 1 and 2 of the ferric pair. Rather, one site of the mixed valence pair (site 4, for example), is switched with one site of the ferric pair (say site 1), and these two sites have statistical combinations of the mixed valence and ferric character, while sites 2 and 3 maintain their original character. Presumably, the two different delocalization schemes have different energies, and we can suppose that there is a thermal mixture at any temperature. Then three types of signals are observed: positive (downfield) shifted with Curie dependence; negative (upfield) shifted with Curie dependence; and, positive shifted with anti-Curie dependence. The number of protons typically observed with these characteristics is 4:2:2 . The anti-Curie peaks are positively shifted because the spin projection coefficients of the mixed valence pair are typically larger than those of the ferric pair. If detrapping is smaller or absent, one expects a 4:4 pattern with no anti-Curie signals, as seen in *E. halophilia* HIPIP II.

We note also that the oxidized $[Fe_4S_4]^{3+}$ systems are in this re spect *not* analogous to the reduced $[Fe_2S_2]^{1+}$ systems. In the 2Fe systems, it is readily shown that while the ground state $| S_1\ S_2\ S \rangle = | 5/2\ 2\ 1/2 \rangle$, (ferric, ferrous, and total spins) have corresponding spin projection coefficients $K_1 = 7/3$, $K_2 = -4/3$. In the first excited state $| 5/2\ 2\ 3/2 \rangle$, both spin projection coefficients are positive, $K_1 = 13/15$, $K_2 = 2/15$, and in the second, $| 5/2\ 2\ 5/2 \rangle$, $K_1 = 23/35$ and $K_2 = 12/35$ [34]. It is then entirely feasible to obtain anti-Curie behavior of protons associated with the ferrous site from occupation of the S=3/2 and 5/2 excited spin states alone, if these are sufficiently low in energy. Detrapping effects may also be present, but these are less apparent than in the oxidized high potential 4Fe4S systems.

A number of important issues remain for further experimental and theoretical examination. Concerning the angular dependence of the proton hyperfine shifts, how general is the correlation with Fe-S-C-H dihedral angle? Does it apply to other cluster oxidation states, or equally to ferric, mixed-valence, and ferrous pairs? How do trapped valence 2Fe2S systems differ from 4Fe4S in the angular and thermal dependence of paramagnetic shifts? There are indications from ENDOR and paramagnetic NMR of a more complicated story when different cluster oxidation states, and different Fe pair types in 4Fe4S proteins and complexes are analyzed. Are other geometric factors involved? What is the behavior of other paramagnetic shifts, either of protons, ^{13}C, $^{14,15}N$ or ^{33}S within the cluster or on adjacent residues? The picture of the temperature dependence of NMR paramagnetic shifts versus those from ENDOR is still very incomplete. The comparative influences of detrapping, thermal occupation of excited spin states, and of electronic structure effects on the energies of low lying spin states across different proteins and synthetic analogue systems still needs to be sorted out.

6. Acknowledgements.

This research was supported by NIH Grant GM39914 and a NATO travel grant CRG-910204.

7. References

1. Bertini, I., Turano, P., and Vila, A.J. (1993) Nuclear magnetic resonance of paramagnetic metalloproteins. *Chem. Rev.* **93,** 2833-2932.

2. Dugad, L. B., La Mar, G.N., Banci, L., and Bertini, I. (1990) Identification of Localized Redox States in Plant-Type Two-Iron Ferredoxins Using the Nuclear Overhauser Effect. *Biochem.* **29,** 2263-2271.

3. Cowan, J.A. and Sola, M. (1990) ^{1}H NMR Studies of oxidized High-Potential Iron Protein from *Chromatium vinosum*. Nuclear Overhauser Effect Measurements. *Biochemistry* **29,** 5633.

4. Bertini, I., Briganti, F., Luchinat, C., Scozzafava, A., and Sola, M. (1991) ^{1}H NMR spectroscopy and the electronic structure of the high potential iron-sulfur protein from *Chromatium vinosum*. *J. Am. Chem. Soc.* **113,** 1237-1245.

5. Bertini, I., Ciurli, S., Dikliy, A., and Luchinat, C. (1993) Electronic structure of the $[Fe_4Se_4]^{3+}$ clusters in *C. vinosum* HiPIP and *Ectothiothodospiza halophila* HiPIP II through NMR and EPR studies. *J. Am. Chem. Soc.* **115,** 12020-12028.

6. Banci, L., Bertini, I., Ciurli, S., Ferretti, S., Luchinat, C., and Piccioli, M. (1993) The electronic structure of $[Fe_4S_4]^{3+}$ clusters in proteins. An investigation of the oxidized high-potential iron-sulfur protein II from *Ectothiorhodospira vacuolata*. *Biochemistry* **32,** 9387-9397.

7. Bertini, I., Capossi, F., Luchinat, C., Piccioli, M., and Vila, A.J. (1994) The Fe_4S_4 centers in ferredoxins studied through proton and carbon hyperfine coupling. Sequence-specific assignments of cyteines in ferredoxins from *Clostridium acidi urici* and *Clostridium pasteurianum*. *J. Am. Chem. Soc.* **116,** 651-660.

8. Ibers, J.A. and Holm, R.H. (1980) Modeling Coordination Sites in Metallo-biomolecules. *Science* **209**, 223-235.

9. Day, E.P., Peterson, J., Bonvoisin, J.J., Moura, I., and Moura, J.J.G. (1988) Magnetization of the Oxidized and Reduced Three-iron Cluster of Desulfovibrio gigas Ferredoxin II. *J. Biol. Chem.* **263**, 3684-3689.

10. Mouesca, J.-M., Rius, G., and Lamotte, B. (1993) Single-crystal proton ENDOR stuides of the $[Fe_4S_4]^{3+}$ cluster: Determination of the spin population distribution and proposal of a model to interpret the ^1H NMR paramagnetic shifts in high-potential ferredoxins. *J. Am. Chem. Soc.* **115**, 4714-4731.

11. Gloux, J., Gloux, P., Lamotte, B., Mouesca, J.-M., and Rius, G. (1994) The different $[Fe_4S_4]^{3+}$ and $[Fe_4S_4]^{+}$ species created by γ irradiation in single crystals of the $(Et_4N)_2[Fe_4S_4(SBenz)_4]$ model compound: Their EPR description and their biological significance. *J. Am. Chem. Soc.* **116**, 1953-1961.

12. Averill, B.A., Herskovitz, T., Holm, R.H., and Ibers, J.A. (1973) Synthetic Analogues of the Active Sites of Iron-Sulfur Proteins. II. Synthesis and Structure of the Tetra[mercapto-μ_3-sulfido-iron] Clusters, $[Fe_4S_4(SR)_4]^{2-}$. *J. Am. Chem. Soc.* **95**, 3523-3534.

13. Middleton, P., Dickson, D.P.E., Johnson, C.E., and Rush, J.D. (1980) Interpretation of the Mossbauer spectra of the high potential iron protein from chromatium. *Eur. J. Biochem.* **104**, 289-296.

14. Papaefthymiou, V., Millar, M.M., and Münck, E. (1986) Mossbauer and EPR Studies of a Synthetic Analogue for the Fe_4S_4 Core of Oxidized and Reduced High-Potential Iron Proteins. *Inorg. Chem.* **25**, 3010-3014.

15. Noodleman, L. and Case, D.A. (1992) Density-functional theory of spin polarization and spin coupling in iron-sulfur clusters. *Adv. Inorg. Chem.* **38**, 423-470.

16. Borshch, S.A., Bominaar, E.L., Blondin, B., and Girerd, J.-J. (1993) Double exchange and vibronic coupling in mixed valence systems. Origin of the broken-symmetry ground state of $[Fe_3S_4]^0$ cores in proteins and models. *J. Am. Chem. Soc.* **115**, 5155-5168.

17. Ziegler, T. (1991) Approximate density functional theory as a practical tool in molecular energetics and dynamics. *Chem. Rev.* **91**, 651-667.

18. Labanowski, J.K. and Andzelm, J.W., eds. (1991) "Density Functional Methods in Chemistry." Springer-Verlag, New York.

19. Sosa, C., Andzelm, J., Elkin, B.C., Wimmer, E., Dobbs, K.D., and Dixon, D.A. (1992) A local density functional study of the structure and vibrational frequencies of molecular transition-metal complexes. *J. Phys. Chem.* **96**, 6630-6636.

20. Andzelm, J., Sosa, C., and Eades, R.A. (1993) Theoretical study of chemical reactions using density functional methods with nonlocal corrections. *J. Phys. Chem.* **97**, 4664-4669.

21. Johnson, B.G., Gill, P.M.W., and Pople, J.A. (1993) The performance of a family of density functional methods. *J. Chem. Phys.* **98**, 5612-5626.

22. te Velde, G. and Baerends, E.J. (1992) Numerical Integration for Polyatomic Systems. *J. Comp. Phys.* **99**, 84-98.

23. Mouesca, J.-M., Noodleman, L., and Case, D.A. (manuscript in preparation).

24. Sands, R.H. and Dunham, W.R. (1975) Spectroscopic Studies on two-iron ferredoxins. *Quart. Rev. Biophys* **7**, 443-504.

25. Antanaitis, B.C. and Moss, T.H. (1975) Magnetic Studies of the four-Iron High-Potential, Non-Heme Protein from Chromatium vinosum. *Biochim Biophys. Acta* **405**, 262.

26. Middleton, P., Dickson, D.P.E., Johnson, C.E., and Rush, J.D. (1978) Interpretation of the Mossbauer spectra of the four-iron ferredoxin from *Bacillus stearothermophilus*. *Eur. J. Biochem.* **88**, 135-141.

27. Papaefthymiou, V., Girerd, J.J., Moura, I., Moura, J.J.G., and Münck, E. (1987) Mossbauer study of D. gigas ferredoxin II and spin-coupling model for the Fe 3 S 4 cluster with valence delocalization. *J. Amer. Chem. Soc.* **109**, 4703-4710.

28. Kent, T.A. and Münck, E. (1986) Structure and Magnetism of iron-sulfur clusters in proteins. *Hyp. Int.* **27**, 161-72.

29. Blondin, G. and Girerd, J.J. (1990) Interplay of Electron Exchange and Electron Transfer in Metal Polynuclear Complexes in Proteins or Chemical Models. *Chem. Rev.* **90**, 1359-1376.

30. Jordanov, J., Roth, E.K.H., Fries, P.H., and Noodleman, L. (1990) Magnetic Studies of the High-Potential Protein Model $[Fe_4S_4(S\text{-}2,4,6\text{-}(i\text{-}Pr)_3C_6H_2)_4]^-$. *Inorg. Chem.* **29**, 4288-4292.

31. Noodleman, L. (1988) A Model for the Spin States of High Potential $[Fe_4S_4]^{3+}$ Proteins. *Inorg. Chem.* **27**, 3677-3679.

32. Banci, L., Bertini, I., Briganti, F., Luchinat, C., Scozzafava, A., and Oliver, M.V. (1991) Iron-Sulfur Protein (HiPIP) from *Rhodocyclus gelatinosus*. A model for Oxidized HiPIPs. *Inorg. Chem.* **30**, 4517-4524.

33. Banci, L., Bertini, I., Ferretti, S., Luchinat, C., and Piccioli, M. (1993) The structure of iron-sulfur clusters in proteins as monitored by NMR, Mossbauer, EPR, and molecular dynamics. *J. Mol. Struct.* **292**, 207-220.

34. Banci, L., Bertini, I., and Luchinat (1990) The ^1H NMR Parameters of Magnetically Coupled Dimers- the Fe_2S_2 Proteins as an Example. *Structure and Bonding* **72**, 113-136.

SPIN DEPENDENT ELECTRON DELOCALIZATION, VIBRONIC AND ANTIFERROMAGNETIC COUPLINGS IN IRON-SULFUR CLUSTERS

G. BLONDIN,[1] E.L. BOMINAAR,[2] J.-J. GIRERD,[1] and
S.A. BORSHCH[3]
1 Laboratoire de Chimie Inorganique, URA CNRS 420, ICMO, Université Paris-Sud, 91405 Orsay, FRANCE
2 Department of Chemistry, Carnegie-Mellon University, 4400 Fifth Avenue, Pittsburgh, PA 15213, USA
3 Institut de Recherche sur la Catalyse, CNRS, 69626 Villeurbanne Cedex, FRANCE

Abstract

Iron-sulfur clusters can contain both high spin Fe(II) and Fe(III) iron. Such mixed-valence systems may exhibit electron delocalization properties. We develop a theory depending on four effects which are in decreasing energetic order: (i) delocalization energy; (ii) vibronic coupling; (iii) antiferromagnetic interaction; (iv) environment effect. This theoretical model allows us to explain the observed electronic ground states for $[Fe_3S_4]^0$ and $[Fe_4S_4]^{3+}$ clusters.

1. Basic Concepts

1.1. EFFECTIVE HAMILTONIAN

Spectroscopy of metal clusters is today the preoccupation of many inorganic and bioinorganic chemists. Basically the problem is to know the nature of the ground state and of the excited states susceptible to be thermally populated. We want to know the energies but also several characteristics of the states as spin value, spin density distribution, charge distribution since these quantities are observables.

Experimentally several clusters can be described by a perturbative theory: the metal ions keep some individuality in the cluster and a zeroth order Hamiltonian is the sum of individual Hamiltonians. The interaction between metal centers is then treated as a perturbation. Formally this means that the Hamiltonian can be written as

$$H = H_0 + V \tag{1}$$

with H_0 being the zeroth order Hamiltonian and V the perturbation.

One supposes that the problem associated with H_0 is completely solved

$$H_0 \phi_i^0 = E_i^0 \phi_i^0 \tag{2}$$

G.N. La Mar (ed.), Nuclear Magnetic Resonance of Paramagnetic Macromolecules, 369-386.

The situation is depicted in Figure 1. For thermal physics experiments the lowest energy levels are sufficient to explain the properties. A theory which would give a good approximation for the description of the levels issued from E_0^0 is sufficient. Such a theory exists and leads to the concept of effective Hamiltonians: an effective Hamiltonian acts in the subspace of energy E_0^0 and reproduces (generally to second order) the exact energy levels (Figure 1).

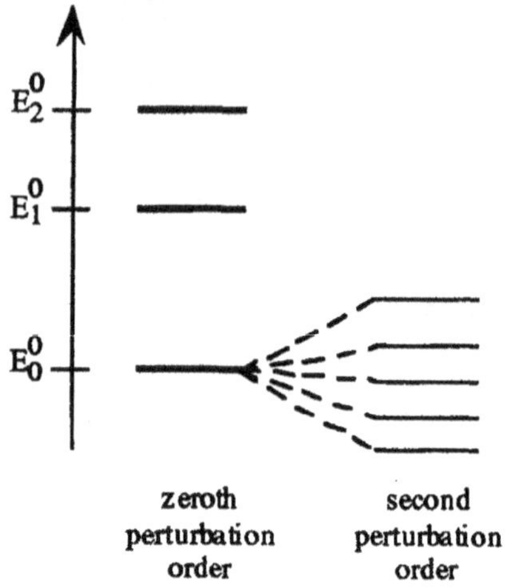

Figure 1. Energy level scheme obtained by treating to zeroth or second order of perturbation metal-metal interaction.

Degenerate perturbation theory gives a simple expression of such an effective Hamiltonian:

$$H_{eff} = P_0 \, V \, P_0 + \sum_m \frac{P_0 \, V \, P_m \, V \, P_0}{E_0^0 - E_m^0} \tag{3}$$

where P_i is the projector in the subspace of energy E_i^0.

As a simple and well-known example of effective Hamiltonian, let us take the Heisenberg Hamiltonian for a Cu^{II}-Cu^{II} dinuclear system.

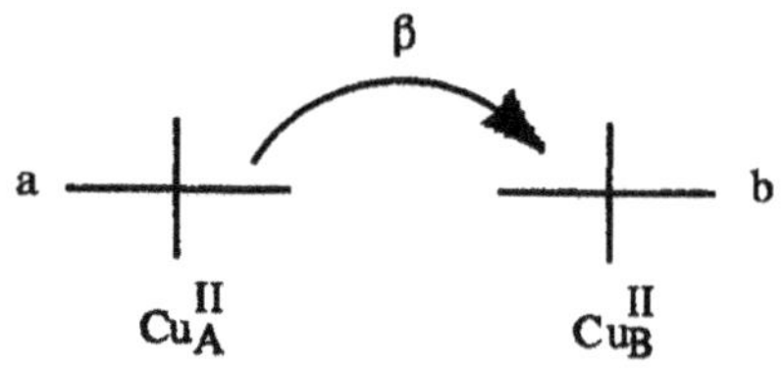

Neglecting intercenter interaction we have two subspaces: one of energy $E_0^0 = 0$ with wavefunctions $|ab\rangle$, $|a\bar{b}\rangle$, $|\bar{a}b\rangle$, $|\bar{a}\bar{b}\rangle$ and the excited one of energy $E_1^0 = U$ with wavefunctions $|a\bar{a}\rangle$, $|b\bar{b}\rangle$. U is the repulsion between two electrons on the same center. The perturbation is $V = \beta \, t_{AB}$ where β is the transfer energy and t_{AB} is the transfer operator.

In this particular case it is easy to recognize that $P_0 \, V \, P_0 = 0$.

Take for instance the wavefunction $| \bar{a}b >$. One has:

$$P_0 \, V \, P_0 \, |\bar{a}b> = P_0 \, V \, | \bar{a}b> = P_0 \, (| \bar{a}a> + | \bar{b}b>) = 0 \qquad (4)$$

This is just a pedantic way to express that transfer from A to B transforms a neutral wavefunction into an ionic one.

It can be shown with some algebra that:

$$P_0 \frac{V \, P_1 \, V}{0 - U} P_0 = \frac{4\,\beta^2}{U} \left[\vec{s}_A \cdot \vec{s}_B - \frac{1}{4} \right] \qquad (5)$$

This in fact constitutes a demonstration of the Heisenberg Hamiltonian

$$H = - J \, \vec{S}_A \cdot \vec{S}_B \qquad (6)$$

with, as a bonus, the expression of the antiferromagnetic component of the exchange parameter J [1].

$$J_{AF} = - \frac{4\,\beta^2}{U} \qquad (7)$$

The ground subspace of dimension 4 is split in two states: one singlet of energy $- 4\beta^2 / U$ and a triplet of energy 0.

More refined theories exist for these phenomena [2] (including ferromagnetic contribution to J) but we do not need to treat these problems here.

The apparition of spin operators in (5) may look amazing. In fact this is due to the peculiar situation studied: in such an exchange problem the charge distribution is the same for every wavefunctions. There is an unequivocal correspondence between complete wavefunctions and a basis where the spin projection is the only apparent information:

$\| ab >$	⟷	$\| + + >$
$\| \bar{a}\bar{b} >$		$\| - - >$
$\| a\bar{b} >$		$\| + - >$
$\| \bar{a}b >$		$\| - + >$

The spin projection is the only degree of freedom. Nevertheless one has not to forget the particularity of the situation and we will see that in cases involving charge distribution variability, other operators than spin operators appear in the effective Hamiltonian.

This brings us to the central question of our talk. What effective Hamiltonian must we use for iron-sulfur clusters ?

We consider that a mixed-valence pair in such a cluster can be equally Fe_A^{II}-Fe_B^{III} or Fe_A^{III}-Fe_B^{II}, i.e. that both situations have the same energy or that the difference in their energy is smaller than the transfer energy β. This means that the corresponding wavefunctions belong to the same subspace E_0. Then it is clear that for this case $P_0 \, \beta \, t_{AB} \, P_0 \neq 0$.

The effective Hamiltonian will contain first order terms in β !

1.2. DOUBLE EXCHANGE

Transfer of one electron between two orbitals is the simplest problem in quantum chemistry. It is well known that the wavefunctions will be $| \varphi_+ >$, $| \bar{\varphi}_+ >$, $| \varphi_- >$, $| \bar{\varphi}_- >$ where

$$\varphi_\pm = \frac{a \pm b}{\sqrt{2}} \tag{8}$$

of energy $\pm \beta$.

Transfer of one electron from a Fe^{II} high spin ion to a Fe^{III} high spin ion is more complicated if one assumes this transfer maintains high spin configurations.

This is more simply understood in the following simple case:

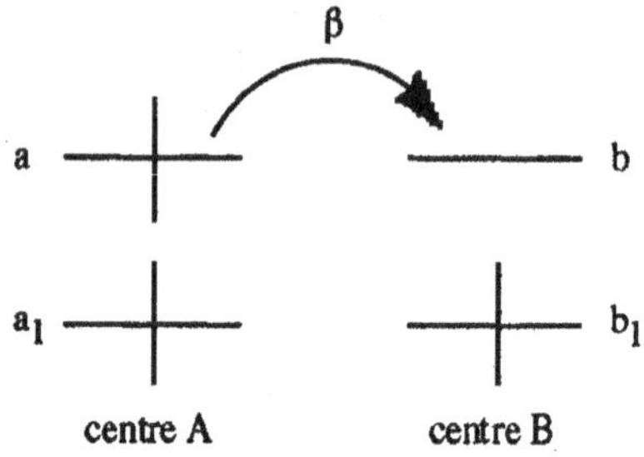

Let us suppose that the extra electron moves from A to B but that it is always ferromagnetically coupled locally to the core spin. The problem has been solved by Anderson and Hasegawa [3]. The spin values for the pair are $S = 3/2$ or $1/2$. In the $S = 3/2$ case, the electron can move from one site to the other, the Hund's rule being always obeyed: the movement of the electron is not impeded and the energy is $\pm \beta$. The wavefunctions can be written $| a_1 b_1 \varphi_\pm >$. In the $S = 1/2$ case, the movement of the electron is inhibited due to non Hund's situations generated by the transfer. The delocalization energy is then only $\pm \beta / 2$.

For situations with half-filled or less than half-filled d shells, Anderson and Hasegawa [3] have shown that the delocalization energy is

$$E_\pm = \pm \beta \frac{S + \frac{1}{2}}{2S_0 + 1} \tag{9}$$

where S is the spin of the pair and S_0 is the core spin. One can check the validity of the formula for the previous d^2-d^1 case.

The phenomenon is called double exchange [4]. Delocalization stabilizes the highest spin state of the pair.

A spectacular experimental illustration of this phenomenon has been found by K. Wieghardt [5]. The compound $[LFe^{II}(\mu\text{-}OH)_3Fe^{III}L]^{2+}$ with L = 1,4,7-trimethyl-1,4,7-triazacyclononane (Figure 2) contains completely delocalized valencies and the spin of the pair is maximum, $S = S_{max} = 9/2$. The value of β is found for this compound by optical spectroscopy: the totally allowed transition between $^{10}\Gamma_u$ and $^{10}\Gamma_g$ gives rise to a strong band with $h\nu_{max} = 13\,000\ cm^{-1}$ and

$\varepsilon_{max} = 5\,200\ M^{-1}\ cm^{-1}$. The transfer energy is $2|\beta|$ so one can deduce that $|\beta| = 6\,500\ cm^{-1}$.

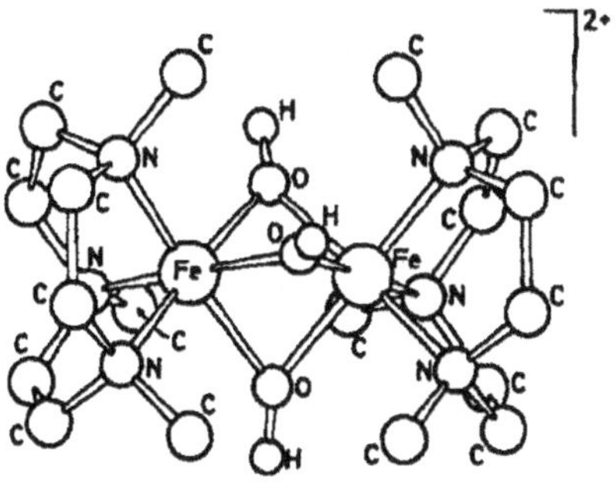

Figure 2. Structure of the cation $[LFe(\mu\text{-}OH)_3FeL]^{2+}$ with L = 1,4,7-trimethyl-1,4,7-triazacyclononane

In iron-sulfur clusters, theoretical calculations by Noodleman et al. [6], give $\beta \approx 3\,000\ cm^{-1}$ which is a sensible value versus the one measured for $[Fe_2(\mu\text{-}OH)_3]^{2+}$, if one takes into account that the Fe—Fe distance is larger in iron-sulfur clusters (2.7 Å) than in the hydroxo system (2.5 Å).

In terms of effective Hamiltonian, double exchange in a dinuclear system is associated with

$$H_{eff} = P_0\ \beta\ t_{AB}\ P_0 \tag{10}$$

If one enlarges to polynuclear mixed-valence system, one ends up with the effective Hamiltonian

$$H_{eff} = P_0 \left[\sum_{i<j} \beta_{ij}\, t_{ij} \right] P_0 \tag{11}$$

where β_{ij} is the transfer energy in the pair center i-center j and t_{ij} is the associated transfer operator.

1.3. VIBRONIC COUPLING

It is well known in the field of mixed-valence complexes in Inorganic Chemistry, that electron transfer between metal centers can be inhibited by vibronic coupling.

Let us consider the symmetric pulsation q_i of each coordination sphere in a pair system. It is clear that the antisymmetric combination

$$q_- = \frac{q_A - q_B}{\sqrt{2}} \tag{12}$$

introduces an asymmetry between metal centers. The energy of the situation with the electron on A will be different from that of the situation with electron on B, as soon as q_- will differ from zero. If this energy difference is larger than the transfer energy, the electron will be trapped on one center [7].

1.4. THE MODEL

We just briefly introduced the basic concepts we propose to describe the electronic structure of iron-sulfur clusters: (i) delocalization energy; (ii) vibronic coupling; (iii) antiferromagnetic coupling. These were ordered by decreasing energy. Transfer energy can be in so compact clusters, easily of several thousands of wave numbers. Vibronic coupling is of the same order (see ref. [8] and references therein). As clearly shown in $[Fe_2S_2]^{2+}$ systems, antiferromagnetic exchange parameter J is of the order of 300 cm^{-1} [9]. This is an order of magnitude smaller than the preceding effects but it may play a role in the fine ordering of the levels.

We developped a theory which corresponds to the exact solution of (11) implemented by vibronic and antiferromagnetic couplings.

At this point we must insist on one idea. We saw that vibronic coupling can trap the electron on one site. But it is important to recognize that at this level of the theory, both sites for a pair are equiprobable.

Recently it has been shown quite clearly by Lamotte et al. [10] and Bertini et al. [11] that in $[Fe_4S_4]^{3+}$ clusters, the excess electron is delocalized in a particular (that is one among six possibilities) Fe^{II}-Fe^{III} pair and they were able to identify the iron atoms belonging to this pair. It is clear that our model, which starts from equivalent sites is unable to explain such an observation. Our model is valid for clusters in vacuum or in solution. Upon insertion of the cluster in a solid or in a protein, iron sites will be distinguished energetically and a charge distribution will be identified versus the environment.

Nevertheless, we believe that the three effects previously mentionned "prepare" the system in a set of equivalent quantum states with precise charge and spin densities. Environment effects will only select one or several of these possibilities.

Environment effects are exemplified in the case of a pair. If the site A has an energy W above that of site B, we will have to compute W < n_A > where n_A is the occupation number of site A. The plain curve with two distinct minima of Figure 3 is obtained. A perturbation treatment gives a very good evaluation of W < n_A >.

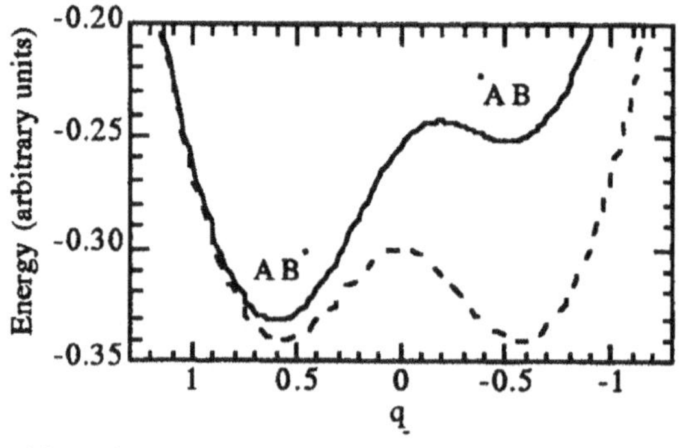

Figure 3. Influence of the environment on the localization of the extra electron for a dinuclear system. The energy in arbitrary units is drawn as a function of the q_- vibration coordinate (arbitrary units). Dashed curve corresponds to $W = 0$, plain curve to $W = 0.1$ ($\beta = 0.3$ for both curves) (W and β are expressed in the same units as the energy)

2. Triangular Systems

2.1. DOUBLE EXCHANGE

The case of one electron moving on a equilateral triangle has a simple solution: a ^2A state of energy 2β and a ^2E state of energy $-\beta$.

These wavefunctions can be built in the following way which will be useful for the iron-sulfur clusters: this equilateral triangle can be considered as made of a pair, AB for instance, connected to a single center C in the appropriate geometry. The pair will have the states φ_+ and φ_-. Action of the transfer operator gives:

$$\beta\,(t_{AB} + t_{AC} + t_{BC})\,\varphi_+ = \beta\,\varphi_+ + \sqrt{2}\,\beta\,c \qquad (13a)$$

$$\beta\,(t_{AB} + t_{AC} + t_{BC})\,\varphi_- = -\beta\,\varphi_- \qquad (13b)$$

It is then clear that φ_- is an eigenstate of the transfer operator with the energy $-\beta$. φ_+ and c will mix to give a state of energy 2β and one of energy $-\beta$.

The archetype of the equilateral triangular double exchange situation is the d^2-d^1-d^1 case. More spin values will occur than in the previous case. Let us suppose that the extra electron is on site A; one has $S_A = 1$, $S_B = 1/2$, $S_C = 1/2$. It is easy to see that the following spin values will be attainable: S = 0 (1), S = 1 (2), S = 2 (1).

Taking into account that the excess electron can be also on B or C, one finds 3 singlet, 6 triplet and 3 quintet states.

Transfer operator (11) leads to the following states and energies:

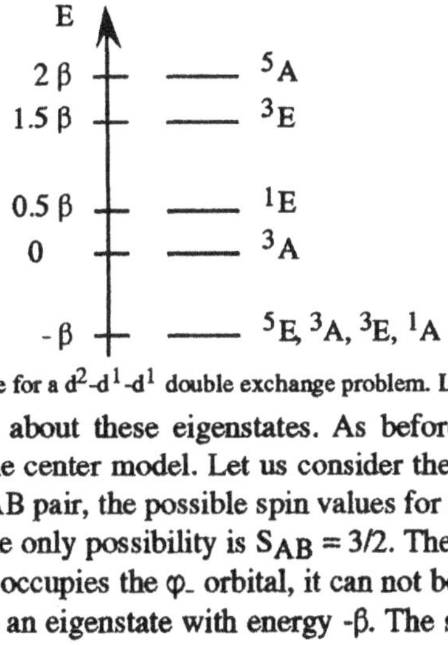

Figure 4. Energy level scheme for a d^2-d^1-d^1 double exchange problem. Labels refer to the C_3 symmetry.

It is useful to think about these eigenstates. As before, we can approach these solutions by the pair + one center model. Let us consider the S = 2 problem. When the excess electron is in the AB pair, the possible spin values for this pair will be 1/2 or 3/2. To get total spin S = 2, the only possibility is $S_{AB} = 3/2$. The extra electron will be in a φ_+ or φ_- orbital. When it occupies the φ_- orbital, it can not be transferred to site C. The state $| a_1 b_1 c_1 \varphi_- >$ will be an eigenstate with energy $-\beta$. The state with φ_+ will combine with a state $S_{AB} = 1$, $S_C = 1$ (extra electron on site C) to give two states, one at 2β, the other one at $-\beta$.

Figure 5 reproduces this construction for every total spin value. One sees that a S = 2 and a S = 1 states will be made of the extra electron delocalized in one pair in φ_- orbital with no probability to occupy the third site.

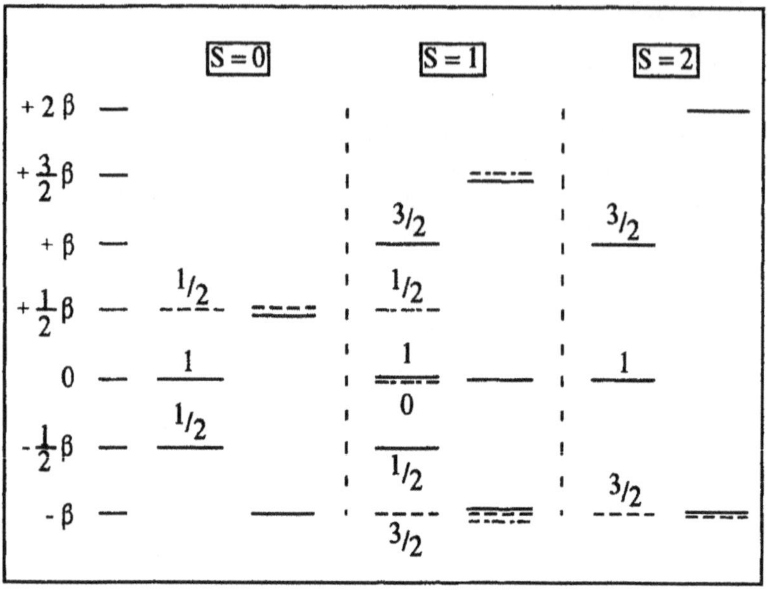

Figure 5. Construction of the double exchange levels for a d^2-d^1-d^1 equilateral triangle by the pair + site method. For each value of total spin S, energies of the C_{3v} eigenstates (right) and of the C_s symmetry adapted basis (left) are shown. Lines of the same style mean that the states mix to give the eigenstates. The numbers indicate subspin S_{AB} of C_s symmetry adapted basis states. S_{AB} is half integer when AB is the mixed-valence d^2-d^1 pair and is integer when AB is the d^1-d^1 pair.

2.2. VIBRONIC COUPLING

As in dimers, vibronic coupling can trap the extra electron. Vibronic coupling for the d^1-d^0-d^0 case has been studied [12].

It can be shown that two combinations of the coordination sphere pulsations are effective; they constitute the q_E vibration.

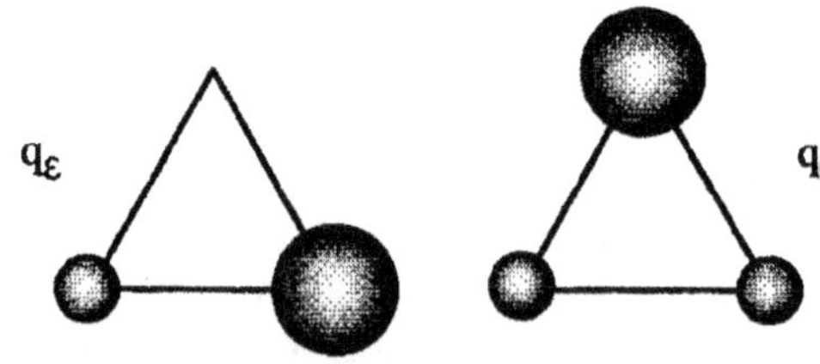

The effect of these vibrations is shown in Figure 6.

For $\beta < 0$ and small (A1) versus energy asymmetry induced by q_E vibration, a minimum corresponding to the trapping of the excess electron on one site occurs. When the transfer energy intensity increases, this minimum changes in nature to finally correspond to a non deformed equilateral triangle (A3).

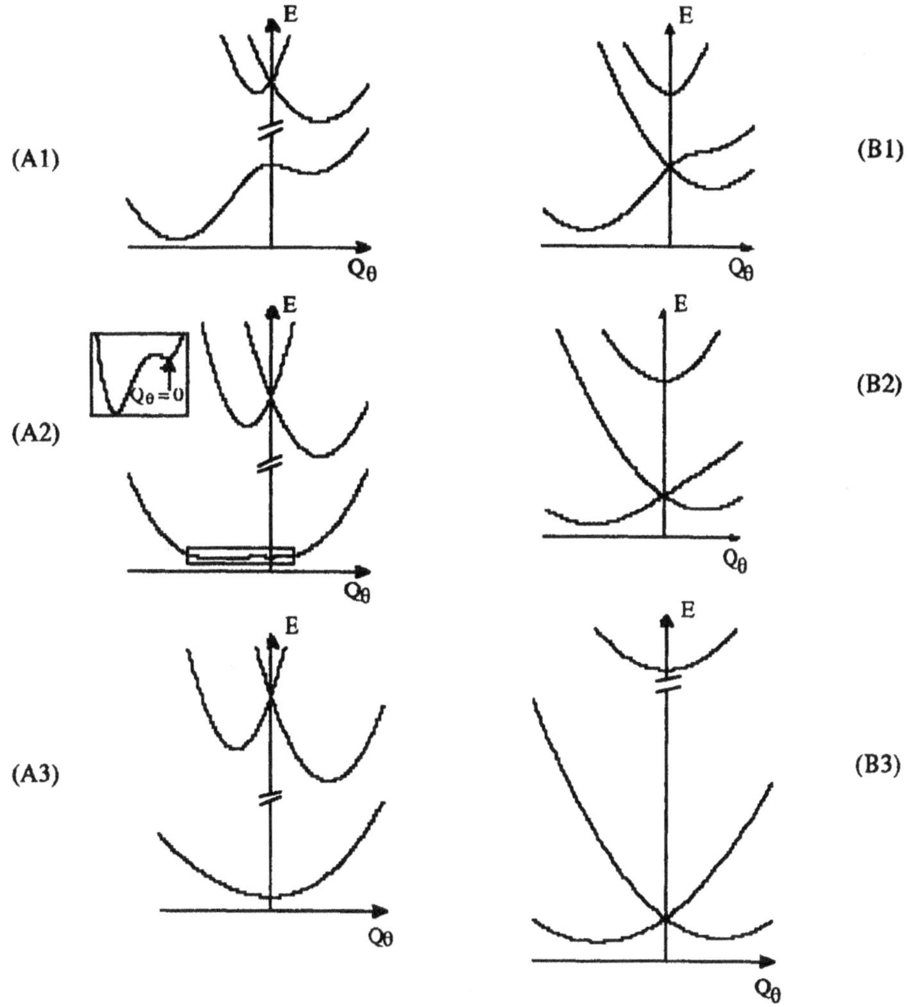

Figure 6. Sections of adiabatic potential surfaces for a monoelectronic mixed-valence trinuclear system along the coordinate Q_θ, obtained for $\beta < 0$ (left) and $\beta > 0$ (right) at small (A1 & B1), intermediate (A2 & B2), and large (A3 & B3) value of transfer energy β versus the vibronic coupling. The inset in A2 displays the bottom of the lowest adiabatic potential surface at a smaller energy scale.

For $\beta > 0$ and small (B1), the same site trapping occurs. For β large (B3), one ends up with a Jahn-Teller effect on the degenerated ^2E state.

3D representation of the potential energy surface is given in Figure 7 which corresponds to a small intensity of delocalization: three minima are observed, each one corresponding to localization of the extra electron on one peculiar site.

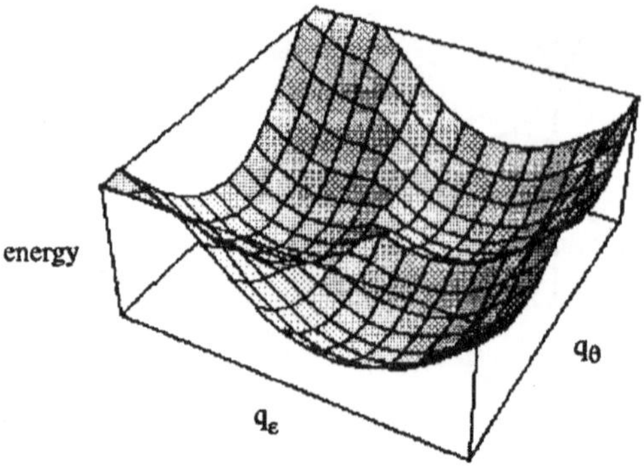

Figure 7. Adiabatic potential surface for the d^1-d^0-d^0 with a small transfer energy.

The study of this effect for the double exchange d^2-d^1-d^1 problem led us [8] to the amazing following result for $\beta > 0$. In this case we saw that the lowest energy levels are ^5E, ^3A, ^3E, ^1A. The singlet and quintet cases are analogous to the d^1-d^0-d^0 case. The triplet case is more instructive (Figure 8).

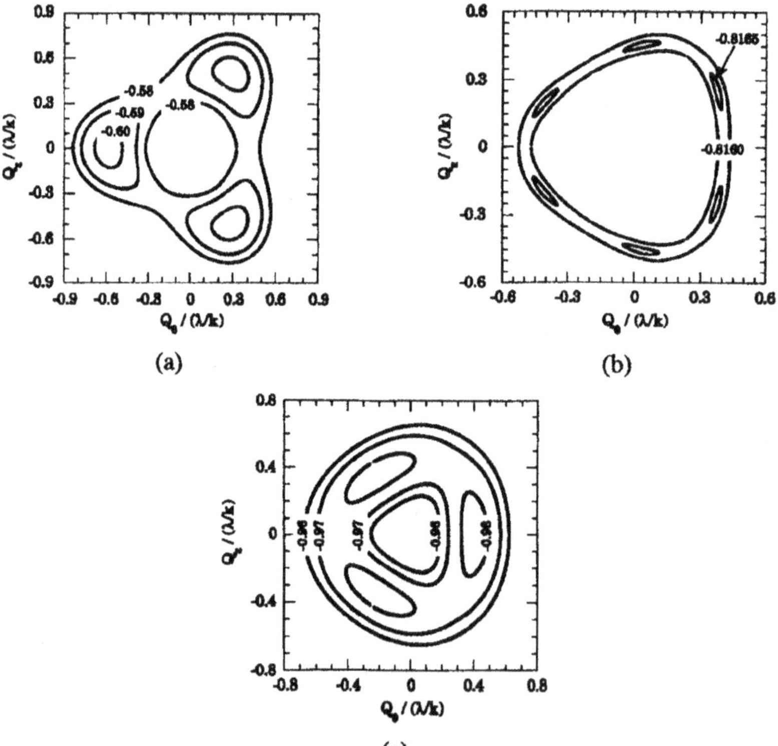

Figure 8. Equipotential curves on lowest adiabatic surface for S = 1 in the system d^2-d^1-d^1, calculated for positive β at different values of the transfer energy versus the vibronic coupling. In (a), the excess electron is localized on one site, in (b), it starts to delocalize and in (c), it is delocalized in a pair.

For β small, three minima are found (a) which correspond to localization of the extra electron on one site. Now when β increases, we see the minima to deform (b) and to move by 60° (c): this precisely means that each minimum corresponds to localization of the extra electron in a pair.

Analysis of one of these minima shows that the associated wavefunction corresponds to the extra electron delocalized in a φ_- orbital for the AB pair with $S_{AB} = 3/2$ antiparallel to the $S_C = 1/2$ spin in order to give a total spin value $S = 1$.

It turns out that the minimum for $S = 1$ has the same energy than that of the $S = 2$ state. The $S = 0$ state, being of symmetry A, gains no vibronic stabilization. Here we see the role of antiferromagnetic coupling: if we suppose for simplicity a uniform interaction ($J < 0$), we understand that the $S = 1$ will be stabilized. If the coupling is too large the $S = 0$ can become the ground state.

In conclusion, we understand with this model calculation that it is possible to explain, based on the three concepts enunciated in the introduction, delocalization of the extra electron in one pair with stabilization of the intermediate $S = 1$ spin state. This was for us an encouragement to apply this approach to iron-sulfur clusters.

At this point we must insist on the idea that the above conclusions explain the localization of the extra electron in one pair but in <u>any</u> of the three pairs. In a solid or in a protein, experiments on tetranuclear clusters by Lamotte [10] and Bertini [11] demonstrated that the pair can be identified. This asks for addition of environmental effects to the previous theory.

We propose that small energy differentiation between the three sites are sufficient to trap the pair without destroying the electronic structure arrived at in vacuum.

The following calculation along this line can be done. Let us express the effect of environment by

$$H_{env} = W < n_C > \tag{14}$$

where $W > 0$ and n_C is the occupation operator of site C by the extra electron. If we consider H_{env} as a perturbation on the previous problem we find:

$$E_{AB} = 0 \qquad E_{AC} = \frac{W}{2} \qquad E_{BC} = \frac{W}{2} \tag{15}$$

The pair AB will be favored by the quantity W/2.

2.3. IRON-SULFUR CASE

The theory exposed in the preceding paragraph was applied to the $[Fe_3S_4]^0$ cluster for which experimentally it was shown that the Fe^{II} oxidation state was confined to a pair with $S_{AB} = 9/2$ and that the total spin value was $S = 2$ ($S_C = 5/2$) [13].

The conclusion exposed in ref. 8 was that indeed our theoretical model leads to the spectroscopically identified ground state. Let us summarize briefly our demonstration.

(i) For a d^5-d^5-d^6 equilateral triangle, double exchange effects lead for $\beta > 0$, to the following manifold ^1A, ^3A, ^3E, ^5A, ^5E, ^5E, ^7A, ^7A, ^7E, ^9A, ^9E, ^{11}E, ^{13}A (energy - $6\beta/5$).

(ii) Vibronic coupling through q_E coordinates lifts the preceding degeneracy: the most stabilized states are those with the lowest occupation of site C, i.e., the ones with the strongest localization in the pair AB. Table VII of ref 8 gives the $< n_C >$ values. For ^1A and ^{13}A states $< n_C > = 1/3$ by symmetry. The ^{11}E state is subject to an ordinary Jahn-Teller effect which leads to a stabilization energy of 0.082 units. The most stabilized triplet state has $< n_C > = 0.257$ (stabilization energy 0.052 units). This is due to the fact that a S = 1 state can be achieved with $S_{AB} = 7/2$ at maximum: this means that double exchange in the AB pair gives a stabilization of $- 4\beta/5$ and that extra stabilization of $- 2\beta/5$ is gained through delocalization onto site C. The triplet state is not the most pair localized state achievable.

This limitation is not encountered for total spin value $2 \leq S \leq 5$. Calculation shows that the smallest $< n_C >$ is realized for a S = 2 state: strict antiferromagnetic order between S_{AB} and S_C minimizes the transfer energy to site C. For this S = 2 state we got $< n_C > = 0.152$ and a stabilization energy of 0.298 units.

The result here is that vibronic coupling alone stabilizes among all the possible states, precisely the one experimentally observed $S = 2$, $S_{AB} = 9/2$, $S_C = 5/2$.

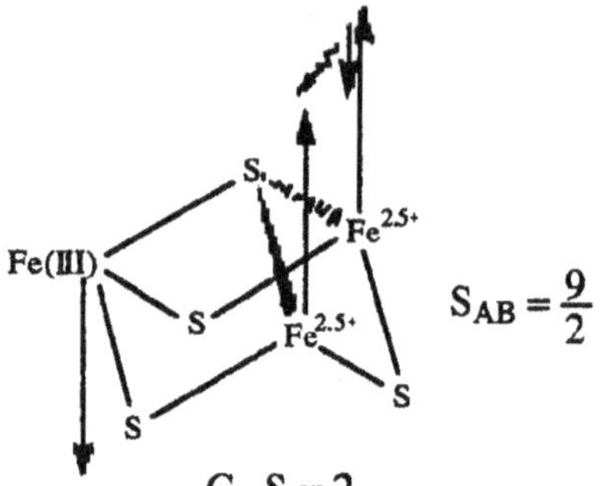

C_s, $S = 2$

Strictly speaking the wave function we got does not correspond to absolute confinement of the extra electron in one pair but the occupation of the third site is so low that it was undetected experimentally.

The effect of antiferromagnetic exchange is to increase the stabilization of this state versus the S > 2 ones. Nevertheless for large enough values of J (uniform) the S = 1 or S = 0 states can become ground states. Introduction of a distinction (physically expected) between $J_{Fe^{II}-Fe^{III}}$ and $J_{Fe^{III}-Fe^{III}}$ can stabilize the S = 2 solution.

The derivation from first principle of the observed ground state for $[Fe_3S_4]^0$ was striking. It has to be noted however that it relies on a positive value for β which we justified by Extended Hückel calculations. This sign could be questioned by more sophisticated methods. This is certainly worthwhile to be discussed in future.

3. Tetranuclear Case

3.1. THE d^2-d^1-d^1-d^1 DOUBLE EXCHANGE PROBLEM

Delocalization of one electron on a T_d core leads to ^2A at the energy 3β and ^2T at the energy $- \beta$.

If we add on each center a S = 1/2 core spin, we end up with the d^2-d^1-d^1-d^1 double exchange problem [14, 15].

For $\beta > 0$, a manifold of ground states is found: ${}^2A_2$, ${}^2T_2$, ${}^2T_1$, ${}^4E$, ${}^4T_1$, ${}^4T_2$, ${}^6T_1$. The interested reader can consult ref 18 to understand the construction of these states from two pairs connected in such a way that the overall T_d symmetry is obeyed.

3.2. THE VIBRONIC PROBLEM

Composition of the four coordination sphere pulsations gives rise to four normal coordinates: A + T. The q_A coordinate does not introduce site differentiation. Only the q_T coordinate is effective.

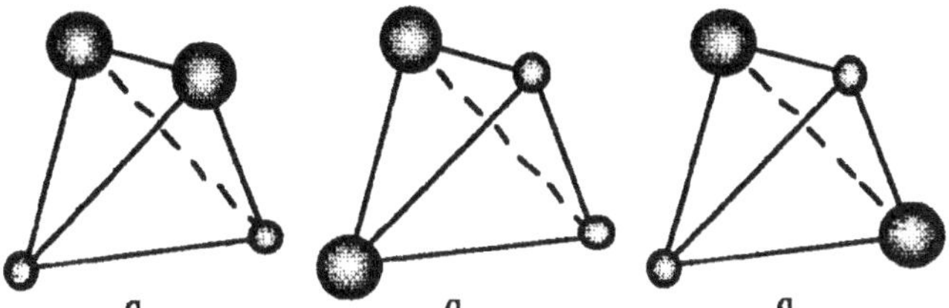

The d^1-d^0-d^0-d^0 T_d vibronic coupling has been studied by Prassides [16]. We propose to illustrate the results in the q_T space in the following manner. Energy for the ground state of the complete double exchange + vibronic problem is represented as the function $E = f(q_x, q_y, q_z)$ where q_x, q_y, q_z are the three components of q_T. To each point in this space corresponds a particular deformation of the tetrahedron (Figure 9).

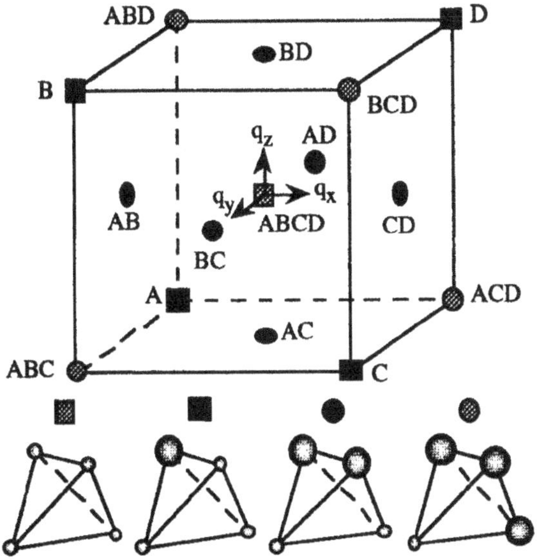

Figure 9. Space of symmetrized vibrational coordinates, q_x, q_y and q_z. The labels, A, ..., AB, ..., ABC, ..., and ABCD mark directions with one, two, three, and four dilated ligand shells, leading to electron delocalization over monomeric, dimeric, trimeric, and tetrameric subunits of the cluster, respectively.

Corners of the cube in the q_T space correspond to expansion of one sphere and contraction of the three others (point -1, -1, -1 and corresponding ones) or expansion of three spheres and contraction of one (point 1, 1, 1 and corresponding ones). A point of type -1, -1, -1 will correspond to a wavefunction with the extra electron localized on one site; a point of type 1, 1, 1 to the extra electron localized on a triad.

The d^2-d^1-d^1-d^1 case is rich when $\beta > 0$ [17]. The $^6T_1$ state exhibits a Jahn-Teller effect. The study of the quartet and doublet states is more amazing. For small values of β versus vibronic energy, trapping of the extra electron on one site has been found: this is a general result in this theory and is physically expected, since in that case the excess electron does not gain enough energy to delocalize.

When the transfer energy and the vibronic coupling are of the same order, we found that the ground state is a S = 5/2 site localized state followed in energy by a S = 3/2 site localized state and three degenerate 1/2, 3/2, 5/2 pair (strictly) localized states (Figure 10).

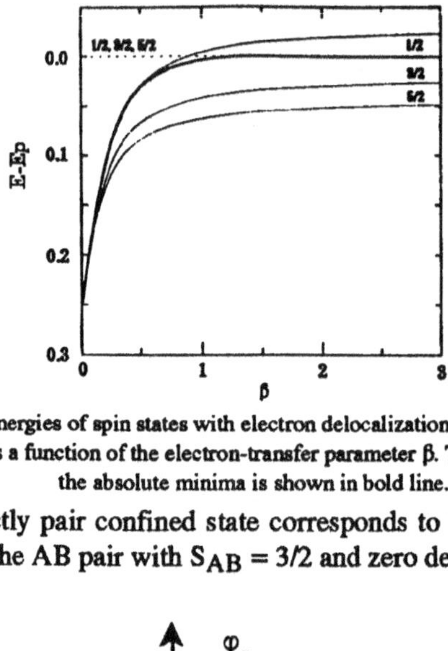

Figure 10. The minimal energies of spin states with electron delocalization (plain curves) and delocalization in a pair (dashed curve) as a function of the electron-transfer parameter β. The energy of the S = 1/2 state at the absolute minima is shown in bold line.

The S = 1/2 strictly pair confined state corresponds to the excess electron in a φ_- molecular orbital on the AB pair with $S_{AB} = 3/2$ and zero delocalization on the CD pair with $S_{CD} = 1$.

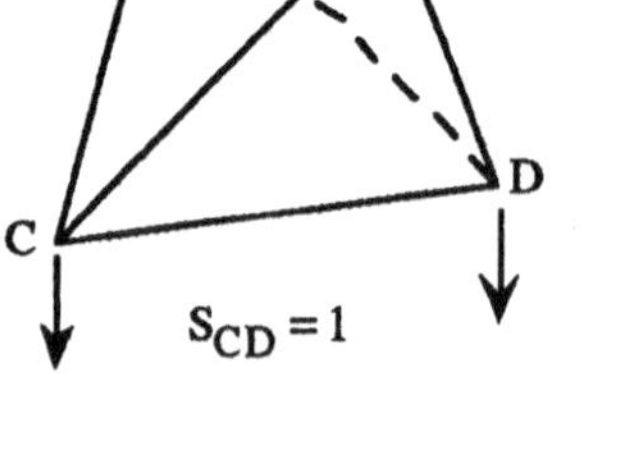

Antiferromagnetic exchange can stabilize the S = 1/2 state.

So again we see that a combination of double exchange, vibronic coupling and antiferromagnetic exchange can lead to a S = 1/2, $S_{AB} = 3/2$, $S_{CD} = 1$ pair localized state strongly reminiscent of what is observed in the simplest $[Fe_4S_4]^{3+}$ cases.

The minima for the d^2-d^1-d^1-d^1 problem are obtained along the q_x, q_y, q_z axes (points underlined in the following scheme). But all these minima have the same energy. So the aforementioned effects prepare the system in one of these six minima but all are equiprobable.

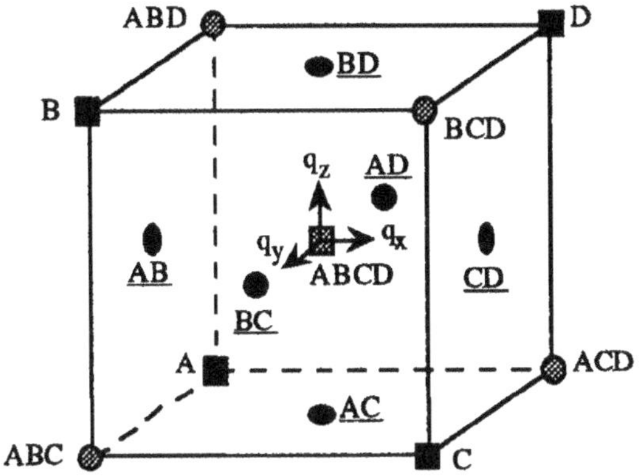

Environment effects have to be taken into account. Let us use as a preliminary answer to this question the perturbative approach delineated for triangles. Certainly in a protein the four sites will have different energies. Let us call them A, B, C, D such that

$$H_{env} = W < n_B > + W' < n_C > + W'' < n_D > \qquad (16)$$

with W" > W' > W > 0.

It is easy to show that two favored pairs will be AB and AC (in that order); the third one will be AD or BC depending on the ratio W"/(W + W'). This can explain why Bertini et al. [11b] observed the extra electron delocalized in two adjacent pairs. The model predicts that AB, AC, AD or AB, AC, BC schemes could also be observed.

3.3. THE IRON CASE

We recently published the results of our study of the $[Fe_4S_4]^{3+}$ case [18].

The d^5-d^5-d^5-d^6 T_d double exchange problem has been solved [19].

Keeping β positive, a manifold of 215 states is found with $1/2 \leq S \leq 17/2$ at the energy -6β/5. Vibronic coupling favors a | 9/2, 5, 1/2 > state (major component) since it has the smallest occupation of the CD pair ($< n_{CD} > = 0.0833$). The following state is | 9/2, 4, 1/2 > (major component) with $< n_{CD} > = 0.0918$ (see Table 2 of ref 18).

The stabilization of the | 9/2, 5, 1/2 > state versus the | 9/2, 4, 1/2 > state can be explained quantitatively as follows.

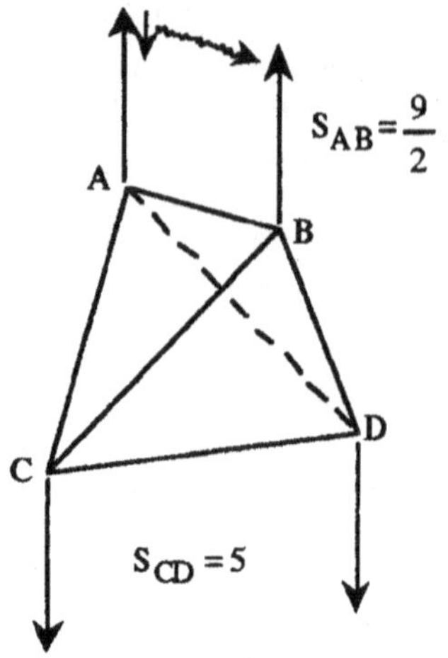

In this case, antiparallelism between S_{AB} and S_{CD} is maximum which minimizes delocalization from AB to CD: localization on the AB pair is maximized.

Experimentally it has been proposed that $S_{CD} = 4$ [11]. The only way to found this result was in this model, to introduce two different antiferromagnetic coupling constants $J_{Fe^{III}\text{-}Fe^{III}}$ and $J_{Fe^{II}\text{-}Fe^{III}}$ with $| J_{Fe^{III}\text{-}Fe^{III}} | > | J_{Fe^{II}\text{-}Fe^{III}} |$. Table 3 of ref. 18 gives the ordering for $\Delta J = 200\ cm^{-1}$. The ground state is | 9/2, 4, 1/2 > followed by | 7/2, 3, 1/2 > (states are named by their main component).

Let us remark that our model does not give a strict localization in a pair but a small delocalization on the other pair: the | 9/2, 4, 1/2 > ground state has $< n_{CD} > = 0.0833$.

4. Conclusion

We showed that a model taking into account double exchange, vibronic coupling and antiferromagnetic exchange can explain several features of the observed electronic structure of the iron-sulfur clusters. It is a three or four parameters model (four if $J_{Fe^{II}\text{-}Fe^{III}}$ is considered different from $J_{Fe^{III}\text{-}Fe^{III}}$).

This model demonstrates the suggestion by Holm that electronic states of these clusters are intrinsic and not principally determined by the environment.

Nevertheless, we noted that our model has to be improved by taking into account environment effects. We propose that a perturbation treatment will be enough in that respect.

Calculations on $[Fe_4S_4]^{n+}$ (n = 1, 2) systems have still to be done.

More ab initio calculations are certainly needed to found some of our hypotheses [6].

As for experimental methods, NMR studies by the Bertini group and ENDOR studies by the Lamotte group have contributed a lot recently in the understanding of these systems (see other chapters in this book).

Charge delocalization could in principle be studied by optical spectroscopy but considering the complexity of the spectra of these systems, it will certainly take some time before complete understanding is achieved.

We have the feeling that more physics is contained in the model delineated here and that fundamental work on this problem deserves on its own right to be done.

It would be interesting to ascertain that iron-sulfur clusters, already so important in biochemistry, bring a physics richer than exchange interaction in the field of metal polynuclear clusters.

5. References

1. Anderson P. W. (1993) *Magnetism*, G. T. Rado and H. Suhl (eds), Academic Press, New York, vol 1, chapter 2, 25-84.
2. Goodenough J. B. (1963) *Magnetism and the Chemical Bond*, Interscience Publishers, New York.
3. Anderson P. W., Hasegawa H. (1955) Considerations on double exchange, *Phys. Rev.*, **100**, 675-681.
4. Zener C. (1951) Interaction between the *d*-shells in the transition metals. II. Ferromagnetic compounds of manganese with perovskite structure, *Phys. Rev.*, **82**, 403-405.
5. Drüeke S., Chaudhuri P., Pohl K., Wieghardt K., Ding X.-Q., Bill E., Sawaryn A., Trautwein A. X., Winkler H., Gurman S. J. (1989) The novel mixed-valence, exchange-coupled, class III dimer $[L_2Fe_2(\mu\text{-}OH)_3]^{2+}$ (L = *N,N',N''*-trimethyl-1,4,7-triazacyclononane), *J. Chem. Soc., Chem. Commun.*, 59-62.
6. see the chapter of Case et al. in this book.
7. Piepho S. B., Krautz E. R., Schatz P. N. (1978) Vibronic coupling model for calculation of mixed valence absorption profiles, *J. Am. Chem. Soc.*, **100**, 2996-3005.
8. Borshch S. A., Bominaar E. L., Blondin G., Girerd J.-J. (1993) Double exchange and vibronic coupling in mixed valence systems. Origin of the broken-symmetry ground state of $[Fe_3S_4]^0$ cores in proteins and models, *J. Am.. Chem. Soc.*, 115, 5155-5168.
9. Bertrand P., Guigliarelli B., More C. (1991) The mixed-valence [Fe(III), Fe(II)] binuclear centers of biological molecules viewed through EPR spectroscopy, *New J. Chem.*, **15**, 445-454 and references therein.
10. (a) Mouesca J.-M., Rius G., Lamotte B. (1993) Single crystal proton ENDOR studies of the $[Fe_4S_4]^{3+}$ cluster: determination of the spin population distribution and proposal of a model to interpret the ^1H NMR paramagnetic shifts in high-potential ferredoxins, *J. Am. Chem. Soc.*, **115**, 4714-4731. (b) Gloux J., Gloux P., Lamotte B., Mouesca J.-M., Rius G. (1994) The different $[Fe_4S_4]^{3+}$ and $[Fe_4S_4]^+$ species created by γ irradiation in single crystals of the $(Et_4N)_2[Fe_4S_4(Sbenz)_4]$ model compound: their EPR description and their biological significance, *J. Am. Chem. Soc.*, **116**, 1953-1961.
11. (a) Bertini I., Turano P., Vila A. J. (1993) Nuclear magnetic resonance of paramagnetic metalloproteins, *Chem. Rev.*, **93**, 2833-2932. (b) Banci L., Bertini I., Ciurli S., Ferretti S. Luchinat C., Piccioli M. (1993) The electronic structure of $[Fe_4S_4]^{3+}$ clusters in proteins. An investigation of the oxidized high-potential iron-sulfur protein II from *Ectothiorhodospira vacuolata*, *Biochemistry*, **32**, 9387-9397.
12. see references 31 to 33 of ref 8.
13. Papaefthymiou V., Girerd J.-J., Moura I., Moura J. J. G., Münck E. (1987) Mössbauer study of *D. gigas* ferredoxin II and spin-coupling model for the Fe_3S_4 cluster with valence delocalization, *J. Am. Chem. Soc.*, **109**, 4703-4710.
14. Palii A. V., Ostrovsky S. M., Tsukerblat B. S. (1992) Double exchange in tetrameric mixed-valence clusters, *New J. Chem.*, 16, 943-952.
15. Marks A. J., Prassides K. (1993) Vibronic effects in a model four-center, one-electron mixed-valence system, *New J. Chem.*, 17, 59-65.
16. Marks A. J., Prassides K (1993) Exchange and vibronic interactions in a tetrahedral, five-electron mixed-valence cluster, *J. Chem. Phys.*, **98**, 4805-4813.

17. Borshch S. A., Bominaar E. L., Girerd J.-J. (1993) Electron delocalization in four-nuclear mixed-valence clusters with paramagnetic ions, *New J. Chem.*, 17, 39-42.
18. Bominaar E. L., Borshch S. A., Girerd J.-J. (1994) Double exchange and vibronic coupling in mixed-valence systems. Electronic structure of $[Fe_4S_4]^{3+}$ clusters in high-potential iron protein and related models, *J. Am. Chem. Soc.*, **116**, 5362-5372.
19. Boras-Almenar J. J., Coronado E., Georges R., Gomez-Garcia C. J. (1992) Electron transfer in tetranuclear mixed-valence iron clusters. Role of the topology on the magnetic properties, *Chem. Phys.*, **166**, 139-144.

INDEX